Process Safety Spotlights 2017

Topical Conference at the 2017 AIChE Spring Meeting and 13th Global Congress on Process Safety

San Antonio, Texas, USA
26 - 30 March 2017

ISBN: 978-1-5108-4367-7

Printed from e-media with permission by:

Curran Associates, Inc.
57 Morehouse Lane
Red Hook, NY 12571

Some format issues inherent in the e-media version may also appear in this print version.

Printed by Curran Associates, Inc. (2017)

For permission requests, please contact AIChE
at the address below.

AIChE
120 Wall Street, FL 23
New York, NY 10005-4020

Phone: (800) 242-4363
Fax: (203) 775-5177

www.aiche.org

Additional copies of this publication are available from:

Curran Associates, Inc.
57 Morehouse Lane
Red Hook, NY 12571 USA
Phone: 845-758-0400
Fax: 845-758-2633
Email: curran@proceedings.com
Web: www.proceedings.com

TABLE OF CONTENTS

482030 AGILE Award Keynote Address

Abstract:

Factors in Selecting AGILE Honoree

1. Overall sustained excellent industry leadership in businesses that employ chemical engineers

2. Recognized for efforts in innovation and creating strategies that ensure success

3. Value of the keynote to the meeting attendees and the members of the Institute Schedule

Event	Time	Location
Continental Breakfast	7:00AM - 8:00AM	Henry Gonzalez Convention Center, Lila Cockrell Theatre
AGILE Keynote Address	8:00AM - 9:15AM	Henry Gonzalez Convention Center, Lila Cockrell Theatre

The 2017 AIChE Spring Meeting and 13th GCPS AGILE Keynote Address will be presented by AGILE award recipient John Y. Televantos, Partner, Arsenal Capital Partners.

John Televantos is a Partner at Arsenal Capital Partners, a middle market private equity firm focused on the specialty industrial, healthcare and specialty financial service sectors. Before joining Arsenal in 2006, Dr. Televantos was President of Aqualon Company, owned by Hercules, Inc.; President and Founder of Helios Chemical Company; and Chief Executive Officer for Foamex International.

Dr. Televantos began his career with Union Carbide in 1977 in silicones research. There he developed two key processes for the manufacture of Methyl Chloride and Siloxane polymers that were commercialized. These processes not only improved efficiency and reduced cost but also dramatically reduced waste. He moved through a variety of management positions there and later at ARCO Chemical, where he became vice president of strategy and development.

At ARCO John co-invented and led a team that developed and commercialized a breakthrough technology in polyol manufacture (a key raw material for polyurethanes) eliminating a whole process step along with waste and emissions, while at the same time opening the door to the creation of new higher-performance polyurethanes. The technology has since been licensed by most of the major global polyol manufacturers with significant waste reduction worldwide.

At Arsenal Capital, Dr. Televantos serves as co-head of the specialty industrial practice. During his tenure, he has been part of several acquisitions and successful exits in the chemicals and materials sectors. In addition to serving on the boards of several Arsenal-owned companies, he is lead director on the board of Balchem Corp.

480703 Big Data Analytics Skills - a Revolution Lacking in Revolutionaries

Abstract:

Since the advent of the common use of the term "Big Data[i]" in 2010 to describe large, *unruly* sets of data, and the artful analysis thereof, industry has felt besieged by opportunity. The chemical and refining industry has been faced with "Big Data" long before it was cachet. Yet within industry there is still a struggle to get hearts and minds around what the Big Data Journey means to an organization and how to take advantage of the data opportunities that lie within. The added complication to the chemical industry is that it must add "contextualization" to the data analysis because it's not enough just to know there is a signal, we need to know what the signal means[ii]. Yet during this data revolution there is an underlying challenge that Academia, Industry and Governmental Agencies must tackle together to ensure the US Chemical industry retains its competitive position in the world. Together we need to provide education for the higher level skills and abilities required across roles in manufacturing to assure that no one gets left behind in the new Big Data age. Indeed, the next great technological "step" made may not be in engineering design or IoT connectedness (reference Industry 4.0), it may be in designing and creating a better human capability in "data acuity" out of our university system; operators, engineers and scientists alike who are better prepared for the data age than their predecessors. This talk will take a look at what Industry requires in skills sets and perspective to be "data ready", and what Academia and Government might do to improve the output of data-ready skills.

---

[i] Big Data, What is it? White, Chemical Engineering Progress, AIChE March 2016.

[ii] Big Data, Getting Started on the Journey. Colegrove, Seasholtz, Khare, Chemical Engineering Progress, AIChE March 2016.

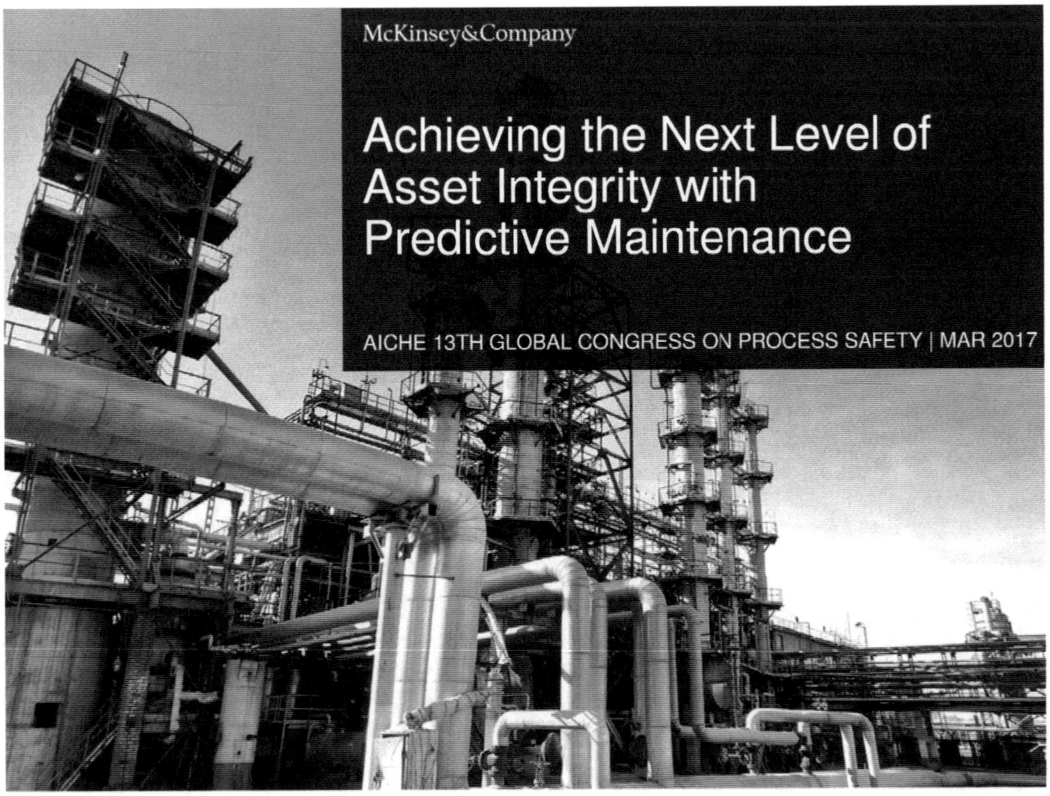

Abstract

Achieving the next level of asset integrity with predictive maintenance

Exceptional organizations manage to deliver superior operational performance without compromising on safety and integrity. On the contrary, high performing players are at the forefront of equipment reliability for critical equipment and safety performance.

In this context, introducing Predictive Maintenance can dramatically help achieve the next level of asset integrity by leveraging data - structured and unstructured, machine and non-machine based – in order to generate insights with advanced analytics methods that would not be possible with conventional techniques
Implementing Predictive Maintenance generates three benefits that reinforce one another:
Predict when equipment will fail and help reduce the safety and integrity threats
Avoid failures and positively contribute to reduce the scale of the value at risk
Enhance specific asset knowledge and learn more about the behavior of critical equipment

In a recent client project for an upstream oil and gas facility, the main gas compressors have been identified as **critical equipment**: they account for many failures and a large portion of production losses. A Weibull analysis showed that out of several known failure modes, some show a random failure pattern, leading to sudden breakdowns releasing high energy and thus representing a safety issue. As per the random failure behavior, classical time- or usage-based maintenance strategies are not fruitful.

Combining 100's of gigabytes of loss data, maintenance work order data and sensor data from 1,200 sensors around the equipment, a data driven model has been created showing that out of the 1,200 sensor sources, 43 explanatory variables predict the time to failure with ~80% accuracy. The model has been tweaked further in order to reduce the number of false negatives while accepting by this a certain number of false positives (i.e. false alarms) due to safety reasons.

As a result of this model, an early warning of an imminent failure means repairs can be planned, spares can be gathered and the equipment can now be shut down in a safe and orderly manner. Overall, this represents an average reduction from 14 to 4 days of downtime per breakdown event.

Outline

- An organization's ability to maintain asset integrity and anticipate equipment failure positively affects its safety performance
- Predictive Maintenance makes it possible to increase asset integrity and improve operational performance using the millions of operational data points that are being collected
- Example of a success story: Early warning of an imminent failure for a gas compressor on a rig

McKinsey & Company 2

There is a strong correlation between equipment reliability and safety performance

EXAMPLE FROM ALUMINUM SMELTING

1993 - 2003 (base 100)

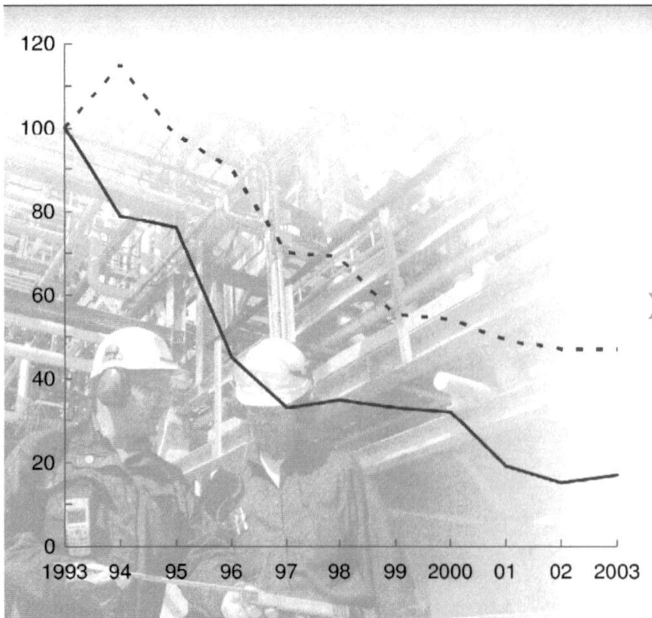

- **Safety** and reliability performance require **control over the plant environment**
- **Unscheduled work** – often in rush – **is by nature much more risky** than planned interventions
- **Reliability and safety performance share multiple common drivers**
  - Importance of **behavioral factors**
  - **Anticipation rather than reaction as a management practice**
  - **Ownership** and care for equipment

SOURCE: "Practices to Reliability Performance" team

McKinsey & Company 4

Industry collects Petabytes of data which can be used for Predictive Maintenance to improve asset integrity and reliability

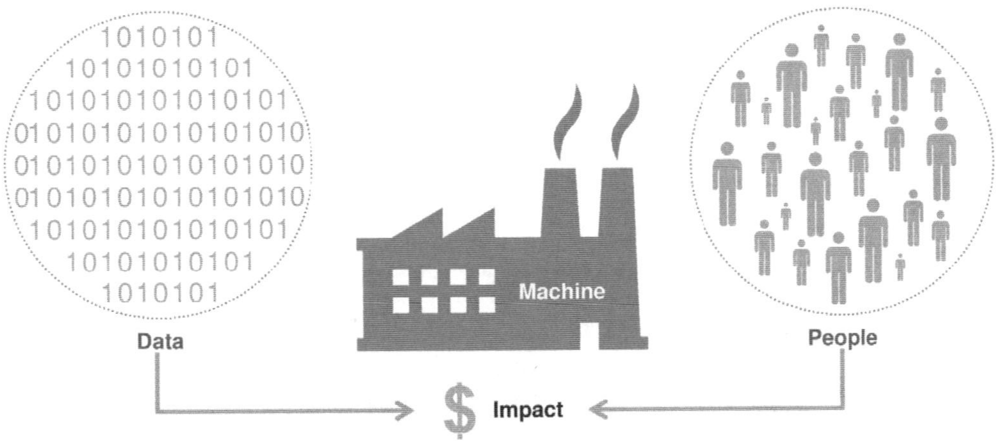

SOURCE: McKinsey

McKinsey & Company 5

Getting from data to impact requires a combination of many different skills

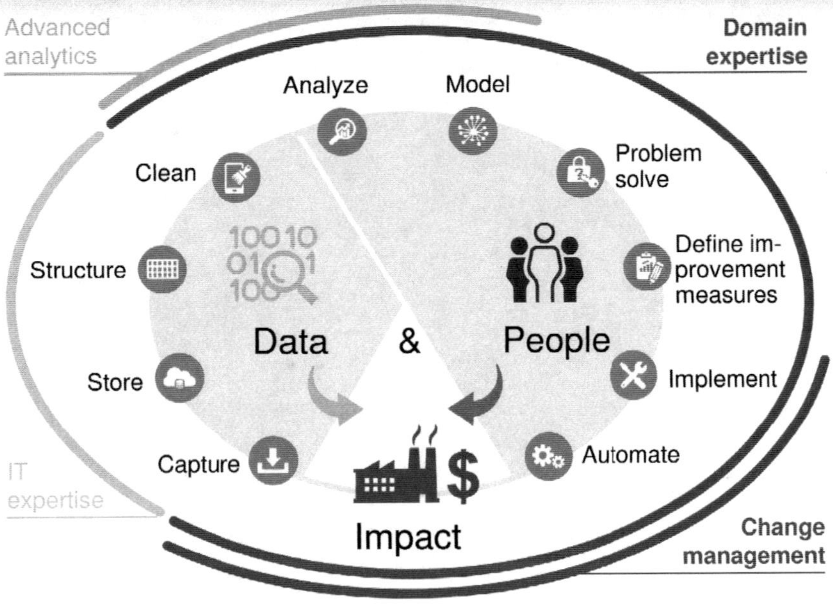

SOURCE: McKinsey

McKinsey & Company 6

Advanced Analytics generates previously untapped insights in Predictive Maintenance from troves of historical data

Predictive Maintenance
is the use of data (structured and unstructured, machine and non-machine based) to generate insights with advanced analytical methods that would not be possible with conventional techniques

Predict

Failure forecasting: Model future asset behavior to anticipate needs *(e.g., HR, spare parts inventory, capital replacement)*
Incipient failure detection: Continuous data acquisition & analysis to detect or predict failure *(e.g., early warnings to allow planned intervention)*

Avoid

Avoid failures
Identification of undesired patterns in data as primary causes of failure (e.g., preventive actions in specific locations of roads with poor conditions to prevent truck failures)

Learn

Leverage specific asset knowledge
Access reliability data from a global equipment database comprised of hundreds of thousands of components across the globe *(e.g. Failure rates, failure mode distributions, causes)* to inform maintenance strategies

SOURCE: McKinsey

The main benefits of implementing predictive maintenance include availability increase and cost reduction

Predictive maintenance aims to ...

... detect early signs of problem **for timely correction** at **minimal costs**

... prioritize and optimize maintenance resources using real-time data

... increase unit availability due to **shorter downtimes**

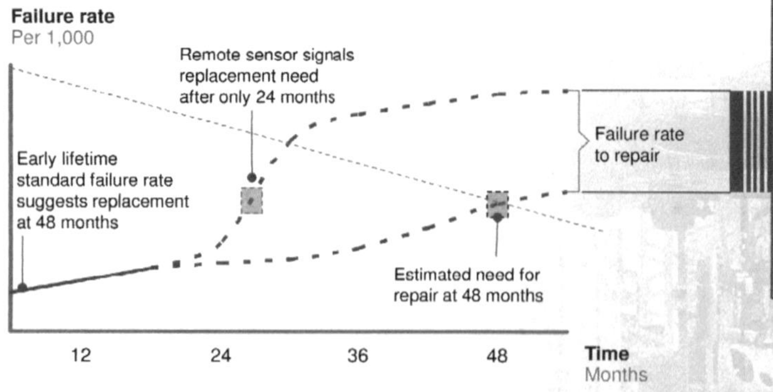

Typical impact of PdM

Machinery breakdowns	-50%
Spare part inventory	-20%
Total machine downtime	-50%
Overtime expenses	-20%
Machine life	+20 - 40%
Overall productivity	**+20 - 30%**

SOURCE: ARC Advisory Group; Department of Defense; expert interviews; GE; Schaeffler; VibroSystM; WiHART Systems

Predictive maintenance is proven to add value in many industries

What was Predictive Maintenance used for

✓ Focus of the engagement
Detailed in following pages

Industry	Reduce operating costs	Reduce maintenance costs	Increase machine availability	Impact
1 Mining			✓	▪ 25% downtime reduction
2 Mining *(Copper mine)*	✓	✓		▪ Savings of ~12% in operational expenditures
3 Oil and gas *(Rig operator)*		✓		▪ Reduced maintenance spend by 18 to 27%
4 Oil and gas			✓	▪ ~70% reduction in downtime on critical component
5 Railroads			✓	▪ Reduced failure cost by 20% to 30%
6 Heavy-Motors *(Cruise operator)*	✓	✓		▪ Reduced operating cost by 7%
7 Semiconductor	✓			▪ Increased failure prediction to 87%

We have more than 20 use cases. The remaining cases cannot be shared due to confidentiality.

SOURCE: McKinsey Asset Productivity Service Line

Summary of the case

Applying advanced analytics in off-shore oil and gas gives early failure warning can reduce production losses due to gas compressor by ~70%

Situation

Sector and site
- Offshore oil and gas operator
- Rig maintenance

Situation
- Gas compressors are **critical equipment**: they account for many failures and a large portion of production losses
- Loss data, maintenance work order data and sensor data is collected

Question
- To what extent can we reduce maintenance spend and improve reliability by predicting when a failure will happen?

Typical failure mode: inlet strainer

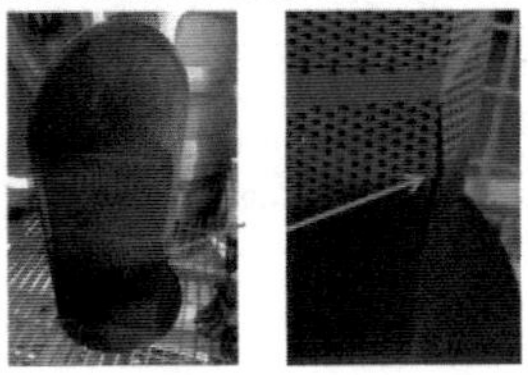

Key Activities

Data collection from several sources
- 120 gigabytes of data
  - ~1,200 sensors
  - 10 years of work orders and deferment data

Identify failures and construct target variable
- Identify if downtime is caused by compressor

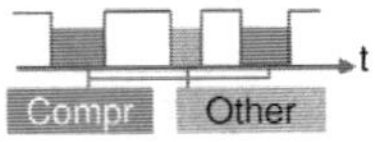

Create a model to predict time to failure
- For any point in time we predict the **time to the next failure**

Inputs 1,200 sensors → Prediction Model → Time to next failure

- **Inputs** read from 1,200 sensors
- **Regression Model** predicts the time to failure with ~80% accuracy

Impact

- Downtime due to gas compressor is expected to reduce by ~**70%** by planning intervention when a failure is imminent
  - **Prediction of time to failure:** if intervened at predicted TTF < 5 days, you would act between 9 and 6 days before the failure actually will happen

Predicted TTF for different failures, Days

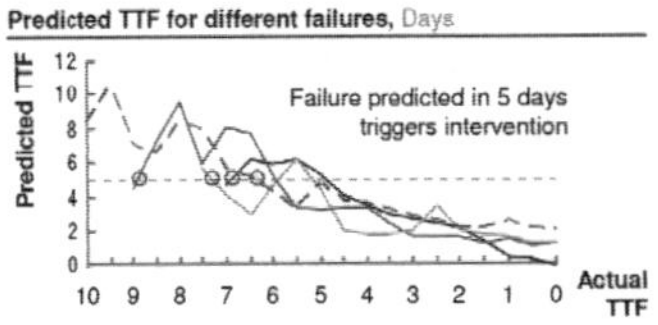

  - **Early warning of an imminent failure** means repairs can be planned. Reduction is on average from 14 to 4 days of downtime

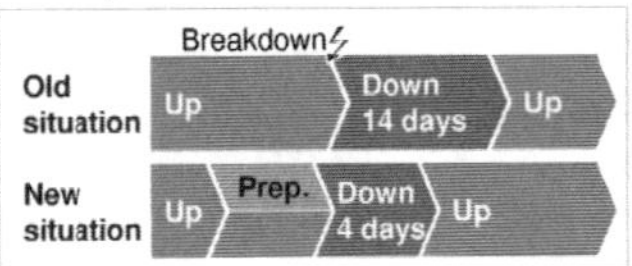

1 TTF = Time To Failure, each of the lines in the graph is one failure case

SOURCE: McKinsey

PREDICTIVE MAINTENANCE IN OIL & GAS

Details of the case

Data collection was based on 120 GB of data – including more than 1200 sensors and 10 years of work orders

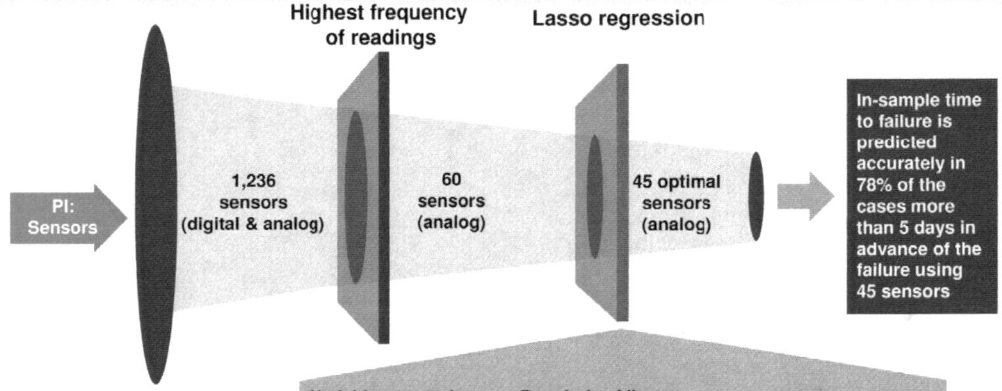

Variables created using sensor data	Regularized linear regression (LASSO)	Time to failure
▪ Moving averages on 6-hour time-windows	▪ Most informative sensors are identified in an automated way ▪ 60 analog sensors with highest frequency of readings reduced to 45 via cross-validation	▪ Binned to 10-min intervals

SOURCE: McKinsey

McKinsey & Company 11

PREDICTIVE MAINTENANCE IN OIL & GAS

Details of the case

In order to predict downtimes, data preparation as well as a clear target variable were keys to success

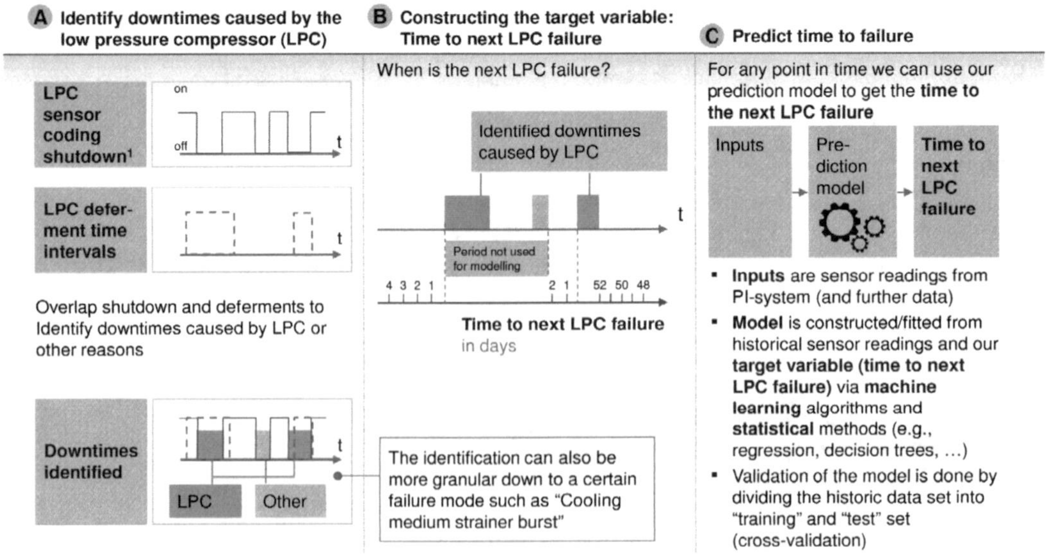

SOURCE: McKinsey

McKinsey & Company 12

Thank you

Carl March

Operations Practice Expert

- Predictive Maintenance

Carl.March@mckinsey.com

McKinsey & Company 13

FROM ANALYSIS TO VALUE CREATION

Top down value assessment for predictive maintenance

PRELIMINARY

Digital asset management

- ↑ **Availability 2-5%**
  - ↓ **Planned downtime 10-20%**
    - Decrease in work order frequency
      - Work order cadence to equipment needs instead of OEM recommendation
    - Decrease in work order duration
      - Word order instructions pushed to mobile devices
      - Improved preparation to increase employee wrench time
      - Fully digital and mobile enable maintenance process and cycle
  - ↓ **Unplanned 30-50%**
    - Actively performing routine work orders based on predictive analytics
    - Monitoring conditions of equipment to accurately predict failures
- ↓ **Maintenance cost, 5-25%**
  - ↓ **Operating maintenance costs**
    - Cheaper, routine work orders (predictive > reactive)
    - Improve part and tool inventory strategy with improved database
  - ↓ **Sustaining capex costs**
    - Longer equipment life through planning component replacement extension
    - Adjust to a smaller fleet size

McKinsey & Company 14

I The predictive model combines sensors in a way that cannot be replicated manually

Most important sensors for prediction

10 most significant out of a total of 45 sensors

Rank	Sensor	Unit of measure	Relative contribution Abs. value of coefficient
1	SEAL VENT	BARG	4.2
2	SEAL GAS VENT DP	BAR	2.3
3	FUEL GAS PRESSURE	BARG	1.9
4	COMPRESSOR FLOW DP	MBAR	1.8
5	INLET FILTER DP	MBAR	1.6
6	SEAL GAS SUP PRESS	BARG	1.3
7	ENG BRG2/3 DRAIN	DEG C	1.3
8	FUEL SYS AIR SUPPLY	BARG	1.2
9	FUEL GAS FLOW DP	MBAR	1.1
10	PRI NDE SEAL VENT	BARG	1.1

- The statistical model has identified a total of **45 sensors** to be significant when predicting compressor failures
- The contribution of the sensors is evenly distributed – there is **no one sensor that can replace the predictive model**, which combines a total of 45 sensors
- The absolute value of a coefficient can only be extracted in a linear regression case, and that other metrics should be used for other models e.g. variable importance in random forests

SOURCE: Team analysis, Sensor data

McKinsey & Company 15

II We are able to predict breakdowns at least 5 days before failure

Predicted time until failure. 12-hour average of prediction

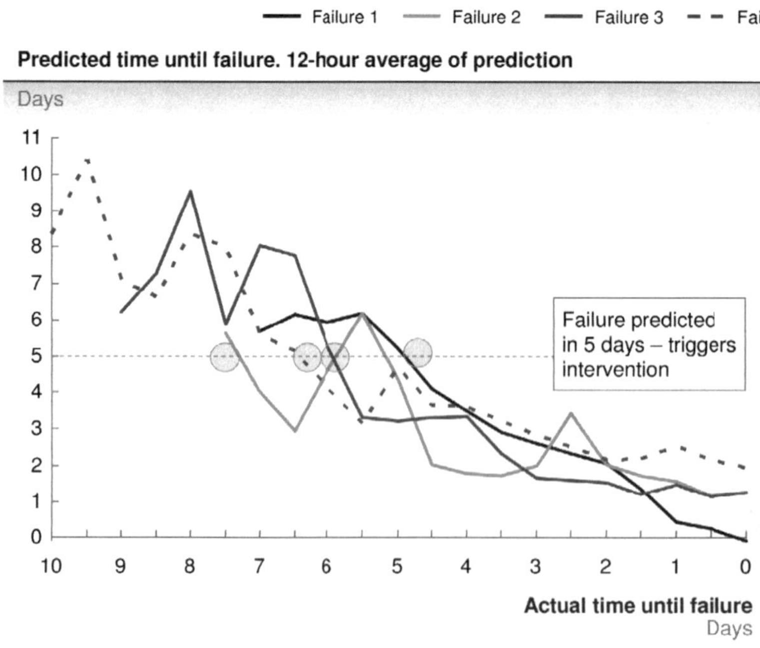

We are able to react prior to failures

- Early indications of failure can be seen 10 days in advance
- In all observed cases an intervention could be triggered more than 5 days prior to failure
- With normal reaction time the maintenance organization will have enough time to inspect and intervene

SOURCE: Team analysis, client data

McKinsey & Company 16

479971 Big Data Vs. Little's Law: A Systems-Based Framework Changing the Risk-Reward of Large Scale Facilities

Abstract:

The current rush to adopt Big Data ignores the fact that *the data is not the system*. Ignoring system dynamics in the interpretation of thedata often produces the opposite of the intended result. Fortunately, Little's Law ($L = \lambda W$) provides a framework to properly interpret the dynamic nature of data and diagnose the true underlying systemic problems. Little's Law is especially useful in identifying and eliminating *hidden factories* in the system that erase profits, increase systemic risk, and erode morale and culture. Dr. Carrier will explore examples from high-risk environments that have saved hundreds of millions of dollars per year while improving safety. In addition, the amount of data required to make and implement informed decisions is dramatically decreased through the use of Little's Law. The approach will be demonstrated with several examples in Oil and Gas, Refining, and Fine Chemicals worked on by the author. The methodology will also be demonstrated for use in maintenance management and the radical improvement of a mid-sized refinery's MoC (management of change process). Finally, the role of culture in actively the dynamic response of our systems will be discussed.

479057 Big Data Doesn't Necessarily Mean Big Intelligence

Abstract:

Big data is the IT phrase of the moment in almost every industry. The concept of collecting huge amounts of data to learn, manage and drive your business forwards is attractive to all managers, especially as without it you may feel that you're being left behind. But without knowing how to collect data effectively, knowing what data has value, and how to effectively and efficiently use the volume of information, Big Data can be a hindrance, not a help. In Process Safety, we rely on data coming from the field in many instances. Collecting this data can be time consuming and costly, so making the most of the data you have is paramount, as is ensuring the data you collect is seen as a valuable asset by everyone in your organisation. But turning that data into intelligence to guide you on your path to Process Excellence is what it's really all about. Data is as valuable an asset as any other. Look after it. Nutrue it. Value it. And make sure you use it. In this paper we examine the core principles of making PSM data work for you to give you the best usable, actionable intelligence possible.

Meeting the Future Organizational and Technical Training Challenges of a Changing Workforce

Dr. Stewart W. Behie
Manager Safety and Process Risk
Occidental Oil and Gas Corporation
Suite 110, 5 Greenway Plaza, Houston, Texas 77046
stewart_behie@oxy.com

Mr. Matthew K. Henwood
Manager Technical Training
Occidental Oil and Gas Corporation
Suite 110, 5 Greenway Plaza, Houston, Texas 77046
matthew_henwood@oxy.com

Prepared for Presentation at
American Institute of Chemical Engineers
2017 Spring Meeting and 13th Global Congress on Process Safety
San Antonio, TX
March 26 – 29, 2017

Meeting the Future Organizational and Technical Training Challenges of a Changing Workforce

Dr. Stewart W. Behie
Manager Safety and Process Risk
Occidental Oil and Gas Corporation
Suite 110, 5 Greenway Plaza, Houston, Texas 77046
stewart_behie@oxy.com

Mr. Matthew K. Henwood
Manager Technical Training
Occidental Oil and Gas Corporation
matthew_henwood@oxy.com

Keywords: technical talent, training and competency assessment, staff retention; multi-generational workforce;

Abstract

The workforce of the future will be dramatically different from that of today in several significant ways. In order to be successful in attracting and retaining technical staff in the future, Companies will have to gain an in-depth understanding of the nature of this technical workforce and be prepared to meet their needs and to match the demands of industry starting now. During most of the 20th century there were three generations actively participating in the workplace during which period older workers were generally the bosses and younger workers toiled under their guidance in a relatively structured manner. The workplace of today has gained a complexity never seen before, as for the first time in history in North America, workers from four distinctive and different generations will be working together. This dynamic will present new challenges and opportunities that need to be managed to achieve success. For example, by 2025, it is reported [2] that seventy-five percent of the workforce will be Millennials with their unique perspectives on the workplace, workplace interactions and their place in society. Millennials or Gen Y workers will be working alongside Baby Boomers, as well as Gen X, and Gen Z (Centennials). The workplace is changing fast and successful companies, which will rely on younger professionals to lead them into this uncertain future, must develop systems and structures that considers trans-generational needs of their workforce. All of these factors must be appreciated if companies are to attract and retain talented staff. Training and development programs will become a differentiator among companies in the future if they are to be successful. Attrition is an additional, significant challenge that can lead to accelerated departure of expertise at a time when the oil and gas industry is already experiencing a declining workforce resulting in part from the difference in focus among the different generations. Learning solutions must provide accelerated development compared to "traditional" methods (used for Boomers and early Gen X) and these must be repeatable, measurable and sustainable. The training environment and overall proposition has changed. Blended approaches to learning is the only solution and must reflect the reality that adult learners learn by doing.

This changing workplace profile will be taking place within a looming manpower shortage and potential crisis in all technical skills areas [1]. For example, high and very high skills gaps are predicted to occur in many developed countries including those in North America, and in the fields of engineering and construction starting as early as 2020 [2]. To be successful in this changing world, companies must meet the additional challenges of developing and promoting new approaches to attract and retain the dwindling technical resource of the future. These approaches could include a range of considerations such as: promoting virtual internships and work schemes; developing innovative ways to integrate Human Resources (HR) considerations into corporate strategy development; creating growth opportunities to retain staff; extending employment opportunities for women and older employees on their terms; and fostering talent mobility. All of these approaches, in and of themselves, will present substantial challenges particularly in face of the recent trend towards nationalization and away from globalization.

This paper will examine all of these factors and discuss ways to use these insights to optimize learning processes and customize training programs to match the needs and requirements of the future workforce. This paper will suggest novel approaches to training, skill set development and competency assessment. Finally, this paper will profile one company's approach to understanding the changing future workforce and reflecting this knowledge in technical training, skills set development and competency assessments in order to maintain a position as a leader in the oil and gas sector through innovation, integrity and prudent investment. The challenges of the future are many and will only be met by those who understand them and put in place plans to address them.

1 Introduction

History has shown that to be successful a company in the energy sector they must be able to attract and retain technical staff on an ongoing basis. The energy sector has faced and will continue to face a period of difficulty in this regard. The demographic gap has effectively produced an associated gap in the talent pipeline around the mid-career sector. A company in the energy sector must benchmark senior talent capability and provide rapid solutions to close the early to mid-career gap while also addressing advanced technical skillsets in order to make the solutions sustainable. In a fast changing world this is a real challenge for a number of reasons including the changing profile of generations over time in terms of their personal focus, interests and overall requirements and demands of the workplace.

During most of the 20th century there were three generations actively participating in the workplace; older workers were generally the bosses and younger workers toiled under their guidance in a relatively structured manner. The workplace of today has gained a complexity never seen before. For the first time in history in North America, workers from four different generations will be working together. This dynamic will present new challenges and opportunities that need to be managed to achieve success. By 2025 [2], it is reported that seventy-five percent of the workforce will be Millennials with their unique perspectives on the workplace, workplace interactions and their place in society. Companies that are successful at attracting and retaining the best and brightest talent have and will continue to have a distinct economic advantage over the competition. However, this advantage will become more difficult to achieve and maintain over time. In order to maintain this advantage. It is critically important for companies not only to understand the dynamics of the changing workforce but also to establish organizational structures and programs that will be able to adjust to meet the changes in the technical workforce of the future. This process should start now since significant change has already started. Statistics from the US Census Bureau[18], indicate that the oil and gas industry will experience the departure of about 22,000 petro-technical professionals over the next five years in North America alone while attracting about only 17,000 professionals. Even allowing for the impact of technological efficiencies of increased productivity and operational capability, the industry will fall 20% short of maintaining status quo compared to current conditions. The solution is not only to drive accelerated development but also to provide a viable career proposition to attract new talent and create a diverse and growing workforce.

The situation will soon be further complicated and severely constrained by the looming crisis globally in manpower shortage in all technical skills areas [1]. For example, high and very high skills gaps are predicted to occur in many developed countries including those in North America in engineering and construction starting as early as 2020 [2].

2 The Profile of the Generations Making up the Workforce of the Future

The diversity of the workforce has never been as great as it is today. Multiple generations are in play in the current workforce and each has their own unique profile of desires, interests and requirements. A generation can be defined as a group of people born around the same time and raised around the same place that exhibit similar characteristics (such as communication, shopping, and motivation preferences) and experienced similar trends at approximately the same life stage and through similar channels (e.g., online, TV, mobile, etc.). Generation-shaping trends are most influential as people come of age, which means that members of a particular generation will develop and share similar values, beliefs and expectations. It is important to remember that at an individual level, everyone is different and people exhibit a wide range of trends. However, by looking at people through a generational lens, one can begin to understand how to reach, inform or influence certain segments of a population.

The three key trends that shape generations are parenting, technology and economics [13]. To be able to engage the workforce of the future, it is helpful to use this "generational lens" to appreciate what it means to become an employer of choice and to maintain this position over time by adjusting strategies to meet the needs of the changing workforce. Employers will need to address the widely divergent characteristics and idiosyncrasies of several generations at the same time, which is new territory for all companies. This is a tremendous challenge that will require significant adjustments to be successful in the not-so-distant future. It begins with understanding the attitudes and requirements of all the generations that will contribute significant numbers to the workforce of the near future.

The generational timeline applicable to North America is described below along with an overall profile of their overall characteristics. The GI generation (born before 1936) has essentially left the workplace and is mentioned here only for the sake of completeness. Referred to as "the greatest generation", the GI generation had the distinct experience of a) growing up during the great depression after the stock market crash in 1929, b) fighting in WWII, 3) experiencing the fear of communism and prohibition, 4) championing the cause of civil rights and e) joining President Franklin Delano Roosevelt's "new deal" to save the civic fabric of the world and rebuild it.

The generations in play in the current workforce are broadly greneralized below. The range in dates that bracket each generation vary a bit from reference to reference

- **Traditionalists or the Silent Generation (born between 1937 and 1945):** Although not strong in numbers, members of this generation are still active in the workplace and will be for a few more years. Traditionalists lived in a pre-civil rights society and went through their formative years during an era of suffocating conformity, but also during the postwar happiness of peace, jobs, suburban living, television, cars, and rock and roll. [15] This generation believed in marriage for life and viewed divorce and having children out of wedlock as unacceptable. The Traditionalists women were pre-feminists who stayed at home to raise children and, if they worked, it was in only certain jobs like teaching, nursing or secretarial. Men pledged loyalty to the company they worked for and generally kept a job for life. Traditionalists had a strong sense of values, were disciplined, cautious and self-sacrificing and were avid readers. Retirement for Traditionalists meant sitting in a rocking chair and living out your life in peace close to home, family and friends.

- **Baby Boomers (born between 1946 and 1964):** Boomers, one of the largest generations in history with about 77 million, defined the American workplace for many years with their workaholic tendencies and work-to-live attitudes. Boomers were optimistic, driven, and team-oriented and tended to be positive about authority, hierarchal structure and tradition. Boomers ushered in the free love and societal, non-violent protests (which did trigger violence). Even though mothers of the Boomer generation were generally housewives, responsible for all child rearing, Boomer women began working outside the home in record numbers, thereby changing the entire nation. This generation was one of firsts: the first generation to have their own children raised in a two-income household where mom was not omnipresent; the first generation with television; the first generation where divorce was beginning to be accepted as a tolerable reality; the first to begin to accept homosexuals; and the first generation to use the word "retirement" to mean being able to enjoy life after the children have left home. Instead of sitting in a rocking chair, Boomer retirees go skydiving, exercise and take up hobbies, which increases their longevity. Boomers are still in the workforce in substantial numbers and their aging has changed and will continue to change America almost incomprehensibly [(15)]. Boomers embrace the notion that "physiology is more important than chronology" [(17)].

- **Generation X (born between 1965 and 1976/1980):** The Gen X generation was raised in the transition phase of written based knowledge to digital based knowledge. Most remember being in school without computers or saw the introduction of computers in middle school or high school. They tend to commit to self rather than an organization or a specific career. This generation averages 7 career changes in their lifetime. It is not normal for a Gen Xer to work for a company for life, unlike previous generations. Gen Xers were raised by the career and money conscious Boomers amidst the societal disappointment over governmental authority and the Vietnam War. They are survivors as individuals, suspicious of all organizations and wary of commitment. Gen Xers were late to marry and quick to divorce [(15)].

- **Millennials or Generation Y (born between 1977 and 1993/96):** This generation is said to be characterized by a short attention span, an instant gratification focus and are technological savvy. They exhibit a sense of entitlement born of the parenting style they were raised under. The generation-defining moment for Millennials was the events of September 11, 2001 ("9/11"). Millennials have very different attitudes toward employment, sales and marketing and these attitudes are challenging many conventional strategies and approaches. Millennials changed every other generation's dependence on social media, text messaging and digital downloading [(14)]. Currently, Millennials have become the largest generation in terms of numbers in the workplace and are the fastest growing generation of customers in the marketplace [(13)]. By 2025, Millennials will make up 75% of the workforce [(13)]. To attract and retain Millennials, employers will be required to make significant changes and investment and provide: state of the art equipment; mobile access and mobile learning (m-learning); workplace flexibility (which is valued over pay) with a teamwork focus; continuous learning opportunities that focus on leadership, team management, self-efficacy building and public speaking [(13)]; ongoing and immediate assessment and feedback on performance; and meaningful mentoring and coaching.

In short, employer-provided technology must meet the expectations of Millennials to enhance retention. Adjustments will need to be made in digital information security and external communication policies as well as cyber security systems to allow the free and open access desired by Millennials.

- **Centennials or Gen Z or iGen (born in 1994/96 or later)**: The start of Gen Z in the North America is closely related to the events of 9/11 in that members of this generation, unlike Millennials, cannot process the significance of 9/11. Members of this generation cannot recall a time when the intranet and mobile devices did not play such a dominant role in society. Gen Z will determine how generations that follow them will ultimately use technology such as wearable devices and the dependence on the "cloud". Gen Z has assumed the Millennial's mantle of the generation to know, understand and engage [14]. Gen Z is the first generation to grow up watching their parents (Millennials or Gen X) glued to their smartphones, which drives their desire to have their own smartphone at an early age. Gen Z children have never known a world without computers and cell phones. They tend to leave their conventional toys behind at a younger and younger age, begin to play with computers at ages 4 or 5 and prefer electronics, such as cell phones and video games, to play with. Gen Z members are currently the majority in college and the trendsetters, particularly in technology. This generation will experience a change in the American melting pot. For example, of the record number of births in the US in 2006, 49% were Hispanic. Gen Z has outnumbered the start of the Boomer generation and will easily be a larger generation than the Boomers. Employers must start now to understand and unlock the power of this entirely new generation [14] by identifying and addressing their specific needs. Some of the characteristics that define Gen Z include access to mobile devices at an early age (and have their own mobile phone as young as 13).

3 Working in a Multi-generational Workplace

For the first time in the history of North America, we now have four different and distinctive generations working side-by-side; namely, Boomers, Gen X, Millennials and Gen Z [16]. For much of the 20th century in North America, (before the year 2000 or so), the workplace was characterized by three generations working side-by-side. During this period, the older workers were generally the bosses and the younger workers toiled under the guidance of an older boss doing what they were asked to do. This workplace was quite structured and most workers followed the established rules and norms of the workplace, both written and implied. Computer technology started to gain a foothold in the early 70's before which time electronic aids and communication devices were largely unavailable. The pace of change was slow until the last 25 years or so.

Roles in the near future will no longer follow the norms and rules of the past. The differences among these generations have started to exert pressures and changes on the workplace and will continue to do at an increased pace in the near future. Generational differences affect all aspects of the workplace including HR processes of recruitment and engagement and remuneration; bonus schemes and timing; working environment; approaches to work; communications both internal and external to the organization; training and development; organizational structure; employee motivation and advancement. The workplace will be changed forever at an unprecedented rate, in many directions and in ways that we cannot imagine today.

One of the major challenges of the workplace in the near future will be communications since it has been shown that how people communicate is based in large part on their generational backgrounds [16]. To be successful, companies will have to implement communications systems that will allow the different generations to communication effectively to avoid unnecessary discourse, misunderstandings and confrontation. Companies will also need to offer flexibility in working times, schedules and places as well as the structure of offices. Even today, some companies have already significantly changed the design of their workplace by introducing an "open office" concept when employees work from the office. Another critical area will be the delivery of employee development programs and plans that offer all styles of high quality, professional and technical training as well as employee assessment and meaningful feedback. To be attractive, these programs must appeal to the advanced technology and communication skills of the younger generations. In addition, programs need to reflect the learning audience and the adoption of adult learning methodologies by covering concepts coupled with simulation and on the job real world applications. To add to the complexity this training needs to be conducted in about half the time that was historically taken to develop technical personnel. By the mid 2020's our Gen Xers will also be departing the industry in large numbers such that by 2015, Millennials will comprise the majority of the workforce [(18)].

Companies are well advised to start to profile their workforce in terms of the numbers of employees in each of the generations and to update this profile on a regular basis to fully appreciate the dynamics at play within the organization. Free and open discussions in group settings will help each group to understand these dynamics and discuss the issues of concern and to keep this dialogue open and ongoing. Creativity and innovation are hallmarks of successful companies and many have already made significant changes to foster these traits by effectively managing their intergenerational groups.

4 The Looming Manpower Shortage in Technical Skills Areas

In addition to the challenges related to managing the complex dynamic of a multi-generational workforce involving four distinctive and different generations, a separate challenge is looming in the very near future. This challenge relates to the availability of employees with the requisite skill sets to meet the future staffing demands of companies in the oil and gas industry. The gap between the number of experienced and inexperienced engineers and other technical professionals in the industry will peak in 2016 and is predicted to continue until about 2022. This shortage of experienced engineers is predicted to increase the probability of costly mistakes being made by novices [(18)].

Publications issued by the World Economic Forum in recent years [(1)] [(2)] have highlighted the looming manpower shortage / crisis in all technical skills areas. In the report of the World Economic Forum published in 2010, entitled "Stimulating Economies through Fostering Talent Mobility" [(1)] the disturbing claim is made that *"The global economy is approaching a demographic shock of a scale not observed since the Middle Ages. The working-age populations of many developed economies will start to decline shortly. Numerous organizations will be unable to find enough employees in their home markets to sustain profitability and growth. By 2050, the global population of those 60 years old and older will exceed the number of young (less than 15 years old) people for the first time in history"*. Figure 1 below, an exhibit from this report, shows predictions of skilled worker availability for a wide range of industries for the period 2020 to 2030 for a number of countries in the Americas, Europe, Asia and Oceania. The blue color coding indicates how countries are positioned to face shortage challenges with the dark blue indicating a high ability to attract migrant workers because of the attractive attributes of the country while the medium and light blue color code indicates medium and low abilities to attract migrants. The skill gaps are color coded as follows: very high skills gaps (red); high skills gaps (orange); and low skills gaps (yellow) and no skills gaps (white).

Focusing on North America, high and very high skills gaps are predicted to occur in Canada and the US in a number of critical fields including engineering and construction, education and IT between 2020 and 2030. This situation has the potential to curtail economic growth around the world and fundamentally change the workforce challenges. Notably 80% of the impending labor shortage will involve skills, not numbers of workers potentially available. Those companies that are successful at attracting and retaining the best and brightest talent from this diminishing skilled labor pool will have a significant economic advantage over the competition.

The ability to attract and retain talent will be dictated, in part, by meeting the needs of new employees, a very important component of which is the growth and development opportunities provided by their employers. Companies need to reflect an awareness of this looming manpower crisis in their Learning and Management Systems not only to stay ahead of the changing needs of new talent but also to reflect the different demands of the various generations that make up the workforce in the near future. For example, to meet the expectations of Millennials, companies must develop achievable development plans that meet the technical requirements of the associated roles along with the talent pipeline. These plans must be readily available to all and also be sustainable, measurable and repeatable while still providing an accelerated and directed path.

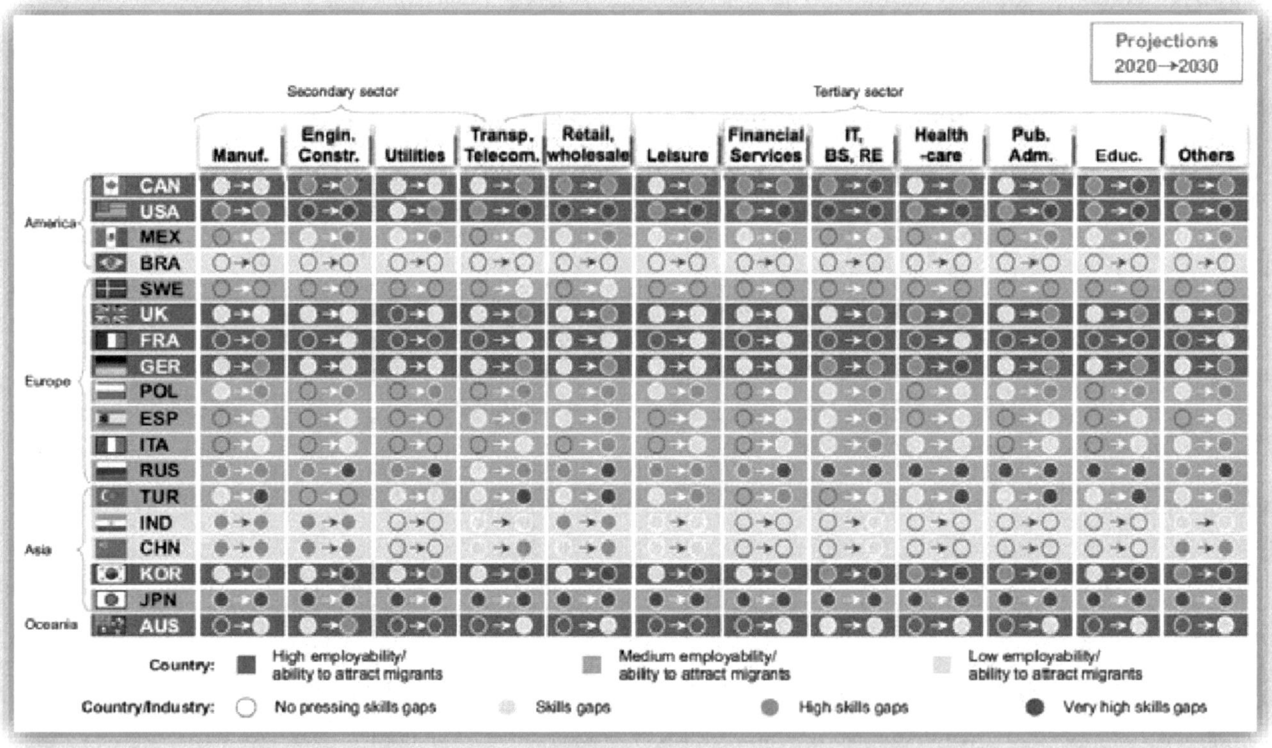

Figure 1 Projections of Skilled Worker Availability 2020 – 2030

4.1 *Importance of Skilled Labor Migration*

Technical manpower shortages in the past have largely been met by the migration of skilled labor from less developed countries to North America and other desirable workplaces. Figure 2 [(1)] below shows a snapshot of the importance of highly skilled immigrant workers to the economic well-being of a number of countries. The significant share of highly skilled migrant workers as the overall percentage of domestic skilled labor is quite significant as shown in Figure 2 as is the average contribution of highly skilled migrants to the GDP of those countries. Maintaining an increasing the flow of highly skilled migrant workers into North America will be very important to companies in the oil and gas sectors if they are to continue to grow and prosper financially.

Talent mobility can pose short-term challenges to the receiving country in terms of integration as well as sending countries in terms of a "brain drain". In the long term, positive benefits can accrue to both receiving and sending countries. However, these positive benefits could be undermined if protectionism policies are implemented by a number of countries. Such policies can compound the challenges that the oil and gas industry faces in meeting their demands of skilled labor in the very near future. One of the long-term benefits to the sending country (1) is the return migration of highly skilled workers to their home countries as conditions and opportunities improve. Planned and mutual cooperation between sending and receiving countries should be developed and implemented to their mutual benefit when the receiving country performs the role of a training ground for skilled workers of the sending country who return on a planned, scheduled basis to repatriate their honed skills. For international companies the needs are largely similar the world over. Career plans, training and development solutions and resources should speak to the global workforce. This drives mobility to get the top talent in the right job at the right time efficiently, consistently and cost effectively to ultimately serve a higher purpose of supporting a desired operating culture.

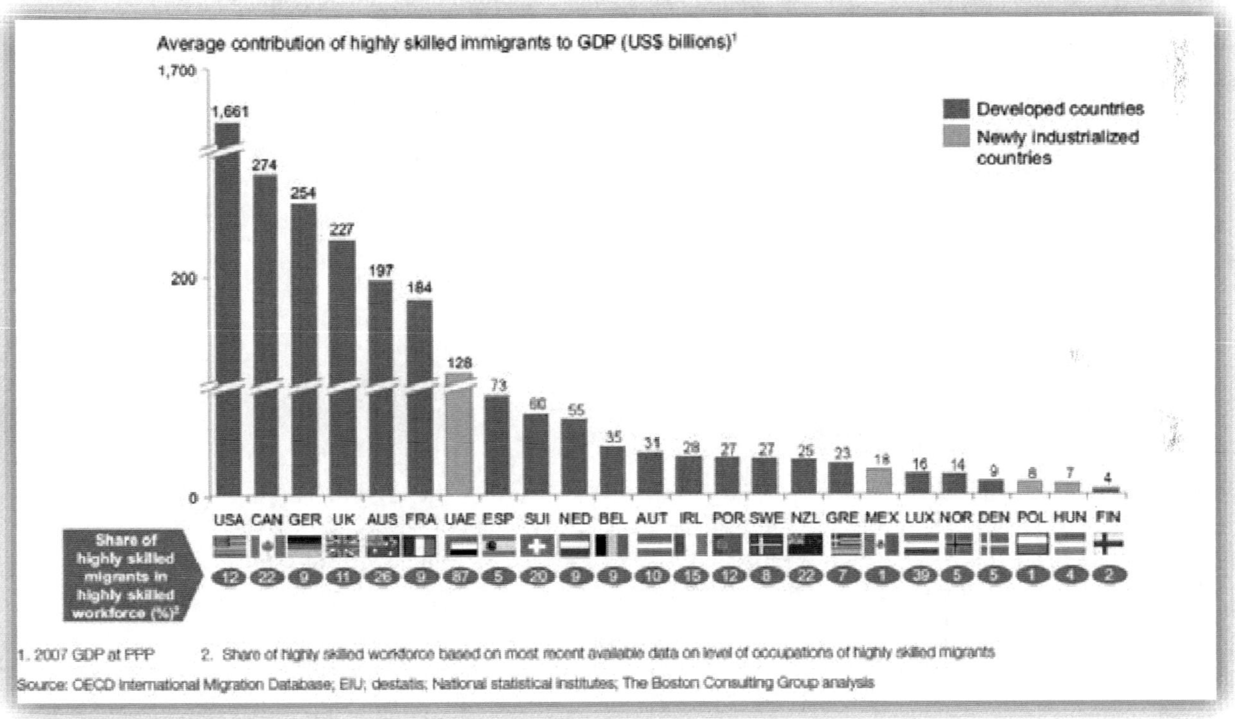

Figure 2 Talent Migration Trends

4.2 Meeting the Global Talent Risk

Figure 3 below, an excerpt from the report published by the in 2011 entitled "Global Talent Risk – Seven Responses" (2), summarizes seven key factors that need to be in addressed in order to increase the chances that the future manpower needs of the oil and gas sector in North America are met. Companies can use these factors in developing strategies to frame the approaches they will take to understand the changing workplace environment and to attract, train and retain new technically skilled staff and to retrain and invigorate existing staff to embrace these substantial changes. Those companies that are willing to invest the energy to prepare for this uncertain future will be better positioned to address the looming deficits in skilled talent.

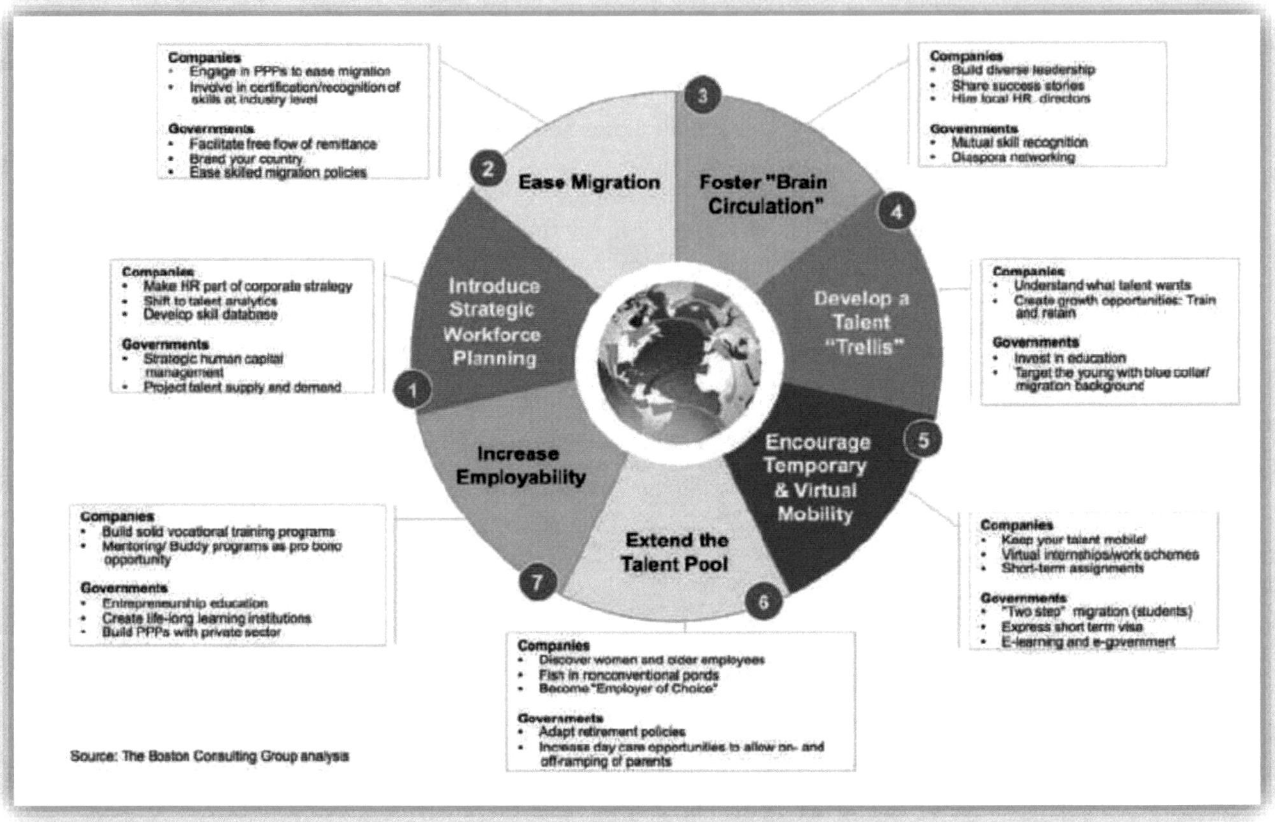

Figure 3 Checklist for Formulating Strategies to Address the Global Talent Shortage Risk

Of the checklist actions outlined in Figure 3, those that are most significant for companies to meet the goal of developing and maintaining a skilled workplace are:

- Build diverse leadership that embraces the HR function as a critical component of corporate strategy, reflects the demographics of the multi-generational workforce at play and fully appreciates and understands what is involved to "become an employer of choice"
- Understanding what their multi-generational staff want in terms of growth opportunities and to cater to these by providing training and retraining suited to each generation including the promotion of virtual internships and work schemes.
- To develop and promote creative ways to extend employment opportunities of older employees in meaningful and unique ways and to encourage the engagement of parents with young children through increased day care opportunities.
- Develop creative and inspiring training programs that embrace the learning attributes of all generations in the workforce and keep these evergreen and exciting and provide ongoing feedback and assessment by way of skills and competency assessments.

Governments have a significant role to play in addressing the looming shortage in skilled labour by supporting company initiatives through investment in education, revised policies related to retirement age and immigration. Only with all parties working in concert will the future demand for skilled labor be met and the negative impacts of skilled labor shortages be avoided.

5 One Company's Approach

In 2014, Occidental Petroleum Corporation (Occidental) identified a demographic and potential competence gap in its skilled labor workforce. Many of its employees were older and approaching retirement age or were younger, newer hires; there were not a lot of employees in between. As older workers retired, there was a risk that the current workforce might lack the experience, skills and numbers to replace them.

To close the gap, Occidental enhanced its training program to focus on technical skills development and competency, using online and in-person training. The company viewed targeted training as a way to retain talent by providing opportunities to promising employees, as well as to facilitate knowledge-sharing by potential retirees within the organization. Key actions focused on training, skills development and competency assessment in all technical areas.

In 2015, the company opened a state-of-the-art training center in Midland, Texas, in the Permian Basin, the base of company's domestic operations. The facility features practical and simulated hands-on training. Operational and field staff had access to customized programs and flexible training schedules, with a focus on providing foundational knowledge and behavioral-based competency. More experienced employees then helped with complementary on-the-job training centered on the "how" and "why" to accelerate the learning process. In the first 12 months of operation, the Midland training center has had approximately 2,300 technical personnel (an average of 190 staff per month) participate in training. The training totaled about 31,000 hours of instruction in a blended environment and involved more than 74 different programs spread over 194 offerings, most of which were integrated with simulation and hands-on activities.

Occidental implemented a new Learning Management System (LMS) in 2016 to support training programs and curricula development with a focus on job role, discipline and position. Employees across the company could now easily access structured training and competency assessments, along with self-directed and just-in-time learning opportunities. The LMS also provided a process and avenue for employees to demonstrate their competence in their current role and qualifications for advancement within the organization.

Beyond the classroom, Occidental is working to create a culture and infrastructure that can support development in a consistent manner inclusive of mentoring programs.

Occidental recently initiated a business transformation program to search for and implement value-added innovative ideas that will propel consistent breakthrough performance. The program seeks to mine and capture the substantial talents and innovative ideas of the entire workforce in a focused and structured manner. The program started in 2016 and will be implemented over the next few years in a multi-phase approach: design, build and deploy, enhance and finally sustain.

All these efforts are facilitated by a mature and integrated technical training team that is well positioned to adjust to and meet the learning styles of all the generations that make up the company's workforce.

6 Conclusions

The workplace in North America is changing very dramatically and within a relatively few years will take on a dynamic never experienced before in history. By 2020, four different and distinctive generations will be working side-by-side. The differences among these generations have started to exert pressures and changes on the workplace and will continue to do so at an increased pace to affect all aspects of the workplace. The workplace will be changed forever in many directions and this change will probably move in directions that are unanticipated today.

To be successful in this new workplace, companies must understand and start to prepare now to meet the challenges that these substantial changes bring. The first step in this process is to understand the detailed profile of the demographics of their current workplace, update this profile on a regular basis, and include an estimate of what the future profile will look like in 3 to 5 years based on the trends of these profiles. These profiles will frame the nature of the multi-generational workplace over time and be the basis of integrated, development plans to meet the future manpower needs of the company.

Critical in this regard is to implement an integrated approach to training and development that:

- reflects the training styles and needs of each generation at play in the workforce
- provides hands-on skills and competency assessment on an ongoing basis
- proposes development programs for highly skilled programs (i.e. risk management, etc.)
- allows cross training and advancement in all directions (the corporate ladder must become a trellis)
- includes a leadership training program that identifies high-potential candidates and provides extensive training and exposure to all aspects of the business,

Working with universities will be important to:

- update technical training program requirements to meet company needs in the short term and long term and actively participate in training, if possible
- develop new programs that speak to the entrepreneurial propensity of the newer generations
- include a substantial number of internships in technical skill areas to promote cooperation and coordination with educational institutions

Active interaction with state and federal governments will be needed to:

- streamline visa processes for foreign students to study at North American universities;
- streamline work visas upon completion of degrees to allow foreign students the opportunity wo work for companies in North America
- develop innovative child care programs that facilitate younger mothers to stay in the workplace;
- revise retirement age restriction where they exist to encourage older workers to keep working after "retirement" age on a basis defined by them

Human Resources departments need to become human capital management departments and an integral component of corporate strategic planning. They should provide training programs that are flexible to encourage retention of staff from all generations. Innovative approaches are needed.

The successful companies of the future will be those companies of today that recognize the nature of the significant changes on the horizon and aggressively put in place plans to embrace the excitement of this new changing dynamic to meet its challenges. Are you ready for the future?

7 References

1. "Stimulating Economies through Fostering Talent Mobility", World Economic Forum in 2010.
2. "Global Talent Risk – Seven Responses", World Economic Forum in 2011
3. "The Coming Neurological Epidemic", Gregory Petsko, Arthur J. Mahon Professor of Neuroscience, Brain and Mind Research Institute, Weill Cornell Medical College, TED talk, February 2008.
4. "How Great Leaders inspire action", TED Talk by Simon Sinek, filmed Sept 2009
5. "The Myth of the Teen Brain", Robert Epstein, Scientific American, April / May 2007
6. "The Amazing Teen Brain", Jay N. Giedd, Scientific American, June 2015
7. "The mysterious workings of the adolescence brain" Sarah-Jayne Blakemore, TED talk, filmed June 2012
8. "The Learning Brain, Lessons for Education", Sarah Jayne Blakemore and Uta Frith, Blackwell Publishing, 2005
9. "The Neurons That Shaped Civilization", Vilayanur Ramachandran, TED India Talk, filmed November 2009.
10. "Growing Evidence of Brain Plasticity", Michael Merzenich, TED Talk, filmed February 2004
11. "Three Clues to Understanding Your Brain", Vilayanur Ramachandran, TED Talk, filmed March 2007
12. "L&D Meets OMG, Learning and Development for Millennials" Bridge, 2015
13. "Generational Breakdown: Info About All of the Generations", Center for Generational Kinetics, Austin, Texas (info@GenHQ.com)
14. "Gen Z Tech Disruption" 2016 National Study on Technology and the Generation after Millennials", The Center for Generational Kinetics and Jason Dorsey, 2016
15. "The Six Generations in America" Dr. Jill Novak, University of Phoenix, Texas A&M University
16. "Mixing and Managing Four Generations of Employees", Greg Hammill, Executive Director for Human Resources Management Studies, Fairleigh Dickinson University, 2005
17. "Making 80 the new 60", Occidental Oil and Gas, internal presentation, Dr. Stewart W. Behie, 2016
18. "Attracting, Retaining & Developing Technical Talent", Matthew Henwood, Technical Training Manager- Growth & Operations Support, Occidental Technical Training Center, September 15, 2016, Marcus Evans Energy Series Conference, Houston.
19. "The Great Generational Shift: The Emerging Post-Boomer Workplace", Bruce Tulgan, Rainmaker Thinking, White Paper, 2016
20. "2016 Trends in the Workplace", CallisonRTKL, 2016
21. "Understanding Generational Differences", Grant Wood Area Education Agency, 2012
22. "Generational Diversity, Teaching and Learning Approaches", Susan A. Johnson and Mary L. Romanello, Nurse Educator, Vol.30, No. 5,September / October 2005
23. "Managing People from 5 Generations" Rebecca Knight, Harvard Business Review, September 25, 2014
24. "Meeting the challenges and opportunities of today's new multi-generational workforce", Fairsail, July 2016
25. "Understanding and managing the 4 Generations in the Workplace", Robert Tanner, Management is a Journey, June 28, 2916

Sustainable Safety – A Case Comprehension

Kumar (Chris) Israni
Principal Consultant - ERM
840 West Sam Houston Parkway North, Suite 600
Houston, TX 77024
Chris.Israni@erm.com

Liam Doyle, CSP
Process Safety Consultant - ERM
3838 North Causeway Boulevard, Suite 3000
Metairie, LA 70002
Liam.Doyle@erm.com

Prepared for Presentation at
American Institute of Chemical Engineers
2017 Spring Meeting and 13th Global Congress on Process Safety
San Antonio, TX
March 26 – 29, 2017

Sustainable Safety – A Case Comprehension

Kumar (Chris) Israni
Principal Consultant - ERM
840 West Sam Houston Parkway North, Suite 600
Houston, TX 77024
Chris.Israni@erm.com

Liam Doyle, CSP
Process Safety Consultant - ERM
3838 North Causeway Boulevard, Suite 3000
Metairie, LA 70002
Liam.Doyle@erm.com

Keywords: Process Safety, Sustainable Safety, Risk Management, Human Factors, Asset Integrity, Accelerated Learning.

Abstract

We have seen a lot of changes in the approach to process safety and risk management over the last four decades, within the Oil and Gas (O&G) and Chemical sectors. Learnings from catastrophic incidents in the industry, led to the evolution and enhancement of risk techniques, as well as global regulatory changes. The last decade has also seen the evolution of the concept of sustainability. One of the key elements for a company in evaluating their sustainability performance is safety; thus tying safety into each company's overall goal to strive towards operation excellence.

This paper elaborates on the sustainable and sequential application of process safety methods in a risk based manner for the entire lifecycle of the facilities. That assures longevity for the facilities and thus operational excellence. The case comprehensions discussed in this paper will highlight the ways we've witnessed/assisted various groups and organizations to achieve success in this area.

1 Introduction

Effective and sustainable process safety programs are vital to the overall health and viability of a company from a business continuity perspective; as evidenced by the devastating business impacts previously incurred by those companies that failed to effectively sustain such programs. Unfortunately, in our experience, we find that even though some companies may spend up to 25% of their operational expenditures on safe production programs, only about 10% of that spending generally ends up going towards programs/activities capable of driving actual improvements in safety and operational performance [3,8]. Sometimes this may be due to inefficient or unwieldly leadership programs. Other times it may be the result of an unbalanced allotment of resources aimed more towards controlling process hazards via personal, compliance-based safety programs, instead of mitigating and/or eliminating process hazards, efficiencies, and/or weak spots via process safety and risk management programs. However, regardless of the cause, such investments are highly inefficient and simply unsustainable. Thus, we recommend such companies redirect their efforts more towards those programs actually capable of driving sustained improvements in process safety and operational performance, such as Sustainable Safety [8].

2 Safety Leadership Programs

Several leadership programs currently in use today focus on encouraging workers to "own" safety as a common practice. Some of the methods used by such leadership programs to encourage workers to "own" safety include: (1) empowering employees to take action when they witness unsafe behaviors; and (2) educating employees about the off duty activities that could possible lead to unsafe acts at work. However, in order for these programs to be effective the management teams facilitating the programs must first develop a work environment that's consistent with the different rationales and expectations present within their multi-cultural workforce [4].

For example, as we've discussed in a previous publication [4], failing to account for the regional customs in the workforce can inhibit the very feedback such leadership programs are desperate to receive. In this publication [4] our colleague, Tianherng (Joshua) Yang, provides an example of a workforce hesitant to report the major safety faux pas being noticed amongst peers and supervisors to the appropriate authorities. In most cases, this hesitation would likely be attributed to a fear of backlash and thus addressed by promoting "whistleblower" protection programs. However, in this workplace, the workforce culture placed an exceptionally high emphasis on respect. Thus, the root cause of this hesitation was also due, in part, to the workforce's fear of disrespecting their peers and supervisors [4]. Failing to account for the respect-based aspects of this workforce's culture would have prevented any employee feedback encouragement programs, such as a "whistleblower" protection program, from achieving its intended objective.

3 What is Sustainable Safety? [3]

Sustainable safety is a concept we've termed to describe how the people working within a typical manufacturing facility utilize the facility's existing processes (management systems, procedures, permits) to operate and maintain the plant (equipment, machines, vehicles, etc.) aspect of the facility's operations. As shown in the figure below, exceptional operational performance typically occurs at the interface between the People, Process, and Plant elements; with Technical Safety (Process Safety/Risk Management) generally occurring at the interface between the Process and Plant elements.

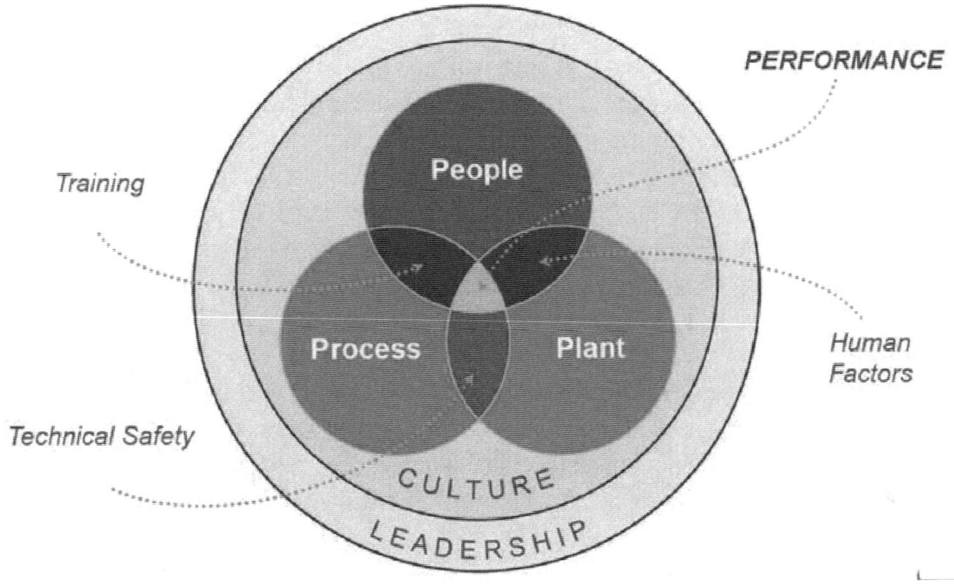

For example, written operating procedures (Process), developed to increase operator productivity, may fail to achieve its intended goal if designed without the input of those (People) by which it's intended to be used. Additionally, such procedures may also fail to realize their intended objective if they're designed without accounting for recent changes made to process equipment (Plant). Consequently, operational performance is generally tied to how well each element interacts with one another.

For the purposes of this paper we have chosen to highlight how we've seen some companies incorporate their existing technical safety management systems into Sustainable Safety to drive considerable improvements in operational performance, risk reduction, and facility longevity.

4 How Does Technical Safety Fit into Sustainable Safety – Case Comprehensions?

In our experience, the best way to promote sustainable safety, from a process, or "technical safety" standpoint, is through the development and implementation of focused, risk-based approaches. Such approaches, when implemented properly, can have significant returns, as further discussed in the examples outlined below pertaining to standard operating procedures, asset integrity, and human factors.

3.1 Standard Operating Procedures

As stated by the Center for Chemical Process Safety (CCPS) in their book, "Guidelines for Risk Based Process Safety" [2], a consistent high level of human performance is a critical aspect of any process safety program. Indeed, a less than adequate level of human performance will adversely impact all aspects of operations. Without written procedures, a facility can have no assurance that an individual operator will consistently execute a particular task in its intended manner [2]; thus leading to an increase in operational waste and a decrease in operational efficiency due to operator error/ignorance.

In 29 CFR 1910.119(f), OSHA outlines its regulatory requirements for standard operating procedures. Such requirements places an emphasis on ensuring employers provide employees with clear, concise, and most importantly, accurate, work instructions for safely conducting their assigned tasks in, on, or near the employer's PSM-covered process. These requirements were put in place, in part, because of OSHA's view that incidents are less likely to occur if written operating procedures are developed so that even a new employee, or someone relatively inexperienced, may perform a task and/or respond to a given event in a pre-considered and prescribed manner [6].

Interacting with many key players in the industry, we have observed that when companies develop and implement new/revised standard operating procedures in a manner that effectively accounts for the "People" and "Process" aspects of Sustainable Safety, they typically realized a 7-10% gain in process productivity, as well as a significant reduction in operational waste because of fewer operator errors and rework [3]. In fact, it's been said that in addition to helping ensure process safety, procedures can also help ensure reliable human performance, which in turn helps ensure predictable product output, yield, and quality [2]. Likewise, less than adequate procedures can result in less than adequate human performance; which can have adverse effects on all aspects of a facility's operation [2].

In our experience, such gains are typically realized when standard operating procedures are developed/implemented in conjunction with effective training programs [3]. Indeed, using operating procedures as a training aid can help: (1) familiarize trainees with the content and format of procedures, (2) reinforce the value of the procedures, and (3) allow trainers or other subject matter experts to identify errors or omissions in procedures [2].

The current trend in the industry, is to use an accelerative learning (AL) approach to training [7]. Enhancement of training programs, using the AL techniques has previously rendered a 30% gain in knowledge retention and on the job implementation of the concepts that are rolled out.

An example of this includes a recent project for an injection molding manufacturing process in which we overhauled the facility's operating procedures and developed a structured on the job training program to help the facility qualify its operators. By incorporating the facility's updated procedures into its revised operator training program, the facility has helped ensure the future accuracy and ongoing use of its written operating procedures [3, 7].

3.2 Asset Integrity

To paraphrase the CCPS [2], asset integrity programs generally consists of a set of written standards, procedures, policies, and related training activities meant to help assure process equipment is fabricated, installed, and maintained in a proper manner. Such programs typically utilize a set of standards, or technical bases, such as recognized and generally accepted good engineering practices (RAGAGEP), to develop/define inspection, testing, and preventative maintenance (ITPM) activities for various types of process equipment. In our experience, the most effective programs typically also utilize a risk-based approach to ITPM activities that adopts a "living" inspection regime based on the actual operating and environmental conditions at the facility, as well as the condition of the facility's equipment when last inspected. Thus, placing an emphasis on ensuring the ITPM activities tied to the areas of greatest concern are always completed.

In fact, the CCPS has indicated that safe operations are directly and closely correlated with reliable operations, and conditions that lead to safe and reliable operations generally lead to high overall plant availability and other business related benefits [2]. Moreover, our experience has shown that implementing a robust asset integrity program can help facilities increase equipment "up-time", extend the useful life of their equipment, and achieve an overall increase in productivity; thus, reducing overall capital expenditures.

For example, we previously implemented an asset integrity and reliability program for a company that lead to an approximate 18% increase in equipment "up-time", as well as a corresponding increase in overall productivity. This implementation included the development of an enhanced mechanical integrity program that better aligned the facility's existing ITPM activities with applicable RAGAGEP [3].

3.3 Human Factors

The term Human Factors represents a concept based on the study of operators, managers, maintenance personnel, and others in the work environment, and of the factors that generally influence humans in their relationship with a technical installation. Thus, human factors related programs and activities generally involve activities meant to help assure machines, operations, and work environments are designed so that they match human capabilities, limitations, and needs [1]. In terms of Sustainable Safety, human factors activities are best represented at the interface between the "People" and "Plant" aspects of a company's operation. Thus, the success of a company's Human Factors program is directly influenced by how well it's able to incorporate the "People" and "Plant" aspects of its operation into its human factors related programs/activities.

For example, a non-regulated OSHA PSM cotton processing pilot plant facility in the eastern United States had recently experienced multiple fire events. The site had previously completed traditional HAZOPs for each of its processes to prevent such events from occurring, but had failed to fully address the human-machine interface of its process. Thus, the facility asked us to help facilitate a critical control study of the site's processes to identify the gaps and disconnections existing in its human factors dependent processes. The resulting corrective actions have since been said to have helped increase operator efficiency and substantially minimize the site's overall risk of fire [3].

5 Conclusion

Effective and sustainable process safety programs are vital to the overall health and viability of a company from a business continuity perspective; as evidenced by the devastating business impacts previously incurred by those companies that failed to effectively sustain such programs. Thus, we feel that it is vital that companies focus much more of their safety/risk reduction-related efforts towards those programs capable of driving sustained improvements in process safety and operational performance, as well as overall facility longevity, such as Sustainable Safety.

6 References (examples given to show format).

(1) CCPS. *Guidelines for Hazard Evaluation Procedures, 3rd Edition.* Center for Chemical Process Safety, American Institute of Chemical Engineers, New York, NY, 2008.

(2) CCPS. *Guidelines for Risk Based Process Safety.* Center for Chemical Process Safety, American Institute of Chemical Engineers, New York, NY, 2007.

(3) "Achieving Operational Excellence Through A Sustainable Approach Within Process Safety", Kumar (Chris) Israni, Steve Hawkins, Presented at GCPS 2016.

(4) "Safety Leadership and Implementation- Independent of Culture", Kumar (Chris) Israni, Sarah Acton, and Tianherng (Joshua) Yang, Presented at GCPS 2014.

(5) "Safety Culture and Leadership - Independent of Language", Kumar (Chris) Israni, Sarah Acton, Yaneira Saud and Sandra Lovell, Presented at Mary Kay 2016.

(6) Occupational Safety and Health Administration. "Process Safety Management of Highly Hazardous Chemicals." U.S. Department of Labor, Occupational Safety and Health Administration. 1910.119

(7) "Applying the "Concept of Accelerative Learning" for Design and Delivery of Process and Personnel Safety Leadership Programs in Oil & Gas Assets", Kumar (Chris) Israni and Collen Watson, Presented at SPE – HSSE New Orleans 2017.

(8) "Achieving Sustainable Safety" Environmental Resources Management (ERM), et al. Internal ERM document, 02-01-2017.

479491 Are Our Current Safety Challenges Different Than 1974 or 1984? Have Wereally Learnt from Accidents? Where Are the Gaps, What Are We Missingand Why? Learn What We Have Not Learnt in Last 40 Years, Keptrepeating the Same Mistakes

Abstract:

Despite the extensive development in technology, a considerable reduction in process safety accidents was expected, but slight improvements are realized. If we look back 50 years, no doubt, the likelihood and severity of consequences for majority of accidents have reduced, however the root causes are still unchanged. So, it is worth debating that why do we keep repeating similar accidents in spite of using latest tools, techniques and learning.

The presentation will cover what has changed and what not over last 50 years. The author will discuss the practical examples about most common gaps, current safety challenges, and most needed process safety improvements to eliminate the learning gaps and prevent the recurrence of accidents. The role of good safety culture and information flow between management and shop floor will also be discussed in relevance to accident prevention.

Several methods and techniques will be presented to identify the opportunities for improvement and tips for effective implementation during design and operating stage of the facility. Emphasis will be given on prevention than mitigation by applying hierarchy of control.

Presentation will conclude by explaining how a robust management system, committed management and motivated employees play a vital role in preventing the recurrence of accidents by eliminating the root causes.

480429 A Data-Driven Early Warning System for Mining Accidents

Abstract:

Often, safety condition of a mine deteriorates before an accident happens. For example, before the 2010 Upper Big Branch Mine disaster, one of the largest in the U.S. history, the mine displayed an alarming rising trend of safety violations. Similar building-up behaviors apply to other major mining accidents in the U.S. According to the Mine Safety and Health Administration (MSHA), on average, there are more than a hundred thousand citations and orders issued yearly. The MSHA accident and inspection databases are an untapped resource for safety analysis. Our model analyzes semi-structured data from MSHA, attempting to uncover a company's safety culture from its public regulatory records. We discuss preliminary results from logistic regression. This work has implications for developing a data-driven early warning system. In the future, we plan to extend this research to inspection data from the Occupational Safety and Health Administration (OSHA) and the Environmental Protection Agency (EPA).

479231 New Ways to Visualize and Consolidate Alarm Management KPIs

Abstract:

Alarm management. We have been dealing with it for the last two decades, but its concepts, despite of been being essentially simple, are still of difficult interpretation by the operators, plant engineers and especially by the managerial levels. One reason for this is that the main clients of alarm management information (operators, plant engineers and site managers) are, as everyone else nowadays, overwhelmed with too much information and subject to pressures to achieve goals and make quick decisions. Thus, it is imperative to translate the main KPIs usually calculated for alarm management systems in a way that is easy, for someone unfamiliar to automation systems, to understand. These KPIs are, according to EEMUA 191: KPI1 - average number of alarm per time, per operator, KPI2 - maximum number of alarm per time, per operator and KPI3 - percent of time that plant are in a burst situation. Our company has adopted a commercial application to deal with the huge amount of information generated by its alarm systems (about millions of messages per month) and calculate those KPIs. This solution has interesting features like a MS Excel® add-in and a web interface to generate graphics and summarize alarm information for reports. However, the daily acquaintanceship with the matter encouraged us to develop more efficient ways to display the alarm results. The objective of this paper is to present the evolution of our company when reporting and analyzing the alarm management situation of each plant. We are going to present how we evolved from just reporting the KPIs generated automatically by the Alarm Management Software we use to presenting a graphical reports that are of easy and fast interpretation by the plant staff and managers. We have developed a highly automated MS Excel® report that makes use of SQL queries and custom store procedures to generate the information required for analysis quickly, thus increasing the productivity of the engineers in charge of generating the monthly reports. Also, we are going to present how to consolidate alarm management KPIs (defined by operator) in plant KPIs. This came up as a requirement to include alarm management practices in the Company´s Quality Management and Production System. At the same time that it expands the alarm management visibility and importance to the company and all level positions it creates the challenge to meaningfully consolidate the data not only for individual operator consoles, but for different levels and over different time periods. We are going to present the questions we were faced with and the tortuous way we have taken (including the mistakes) to arrive to a reasonably answer.

479232 A Simple Way to Calculate Control Loop Service Factor Taking in Account Plant Conditionals

Abstract:

Process safety considers several layers of protection to avoid that our process achieves an undesirable and dangerous situation. One of the most important of these protection layers is the basic process control system (BPCS). Nowadays, this protection layers is usually performed by distributed control systems (DCS) or programmable logic controllers (PLC), where each layer is configured as a control loop. Nevertheless, having these control loops configured is not enough. It is also necessary that the loop is closed and performing its job well. Although there are in the market several systems that provide means of analyzing control loops, they compete with the advanced process control (APC) and Process Information Management Systems (PIMS) in the acquisition of data from the BPCS, requiring a dedicated OPC server to get data in the required frequency. Secondly, many of the indexes calculated by them are of difficult interpretation by people outside the automation and process control environment. Thirdly, the information needed to calculate many of that index are already stored in PIMS systems. We have learned by experience that if a control loop is helping the operators to do their job, they will keep them closed on its correct operating mode (AUTOMATIC, CASCATE, REMOTE CASCATE, etc.). Based on this assumption, we decided that one of the simplest ways to track the performance of our control loops was to calculate the percent of time that control loops work on their target operating mode. This KPI is called by different name by different companies. We have chosen to call it service factor. All the information needed to calculate it is already stored in our process management information system (PIMS). To calculate this KPI is simple but there is a complication factor for continuous plants: conditionals. As soon as we started presenting this KPI for the plant engineers, we started receiving tons of justifications of why some loops were having a poor service factor. The problem was that the plant operation is not a steady state process. There are many situations when operators must break a loop (e.g.: during a reactor regeneration, unit start-up, etc...). In this paper we are going to present a simple approach using MS Excel and some PIMS store procedures to automatically take in account these conditionals and how the control loops are managed.

478550 Thermal Decomposition of Mono-Nitrated Toluene (MNT) with Additives

Abstract:

Abstract: On September 21st 1992, an explosion happened in Castleford, UK, at the distillation unit, owned by Hickson & Welch Ltd. The incident killed 5 people and caused nearly 200 injuries. The main reason is due to the decomposition of mono-nitrated toluene (MNT). Nitrated aromatic compounds present a potential run-away reaction threat to the chemical process industry, due to its unstable structure and extensive heat release during upset situations. This study employs the Advanced Reactive System Screening Tool (ARSST) to study the effects of various common contaminants on MNT's thermal decomposition, including sodium nitrate, sodium hydroxide, and sodium carbonate. Experimental parameters and results such as 'onset' temperature, pressure increase rate, temperature increase rate and maximum temperature are present. Sodium hydroxide and sodium nitrate can lower the decomposition onset temperature of MNT and make the consequence more severe. Thermokinetics of the decomposition reaction is shown and SADT prediction of UN25 kg package storage is calculated.

479379 Addressing Challenges of Controlling Change By Implementing Best MOC Practices

Abstract:

Effectively controlling change in a plant environment results in a cumbersome process with an inefficient paper-based Management of Change process or use of a traditional and complex electronic Management of Change (MOC) system. Recent research indicates that the majority of chemical incidents are the outcome of poor implementation of Process Safety Management (PSM) and the lack of efficient MOC processes. As a result, MOC workflow steps are often skipped or prolonged resulting in an ineffective process that does not capture the change fully or safely; approvals are delayed, Process Safety Information (PSI) is not being updated, Pre-Startup Safety Review (PSSR's) are not being performed, and MOCs are not being closed in a timely manner.

The goal of this paper is to address industry challenges with controlling change, and how Siemens implements best practices for an effective MOC process.

479243 Process Change Impact on Relief System Design

Abstract:

Process change impact on relief system design

(Understanding equipment impact and the evaluation warranted)

Various scenarios can cause the pressure within equipment to exceed design pressure. Appropriately sized relief devices are one of the most common methods of protecting equipment and personnel operating this equipment. A key challenge to being in compliance is the dynamic nature of processing plants. Numerous changes can be expected to happen over the lifecycle of a processing plant. Management of Change, one of the Process Safety Management (PSM) elements, requires equipment impact evaluation when these changes occur to ensure adequate pressure relief. This presentation talks about numerous types of changes which will impact equipment and warrant overpressure protection re-evaluation.

479152 Identification and Local Impact Analysis of Projectile Hazard in the LNG Industry

Abstract:

With a strong growth in natural gas production in the United States, a number of liquefied natural gas (LNG) facilities have been proposed and authorized. These LNG facilities have sufficient safety, security, risk assessment and mitigation plans developed in close coordination with local, state, and federal authorities. However, to continue to ensure safe operation, it is vital to continue to improve the scope of risk assessment and management strategies. Recent incidents have highlighted the need to consider the potential hazard and risk from projectiles. This paper studies the sequence of phenomena that occur as part of the local impact of a projectile. The main objectives of this study were to: 1) list projectile (missile/fragment) parameters such as source, type, and characteristics (mass, diameter, velocity) and 2) conduct a local impact analysis for identified potential projectile hazard scenarios at an LNG facility. The paper concludes that the five different calculated thicknesses for concrete penetration, scabbing, and perforation, and steel penetration and perforation, depend directly on the kinetic energy and diameter of the projectile. Furthermore, the authors recommend that operators conduct a projectile hazard analyses for LNG facility projects, which would be useful in making recommendations for the thickness and type of storage tanks and other equipment and for other appropriate mitigation measures.

478563 Thermal Hazard Analysis and Chemical Incompatibility Tests with Custom High-Pressure Crucibles Made from Commonly Used Metals and Alloys

Abstract:

The assessment of reactive hazards and chemical incompatibilities is necessary for safe operations in the chemical and pharmaceutical industries. Commercial ARC (Accelerating Rate Calorimeter), RSST (Reactive System Screening Tools) and high pressure crucibles for Differential Scanning Calorimetry (DSC) can be used for this purpose [1-2]. However, ARC and RSST require significant capital investment and require relatively large amounts of sample (at least 1 g).The alternative, commercial high pressure crucibles for DSC, can also be expensive (e.g., several hundred U.S. dollars for each pair of gold-plated high-pressure crucibles) when multiple samples need to be investigated. These requirements make thermal hazard analysis prohibitively expensive for research laboratories that do not focus on chemical process safety, and where materials under development may be available in limited quantities (e.g., milligrams).

To address these short-comings of current hazard screening methods, inexpensive custom DSC high-pressure crucibles made from different commonly used metals and alloys (stainless steel 304, stainless steel 316, titanium, brass and tantalum) have been constructed in-house using a micro-welding system containing a jewelry spot welder and a microscope. Chemical compounds are tested in different crucibles made from different metal/alloys, which simulates different real-world storage, transportation, and process operations in the chemical and pharmaceutical industry. The cost for each pair of crucibles is less than two U.S. dollars. In a typical experiment, only 1-5 mg of sample is required. Kinetic parameters can be extracted from only a few DSC runs with these crucibles, allowing for prediction of the chemical's thermal behavior at larger scales. This method provides an inexpensive and efficient way to perform thermal hazard analysis and incompatibility tests. The application of these crucibles is demonstrated with case studies of ammonium nitrate (AN) and ionic liquids. Predictions of thermal behavior at larger scales have been verified with ARC tests.

References

1. Oxley, Jimmie C. 2003. "Handbook of Thermal Analysis and Calorimetry. Vol. 2: Applications to Inorganic and Miscellaneous Materials M.E. Brown and P.K. Gallagher, Editors 9 2003 Elsevier B.V. All Right Reserved. 349" 2: 349–69.

2. Han, Zhe, Sonny Sachdeva, Maria I. Papadaki, and M. Sam Mannan. 2014. "Ammonium Nitrate Thermal Decomposition with Additives." Journal of Loss Prevention in the Process Industries. Elsevier Ltd, 1–9. doi:10.1016/j.jlp.2014.10.011.

13TH GLOBAL CONGRESS ON PROCESS SAFETY

Do Dust Hazards Analysis (DHA) Really Differ from a Process Hazards Analysis (PHA) or a Process Analysis (PA)?

Michael E. Robertson

4094 Roslindale Ln
Buford, GA 30519
processesmatter@gmail.com

Prepared for Presentation at
American Institute of Chemical Engineers
2017 Spring Meeting and 13th Global Congress on Process Safety
San Antonio, TX
March 26 – 29, 2017

Do Dust Hazards Analysis (DHA) Really Differ from a Process Hazards Analysis (PHA) or a Process Analysis (PA)?

Michael E. Robertson

4094 Roslindale Ln
Buford, GA 30519
processesmatter@gmail.com

Keywords: NFPA 652, Dust Hazards Analysis, DHA

Abstract

Do Dust Hazards Analysis (DHA) Really Differ From a Process Hazards Analysis (PHA) or a Process Analysis (PA)?

This author reviews the latest NFPA 652 requirement on Dust Hazards Analyses (DHA) and compares it to OSHA PSM's Process Hazards Analyses (PHA) requirement and other process analyses required by industry specific NFPA dust regulations. Though process analyses have been a part of NFPA regulations for several years and in the OSHA PSM regulation for 25 years, the 2016 version of NFPA 652 has clearly defined that a process analysis called Dust Hazards Analysis must be done where there is a combustible dust risk. The subject of this paper is the Dust Hazards Analysis.

The aim of the paper is to find commonality in dust hazards analysis, process hazards analysis and process analysis for tips that may limit the number of analyses a company may have to perform. These tips provided will assist a PHA leader in doing their DHA's. This will not be an exhaustive review of standards or regulations, but an overview and guide for those starting DHA's. It has become obvious that NFPA regulations are modeling a PSM framework and some discussion will be made on the subject in the area of risk assessment and management systems. Some comments will be made concerning the software that can be used, but no software will be sold or marketed. The author is not sharing examples of DHA's from his company due to confidentiality and timing of his DHA's.

Some of the Dust Hazards Analysis differences detailed are the emphasis on dust inside and outside of equipment, the use of NFPA dust thickness calculations, building reviews, and frequency of re-validating the analyses. How does a Dust Hazards Analyses affect the process safety information required, team make-up or report after the review? What are the real requirements for a DHA leader?

1 Introduction

There have been several dust related incidents that resulted in loss of life and severe damage to industrial properties. Imperial Sugar in Savannah, GA (2008) and West Pharmaceutical in North Carolina (2003) are two that top the list. Since the early 2000's, the Chemical Safety Board (CSB) has been recommending the Occupational Safety and Health Administration (OSHA) develop standards surrounding combustible dust. The National Fire Protection Association (NFPA) has had guidelines in several of its standards concerning dust for years.

Table 1. Industry Specific NFPA Guidelines Pertaining to Dust

61 – Agriculture
484 - Metal
654 - General
664 - Wood

NFPA 654, 664, 61, 484, 86 and 499 have dealt with combustible dust for years.

It is 2017 and while OSHA still has not set forth comprehensive dust standards, NFPA has published a comprehensive guideline on the management of combustible dust as 2016 NFPA 652 Standards on the Fundamentals of Combustible Dust. This guideline requires dust hazards analysis (DHA) to be complete by September 7, 2018. long with renewed regulations, risk assessments are now becoming part of every safety performance system. NFPA has required process analysis and hazards analysis in the industry specific guidelines..

A DHA is similar to a Process Hazards Analysis (PHA), but there are some key differences that need to be understood by study leaders. PHA methodology is well- defined especially from the OSHA 1910 PSM regulation, the CCPS book *Guidelines for Hazards Evaluations* **[2]**,and other information from new and vintage PHA courses like Primatech's Process Hazards Analysis Leaders' Training and Process Improvement Institute's PHA Leader's Training.

The similarities between PHA's and DHA's are in the methodology. The methodologies used to conduct a PHA, such as HAZOP, Checklist, What-If, are applicable for identifying and assessing hazards in dust handling processes. Expertise with PHA's can assist with the completion of DHA's. The differences begin when the building has to be evaluated for combustible dust hazards. Process Analysis is defined in NFPA 664-12 – Standard for the Prevention of Fires and Explosions in Wood Processing and Woodworking Facilities. [3]. However, NFPA did not offer much instruction on how to do a Process Analysis (PA).

There are other tools that have to be employed for the building evaluation. Pictures, audits, root cause analysis and cause and effect diagrams can be used for the out of the process review.

Detailed building structure evaluations may require a code inspection or some other engineering tools.

This paper evaluates the elements of dust hazards analysis. Comparisons are made between dust hazards analysis, a process hazards analysis methodology called HAZOP (HAZard and OPerability Study), and a process analysis as described in older NFPA guidelines. These comparisons are made to assist PSM companies in aligning PHA's and DHA's and to assist non-PSM companies on resources to assist them in conducting DHA's.

2 What Is a Dust Hazards Analysis (DHA)

NFPA 652 requires that all facilities handling or creating combustible dust must do a Dust Hazards Analysis. A dust hazards analysis is a systematic review of processes and buildings where dust is created or handled and could create a deflagration.

Per the CCPS process safety glossary: A combustion that propagates by heat and mass transfer through the un-reacted medium at a velocity less than the speed of sound.

This requirement has many businesses scrambling in a similar way to the release of the PSM standard in 1992, which left many oil and gas and chemical companies reeling. The dust hazards analysis is not a PSM requirement nor is it a type of process hazards analysis. While dust hazards analyses are new, analyzing your process for dust hazards has been a part of industrial specific NFPA guidelines for a few years now. Now with NPFA 652, OSHA through Recognized and Generally Accepted Good Engineering Practice(RAGAGEP) has the standards it needs for inspection and enforcement to address the potential for a dust deflagration in a facility.

The dust hazards analysis must be complete by September 2018. The DHA requirement is set forth in Element 7.0 of the 2016 version of NFPA 652. There is a good example of the DHA in Appendix B of NFPA 652. Operating areas are reviewed for the probability of deflagration. Classifications are 'Not A Hazard', 'Maybe a Hazard' and 'Deflagration Hazard'. If the probability exists, then the owner must document how they manage or mitigate that risk. If the owner does not have current controls to prevent the deflagration, steps must be taken to document a hazards management plan with steps to reduce the risk.

Within a process system or building, when there is a sufficient quantity of combustible dust, the following conditions must be considered and documented:

1) Oxidizing atmosphere
2) Credible ignition source
3) Credible suspension mechanism

These are items from the fire triangle used in most fire prevention training. These are key in the dust hazards analysis and must be treated with high importance. Housekeeping also must be properly reviewed and documented to prevent the third item mentioned above – credible suspension mechanism.

3 How Do Dust Hazards Analysis (DHA) Compared to Process Hazards Analysis (PHA)

A major difference between the DHA and a PHA is the consideration of dust risk in a building. PHA's do evaluate the risk to occupied buildings in relation to the potential for releases of highly hazardous chemicals covered by the PSM regulation. This risk is evaluated through facility siting studies and more detailed reviews like a Quantitative Risk Analysis (QRA) or Occupied Building Risk Assessment (OBRA).

Table 2. PHA and DHA Methodology Comparison

Process Hazards Analysis (HAZOP)	Dust Hazards Analysis	Process Analysis
Source: OSHA 1910.119 (e)	Source: NFPA 652 7.0 and Appendix B – Dust Hazards Analysis	Source: NFPA 664 4.0 and Explanatory Material A4.3
Covers the process system	Covers the process system and the building construction	A methodical review of the facility, each operation housed within, and the identification of where a hazard exists.
Gather Data – Drawings and Process Information	Gather Data – Lab Data, Drawings and Process Information	
Have a Qualified PHA Leader a Team with an Operator and a Process Expert, along with other Subject Matter Experts as needed	Have a Qualified DHA Review Leader and at least one other for a team.	
Set Objectives and Boundaries		
Determine the Study Nodes by marking up the Piping and Instrument Diagrams	Determine the nodes on the drawings and sketches	
Review HAZOP Guidewords	Review DHA questions from Appendix B of NFPA 652	
Describe Process Node	Describe Process Node	
Use all Guidewords in all Nodes	Use all questions in all Nodes	
Review Global Node, Human Factors, Management of Change and Incidents		

With the PSM regulation, PHA's are required in facilities with highly hazardous chemicals. These chemicals are listed in the PSM requirement. PHA's are a group of tools that are used to review processes handling highly hazardous chemicals. PHA's include HAZOP, Failure Mode and Effect Analysis (FMEA), What If Studies, and Checklists.

The DHA requires buildings to be reviewed for the possibility of a fire from dust build-up. Building construction must minimize the effects of a deflagration or explosion from a combustible dust.

From the table above, you can see some differences in PHA's and DHA's. DHA's do not require a review of management of change and incident review history. The team composition is not set in a DHA. P&ID's are the basis, but DHA's do not have a required standard drawing type although P&ID's are applicable for some aspects of dust handling.

4 DHA Toolbox

Similar to PHA's, multiple tools are needed to complete a DHA.A PHA method could be used to evaluate the risk associated with combustible dust in a process. However, beyond typical PHA methods, field audits, pictures and video recordings of the areas may be need of areas where dust can build up.Alternatively, a root cause analysis method like 5 Why method might also be applied in areas where a dust build-up to determine sources of dust leakage.. The key is to locate where dust gets out of a process or accumulates beyond its minimum explosive concentration (MEC) in a process, why the dust gets outside the confines of the process and what can be done to prevent the leakage.

Whereas chemical analysis and laboratory data are helpful in a PHA, parameters like Maximum Explosion Concentration (MEC), Minimum Ignition Temperature (MIT) and Kst are very key in determining dust combustibility and risk for a process and building. NFPA 652 has information in the use of these parameters. There is also good information in NFPA's Guide to Combustible Dust and the AIChE Book on Solids Handling Equipment. This paper will not cover in detail any laboratory or chemical analysis of dust. [4], [5]

5 Documentation for Review

Piping and instrument diagrams (P&ID's) have long been the basis of the PHA techniques, including HAZOP, Checklist, and What-If. Supporting documentation may include SDS's, process descriptions, management of change documents, incident history and standard operating procedures. Plant layouts and general arrangement drawings will help orient the team and provide information for facility siting and items in the global node.

P&ID's will help for the process analysis in the DHA. You may also need general arrangement drawings and assembly drawings where machines are involved. Plant layouts and 3D renderings may help with the building analysis portion of the DHA. Pictures and video of areas where dust buildup can occur will also be helpful.

Lastly, plant conveying, processing and finishing OEM information such as screen sizes, motor power, pneumatic flow rates, and duct sizing are key in understanding out the equipment affects the creating or lack of creation of combustible dust. Information on seals and design pressures must be available for comparison with operating parameters.

The bottom line is DHA documentation is as important to the process as Process Safety Information is for PHA. However, it might not be as standard or available as PSI is for a PHA. Some of the information may also be very old since plant layouts and renderings might come from the original start of the plant.

6 The Dust Hazards Analysis Process

The DHA process can start like a PHA with an overview, process description and agenda. The DHA must cover the fugitive combustible dust that gets out into the building and the building's ability to withstand deflagrations and explosions.

Documentation must be available to follow the flow of the materials that contribute to creation of combustible dust. Equipment like fans and screens are especially sensitive since those components can create dust suspensions and impact the Minimum Explosion Concentration of the system. Critical laboratory must be available for discussion since it will provide the data to determine combustibility.
As with a PHA, a tour of the area would be very critical to see the areas were combustible dust accumulates the most. A start to the process is chosen and each part is reviewed by asking a series of question. Below are the questions from Appendix B of NFPA 652 used in a DHA.

Table 3. DHA Process Questions ([1] – B.4.5.1.1 – B.4.5.1.5)

I.	Is the particulate deflagrable (explosible)?
II.	Is the particulate suspended in air?
III.	Is there sufficient concentration to propagate a flame front?
IV.	Are there competent igniters available?
V.	What hazard management is in place?
VI.	Is there metal detection, spark detection, bonding and grounding, or other hazard management means in place?

These questions are asked for each piece of equipment, piping and ducting. The answers will should lead the team or competent person to the hazard management system measures in place and the gaps that will need attention from site management. The process must continue until the entire process is analyzed and all the outcomes have been reached.

As mentioned previously, Appendix B of NFPA 652 (2016) provides a good example to do a DHA. For those in the wood industry, the example is actually of a wood processing plant.

7 Outcomes

NFPA 652 requires that results from the Dust Hazards Analysis be documented. This includes necessary action items requiring change to the process materials, process itself, operations or facilities. NFPA 652 has requirements for managing the hazards under the heading of Hazards Management Program.

The results from a Process Hazards Analysis are the hazard scenarios, a list of safeguards currently in place, and any action items required to ensure that risk is effectively managed within the process handling highly hazardous materials. This usually involves a risk ranking system as well.

The Dust Hazards Analysis does not require a risk ranking system. Where a hazard exists, identification and evaluation of the fire scenario shall include safe operating ranges, safeguard identification, recommendations for additional safeguards as needed to properly manage the fire event, and lastly a plan for implementation.

8 Leaders

A meeting has a leader and teams have leaders. DHA's and PHA's are no different. Each review must have a leader. Her qualifications are listed in Table 2. There are no certifications or extensive training requirements. Many believe that certification is coming for PHA leaders because the competency of those who lead PSM is being discussed in government and industry especially as we enter the next 25 years of PSM.

Many experts agree that an experienced PHA leader can also make an effective DHA leader. This is a good start, but the DHA leader must also be familiar with the hazards of combustible dust. Additionally, it would be prudent to have a leader who is experienced in the process being reviewed or ensure that a process expert is present as a team member. I would review the qualifications listed in CCPS's book on Hazards Evaluation and information.

Table 4. Comparison of Leader Qualifications

Dust Hazards Analysis Leader	Process Hazards Analysis Leader
Must be able to lead DHA's Must be familiar with the hazards of combustible dust	Must be trained in PHA techniques Must be re-qualified every 5 years

9 Frequency

PHA's are done initially to establish a base evaluation and then every five years after the initial PHA to address changes or revisions to covered processes. . DHA's are done once for the original process and for any major upgrades. There is no renewal period, but the initial reviews have to be done by September 7, 2018.

10 Management Systems

What is a dust hazards management system? How does a dust hazards management system compare to the standard elements of PSM? Chapter 9 of NFPA 652 mentions its elements as listed in Table 6. Both PSM and NFPA 652 have the requirement of risk assessment that has been the subject of this paper.

PSM is embedded in many of the NFPA guidelines. Though the thought of PSM to many firms is frightening, PSM offers a good framework for managing risk whether it is combustible dust or other system or process risks.

Table 5. Management Systems Comparison

Dust Hazards Analysis Management System (Ch. 9)	Elements of Process Safety Management
1. Operating procedures and practices	1. Standard Operating Procedures
2. Inspection testing and maintenance	2. Mechanical Integrity
3. Training and Hazard Awareness	3. Training
4. Emergency Planning and Response	4. Emergency Planning and Response
5. Incident investigations	5. Incident investigations
6. Management of change	6. Management of change
7. Management system review,	7. Compliance Auditing
8. Contractor training	8. Contractor Safety
9. Employee participation	9. Employee Participation
10. Document retention	10. Process Safety Information
11. Ignition Control (Covered in Chapter 8)	11. Hot Work Permits
12. DHA	12. Trade Secrets
	13. Pre-Start Up Safety Review
	14. PHA

11 Conclusions

DHA's must be completed. It is similar but different than a PHA. Expertise with PHA's can assist with the completion of DHA's. The similarity between PHA's and DHA's is in the methodology of reviewing processes for dust. The differences start when the building has to be evaluated for combustible dust. Here other tools have to be employed. These tools as previously discussed may vary from audits and inspections to root cause analysis tools. Detailed building structure evaluations may require a code inspection or some other engineering tools.

12 References

[1] NFPA 652-15 – Standards on the Fundamentals of Combustible Dust, National Fire Protection Association, September 7, 2015.

[2] CCPS. *Guidelines for Hazards Evaluations.* Center for Chemical Process Safety, American Institute of Chemical Engineers, New York, NY, 2007.

[3] NFPA 664-12 – Standards on the Fundamentals of Combustible Dust, National Fire Protection Association, 2011.

[4] NFPA– Guide to Combustible Dusts, National Fire Protection Association, 2012.

[5] CCPS. *Guidelines for Safe Handling of Powders and Bulk Solids.* Center for Chemical Process Safety, American Institute of Chemical Engineers, New York, NY, 2005.

478758 A Leading Indicators Based Decision Support Tool to Predict Blowout Events

Abstract:

Predicting a potential blowout scenario timely and efficiently is often a challenging task due to the complexity of drilling or other well intervention activities. Blowouts are usually preceded by gas kicks and early prediction of gas kick events can provide some precious time window to take control of the well in concern. A leading risk indicators based approach for predicting gas kick or blowout events has been introduced in this work. A causal factors based step-by-step approach for identifying comprehensive sets of leading indicators are presented and discussed. Leading indicators were divided into two broad sections - real-time indicators and long-term organizational safety performance indicators. With the real-time indicators, different decision support algorithms are developed which can be used as training tools. A real-time indicator may not always successfully lead to a well control event but at the same time the situation needs to be assessed carefully considering other factors and indicators. Such uncertainties are taken into account while developing the algorithms. For evaluating the relative importance of different leading indicators Bayesian Network models are developed. The key causal factors for well control barriers failure are identified by conducting fault tree analysis and appropriate indicators are linked with relevant causal factors in a Bayesian model. At the end, a list of efficient leading risk indicators are presented for predicting primary well control barrier failure which can initiate gas kicks.

Consequence Modeling at the Eastman Kingsport Site

David Skelton
Principal Engineer
Eastman Chemical Company
P.O. Box 511, Building 18
Kingsport, TN 37662
davidskelton@eastman.com

Michael James
Senior Chemical Engineer
Eastman Chemical Company
P.O. Box 511, Building 18
Kingsport, TN 37662
mjames@eastman.com

Prepared for Presentation at
13th Global Congress on Process Safety
San Antonio, TX
March 26-29, 2017

UNPUBLISHED

Consequence Modeling at the Eastman Kingsport Site

Michael James
Senior Chemical Engineer
Eastman Chemical Company
P.O. Box 511, Building18
Kingsport, TN 37662
mjames@eastman.com

David Skelton
Principal Engineer
Eastman Chemical Company
P.O. Box 511, Building 18
Kingsport, TN 37662
davidskelton@eastman.com

Keywords: Dispersion modeling, consequence analysis, risk assessment, dosage, shelter-in-place, temporary haven, evacuation, toxicity, toxic chemical release, toxic load, downwind concentration, blast overpressure, vapor cloud explosion, pressure vessel burst

Abstract

The intent of this paper will be to review the methods utilized at the Eastman Kingsport site in determining magnitude of consequences of toxic, thermal and blast overpressure scenarios. This will include discussion on dispersion modeling, dose-response methods for toxic and thermal probits as well as assessing occupant vulnerability through the use of BakerRisk's Explosion Risk Cooperative tools.

1. Potential Consequences of Loss of Containment

Figure 1 is a typical event tree for potential losses of containment [1].

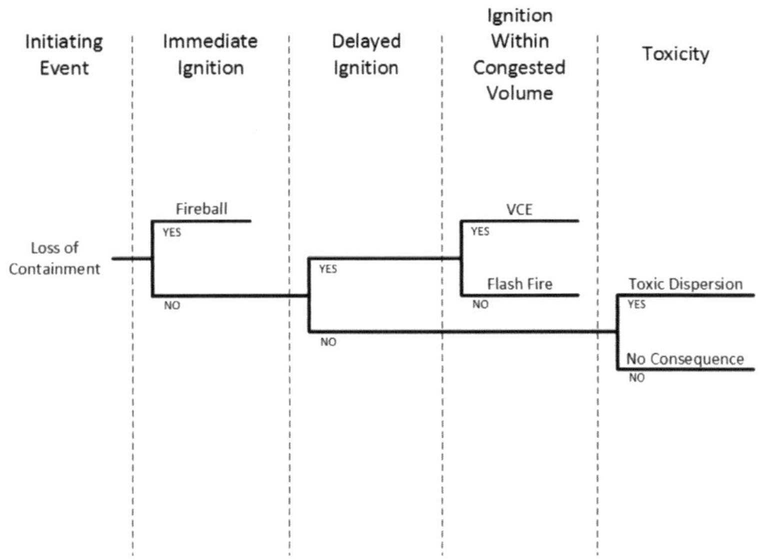

Figure 1. Potential Losses of Containment

The consequences of loss of containment in Figure 1 detail the potential release scenarios that are chiefly of concern. The method for consequence modeling for each type of scenario is described in this paper.

2. Source Terms

Figure 2 details the potential source terms that will need to be developed for each type of scenario.

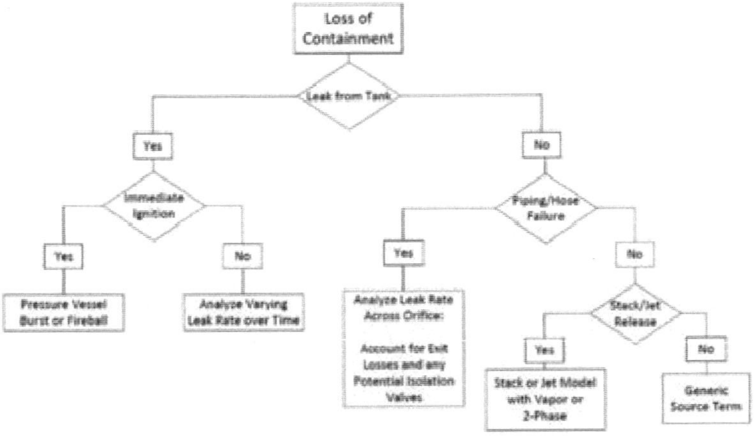

Figure 2. Potential Source Terms.

The source term may have any of the following characteristics:

- Initial momentum
    - Based on the linear velocity across the orifice and other physical properties of the release, the source term may need to account for significant initial momentum of the cloud.
    - In addition, if the linear velocity is sufficiently high, the source of the release may ingest air, resulting in a jetting condition. As a rule of thumb, 10 m/s is a lower limit for potentially jetting releases [2]. There are specific dispersion models to account for jetting conditions during a release.

- Pool Formation
    - Most dispersion models have an integrated pool evaporation model to determine the dynamic evolution of vapor from a liquid spill.
    - If the preferred evaporation model is outside of the dispersion model, the evaporation rate results will need to be analyzed for integration into the dispersion modeling tool.
- Flashing
    - Leaks across an orifice may partially or completely flash from liquid to vapor based on the pressure differential.
    - Modeling tools may report choke conditions at the outlet.
    - Further flash analysis may be required to determine the temperature and liquid/vapor composition from choke to atmospheric conditions.

- Piping Layout between Pressure Source and Leak Point
  - Ensure that all pressure losses are accounted for. This includes entrances, exits, elevation changes, elbows, pipe diameter changes, and other fittings.

- Indoor Release
  - If the release occurs indoors, analyze the concentration changes over time inside the structure based on the source term calculations of the release.
  - Concentration gradients throughout the structure may need to be ignored and perfect mixing assumed.
  - If the vapor portion of the leak will cause overpressuring of the indoor area, determine the amount of vapor required to be vented in order to balance pressure in the room.
  - A flash block can be employed to assess the amount of vapor generated from a leak that will need to be relieved from an indoor area.
  - Flash blocks can also be used to account for vapor suppression systems inside of buildings.
  - This level of modeling is beyond the scope of this paper. However, methods for assessing indoor releases can be referenced from "Consequence Modeling of Dynamic Source Terms" [3].

Presence of the above factors may result in models that require expert judgment and the development of conservative estimates. Most dispersion modeling tools do not have the levels of sophistication required to accurately model these types of source terms. However, the results of a dispersion model employed with conservative estimates may be suitable for site risk assessment.

The following section will discuss the potential leak scenarios and how source terms can be developed, accounting for each of the above factors.

3. Blast Modeling

Simple blast models can be used to predict potential consequences of explosions in lieu of more accurate but complicated methods such as computational fluid dynamic (CFD) models. The TNT equivalency model developed by the U. S. Navy and Air Force was largely used for years, even in industrial settings. Over time it became clear that the TNT model did not predict the behavior of all explosions accurately enough, particularly in regard to vapor cloud explosions more commonly occurring in refining and chemical producing industries. Today several simplified are available to estimate the potential of consequences of explosions, each method with its own strengths and weaknesses.

3.1 Vapor Cloud Explosion (VCE) Modeling

Baker Strehlow Tang (BST) is the simplified method predominantly used at Eastman Kingsport site to model vapor cloud explosions [4][5][7][8]. The BST model is based on the premise that only the portion of a flammable vapor cloud that is confined or contained within a congested area will significantly contribute energy to an explosion. Unconfined or uncongested gas clouds are more likely to result in flash fires or deflagrations. The first step to perform a BST analysis is to identify the confined/congested volume of the plant that has the potential to harbor a flammable cloud of vapor. Examples might include piping and equipment, closely spaced railcars or cars, solid walls, ceilings and floors. A flammable cloud is most often generated from a release of material as a result of an equipment failure, a leak, or a spill. Flammable clouds can drift a considerable distance under the right conditions, so it's important to account for all the equipment within a reasonable radius of the congestion. Once the confined/congested volume is calculated, two parameters contribute the greatest influence on explosion strength using BST methodology; the energy contained within the flammable vapor cloud and flame speed as it burns through the available fuel in the congested region. The first parameter is calculated using the spreadsheet shown in Figure 3. It is a simple tool developed by Eastman to calculate volume based on dimensions of the confined cloud, and then estimate the energy contained with the calculated volume.

ESTIMATION OF VAPOR CLOUD EXPLOSION ENERGY

The user enters dimensions (in ft.) of vapor cloud subject to confinement & obstructions, ambient temperature (F), and the chemical involved in the designated boxes. Thermodynamic data listed in the lookup table to the right is used, or if not available, an approximation is made based on typical hydrocarbons.

The spreadsheet will output the participating energy to be used in V-CLOUD for prediction of pressure and impulse by the Baker-Strehlow method. The methodology is based on the paper "Vapor Cloud Explosion Analysis" by Baker, Tang, and Scheier; Prepared for the AIChE Loss Prevention Symposium, April 1994.

Width (ft)	200	
Height (ft)	50	
Depth (ft)	100	
Volume (ft3)	1000000	
Volume (m3)	2.83E+04	
Ambient Temp (F)	90	
Ambient Temp (C)	32	
Chemical	Acetic Acid	
MW	60.05	
Hc(kJ/mole)	919.73	(heat of combustion in KJ/mole)
Hc(J/m3)	3.67E+07	(heat of combustion in J/m3)
Hc (BTU/lb)	6.59E+03	(heat of combustion BTU/lb for ref.)
Hcs(J/M3)	3.48E+06	(heat of combustion of a stoichiometric mixture with air)
Energy (J)	9.86E+10	
Energy (in-lb.)	8.72E+11	(total available energy in confined / obstructed volume)
% Participation	100	
Equiv Energy in-lb.	8.72E+11	(net energy of explosion in-lb.)

Figure 3. Spreadsheet used to determine energy content of vapor cloud

For a single congested area or multiple areas with little separation distance so they can be effectively considered a single area of congestion, the results from the spreadsheet are used as an input in VCloud®. VCloud® is a simple software tool developed by the BakerRisk Engineering Consultant to produce Pressure/Impulse (P-I) diagrams (this is not to be confused with PI Diagrams representing the response curve for structures and might better be referred to as the side-on blast pressure and impulse as a function of standoff distance). Notice the check box denoted "Plot Using Co-op Energy Factor." Membership in the BakerRisk Engineering Consultant Explosion Research Cooperative (ERC) includes access to experimentally verified energy factors that can reduce conservatism and provide a more accurate prediction of explosion effects. The first page of inputs for VCloud® is illustrated in Figure 4.

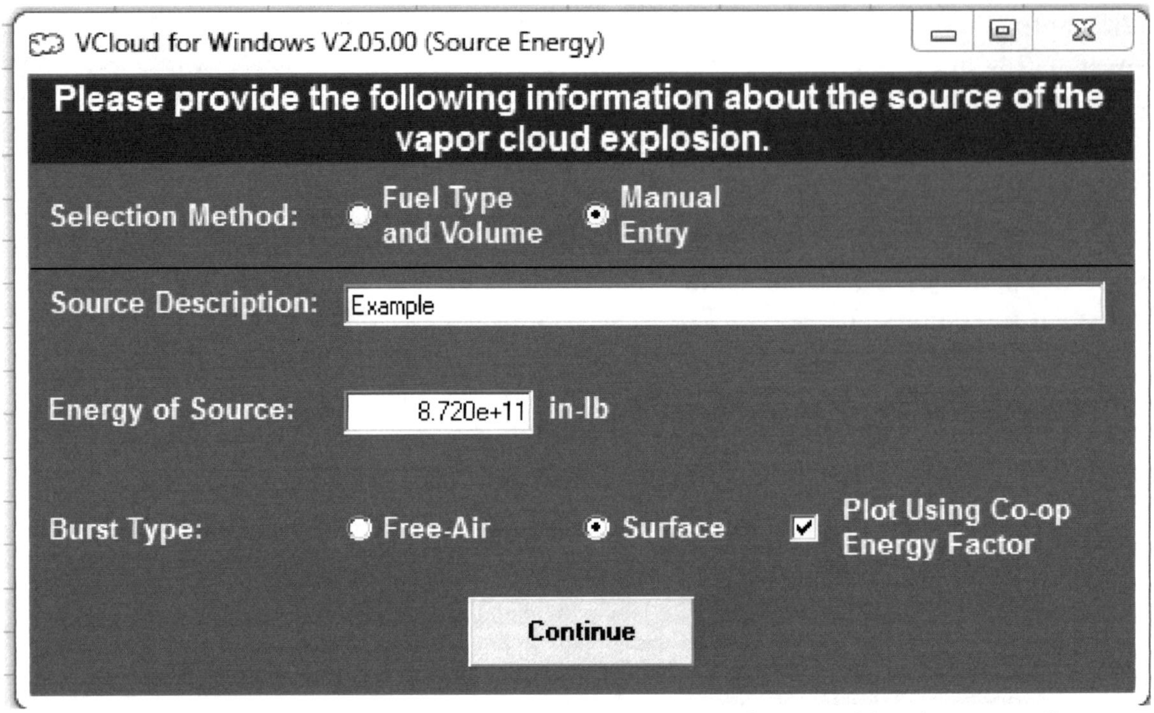

Figure 4. VCloud® Input to Determine Energy Content of Vapor Cloud. Image Courtesy of BakerRisk Engineering Consultant Explosion Research Cooperative.

The second parameter greatly influencing explosive strength when using BST methodology is flame speed, measured as a ratio to the speed of sound or Mach number. Four criterion most influence how fast a flame front burns through fuel: congestion, confinement, fuel reactivity, and the distance to free venting.

- Congestion
  - Burning rate is greatly dependent on the surface area of the flame as it travels through the fuel source. Turbulence causes the flame front to furrow and can greatly increase the surface area which contributes to increasing the speed of the advancing flame. In an industrial setting, turbulence can be generated by piping, equipment, gravel surfaces, vehicles, or any obstacle(s) that inhibits simple linear progression of the burning flame front.
  - The greater the induced turbulence, the faster a flame front will progress increasing the likelihood of detonation. BST method calculations qualify congestion into three categories; high, medium or low congestion. Congestion categorization is the most subjective parameter in BST methodology. Experience is probably the most important tool used to determine degree of congestion.
  - A good rule of thumb is if process equipment occupies an area but can be easily walked through, it is low congestion. If you can manage to slip through an area with difficulty, it's medium congestion. If it is impossible to physically pass through an area, it's high congestion.

- Confinement
  - Any solid surface that prevents the propagation of the flame in a given direction is considered confinement, with the exception of the ground. The BST method models a spherical, free-air vapor cloud explosion, but multiplies the energy of the explosion by a factor of 2 to account for ground reflection of the blast wave from its origin surface.
  - BST calculations categorize confinement in four ways; 3D, 2D, and 2½ D, and 1D.
    - 3D Confinement: A flame front burning through fuel is capable of expanding in all three dimensions with the exception of the ground. It applies to all free-field processing areas not bound by solid walls or ceilings.
    - *2D Confinement:* The propagation of the flame front is free to expand in only two axes because one dimension is blocked by a solid obstacle. A common example is a processing unit located between a solid concrete ceiling with open walls. It also applies if a processing unit is located between two solid walls with no roof. It is important to consider the aspect ratio of 2D confinement, which is the ratio of total width to height of the confined volume. Higher aspect ratios constrict venting capacity and allow additional pressure accumulation over those with a shorter distance to free venting. Typical values are available in a drop down menu in VCloud®, ranging as high as a 25:1 aspect ratio. Anything higher than 25:1 will not provide significantly more pressure accumulation, nor are aspect ratios greater than this commonly found in the field. Any aspect ratio greater than 4:1 should be represented in VCloud as 25:1 to maintain a conservative estimate.
    - 2½ D Confinement: This is a hybrid condition where neither 3D nor 2D accurately describe confinement. For instance, processing equipment contained within a pre-engineered building with lightweight metal walls and roof that will confine the flame front for a time but release at low pressures, before the impending explosion is fully realized. The roof and walls provide some containment, but in short order the blast wave is capable of 3D expansion. Tightly packed equipment on the boundaries of the congested area generating an explosion can also be considered 2½ D confinement, since it provides a small amount of venting, but not enough to consider it 3D expansion.
    - 1 D Confinement: Flame burn in a linear direction, such as inside a pipe.

- Fuel Reactivity
  - The BST method categorizes fuel reactivity in only three ways; low, medium, and high reactivity. Fuel reactivity is based on the laminar burning velocity at a specific concentration, determined by a standard testing procedure in stagnant conditions.
  - Fuels with a laminar burning velocity less than 45 cm/s are considered to have a low reactivity. Fuels with laminar burning velocities above 75 cm/s are considered highly reactive. Everything between is considered to be of medium reactivity.
  - Table 1 illustrates reactivity of a few common chemicals for reference here. NFPA 68 is a good reference for finding laminar burning velocities of many more chemicals so the reactivity category can be determined.

Table 1. Relative Reactivity

Reactivity	Fuel	Group	Relative Reactivity	Formula
Low	Methane	Alkanes	1	CH_4
Low	Formic Acid	Acids	1	HCOOH
Low	Acetic Acid	Acids	1	CH_3COOH
Medium	Ethane	Alkanes	3	C_2H_6
Medium	Propane	Alkanes	3	C_3H_8
Medium	Butane	Alkanes	3	C_4H_{10}
High	Ethylene	Alkenes	6	C_2H_4
High	Acetylene	Alkynes	7	C_2H_2
High	Hydrogen	Inorganics	7	H_2

(Courtesy Eastman Chemical Company)

- Distance to Free Venting
  - Distance to free venting is simply the shortest distance a flame must travel to exit the area of congestion/confinement to find free space.
  - After free venting, turbulence is less likely to accelerate the flame front and under most circumstances it will begin to slow. This reduces the possibility of a transition from deflagration to detonation (DDT).

These four criterion are now input into VCloud® in the format shown in Figure 5. From the provided data, a Mach number (M_f) is assigned to the flame front.

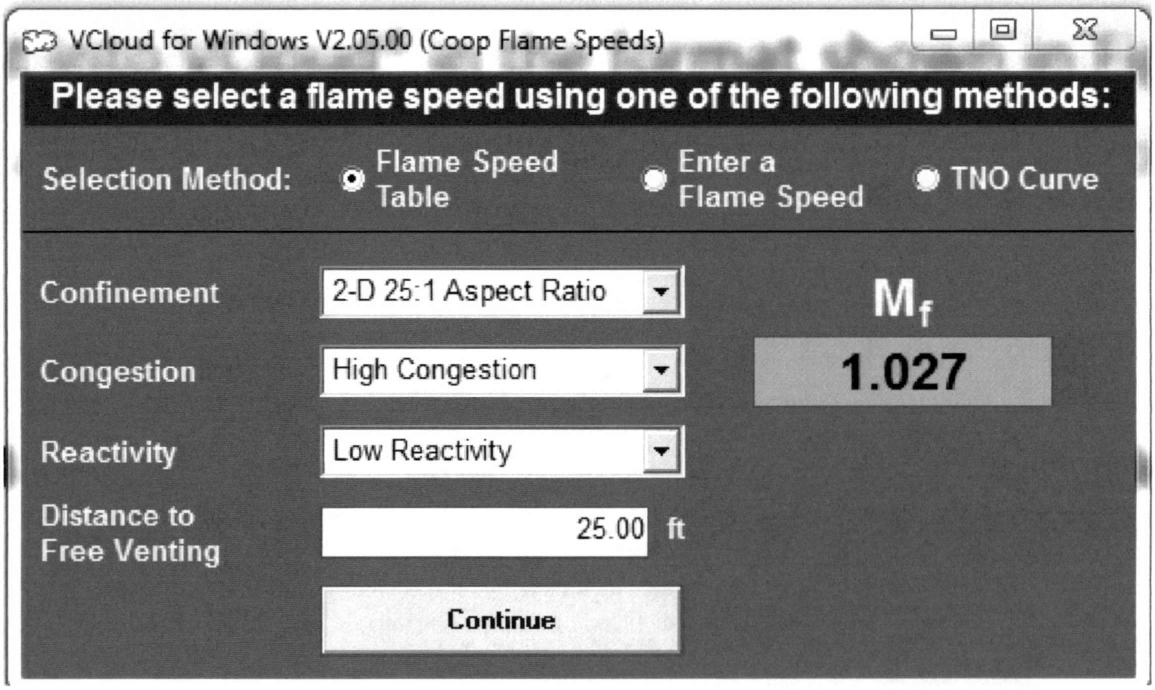

Figure 5. Vcloud® Input to Determine Mach Number. Image Courtesy of BakerRisk Engineering Consultant Explosion Research Cooperative.

A Mach number greater than one indicates the flame front accelerated faster than the speed of sound and beyond, meaning a detonation is more likely to occur. Some typical flame speeds in Mach numbers are provided in Table 2 for reference, with respect to fuel reactivity and degree of congestion.

Table 2. Flame Speeds for Soft Ignition Sources [9]. Provided as a reference but not relevant to the example calculation.

Flame Speed in Mach Numbers (M_w) for Soft Ignition Sources

Flame Expansion	Fuel Reactivity	Obstacle Density High	Obstacle Density Medium	Obstacle Density Low
ID	High	5.2	5.2	5.2
	Medium	2.27	1.77	1.03
	Low	2.27	1.03	0.294
2D	High	1.77	1.03	0.588
	Medium	1.24	0.662	0.118
	Low	0.662	0.471	0.079
3D	High	0.588	0.153	0.071
	Medium	0.206	0.100	0.037
	Low	0.147	0.100	0.037

VCloud® uses the provided inputs to produce a Pressure/Impulse (PI) diagram. The PI diagram is used to determine the overpressure and impulse experienced by a target at a given distance. Figure 6 is an example of a PI diagram produced by VCloud®. The distance between the target and the explosion is manually input to determine the blast effects at that specified distance. The target at 150 feet from the explosion will experience a 10.12 psig side-on overpressure with a 316.05 psi-ms impulse. This is more than enough to completely demolish an unreinforced masonry building. A key attribute of vapor cloud explosions that differs from high explosives is a much higher impulse. Impulse is more threatening to rigid construction than overpressure, because it tends to distort supporting structural members beyond their plastic limits.

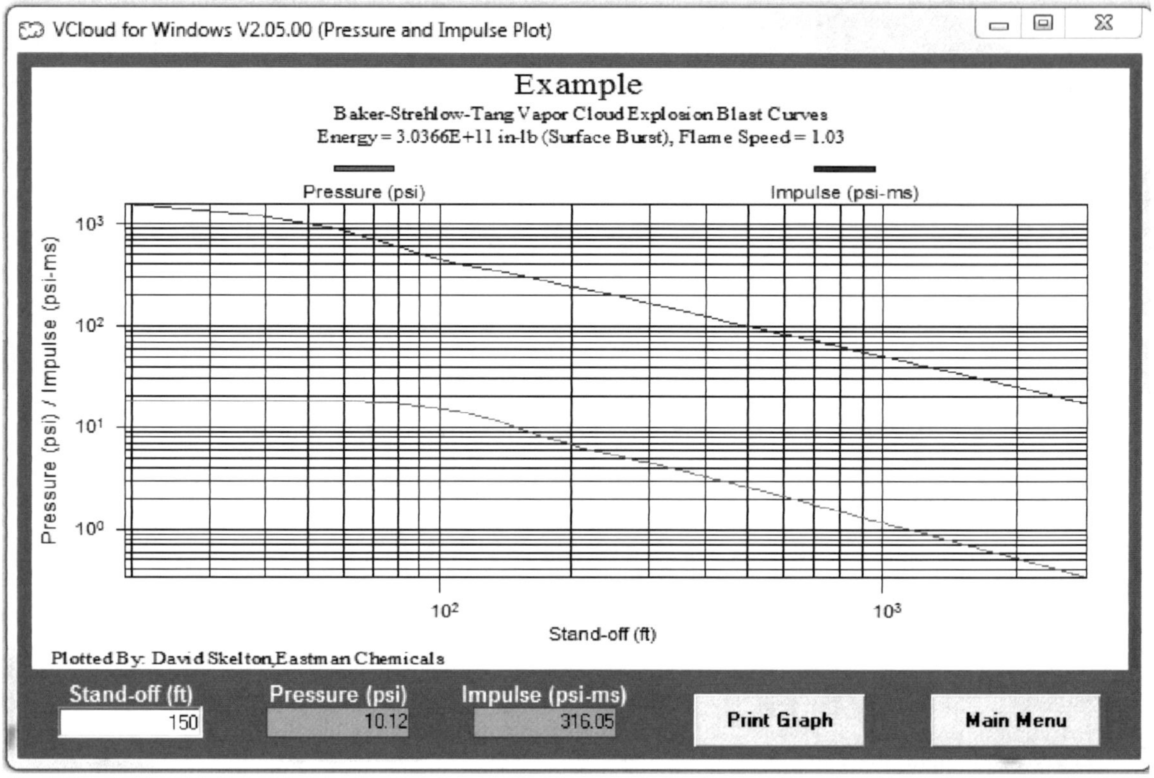

Figure 6. P-I Diagram Output From VCloud®. Image Courtesy of BakerRisk Engineering Consultant Explosion Research Cooperative.

Once a PI diagram is prepared, it can be used to predict the potential damage to conventional industrial buildings from the modeled blast event. Another software tool developed by the ERC can be used for this purpose, the Blast Evaluation and Screening Tool, affectionately known as BEAST®. BEAST® uses empirically based correlations from accidental explosions, shock tube tests, and government sponsored high explosive tests to predict the level of damage imparted by an explosion to a given building or structural component. Originally developed in 1995, BEAST® has been updated over the years to include more building and component damage curves, improve the user interface, and add more building types. The most recent version of BEAST® can evaluate 19 unique building types and many types of individual structural components, such as windows and doors. The first step is to define the building for evaluation. An example using the current BEAST® interface is included in Figures 7a and 7b. Required inputs are type of construction (drop down menu), building dimensions, blast overpressure and impulse (from PI Diagram), and whether wall and roof loading are side-on or reflected.

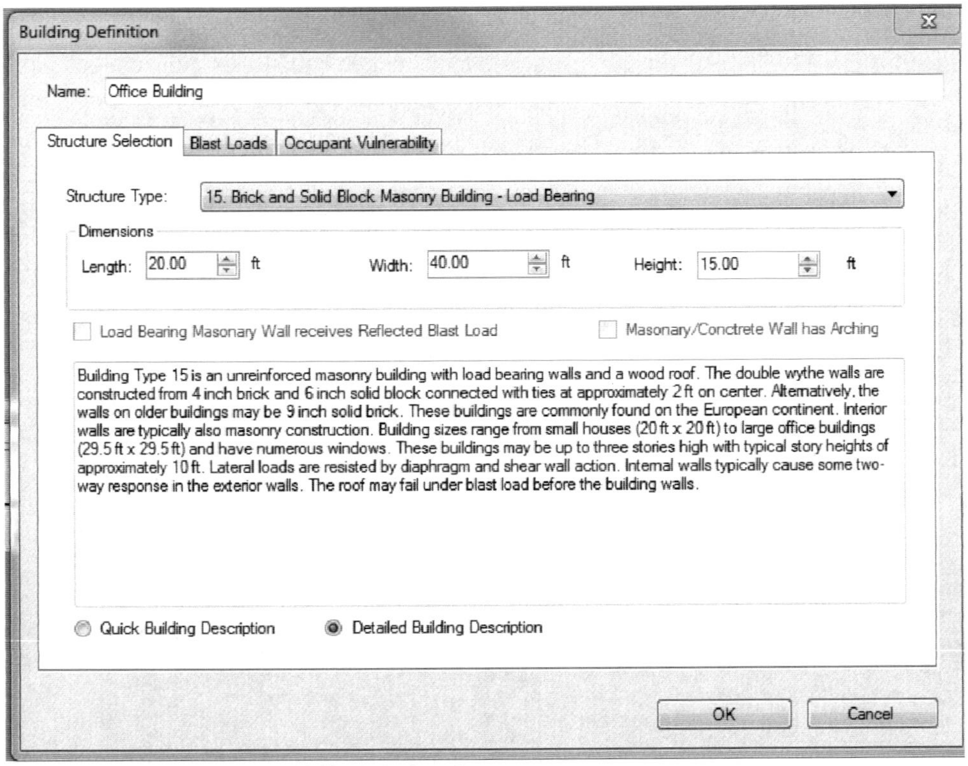

Figure 7a. Beast® Building Selection.

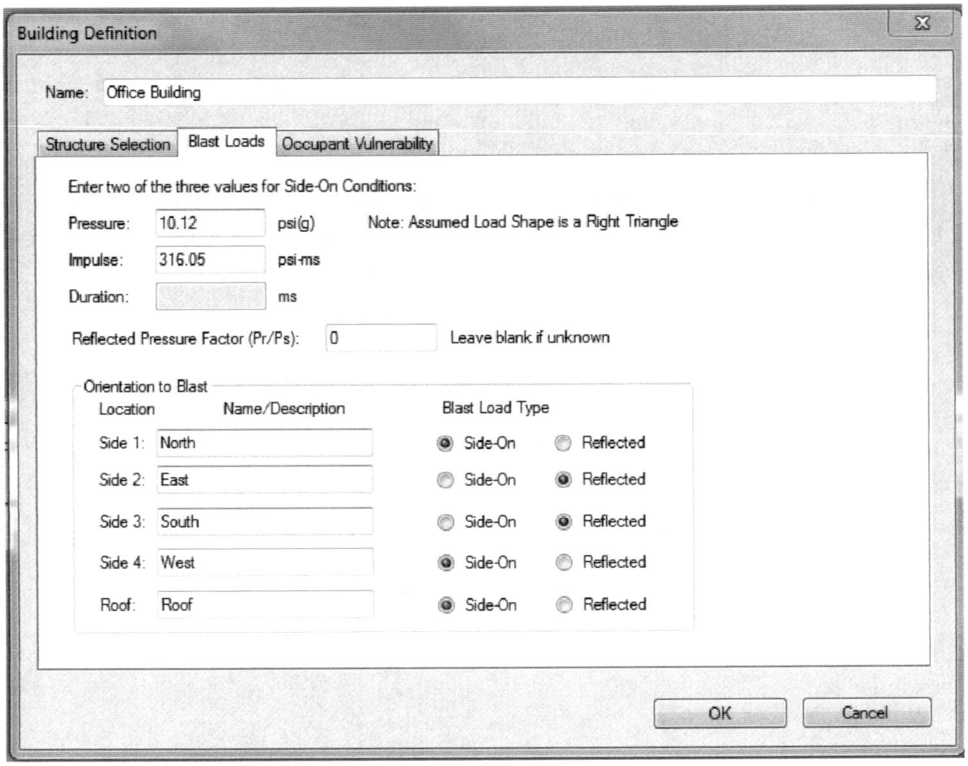

Figure 7b. Beast® Building Selection (cont'd).

Images Courtesy of BakerRisk Engineering Consultant Explosion Research Cooperative.

Given the appropriate inputs, BEAST® predicts a damage level to the building and an occupant vulnerability. Figure 8 illustrates the output provided from BEAST®.

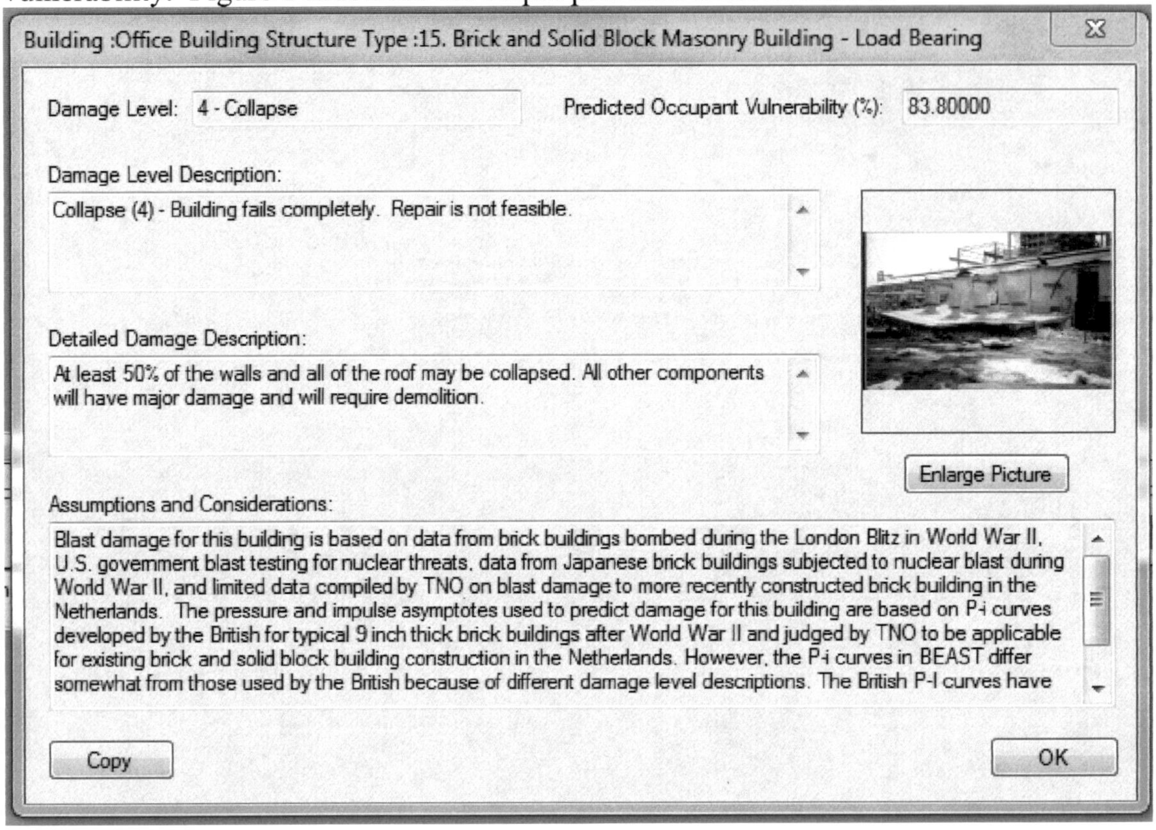

Figure 8. BEAST® Output

Of course, not all explosions result in the complete destruction of a building like the extreme example provided. BEAST® categorizes building damage using the criteria found in Table 3 [6]. A Building Damage Level up to 2B is still survivable by most occupants.

Table 3. BEAST® Building Damage Levels

Building Damage Level	Damage Description
1	Onset of visible damage to building walls facing explosive source. Repairs only needed for cosmetic reasons. Building is reusable following an explosion.
2A	Localized building damage, primarily to components on building walls facing the explosive source. Building performs function and can be used; however, major repairs are required to restore integrity of structural envelope. Total cost of repairs is moderate.
2B	Widespread building damage. Building cannot be used until major repairs are completed. Total cost of repairs is significant, approaching replacement cost of building.
3	Building has lost structural integrity and may collapse due to environmental conditions (i.e. snow or wind). Total repair cost exceeds replacement cost of building.
4	Building fails completely. Repair is not feasible.

Courtesy of BakerRisk Engineering Consultant Explosion Research Cooperative

4. Development of Source Terms for Potential Vapor Dispersion Release Scenarios

Methods for determining the source term for releases from tanks, liquid pools and piping systems are discussed here. All examples provided here do not reflect actual potential release conditions. For all examples, any relation to true conditions is incidental.

Prior to the modeling methods discussion, it is important to review a tool for determining losses due to piping layout.

IoMosaic SuperChems® v 8.21 (2016) allows for development of piping layouts to detail the piping diameters, elbows, and other potential connections. In simple models, as will be discussed in this paper, this can be used as the outlet piping for a pressure source (e.g., tanks) to a release point.

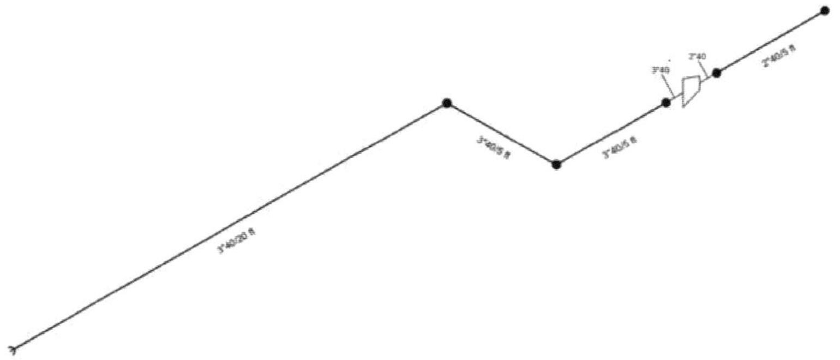

Figure 9. Example Piping Layout.

Figure 3 details a layout from the source at 3 inches with 2 90-degree elbows and a reduction to the 2-inch leak point. The release will be from a guillotine break on the 2-inch line.

4.1 Release from Tank

A reliable model for release from vessels is contained in ioMosaic SuperChems® v8.21 (2016).

This example will model the release of a liquid blanketed by nitrogen. The "Vessels Containing 2-Phases Model" allows for estimation of the vapor and liquid contents of a vessel, accounting for nitrogen blanketing or air in the head space. This same model can be utilized for relief from the vapor phase. There are alternative models for gas-filled vessels.

The overall process for using SuperChems® to model this is as follows:

1. Development of objects for mixture composition, vessel dimensions, piping layouts and other objects. Ensure piping layouts are properly applied.

   The following steps are within the "Vessels Containing 2-Phases" Model.

2. Input the vessel pressure and an initial estimate for overall composition inside of the vessel. The overall composition includes both the vapor and liquid phases. In this example, the modeled vessel is filled to 80% with methyl isocyanate and has a nitrogen blanket.

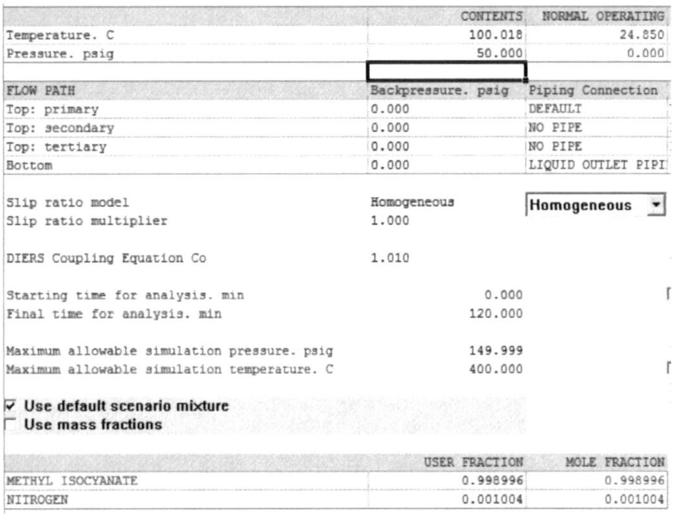

Figure 10. Vessels Containing 2-Phase Inputs

   Uncheck the "Use default scenario mixture" as defining this model is an iterative process.

3. Use "Specify Fill Fraction" to input the approximate vessel fill level. This example uses an estimate 80% fill level.

   The model it calculates the vessel temperature at the specified overall composition.

```
Fill Fraction = [illegible] %, T/F= [illegible], T = [illegible] C, Liq = [illegible] lb, Vap = [illegible] lb, Total =
[illegible] lb
Fill Fraction = [illegible] %, T/F= [illegible], T = [illegible] C, Liq = [illegible] lb, Vap = [illegible] lb, Total =
[illegible] lb
Fill Fraction = [illegible] %, T/F= [illegible], T = [illegible] C, Liq = [illegible] lb, Vap = [illegible] lb, Total =
[illegible] lb
Fill Fraction = [illegible] %, T/F= [illegible], T = [illegible] C, Liq = [illegible] lb, Vap = [illegible] lb, Total =
[illegible] lb
Fill Fraction = [illegible] %, T/F= [illegible], T = [illegible] C, Liq = [illegible] lb, Vap = [illegible] lb, Total =
[illegible] lb
Fill Fraction = [illegible] %, T/F= [illegible], T = [illegible] C, Liq = [illegible] lb, Vap = [illegible] lb, Total =
[illegible] lb
Fill Fraction = [illegible] %, T/F= [illegible], T = [illegible] C, Liq = [illegible] lb, Vap = [illegible] lb, Total =
[illegible] lb
Invalidated Results for [illegible]
Updating Downstream Vessel Connections
```

Figure 11. Model Iterations to determine vessel temperature at 80% liquid fill level.

4. Iterate over steps 2 and 3, varying the composition and re-entering the fill fraction until the modeled vessel temperature matches the true conditions. Increasing content of lower boilers will decrease the modeled vessel temperature.

   In the example, the composition was iterated until the storage temperature was approximately 25 deg C.

	CONTENTS	NORMAL OPERATING
Temperature. C	24.918	24.850
Pressure. psig	50.000	0.000

FLOW PATH	Backpressure. psig	Piping Connection
Top: primary	0.000	DEFAULT
Top: secondary	0.000	NO PIPE
Top: tertiary	0.000	NO PIPE
Bottom	0.000	LIQUID OUTLET PIPI

Slip ratio model	Fauske	Fauske
Slip ratio multiplier	1.000	
DIERS Coupling Equation Co	1.010	
Starting time for analysis. min	0.000	
Final time for analysis. min	120.000	
Maximum allowable simulation pressure. psig	149.999	
Maximum allowable simulation temperature. C	400.000	

☐ **Use default scenario mixture**
☐ **Use mass fractions**

	USER FRACTION	MOLE FRACTION
METHYL ISOCYANATE	0.991100	0.991100
NITROGEN	0.008900	0.008900

Figure 12. Final composition.

5. Running the model will determine the flow rate through the piping layout and guillotine break.

6. When the run is complete, a report will appear. The "Profiles-Bottom" tab is the most useful for this source term model.

TIME. min	VESSEL CONTENTS TEMPERATURE. C	VESSEL CONTENTS PRESSURE. psig	PRESSURE AT VESSEL BOTTOM. psig	BOTTOM PIPING LIQUID MASS FLOW RATE. lb/min	BOTTOM PIPING GAS/VAPOR MASS FLOW RATE. lb/min	Bottom flow reduction factor (Actual flux with pipe/Ideal Flux)	BOTTOM FLOW NOZZLE PRESSURE. psig	BOTTOM FLOW DISCHARGE PRESSURE. psig	BOTTOM FLOW DISCHARGE TEMPERATURE. C
0.0000	24.92	50.000	56.394	2.005E+03	0.000	0.405	56.392	-0.001	25.094
0.0246	24.88	47.142	53.517	1.966E+03	0.000	0.394	53.516	0.000	25.051
0.0506	24.85	44.305	50.662	1.871E+03	0.000	0.383	50.660	-0.001	25.006
0.0782	24.81	41.472	47.809	1.775E+03	0.000	0.372	47.808	-0.001	24.960
0.1077	24.77	38.642	44.961	1.679E+03	0.000	0.360	44.959	-0.001	24.913
0.1394	24.73	35.817	42.116	1.583E+03	0.000	0.347	42.115	-0.001	24.864
0.1735	24.69	32.996	39.275	1.485E+03	0.000	0.334	39.274	-0.001	24.814
0.2104	24.64	30.181	36.440	1.387E+03	0.000	0.321	36.439	0.000	24.761
0.2506	24.60	27.371	33.610	1.289E+03	0.000	0.306	33.609	0.000	24.706
0.2946	24.55	24.568	30.787	1.190E+03	0.000	0.291	30.786	0.000	24.649
0.3433	24.50	21.773	27.971	1.090E+03	0.000	0.276	27.970	0.000	24.588

Figure 13. Vessels Containing 2-Phases Output.

The liquid and vapor release rates out of the piping as well as the discharge temperature and pressure are reported. The discharge pressure will be above atmospheric if there is choked flow inside the piping.

If the choke pressure is significantly above atmospheric, running some select points through the "2-Phase Expansion and Aerosol Model" to determine the final vapor/liquid composition and temperature is recommended.

7. As the release is liquid in this case, evaporation from a liquid forming pool will next be modeled. As stated previously, several dispersion models have integrated pool models. The SuperChems® pool evaporation will be employed in this case, and the output from the model will be input to an external dispersion model.

4.2 Liquid Pools

As with tank leaks, SuperChems® will be utilized for determining the source term of a liquid pool formation.

An important consideration is the variety of weather conditions over which a release could take place. As solar incidence is a major factor in pool evaporation, ensure that the most probable conditions for both daytime and nighttime are employed. Also, to account for solar incidence, ensure that the appropriate time of day and site coordinates are input for the scenario in SuperChems®. In addition, diked area dimensions and substrate composition can be specified.

SuperChems® pool evaporation model allows for a static release rate over a set period of time. As most releases will be dynamic in nature, this may involve expert judgment.

Once a characteristic release rate, duration and temperature have been determined, this information can be entered into the pool evaporation model.

Liquid discharge temperature. C	24.209
Liquid discharge rate. lb/min	583.903
Liquid discharge duration. min	1.949

Use default scenario mixture
Use mass fractions

COMPONENT	USER FRACTION
METHYL ISOCYANATE	1.0000
NITROGEN	0.0000

Average molecular weight	57.0520

Figure 14. Pool Evaporation Model Inputs

The model can perform evaporation on multi-component mixtures. In this case, the evolution of any dissolved nitrogen is not of concern. The liquid rate has been imported from the Vessels Containing 2-Phases model, and the source concentration has been set to pure methyl isocyanate.

Based on the inputs, the model will provide a conservative estimate for the evaporation of methyl isocyanate from the formed pool.

Time. min	OVERALL POOL EVAPORATION RATE. lb/min	Pool diameter. ft	Pool surface temperature. C
0.017	0.000	3.146	4.209
0.017	0.258	3.146	-5.182
0.017	0.257	3.146	-5.235
0.017	0.257	3.147	-5.287
0.017	0.256	3.147	-5.339
0.018	0.257	3.147	-5.389
0.028	0.249	3.171	-5.439
0.038	0.242	3.241	-7.525
0.049	0.252	3.371	-8.514
0.059	0.273	3.553	-8.996
0.070	0.302	3.765	-9.216
0.080	0.335	3.988	-9.306
0.090	0.371	4.212	-9.334
0.101	0.409	4.431	-9.335
0.111	0.447	4.644	-9.322
0.122	0.485	4.849	-9.305

Figure 15. Pool Evaporation Model Output.

Figure 15 details the most critical data points in the pool evaporation model. Understanding these will allow for straightforward input to a dispersion model.

While SuperChems® does have an integrated dispersion model, the author's preference is SAFER Trace® v10.2 (2012). Implementation of pool models into dispersion models is discussed later in this paper.

4.3 Leaks from Piping

For lengths of piping with isolation, it may be appropriate to treat the isolated volume as a vessel and analyze the varying leak rate over time. This section considers the situation where the pressure differential across the orifice may be assumed constant. In addition to the previous source terms, SuperChems® is utilized in determining the source term.

The "Source Term Wizard" can be employed to specify an upstream pressure and temperature to determine the flow rate across a specified piping layout or idealized nozzle. The example here will determine the release rate of methyl isocyanate pressurized to 50 psig and released through a guillotine break in the 2" segment of piping from Figure 9.

OPTIONS	
Liquid Density	Databank
PVT Behavior	Equation of state
VLE Search Direction	Dew point

- [ ] Use mass basis data to compute fractions
- [ ] Mixture is soluble in water
- [ ] Use an existing mixture

COMPOUND	FORMULA
METHYL ISOCYANATE	C2H3NO

Inlet temperature. C	25.00000
Inlet pressure. psig	50.00000
Back pressure. psig	0.00000

- [x] Initial state is subcooled liquid

Figure 16. Liquid Source Term Wizard Inputs

The release is treated as a subcooled liquid in order to allow the model to estimate if any portion of the release will flash across the opening. It is not expected to have any flashing due to the release conditions in comparison to the chemical's vapor pressure at 50 psig. However, this model is employed for consistency.

** INLET CONDITIONS	
Inlet temperature. C	25.0
Inlet pressure. psig	50.0
Surroundings/back pressure. psig	-3.974E-06
Slip ratio exponent eta	0.001
Vapor quality	1.E-04

** EXIT CONDITIONS	
Exit temperature. C	25.160
Exit pressure. psig	-1.7331E-04
Barometric pressure at site elevation. psig	-3.974E-06
Mixture exit velocity. ft/s	21.213
Exit Mach Number	0.006
Maximum Mach Number	0.006
Exit rho*u*u in kg/m/s2 or in Pascals	38695.90
Maximum rho*u*u in kg/m/s2 or in Pascals	38695.90
Exit steady state impulse. lbf	68.146
Maximum steady state impulse difference. lbf	418.675
Pipe length travelled / total piping length	0.998
Flow rate. lb/min	1714.327

COMPONENT	MW	TOTAL FLOW. lb/min	LIQUID FLOW IN. lb/min	LIQUID FLOW OUT. lb/min	VAPOR FLOW IN. lb/min	VAPOR FLOW OUT. lb/min
METHYL ISOCYANATE	57.052	1713.817	1713.817	1713.817	0	0

Figure 17. Source Term Wizard Outputs.

The results from this model include liquid and vapor release rates as well as exit pressure and temperature. Had flow been choked in the piping, the exit pressure would be above atmospheric. As with previous models, it is recommended to utilize a flash block to determine conditions at atmospheric pressure for releases from systems with choked flow.

For releases from holes or cracks in piping, the piping layout can be developed with an orifice at the end of the layout. The orifice area would be equal to the area of the hole, and the orifice length would be equal to the width of the piping. In addition, due to constraints with the software, a short pipe segment of with the same diameter as the leak should be placed at the exit of the segment.

The results for the above cases can directly be input to a dispersion model, as described in the next section.

5. Application of Source Terms to Dispersion Models

This section will describe the inputs to dispersion models based on the results from the two-phase leak from the sour stream.

Section 5 will review how to utilize the data from the dispersion models in order to assess the hazard posed to a hypothetical shelter 500 ft downwind from the release.

5.1 Test for Jetting

In two-phase or vapor release cases, the linear velocity across an opening may be sufficient to cause jetting. As described previously, there are specific dispersion models for jetting conditions. The rule of thumb for jetting regime is 10 m/s [1]. A simple ideal gas calculation can provide an estimate for linear velocity, given a known release rate, hole size and temperature. This can be expanded to a real gas calculation for highly non-ideal cases. The temperature used in this calculation should be the flashed temperature to atmospheric pressure. The rate employed should be that of the vapor phase for two-phase releases as this will be the largest volumetric component once flashed to atmosphere.

In this example, a 200 psig stream of sour hydrocarbons at 150 deg C is released across a 1-inch hole in 6-inch piping. While several VLE models are available for hydrocarbon mixtures, ideality is assumed in this example.

Piping Isometric / Layout	6-INCH LINE WITH 1-INCH CRACK
Site Name	DEFAULT
Liquid Density Method	Databank
VLE/PVT Method	Equation of state
**** INLET CONDITIONS**	
Inlet temperature. C	150.691
Inlet pressure. psig	200.0
Surroundings/back pressure. psig	-3.974E-06
Slip ratio exponent eta	0.001
**** EXIT CONDITIONS**	
Exit temperature. C	133.870
Exit pressure. psig	138.432
Barometric pressure at site elevation. psig	-3.974E-06
Mixture exit velocity. ft/s	190.056
Exit Mach Number	1.0
Maximum Mach Number	1.0
Exit rho*u*u in kg/m/s2 or in Pascals	393239.785
Maximum rho*u*u in kg/m/s2 or in Pascals	393239.785
Exit steady state impulse. lbf	165.061
Maximum steady state impulse difference. lbf	6037.864
Pipe length travelled / total piping length	0.998
Flow rate. lb/min	455.119

COMPONENT	MW	TOTAL FLOW. lb/min	LIQUID FLOW IN. lb/min	LIQUID FLOW OUT. lb/min	VAPOR FLOW IN. lb/min	VAPOR FLOW OUT. lb/min
HYDROGEN SULFIDE	34.082	2.052	2.052	0.866	0.001	1.186
n-PENTANE	72.150	347.584	347.547	268.991	0.037	78.593
HEXANE	86.175	77.840	77.835	66.892	0.005	10.949
n-OCTANE	114.231	27.515	27.515	26.078	0.001	1.438

Figure 18. Flow of stream across 1-inch crack.

The exit conditions show that the flow has choked. As previously discussed, the "2-Phase Expansion and Aerosol Model" will be employed to determine the composition and temperature at atmospheric pressure. In addition, this model will report the linear velocity and cloud diameter when flashed to atmospheric pressure. The linear velocity can be used to determine if jetting is possible.

```
**TWO-PHASE EXPANSION / DEFAULT

                                                     INITIAL            FINAL
Flow diameter. ft                                 8.3333E-02       2.3754E-01
Velocity. ft/s                                        190.42           651.59

Vapor density. lb/gal                                   0.24             0.02
Liquid density. lb/gal                                  4.20             5.19
Mixture density. lb/gal                                 0.98             0.04

Temperature. C                                        133.87            45.56
Pressure. psig                                        138.43             0.00
Vapor/Feed (V/F) mole fraction                          0.21             0.71
Mass flow rate. lb/min                                455.12           455.12

Available energy. BTU/lb                               26.45
Modified Weber number for thermal breakup             392.30
Critical Weber number for mechanical breakup           12.01
Viscosity group number                                  0.01

Stable droplet diameter. Microns                       39.50
Thermal breakup droplet diameter. Microns              39.50
Mechanical breakup droplet diameter. Microns           44.18
Number of droplets                                4.9421E+10
Droplet mass. lb                                  4.4271E-11

COMPONENT                                         TOTAL FLOW      INITIAL LIQ        FINAL LIQ      INITIAL VAP        FINAL VAP
HYDROGEN SULFIDE. lb/min                               2.053            0.866            0.038            1.186            2.015
n-PENTANE. lb/min                                    347.681          269.076           80.158           78.605          267.523
HEXANE. lb/min                                        77.862           66.912           35.926           10.950           41.936
n-OCTANE. lb/min                                      27.523           26.085           23.812            1.438            3.711
```

Figure 19. Sour Stream Contents after Expansion.

Per Figure 19, the temperature of the release cloud will be 46 deg C at atmospheric pressure and will have a linear velocity of 652 ft/s (199 m/s). This is an order of magnitude above the 10 m/s guidance, and should result in significant jetting.

5.2 Input to Jet Model

The dispersion modeling tool utilized for these examples is SAFER Trace® 10.2 (2012).

Figure 20 details the inputs to the jet model based on our example. This is focusing on the vapor portion of the model, ignoring liquid entrainment.

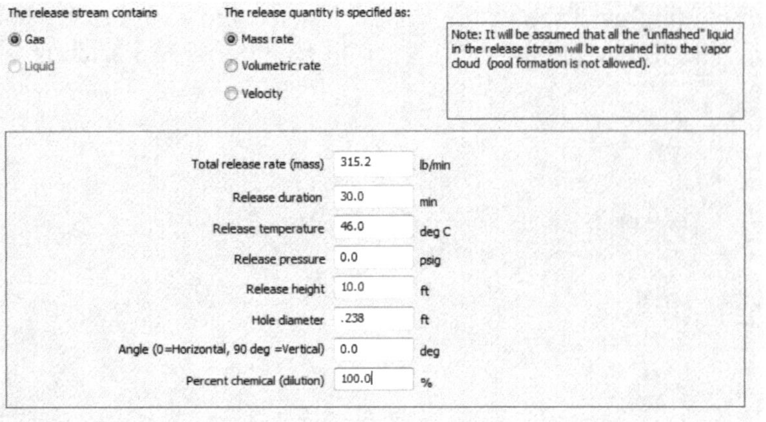

Figure 20. Jet Model Inputs.

The dispersion outputs will report concentrations and plots related to concentration of H2S in air.

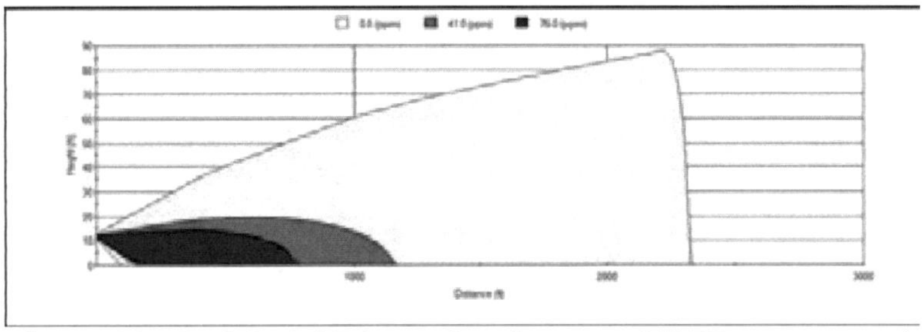

Figure 21. Jet Model Dispersion Output. AEGL concentration limits based on 10-minute exposure to H2S [10].

The H2S concentration profile at the outside of a shelter 500 ft downwind from the release point is shown in Figure 22.

Figure 22. H2S Concentration Outside Shelter

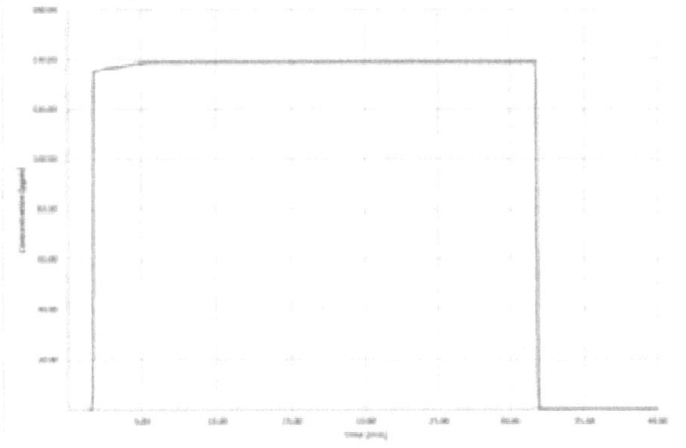

Consequence to personnel in this shelter will be analyzed using a probit function, detailed in Section 5.

5.3 Modeling Pool Evaporation

The liquid portion of the example two-phase release will be treated as the source of a liquid pool.

Spill Surface	DEFAULT
Liquid discharge temperature. C	45.560 <<<
Liquid discharge rate. lb/min	139.934
Liquid discharge duration. min	15.000

☑ **Use default scenario mixture**
☐ **Use mass fractions**

COMPONENT	USER FRACTION
HYDROGEN SULFIDE	0.0006
n-PENTANE	0.6394
HEXANE	0.2399
n-OCTANE	0.1200
Average molecular weight	80.5399

RUN OPTIONS

Figure 23. Liquid Pool Model Inputs

The concentration profile of evaporate is highly dynamic over the studied period of the release.

Figure 24. Pool Evaporate Concentration Profile

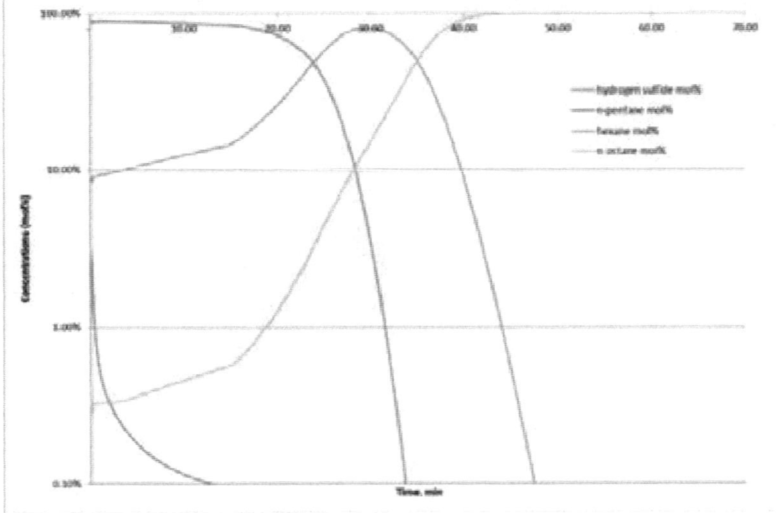

While the concentration is highly dynamic, the main toxicity concern is with H2S. The majority of H2S appears to evaporate in the first 15 minutes. An average rate will be developed based on the total evaporated mass of each component over that time. The effective concentrations and dispersion parameters are shown in Table 4.

Table 4. Molar Concentrations and Dispersion Parameters

Chemical	Mol%
Hydrogen Sulfide	0.1%
Pentane	87.0%
Hexane	12.4%
Octane	0.5%
Average Evaporation Rate	57.0 lb/min
Temperature	2°C
Duration	15 min

The data from this model is input to the SAFER Trace® "Single Phase" steady state model, per Figure 25.

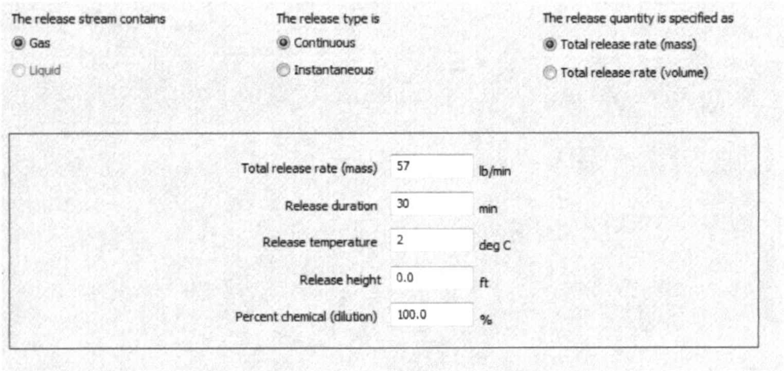

Figure 25. Steady State Model Inputs.

The downwind dispersion of the cloud, focusing on H2S concentrations in air is shown in Figure 26. The concentration is negligible at the 500-ft shelter, so the contribution from evaporation to toxic impact can be ignored.

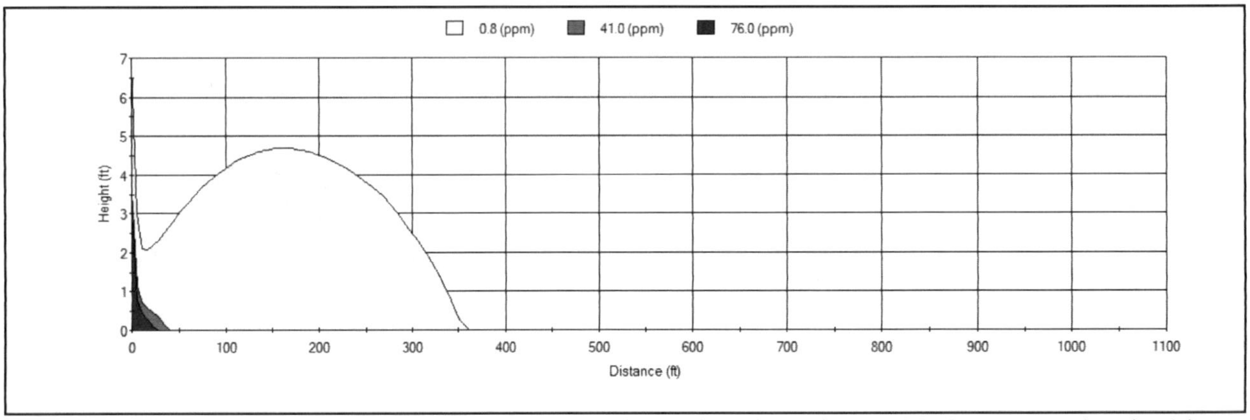

Figure 26. Downwind Impact from Evaporate Dispersion, based on H2S Concentration in Air.

6. Magnitude of Consequence Determination Using Probits

The methods utilized here are described in detail in "Simplified Methods of Using Probit Analysis in Consequence Analysis" [1].

To summarize, the impact to individuals may calculated using dose [5].

$$Dose = \int C^n dt \qquad \text{[Eq. 1]}$$

Dose may be transformed to probability of fatality using a probit transformation [5].

$$Y = A + B\log(Dose) \qquad \text{[Eq. 2]}$$

The constants "A", "B", and "n" are specific to the chemical. The Y value, or probit score is referenced on a probit table to determine probability of studied consequence.

Table 5. Probit table.

%	0	1	2	3	4	5	6	7	8	9
0	—	2.67	2.95	3.12	3.25	3.36	3.45	3.52	3.59	3.66
10	3.72	3.77	3.82	3.87	3.92	3.96	4.01	4.05	4.08	4.12
20	4.16	4.19	4.23	4.26	4.20	4.33	4.36	4.39	4.42	4.45
30	4.48	4.50	4.53	4.56	4.59	4.61	4.64	4.67	4.69	4.72
40	4.75	4.77	4.80	4.82	4.85	4.87	4.90	4.92	4.95	4.97
50	5.00	5.03	5.05	5.08	5.10	5.13	5.15	5.18	5.20	5.23
60	5.25	5.28	5.31	5.33	5.36	5.39	5.41	5.44	5.47	5.50
70	5.52	5.55	5.58	5.61	5.64	5.67	5.71	5.74	5.77	5.81
80	5.84	5.88	5.92	5.95	5.99	6.04	6.08	6.13	6.18	6.23
90	6.28	6.34	6.41	6.48	6.55	6.64	6.75	6.88	7.05	7.33
—	0.0	0.1	0.2	0.3	0.4	0.5	0.6	0.7	0.8	0.9
99	7.33	7.37	7.41	7.46	7.51	7.58	7.65	7.75	7.88	8.09

Source: Finney, 1952 [11]

The Microsoft Excel (2016) function "NORMSDIST" will convert the probit score to a probability.

The Equation 1 is combined with the "NORMSDIST" function to determine impact to occupants of the shelter, per Table 6. This is assuming that the air handling provides 10 air changes per hour to the shelter.

Table 6. Probability of Consequence to Sheltered Occupants.

Time (min)	Dosage ($ppm^{1.43}$-min)	% Probability
1.45		0%
1.46	-	0%
1.47	-	0%
1.48	-	0%
1.49	-	0%
1.50	-	0%
...	...	...
57.00	44,794	0%
58.00	45,354	0%
59.00	45,902	0%
60.00	46,436	0%
61.00	46,957	0%
62.00	47,466	0%
63.00	47,963	0%
64.00	48,448	0%
65.00	48,922	0%

In this example, the occupants can shelter for over an hour with 0% probability of consequence. Probit transformations can be utilized to study probability of fatal exposure from toxicity, thermal radiation, and blast overpressure. In addition, they can analyze probability of burn severity.

7. Conclusion

This paper has summarized the methods utilized in determining impact to personnel from toxic exposure and vapor cloud explosions. The covered models include pipe breach, tank release, and jetting models. These are highly utilized models for consequence analysis. In addition, the paper summarized methods to utilize dispersion modeling data to fully quantify impact to occupants of shelters downwind from releases.

For blast modeling, this paper provided an overview of one simplified method to predict explosion overpressure, impulse, and damage to buildings within the radius affected by the explosion. The method utilized is Baker-Strehlow-Tang (BST), one of several methods engaged by industry for this purpose. Most of the software tools used to perform the analysis were developed by the BakerRisk Engineering Consultant Explosion Research Cooperative, including VCloud® and BEAST®.

For all consequence models, more rigorous methods such as computational fluid dynamic (CFD) modeling will produce more accurate results. However, the methods discussed in this paper are excellent screening tools whose value lies in their ease of use and are suitable for risk analysis.

References

[1] Baybutt, Paul, "Using Layers of Protection Analysis to Evaluate Fire and Gas Systems", Process Safety Progress, Volume 31, Issue 3, 2011

[2] "SAFER Trace® Online Help", SAFER Systems, 2003

[3] James, Michael D., "Consequence Modeling of Dynamic Source Terms", Process Safety Progress, Volume 35, Issue 4, 2016

[4] Baker, Q. A., Tang, M. J., E. A. Scheier, and G. J. Silva, "Vapor Cloud Explosion Analysis," Process Safety Progress, Vol. 15, No. 2, 1996, pp. 106-109

[5] A. Sari, "Comparison of TNO Multienergy and Baker-Strehlow-Tang Models," Process Safety Progress 30 (2011), pp. 23-26

[6] BEAST® Help File, 2000 BEAST® Report, 2000

[7] "VCE Evaluation Method Comparison," The Quest Quarterly, Volume 4, Issue 1, Spring 1999

[8] Baker-Strehlow-Tang Methodology, Guidelines for Evaluating the Characteristics of Vapor Cloud Explosions, Flash Fires, and BLEVEs, CCPS, New York, 1994

[9] Pierorazio, A.J., J.K. Thomas, Q.A. Baker and D.E. Ketchum (2005) "An Update to the Baker-Strehlow-Tang Vapor Cloud Explosion Prediction Methodology Flame Speed Table," Process Safety Progress, 24(1): 59-65

[10] "About Acute Exposure Guideline Levels (AEGLs)", 2016, https://www.epa.gov/aegl/about-acute-exposure-guideline-levels-aegls (Accessed Mar 2, 2017)

[11] D. J. Finney, *Probit Analysis*, Cambridge University Press, New York 1952

Relief Systems Design: The Importance of Accounting for Static Head

Nicholas N. Cristea
Smith & Burgess, LLC
7600 W Tidwell Rd, Ste 600, Houston, TX 77040
Nicholas.cristea@smithburgess.com

Zubin Kumana
Smith & Burgess, LLC
7600 W Tidwell Rd, Ste 600, Houston, TX 77040
Zubin.kumana@smithburgess.com

Guomin Wang
Smith & Burgess, LLC
7600 W Tidwell Rd, Ste 600, Houston, TX 77040
Guomin.wang@smithburgess.com

Prepared for Presentation at
American Institute of Chemical Engineers
2017 Spring Meeting
13th Global Congress on Process Safety
San Antonio, Texas
March 26-29, 2017

Relief Systems Design: The Importance of Accounting for Static Head

Nicholas N. Cristea
Smith & Burgess, LLC
7600 W Tidwell Rd, Ste 600, Houston, TX 77040
Nicholas.cristea@smithburgess.com

Zubin Kumana
Smith & Burgess, LLC

Guomin Wang
Smith & Burgess, LLC

Keywords: Relief Systems Design, Static Head, Upstream, Downstream

Abstract

The process of evaluating an entire facility's relief and flare system design can be time consuming and costly. During the process, there will be systems to be evaluated that have liquid scenarios that may be protecting other assets within the facility. If the static head for the liquid scenarios is not accounted for, then the analysis can become even more costly due to the analysis not being performed correctly the first time. The result could be not meeting code by not adequately protecting system. This paper will look at two different cases where the static head was not accounted for, and how the issue was mitigated in the most cost-effective way. First, when dealing with the liquid heading on an offshore platform, there are many different levels on the platform and ensuring that the static head is accounted for is important to protect the different assets in the liquid handling train. Second, when evaluating a column system, it is important to ensure that the relief device provides protection for every ASME Section VIII Div 1 stamped equipment within the system. These two case studies highlight the necessity of accounting for static head in relief systems design.

1 Introduction

In the process safety industry, relief devices are set such that at least one relief device set pressure is equal to the lowest system MAWP within the protected system, to meet the requirement of ASME Section VIII Div 1 to provide adequate protection against overpressure. ASME Section VIII Div 1 UG-125 states "All pressure vessels within the scope of this Division, irrespective of size or pressure, shall be provided with overpressure protection in accordance with the requirements of UG-125 through UG-138, or with overpressure protection by system design in accordance with requirements of UG-140, or a combination of the two". ASME Section VIII Division VIII UG-125(c) and (1) also state that "when a pressure relief device is provided, it shall prevent the pressure from rising more than 10% or 3 psi (20 kpa), whichever is greater, above maximum allowable working pressure" and "when multiple pressure relief devices are provided and set in accordance with UG-134(a), they shall prevent the pressure from rising more than 16% or 4 psi (30 kpa), whichever is greater, above the maximum allowable working pressure". The vessel is normally hydrotested without accounting for the potential impact of inlet piping to the relief devices. Therefore, when the elevation difference between the top of the vessel and the relief device becomes significant [2], the vessel may experience a pressure higher than its allowable accumulation due to the static head, especially during liquid relief scenarios. The equation below is a conversion from the liquid height to the equivalent liquid head (psig).

$$P = 0.433 \; x \; SG \; x \; H$$

P: liquid head in psi

SG: Specific Gravity

H: liquid head in feet

Equation 1: Liquid height to pressure conversion

This paper will look at two different cases where the static head was not accounted for and how the issue was mitigated for each case. The first case will look at a blocked system liquid outlet scenario on a platform where the vessel of interest has the same MAWP as the protecting vessel on an upper level. The second case will look at the liquid relief case (i.e. overfilling) and the mitigation for a column system that contains several vessels. These two cases will show the importance of accounting for static head and the potential mitigation options for these types of concerns.

2 Case 1 – Produced Water Desanders and Accumulators on an Offshore Platform

2.1 System overview

Figure 1 below is an illustration of case study 1. A produced water desander on the upper deck of a platform is fed by a booster pump. The desander then feeds the desander accumulator. The desander and the accumulator both have the same MAWP of 350 psig. The maximum discharge pressure of the booster pump is 415 psig. In the event of a blocked outlet, the relief device PSV-1 (4" P 6") on the desander is intended to provide adequate protection for the desander and the downstream accumulator. The relief device PSV-2 (1" E 2") on the accumulator is intended only for the fire scenario, based on the assumption that the upstream relief device (PSV-1) would limit the maximum deliverable pressure to the accumulator to 350 psig for the blocked outlet scenario. However, the vessels are located on different levels of the platform. The elevation difference was never accounted for in the choice of relief device set pressure. When PSV-1 is relieving at 385 psig, the pressure inside the accumulator would be the relief pressure (385 psig) plus the static head due to the difference in elevation. If the difference in elevation is large, the downstream accumulator may no longer be adequately protected by PSV-1, because the vessel would experience a pressure higher than its allowable accumulation. The elevation difference between the two relief devices illustrated in Figure 1 is roughly 27 feet. The static head from the elevation difference is about 12 psig. This means that when the desander is at the relief pressure, the pressure in the accumulator would be 397 psig, which exceeds the 10% allowable accumulation [2] (385 psig) of the accumulator.

2.2 Case 1 mitigation

The common mitigation option to account for the static head is to lower the relief device set pressure on the desander to account for the static head, so that the downstream accumulator would be adequately protected during the liquid relief event. However, for this particular case study, a more detailed engineering analysis revealed that the installation was acceptable as-is.

However, it was determined that there are technically two relief devices protecting the system in the event a blocked outlet on the accumulator. The allowable accumulation for the accumulator can go up to 16% of the system (406 psig) because of the multiple relief devices. Since the manual valve between the two systems will remain open (the initiating event is the closure of the downstream block valve), the accumulator is still protected. If the manual valve between the two systems were to be closed, then the blocked outlet case on the accumulator would no longer be applicable as the source of pressure would be lost as well.

If the static head is greater than 21 psig (~49 feet), the accumulator may experience overpressure in excess of the allowable (a pressure of 406 psig). The potential mitigation would be to lower the set pressure of the desander relief device to account for the static head such that the maximum pressure did not exceed this limit. For this case study, PSV-1 set pressure is not required to be lowered because the accumulator is still within the 16% allowable accumulation after accounting for the static head. Additionally, if the relief device set pressure were lowered, the potential exists that the relief device may be undersized, and additional mitigations may be required to resolve the concerns.

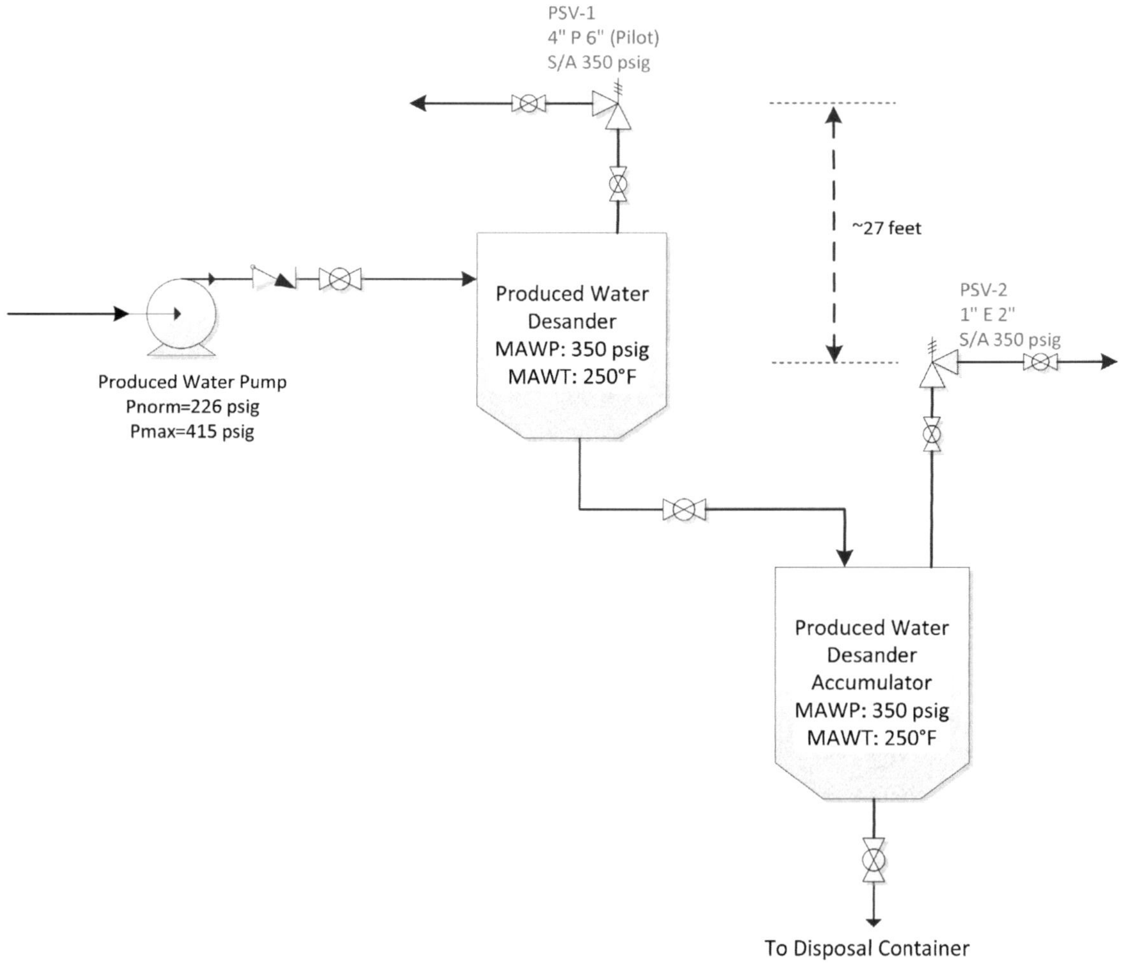

Figure 1, Case 1 – Produced Water System Overview

3 Case 2 – Common Column System

3.1 System Overview

Figure 2 below is a simplified column system that contains several pieces of equipment. The top of column is 50 feet from grade. The reboiler is located at grade. The relief device (PSV-3) is 5 feet above the top of the column and set at 330 psig. The lowest system MAWP is 330 psig based on the MAWPs of the column and the reboiler (shell side). This case study only looks at the "overfilling" scenario for the column. As the lowest system MAWP and the set pressure are the same, it may appear at first glance that the relief device is set correctly to protect the system. However, during the overfilling scenario, there are 55 feet of static head on top of the reboiler (assuming the reboiler is flooded). Therefore, the pressure that the reboiler (shell side) experiences during the liquid relieving event is higher than the allowable accumulation. For this case study, the specific gravity of the fluid is assumed to be 0.7, which would result in roughly 17 psig of static head. This means that during the relief event, the pressure in the reboiler (shell side) is 380 psig (~15% accumulation), which exceeds the 10% allowable accumulation. Therefore, mitigations may be required as the reboiler is not protected due to the static head.

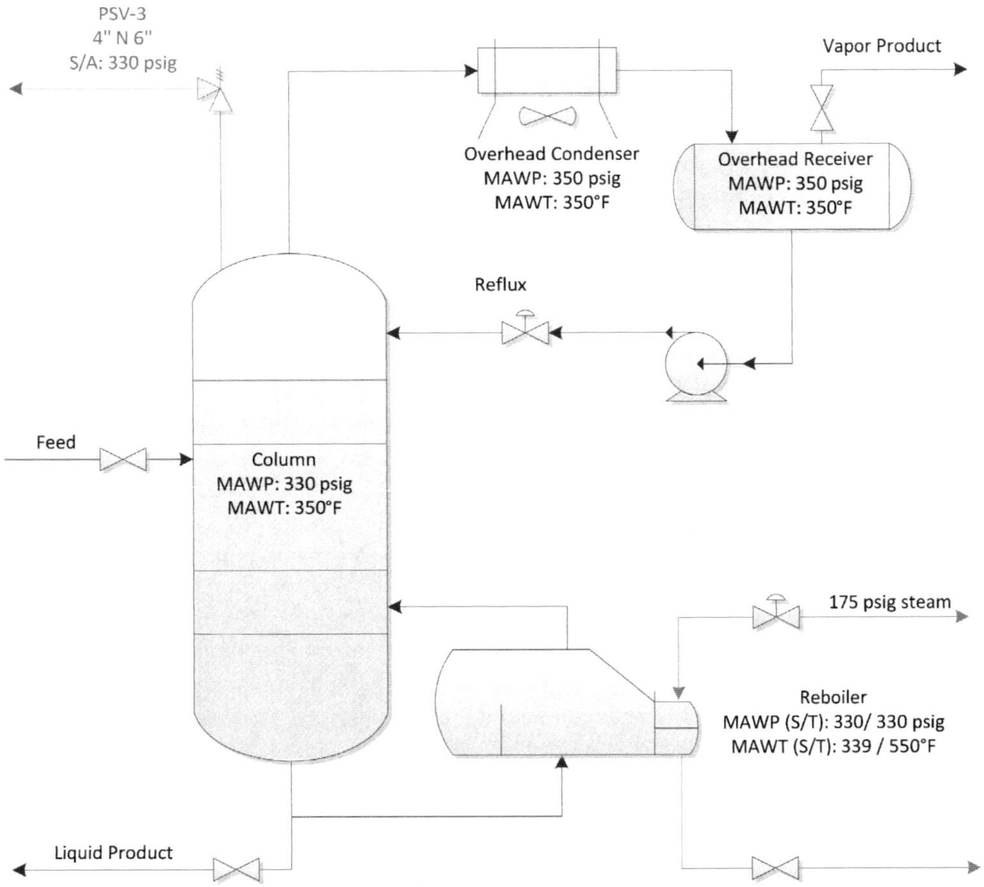

Figure 2, Case 2 - Common Column System Overview

3.2 Case 2 mitigation

There are multiple mitigation options for this type of concern. One mitigation option is to re-rate the reboiler to a higher MAWP to account for the static head. Another mitigation is to lower the set pressure of the relief device to account for the static head. Therefore, the new set pressure for this case study can be 314.5 psig or lower if the reboiler MAWP plus allowable accumulation is not exceeded by the relief pressure plus static head. When the relief device set pressure is lowered to account for the static head during the liquid relief events, engineers must confirm the relief device can still provide adequate capacity at the reduced pressure for all applicable overpressure scenarios. Also, the operating pressure of the column needs to be considered to ensure that the lower set pressure would not result in simmering of the relief device during normal operation. In practice, many plants would also have high independent liquid level alarms and fill time criteria to remove the potential liquid relief scenarios by taking credit for operator responses. Another option is to look at risk assessing the concern as the MAWP is not exceeded by 16% accumulation.

There are many ways to mitigate this concern; this case shows one of the most common mitigation option.

4 Conclusion

The two case studies discussed in this paper show the need to consider the static head when designing the relief system. The issue of static head should be considered during the design phase to ensure the system is code compliant. The cases studies show that the system may not be protected by code, but may be able to be risk accepted based on what the resulting pressure at the vessel is question becomes. It is important during the design phase to be cognizant on the elevation of the systems and the protecting relief devices to ensure the static head has been considered. This can be especially true when designing an offshore platform with multiple levels. Also, when commissioning columns or other vessels, it is important to consider the MAWP of all the vessels in the system to ensure all aspects of the system are protected by the relief device. The case studies give a brief overview of some of the possible mitigation options for static head concerns. It is the responsibility of the engineer reviewing the information to select the option that best fits the facility's needs.

5 References

[1] ASME Section VIII, "Boiler and Pressure Vessel Code," American Society of Mechanical Engineers, 2013.

[2] ANSI/API Standard 520 Part 2, "Sizing, Selection, and Installation of Pressure-relieving Devices," American Petroleum Institute, 6th ed., 2015.

[3] ANSI/API Standard 521, "Pressure-relieving and Depressuring Systems," American Petroleum Institute, 6th ed., 2014.

478548 Study of Factors That Affect Dust Explosion Characteristics of Fibrous Dust Materials

Abstract:

Abstract for 13th Global Congress on Process Safety

Track: Annual Loss Prevention Symposium (LPS)

Study of factors that affect dust explosion characteristics of fibrous dust materials

Bharatvaaj Ravi, Yan-Ru Lin, Chad Mashuga, M. Sam Mannan

Mary Kay O'Connor Process Safety Center

Artie McFerrin Department of Chemical Engineering

Texas A&M University

College Station, TX-77843-3122, U.S.A

+ 1 (774)-262-5215, vaaj26@tamu.edu

Combustible dust hazard is a serious threat to the process industries. Out of the 281 combustible dust incidents reported by the U.S. Chemical Safety and Health Investigation Board (CSB), 66 incidents have been caused by wood dust. In spite of the National Fire Protection Agency's (NFPA) efforts in specifically catering to the wood working and wood-based products manufacturing industries and Occupational Safety and Health Administration's (OSHA) National Emphasis Program on combustible dust hazards, to create awareness and minimize wood dust hazards, the wood dust incidents continue to occur. The fundamental difference between traditional dusts and wood dusts is that the latter are flocculent and fibrous. They are characterized effectively by an aspect ratio (length to diameter ratio) rather than a single dimension, such as diameter, as in the case of traditional - spherical or near spherical - dusts. Various studies in the literature have revealed that the standard dust explosion testing procedures and the categorization of these non-traditional dust based on particle diameter (mesh sizes) might not be valid. This research is concerned with similar line of thought and exploring the effect of aspect ratio on the time of suspension of these fibrous wood dusts. The likelihood of ignition increases with the time of suspension given other conditions to form a combustible dust hazard. The study also deals with characterizing the participating and non-participating portions when a dust explosion occurs through analysis of samples before and after explosion testing. Does the dust concentration give the entire picture of explosion characteristics such as maximum explosion pressure and explosion pressure rise rate? Or is there a better indicator of these characteristics when it comes to flocculent dust? These are the questions that the current research is trying to answer.

Key words: Combustible dust, Fibrous dust, Dust explosion testing, Dust concentration, Dust explosion characteristics, Non-traditional dust.

High Integrity Pressure Protection System (HIPPS) and Implementation Challenges

Janardhanan Kallambettu P.Eng.
Principal Engineer-Control Systems
Bechtel Corporation
3000 Post Oak Blvd, Houston, TX 77459
jkallamb@bechtel.com

Prepared for Presentation at
American Institute of Chemical Engineers
2017 Spring Meeting and 13th Global Congress on Process Safety
San Antonio, TX
March 26 – 29, 2017

High Integrity Pressure Protection System (HIPPS) and Implementation Challenges

Janardhanan Kallambettu P.Eng.
Principal Engineer-Control Systems
Bechtel Corporation
3000 Post Oak Blvd, Houston, TX 77459
jkallamb@bechtel.com

Keywords: PHA, HIPPS, SIS, SIF, SIL, PLC

Abstract

Over the years, the process industries have heavily utilized pressure relief devices (PRDs) to protect the mechanical integrity of system components. While PRDs offer a straightforward and practical way of controlling overpressure in process plants, they do have some inherent limitations such as impacts to flare system design, environmental concerns, and some physical limitations in PRD sizing. In these cases, where the use of PRDs is limited, High Integrity Pressure Protection System (HIPPS) can be utilized as a substitute for PRDs.

While HIPPS offers a solution to address the PRD limitations, there are some challenges associated with a successful HIPPS implementation. The HIPPS are classified as Safety Instrumented Systems (SIS) and require compliance with the functional safety standard IEC 61511[**1**]. The aim of this paper is to guide the reader with the necessary tools to successfully perform all the safety lifecycle activities of HIPPS, per functional safety standards and codes. Challenges in the various phases of the safety lifecycle, from process hazard analysis (PHA) and Safety Integrity Level (SIL) selection to operation and maintenance of HIPPS, are discussed in this paper. Additionally, this article describes many types of HIPPS, their pros and cons, to choose the right system for a given application. The paper concludes with an example HIPPS to achieve high-level risk reductions and demonstrates the appropriate use of quantitative tools, such as fault tree analysis (FTA), during the HIPPS design process.

1 Introduction

HIPPS is an acceptable solution instead of PRDs as a means of overpressure protection for vessels and pipelines. Codes such as ASME Section VIII, Division 1[**2**] and recommended practices and standards such as API Standard 521[**3**], API 17O [**4**] and HSE, U.K [**5**] provide criteria for the usage of HIPPS. The reader is advised to refer to the cited references to understand in detail the selection criteria for HIPPS.

The HIPPS is a safety instrumented function (SIF) implemented in a SIS and, therefore, the codes and the standards referenced above require compliance with the functional safety standards IEC 61508[**6**] and IEC 61511.

The functional safety standards require adherence to a safety lifecycle model. The various safety lifecycle activities are:

- process hazard analysis;
- SIL selection;
- safety requirements specification (SRS);
- SIS design per SRS;
- design verification meeting the target SIL including systematic capability, and architectural constraints (redundancy);
- installation, commissioning, validation, operation, and maintenance per SRS; and
- change management.

The intent of this paper is to expose particularly process engineers and other non-control systems engineers, about specific aspects of the HIPPS in a straightforward language.

2 PHA and SIL Selection

Whenever a HIPPS is indicated instead of PRDs, process and process safety engineers assign an SIL 3 target to the HIPPS as supported by Lawley and Kletz's [7] requirement that the reliability is ten times better than that of a relief device. For clean services, the probability of failure on demand (PFD) of a relief device is typically two orders of magnitude, hence the SIL 3 target for a HIPPS. API Standard 521, while emphasizing the need for an appropriate SIL selection process per established risk tolerance criteria, also suggests SIL 3 target for HIPPS in the absence of a documented risk tolerance criteria.

An assumption, however, that the HIPPS's SIL target is always SIL 3 may not be correct in some circumstances. Applicable codes and standards mandate a requirement for a documented PHA to identify and examine all potential overpressure scenarios. The severity of consequences typically ranges from moderate to catastrophic and, based on the owner's risk tolerance criteria, the HIPPS target can vary from SIL 2 to SIL 4. For example, if the likelihood of an initiating event causing the overpressure is 1 in 10 years and the owner's risk criteria for catastrophic severity is 1in 100,000 years, and in the absence of any other independent protection layers or conditional modifiers, the required risk reduction is 10,000, a SIL 4 target.

Designing a SIL 4 HIPPS is a significant challenge and not recommended. The functional safety standards discourage such high-level risk reduction measures by a single SIF. Every effort should be made to remove or reduce the hazard at the source, introduce other risk reduction measures which are not based on the SIS or minimize the likelihood of the causes of the hazardous event. If the above steps do not help, then use of multiple independent HIPPS, each with a lower integrity, to meet the overall risk reduction requirements should be considered. See section 6 for a hybrid HIPPS design solution meeting an overall risk reduction of 10,000 or more.

3 Types of HIPPS

Many in the process industry believe that HIPPS always includes 2oo3-voted pressure transmitters (sensors), Triple Modular Redundant (TMR) safety PLC (logic solver), and 1oo2-voted ball valves (final elements). In reality, there are many types of HIPPS available to choose from based on the specific application.

HIPPS can be broadly classified as **Electronic**, **Mechanical** or **Hybrid** based on the technology used for the logic solver as shown in **Figure 1**.

A brief description, pros and cons, and applications for each of the above types are given in the following subsections.

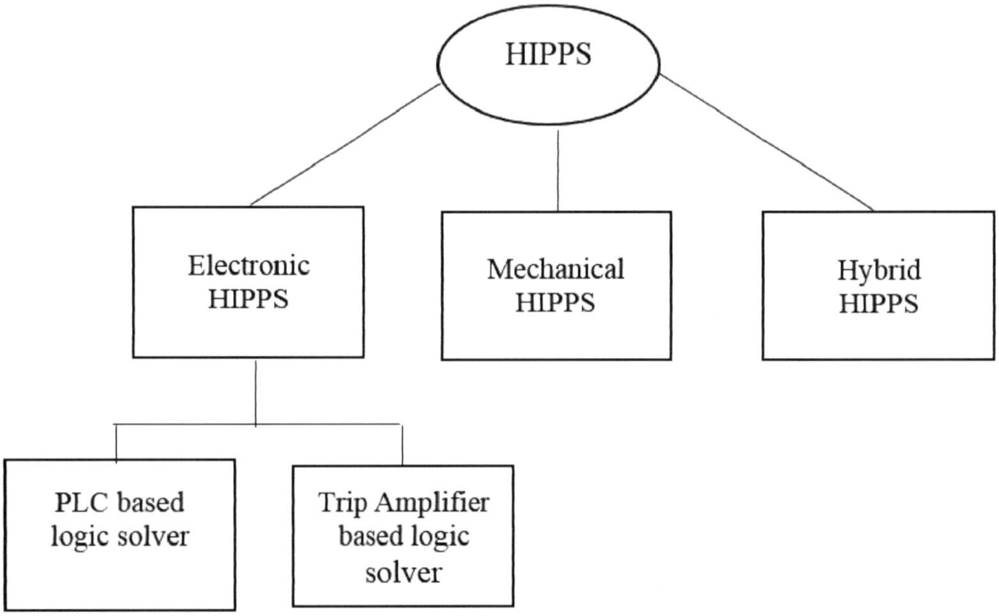

Figure1. Types of HIPPS

3.1 Electronic-HIPPS

An electronic-HIPPS uses either a programmable logic controller (PLC) or a trip amplifier system as a logic solver.

3.1.1 PLC based HIPPS

There are multiple varieties of PLCs equipped with microprocessor or solid state technology in a multitude of architectures such 1oo2, 1oo2D, 2oo3, 2oo4, and 2oo4D. Typically, all these PLCs will meet SIL 3 reliability. These PLCs are further categorized as complex devices and use application programming languages such as ladder logic or functional block diagrams. Refer to **Figure 2**.

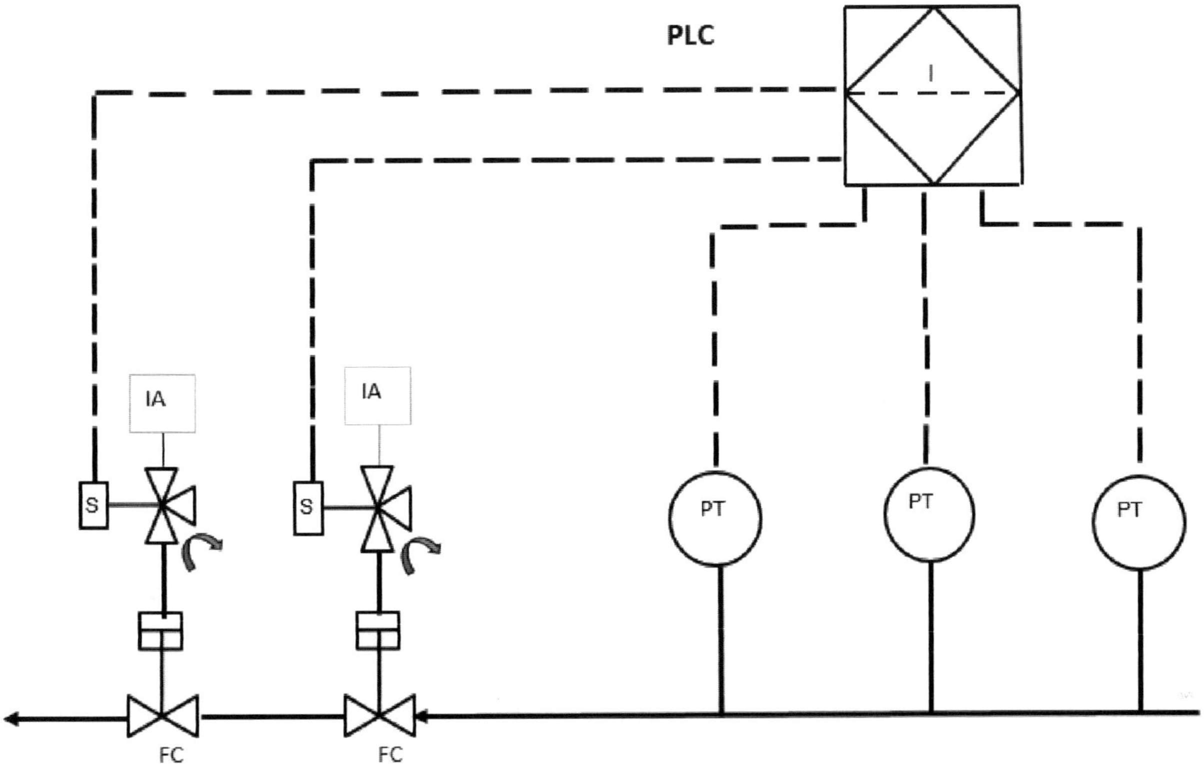

Figure 2. PLC based HIPPS

Pros:

- Flexible in developing the logic program and accommodates higher number of field inputs and outputs (I/Os);
- Equipped with special features for application programming such as variable valve closure speeds while closing the valve to mitigate consequences associated with slamming shut.
- Accommodate multiple HIPPS in a single logic solver within a plant facility;
- Facilitates partial stroke testing to meet target SILs or extend the testing interval, when practicable;
- Easily interfaces with the plant operator stations (human machine interfaces, HMIs).

Cons:

- Often needs specialized application programming expertise, in addition to vendor-specific training for selection of hardware and complex control strategies;
- Validation of the vendor claimed SIL against the safety manual is not a trivial exercise;
- Requires protection against cyber security threats;
- Potential for introducing systematic errors in application program;
- Application program using a low variability language (LVL) calls for special considerations per IEC 61508/IEC 61511 standards including application program safety requirements specification;
- Requires rigorous Factory Acceptance Test (FAT);
- Large footprint is needed -typically 800mmX600mmX1000mm panel;

- Extensive installation requirements such as controlled environment and special grounding;
- Requires complex, and redundant large capacity power supply;
- Voluminous documentation is involved;
- Requires a high level of competence in commissioning, validation, maintenance and operation; and
- Expensive for simple HIPPS applications ($50,000 for a logic solver alone).

Applications:

In a large facility, where many HIPPS are to be implemented for different hazardous event scenarios requiring complex functions or features, a PLC based HIPPS is recommended. For example, over pressure protection of multiple vessels, various piping segments, and process equipment within a facility to avoid loading of the flare system.

3.1.2 Trip-Amplifier based HIPPS

Trip-amplifiers are discrete logic solvers of low complexity. A trip amplifier accepts analog 4-20mA input from a field device such as a pressure transmitter and de-energizes contact output(s) when the process value exceeds the trip set point. Each sensor needs a dedicated trip amplifier. Refer to **Figure 3**.

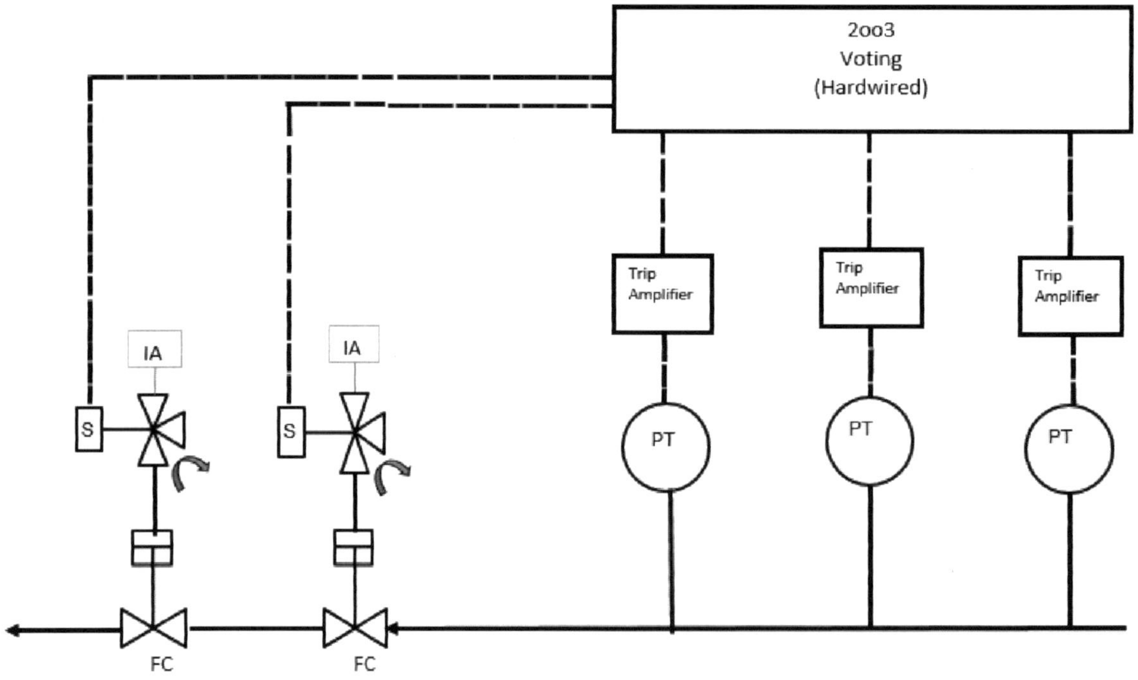

Figure 3. Trip Amplifier based HIPPS

Pros:

- Simple, reliable, with low design complexity;
- Meets SIL 2 reliability without redundancy for Trip Amplifiers and SIL 3 reliability with redundant architecture;
- Does not require application programming; classified as a device with fixed programming language (FPL) per IEC 61508/IEC 61511and, therefore, eliminates the need for application program safety requirements specification;
- No cybersecurity related issues;
- Less potential for systematic errors;
- Well suited for simple logic such as HIPPS logic;
- Low space requirement-typically DIN rail mounted;
- Low power device with no special requirements;
- No need for a controlled environment;
- Being a simple device FAT is not required for the logic solver;
- Easy to install and maintain; and
- Low cost; (logic solver range less than $10000)

Cons:

- Requires external devices for complex logic such as variable speed closing times for valves to avoid slam shut scenarios;
- Interfacing with operator stations requires additional I/Os in the plant DCS;
- Difficult to interface and facilitate partial stroke testing;
- Voting among the trip-amplifiers are external to the trip-amplifiers using the trip contacts and physical wiring which is less straightforward; and
- Implementation of maintenance bypasses is not practical;

Applications:

Well suited for applications involving straightforward, simple logic and for facilities which do not use PLCs. Also, appropriate for applications requiring a small footprint, low lifecycle cost and facilities not having multiple HIPPS. For example, remote pipeline compressor stations or storage tanks.

3.2 Mechanical-HIPPS

A mechanical HIPPS uses mechanical initiators. The mechanical initiator not only senses the over pressure but also serves as a logic solver to detect high pressure and actuate the valve. Mechanical HIPPS do not use any electrical or electronic components and, therefore, are inherently safer. Standalone versions of mechanical HIPPS are available which do not require any energy sources, such as instrument air supplies. Refer to **Figure 4**.

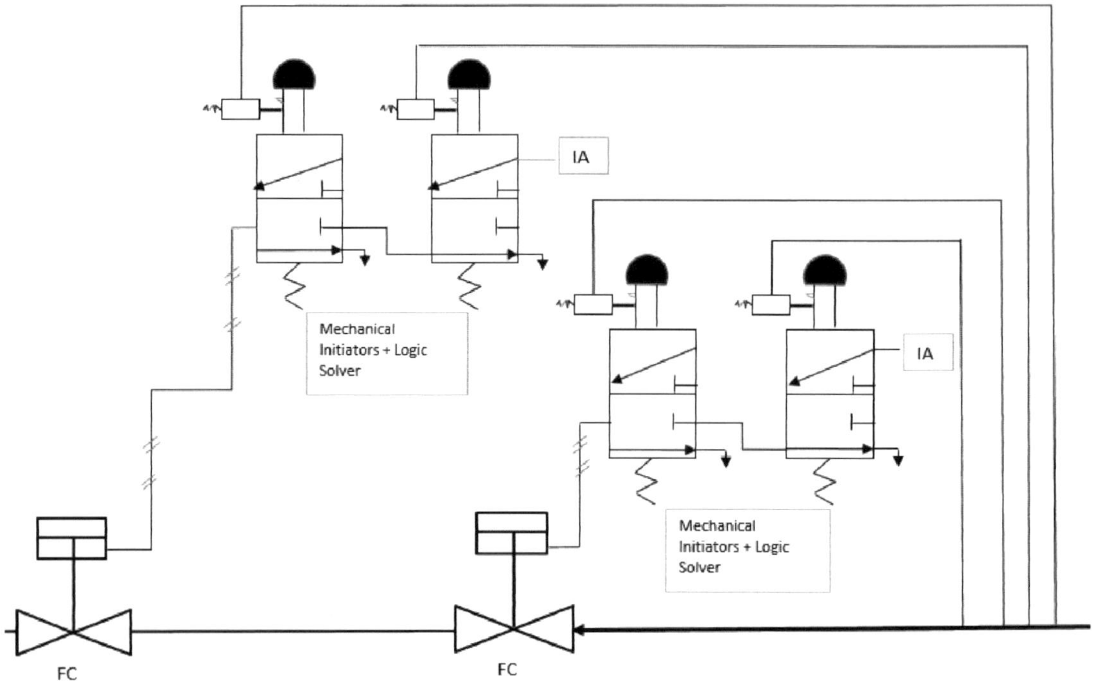

Figure 4. Mechanical HIPPS

Pro:

- Simple, reliable and minimally complex;
- Does not involve wiring and programming;
- Less susceptible to systematic errors;
- Cables, cable trays, conduits, etc. are not required;
- Can be made standalone without external energy sources;
- Single valve system can meet SIL 2 targets, redundant valve configuration can even meet SIL 3;
- Third party agencies certify an entire system in contrast to individual elements/subsystems of electronic HIPPS;
- Fewer components to fail;
- Easy to install, commission, validate, maintain and operate;
- Minimal documentation: no wiring diagrams or, installation drawings for logic solvers and initiators, application program SRS required;
- Competence in programming or maintaining a complex electronic system is not required;
- There are no EMC interference issues;
- There are no impacts due to cyber security threats;
- Suitable for harsh environments-no environmental climate control is needed;
- Easy to comprehend by operators and maintenance mechanics;
- Lowest life cycle cost among all the types of HIPPS;

Cons:

- Limited distance between the pressure sensing points, mechanical initiators and the valve- all components must be in the close proximities to avoid transportation time lags impacting the response times;
- Mechanical initiators need to be carefully selected as they are sensitive to setpoints and their accuracies;
- Interfacing with operator stations requires additional transmitters in the field and Inputs/outputs (I/Os) in the plant DCS;
- Requires external devices to facilitate partial stroke testing;
- Implementing maintenance bypasses consistent with the overall facility philosophy is not practical;

Applications:

Mechanical HIPPS are attractive for standalone applications in remote plant locations, for areas where electrical cabling is impractical and in facilities, like pipeline pump/compressor stations, where PLCs are not commonly used.

3.3 Hybrid-HIPPS

Hybrid-HIPPS are a combination of electronic and mechanical HIPPS.

Pros:

- High level of risk reduction- 10,000 (SIL 4) or higher can be obtained;
- Significant reduction of common cause and common mode failures due to diversity (mechanical and electronic);
- Low-level of systematic errors;

Cons:

- Complex system;
- Requires high level of competence in the installation, commissioning, validation, maintenance and operation;
- Requires a sizeable spare parts inventory;
- Many of the Cons of electronic and mechanical HIPPSs described above are applicable for the hybrid HIPPS as well.

Application:

High-pressure feed gas to a facility typically requires a high-risk reduction in the absence of non-instrumented IPLs. When a high level of risk reduction (greater than 10,000) is required, hybrid systems are well-suited as they are diverse systems with minimum common cause and common mode failures.

4 Safety requirements specification (SRS)

Functional safety standard IEC 61511 requires the development of a SIS safety requirements specification (SRS), including application program safety requirements, when applicable. HIPPS as a functional SIS, requires an SRS. There are a minimum of 29 SRS requirements specified by the standard. While every element is important, a few present unique challenges that can significantly impact the selection, design, and quantification of random hardware failures of the HIPPS.

4.1 Process safety time (PST)

Process safety time, per clause 3.2.52.1, IEC 61511, is the period between a failure occurring in the process or the basic process control system and the occurrence of a hazardous event, if the SIF is not performed. The SIF has to detect the failure and complete its action soon enough to prevent the hazardous event taking into account any lag in the process.

HIPPS that implements the overpressure protection function must complete the action to bring the process to a safe state within the process safety time. Quite often, the process safety times for HIPPS are less than few seconds. The valve subsystems may need quick exhaust valves, larger size air supply tubing, or solenoid valves to meet the response times. Therefore, the PST determination is an important specification which in turn determines various components within the HIPPS to achieve the response time.

4.2 HIPPS output action success criteria

Output actions of a SIF will prevent a particular consequence of concern only when the action is completed with certain performance criteria. For example, stating an action as to close a valve is inadequate. The criteria for the successful operation should be specified. Many HIPPS applications do not require tight shut-off requirements for valves. However, there are situations where larger leakage rates may impact the process safe states. In such scenarios, the process engineer must specify an acceptable leakage rate for the valves, and ensure the design shall have means to achieve such low leakage rates. The design may require unique valve types and trims. Also, additional pressure transmitters and isolation/bypass valves are required for measuring the extent of leakages.

Low leakage rate specification may impact the proof test coverages for the final elements in the quantification of random hardware failures (PFD_{avg}). Achieving a SIL 3 target with tight shut-off requirements is a significant challenge.

4.3 Proof test interval and overhaul interval (mission time)

IEC 61511 defines the proof test as a periodic test performed to detect dangerous (hidden) failures in a SIS so that, if necessary, a repair can restore the system to an “as new” condition or as close as practical to this condition.

The frequency of testing plays a significant role in determining the HIPPS’s availability to perform intended safety function on demand. However, frequent testing may affect the production, introduce spurious shutdowns due to human error or affect the life of a device. If only a full stroke test is performed for a valve, then the test is imperfect since it will not detect all the covert failures. The undetected, hidden failures are detected and repaired during overhaul periods (mission time). The longer the mission time, the higher the PFDavg; therefore, it will be difficult to achieve SIL 3 targets with longer mission times.

Bypass valves and isolation valves are required for proof testing intervals shorter than the overhaul periods to avoid process interruption. Provision of bypass and isolation valves increases the complexity of the design, footprint, capital and the life cycle costs of the facilities.

4.4 Survivability

As per IEC 61511, the SRS shall specify any requirements for a SIF to survive a major accident event, e.g. the time required for the HIPPS valve to remain in operation in the event of fire shall be specified. Survivability specification impacts the selection of valve trims and additional fireproofing requirements.

4.5 Common cause failures

The SRS must specify the measures to identify and minimize common cause failures. Common cause failures will significantly impact PFD_{avg} for high SILs, such as SIL 3 and SIL 4. Common cause failures can be due to process conditions or similar redundant equipment configurations or systematic failures.

Physical and electrical separation and use of diverse devices which respond differently to the common stresses will minimize common cause failures. An effective functional safety management plan and verification processes will reduce systematic failures.

5 Quantification of random failure of HIPPS

The HIPPS design must be evaluated to ensure that the calculated failure measure (PFDavg) is less than the target related to the SIL as specified in the SRS.

The functional safety standard mandates assessment of the following as part of the SIL verification:

- Systematic capability of all the elements associated with a HIPPS meets the specified SIL target
- Calculated PFD_{avg} is less than the target PFD_{avg} of the selected SIL
- Hardware fault tolerance (HFT) complies with the required SIL target

5.1 Systematic capability

IEC 61511 defines the "systematic capability" as a measure (expressed on a scale of SC 1 to SC 4) of the confidence that the systematic safety integrity of a device meets the requirements of the specified SIL, in respect of the specified safety function, when the device is applied in accordance with the instructions specified in the device safety manual.

Systematic capability assessment is for addressing systematic failures which are not quantified and hence not included in the SIL verification calculations. IEC 61508 and IEC 61511explicitly identify a systematic capability parameter (SC). A systematic capability of "SC N" for a device means that the systematic safety integrity of "SC N" has been met when the device is applied following the instructions specified in the instrument safety manual. If a device is to be used in an SIL 3 application, then it must be assessed for the systematic capability of SC 3.

The systematic capability of a device depends on the nature of the device. If a device has no software, then only hardware systematic failure mechanisms are considered. If a device has both hardware and software, then both hardware and software failure mechanisms shall be considered.

There are agencies, such as TUV, Exida, and SIRA that perform systematic capability assessment of a product as per IEC 61508 and provide a certificate with a report specifying an SC parameter. Following a well-established functional safety management plan during detailed engineering, installation, operation and maintenance phases ensures that the systematic safety integrity of a device meets the requirements of the specified SIL.

5.2 PFD_{avg}

Many factors influence the PFD_{avg} of HIPPS function. A few of the factors that pose unique challenges are considered in this section.

5.2.1 Failure rates

There is always an uncertainty in the quantification of random failures. The failure rate is, probably, the least precise parameter used in performing SIL verifications [**8**]. Therefore, one must acknowledge that the results of the quantification are not real values, but rather a basis for comparing different design options and for monitoring the reliability performance during the operational phase of the safety lifecycle. The analysis is only a prediction. Historical performance is not the same as future performance.

If the failure rates are conservative, then achieving target SIL will be demanding. However, optimistic failure rates will result in inadequate protection integrity. Therefore, the failure rates of components used in the analysis should be sufficiently realistic.

There are many sources of failure rate data, such as application-specific, site-specific, company-specific, manufacturer-specific, industry-specific, and generic data. These failure rates can vary significantly by several orders of magnitude. While application specific or site specific data is the best option, but it can often be difficult to obtain. A detailed discussion on the selection of failure rates to use in the analysis is beyond the scope of this paper, and the reader is advised to refer to other sources [**9**].

5.2.2 Proof test interval, test coverage and mission time

As explained in section 4.3, a proof test may be imperfect based on the following:

- The extent of tests that could be performed to reveal the hidden failures. For example, if the valve is only full stroke tested without examining the internals then the proof test is imperfect;
- The conditions under which the proof test is carried out may differ from a real demand;
- Lack of competence of testing crew;
- Inadequate procedures.

Perfect testing is not a realistic assumption. IEC 61511/IEC 61508 requires consideration of the impact of imperfect testing in the analysis. In the SIL verification calculations, a proof test coverage (PTC) is considered for incomplete testing. The incomplete testing will increase the PFDavg. The lower the proof test coverage, the higher the PFD_{avg}. Undetected dangerous failures due to imperfect proof testing will be detected either during the complete overhaul period (when repaired or refurbished) or when the component is replaced or during an actual demand. Mission time is the terminology used by many reliability specialists for the above overhaul period.

For example, if an overhaul period or demand interval for HIPPS is ten years and the imperfect proof test interval for a valve is one year with a proof test coverage of 90%, then achieving an SIL 3 target for HIPPS is an arduous task. If tight shut-off requirement is specified, then the proof test coverage could be even less than 30%.

From the above, it is evident, that the significance of proof test interval, the effectiveness of proof test, and the mission time cannot be ignored and must be taken into account in the analysis.

5.2.3 Partial stroke testing

Partial stroke test will help to improve the PFD_{avg}, extend the proof test interval, or increase the safe failure fraction. A single valve with partial stroke test may achieve an SIL 2 integrity. However, for many HIPPS with short process safety times, partial stroke test may not be useful. Moreover, partial stroke tests may result in spurious trips and cause potential damage/wear to the valve seat due to the frequent testing.

A perusal of literature reveals that the partial stroke testing coverage is in the range of 40-80% [10]. However, the above-stated coverage is widely debated.

If the partial stroke testing is not automated, then it is not considered as the diagnostic testing per functional safety standards. Also, the analyst shall take into consideration the partial stroke testing interval with the repair time in the PFD calculations.

5.2.4 Common cause failures

Common cause failures impact the redundant components such as valves and sensors. The common cause failures occur simultaneously or, at least, within a short interval as to be perceived as simultaneous.

The following are the defenses against the common cause failures:

- Diversity (different types of valves and systems);
- Separation (physically and electrically);
- Staggered testing;
- Use of low complexity devices- simpler equipment less prone to common cause failures;
- Well administered and complied procedures;
- Environmental controls;

When diverse components are used, the common cause factor (Beta factor) is minimal. Using IEC 61508, Part 6 [5], Annex D, Tables, D.1, D.2 and D.3, the Beta factor can be estimated to be as low as 2% with diverse components. However, the diverse components must be suitable for the application with comparable performance without any degradation.

Also, staggered testing reduces the overall PFD_{avg} by a factor of 30 to 50% depending upon the voting configurations [**10**].

5.2.5 Modeling of components

Modeling of HIPPS elements in the analysis must include all the components that are involved in bringing the process to a safe state within the process safety time.

Many HIPPS valves are equipped with quick exhaust valves (QEVs) to reduce the valve closing time to meet the process safety time. Sometimes, redundant QEVs are provided in various voting configurations such 1oo2 or 3oo4, etc. The contribution of QEV failures to the PFD_{avg} is significant and cannot be ignored. Therefore, the reliability analyst must consider the QEVs with their appropriate voting architecture in the PFD_{avg} calculations. Similarly, the likelihood of plugging of sensing lines must be taken into account in freezing applications.

Any interfacing components such as barriers, isolators, interposing relays, and poppet valves, which are part of the safety instrumented function to achieve or maintain the process to a safe state must be modeled and analyzed.

5.3 Hardware fault tolerance (HFT)

Irrespective of the hardware reliability calculated (PFD_{avg}) for the design, the functional safety standard IEC 61511 specifies minimum levels of redundancy for each safety integrity level.

There are two approaches:

- Route 1_H: Determination of HFT based on the calculated safe failure fraction (SFF);
- Route 2_H: Determination of HFT based on field experience and field data (proven-in-use)

5.3.1 Route1$_H$

The Route1$_H$ approach uses SFF and the SIL target to determine the hardware fault tolerance. The SFF is defined as the sum of the potentially dangerous failures revealed by auto-test and those failures which result in a safe state, as a fraction of a total number of failures.

$$SFF = \frac{[dangerous\ detected\ failures(self\ test) + safe\ failures]}{Total\ failures}$$

The above is calculated for each element (sensor, complete valve assembly, logic solver). Two tables, Table 2 and Table 3, are provided in IEC 61508- Part 2. Table 2 is for "Type A" devices which are simple devices such as valves (no microprocessors and programming are involved), and Table 3 for "Type B" devices which are complex devices such as electronic transmitters, trip amplifiers, and PLCs. Based on the SFF, type of the device, and the SIL target, the hardware fault tolerance is determined using Table 2 and Table 3.

5.3.2 Route2$_H$

In Route 2$_H$, the hardware fault tolerance is specified by the functional safety standard for each SIL target if the device is selected based on well documented and verified field based failure rate data with 90% statistical confidence. For low demand application with SIL 1 or SIL 2 target, there is no need for any redundancies (HFT=0), and SIL 3 requires one hardware fault tolerance.

Unless the field failure data with 90% confidence levels are available, the Route 2$_H$ cannot be justified. There are third party agencies who provide Route 2$_H$ certifications and should be evaluated carefully before adopting the Route 2$_H$ approach based on those certifications.

6 High level risk reductions (RRF> 10,000)

As discussed in section 2, sometimes, PHA/SIL selection may indicate a required risk reduction greater than 10,000 in the absence of any non-instrumented protection layer or, human intervention is impractical due to process safety time considerations or safety reasons.

If there are no options other than providing HIPPS with a risk reduction greater than 10,000, then the functional safety standard IEC 61511 mandates certain requirements. The following are a few of the clauses and their requirements:

Clause 9.2.6: *If further consideration of application and confirmation that a risk reduction requirement >10,000....is still required, then* consideration *should be given to achieving the safety integrity requirement using a* <u>*number of protection layers with lower risk reduction requirements*</u>. *If the risk reduction is allocated to multiple protection layers, then such protection layers shall be independent from each other or the lack of independence shall be assessed and shown to be sufficiently low compared to the risk reduction requirements.*

Clause 9.2.7: *If a risk reduction greater than 10,000....is to be implemented, whether allocated to a single SIS or multiple SIS or SIS in conjunction with a BPCS protection layer, then a further risk assessment shall be carried out using a quantitative methodology to confirm that the safety integrity requirements are achieved.*

Clause 9.2.8: *If the risk reduction required for a hazardous event is allocated to multiple SIFs in a single SIS, then the SIS shall meet the overall risk reduction requirement.*

A hybrid HIPPS using the combination of electronic and mechanical HIPPS is a solution to achieve risk reductions greater than 10,000. Each HIPPS can be designed to be better than SIL 2, and the overall risk reduction will be better than 10,000. The performance of electronic and mechanical HIPPS are comparable.

High level of risk reduction is achieved due to the diversity and very low common cause failures between the electronic and mechanical HIPPS. All the subsystems of the HIPPSs are diverse with different technologies. Also, systematic failures are minimal. The integrity of such multiple HIPPS is best analyzed using a fault tree analysis (FTA).

An example fault tree analysis for a hybrid HIPPS using ISOGRAPH's Reliability Workbench™ is shown in **Figure 5**.

The example hybrid HIPPS achieved a risk reduction of 12,815(PFD_{avg} = 7.80E-05).

The FTA shows an approach for modeling of various components including the common cause failures. The failure rates assumed are generic for the purpose of demonstration. The actual failure rates and other factors are application specific and cannot be generalized.

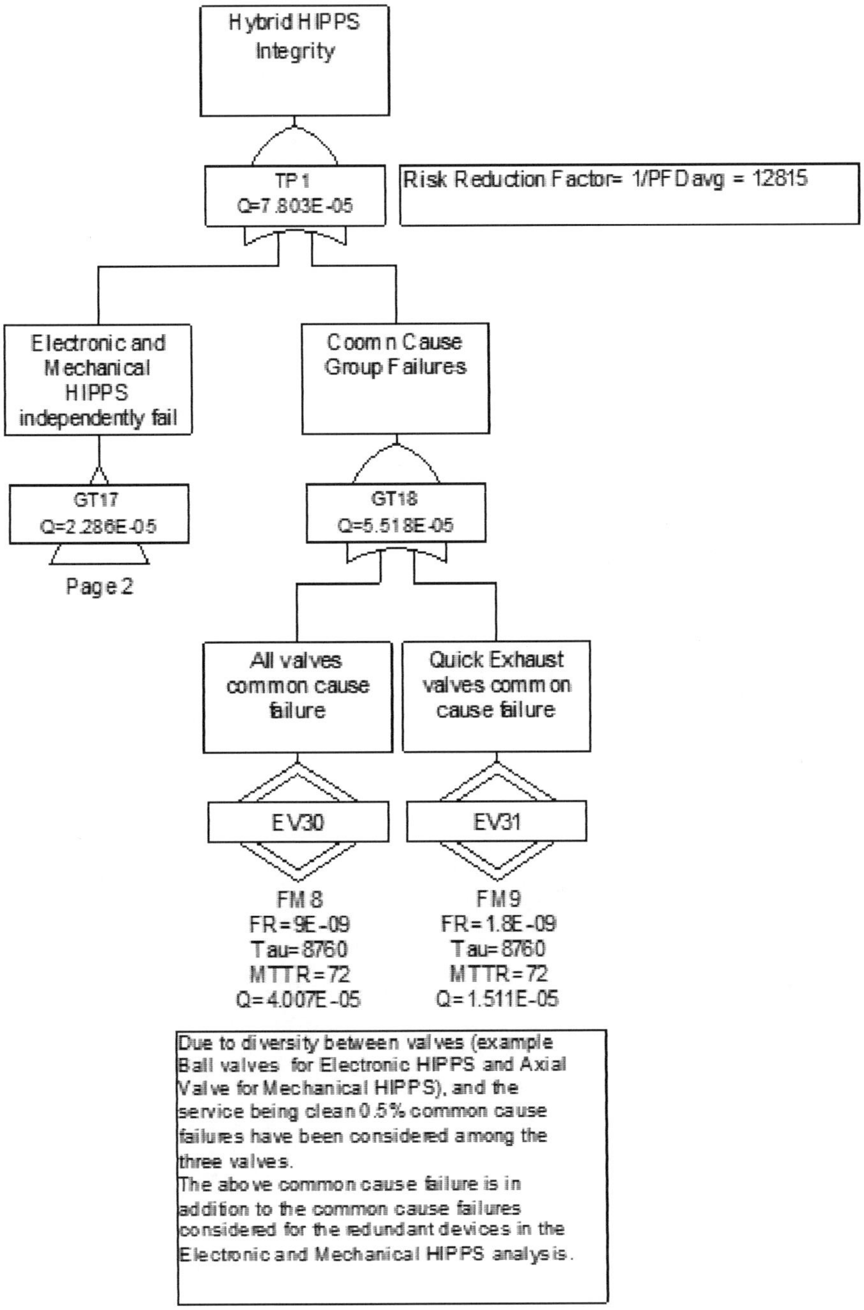

Figure 5. FTA for multiple HIPPS to achieve high risk reduction

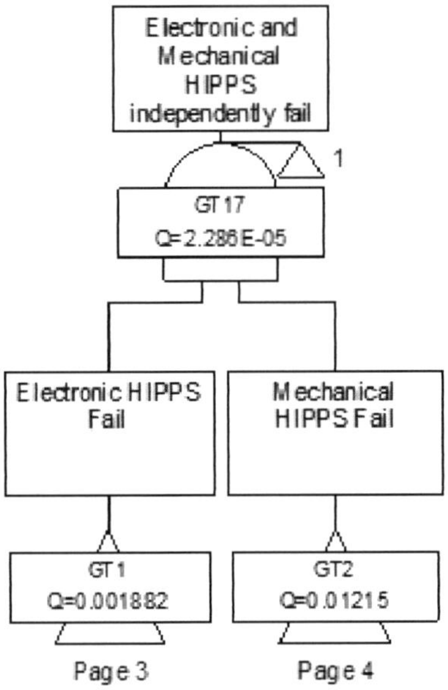

Figure 5. FTA for multiple HIPPS to achieve high risk reduction (contd..)

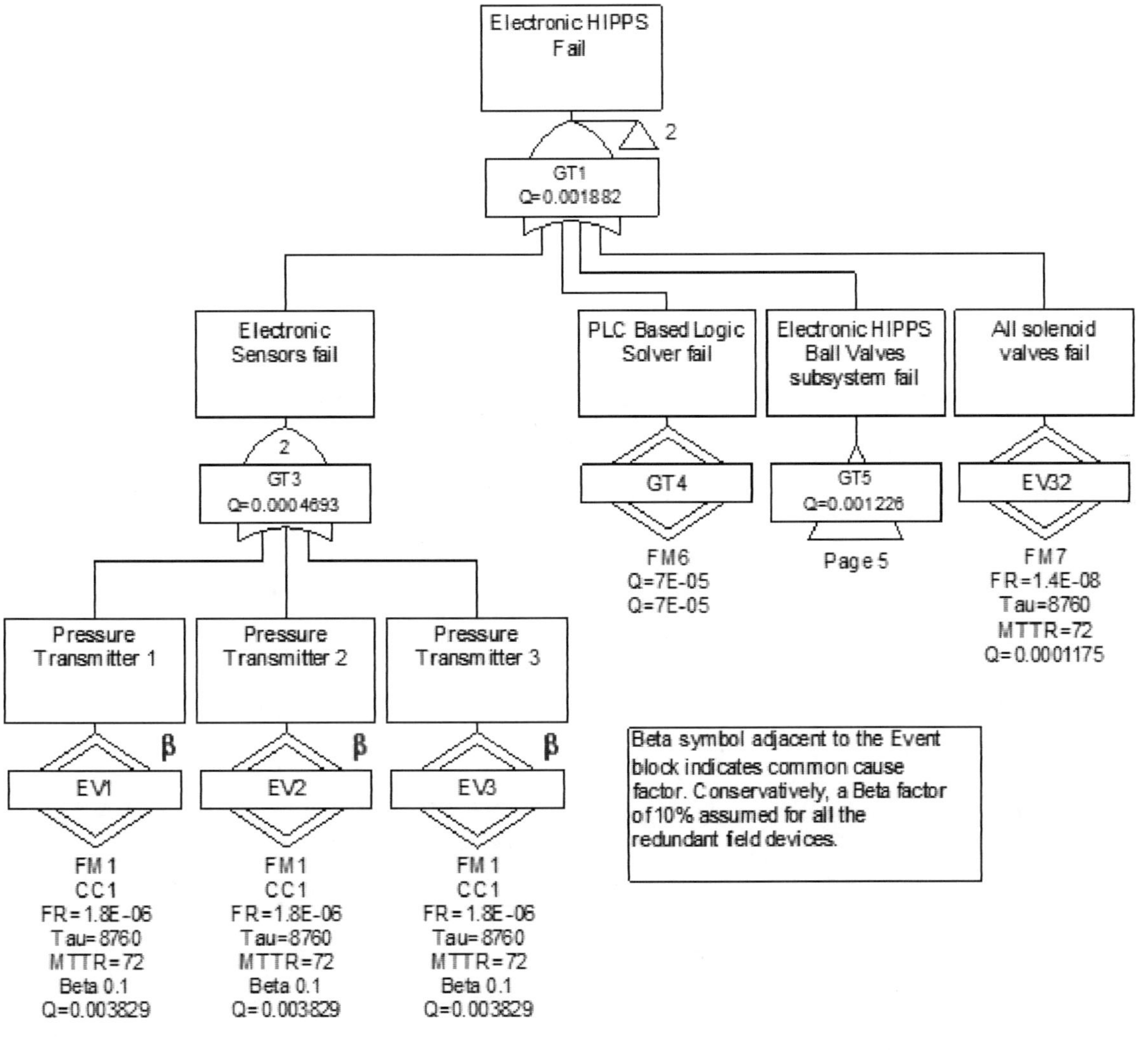

Figure 5. FTA for multiple HIPPS to achieve high risk reduction (contd..)

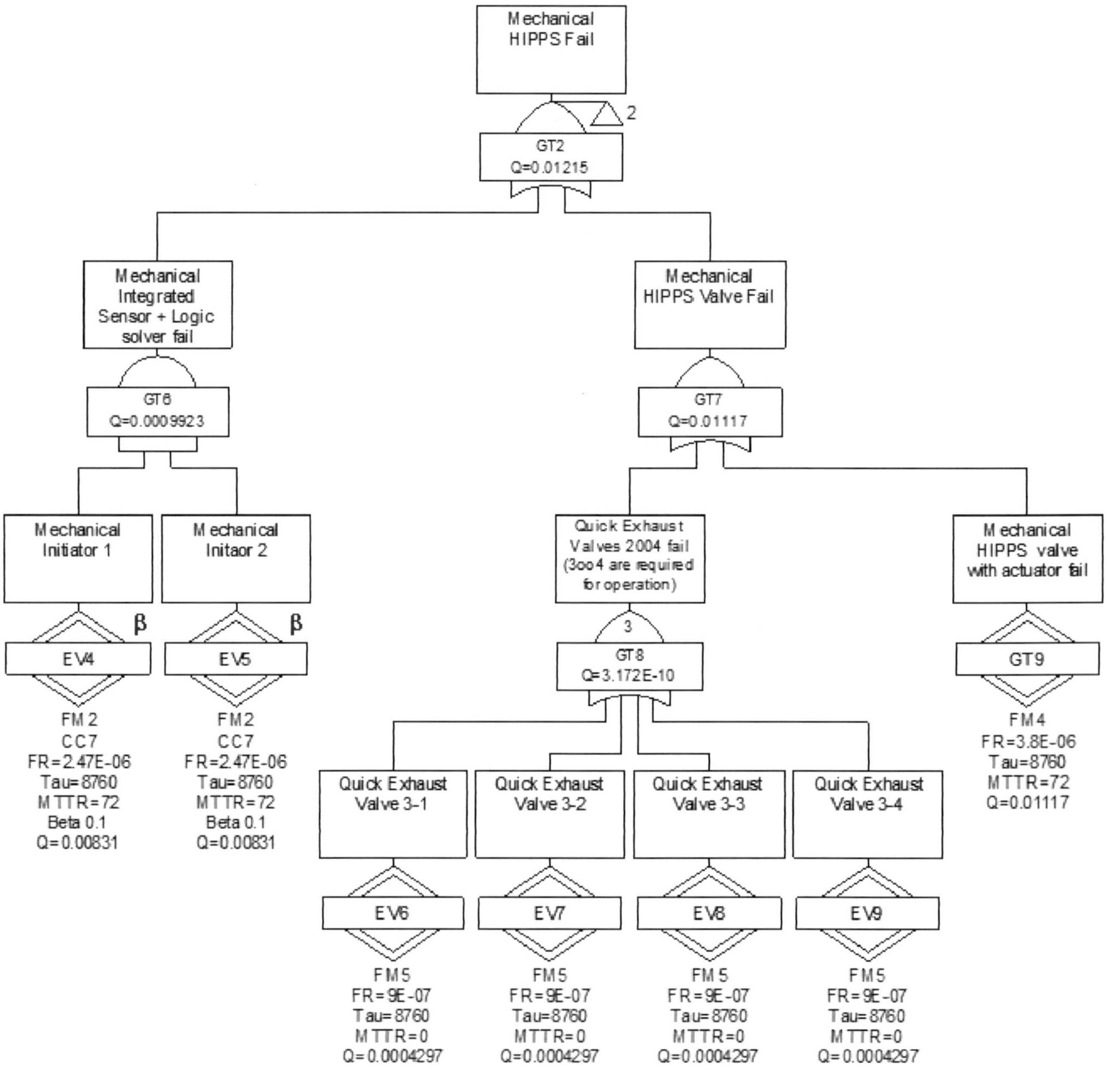

Figure 5. FTA for multiple HIPPS to achieve high risk reduction (contd..)

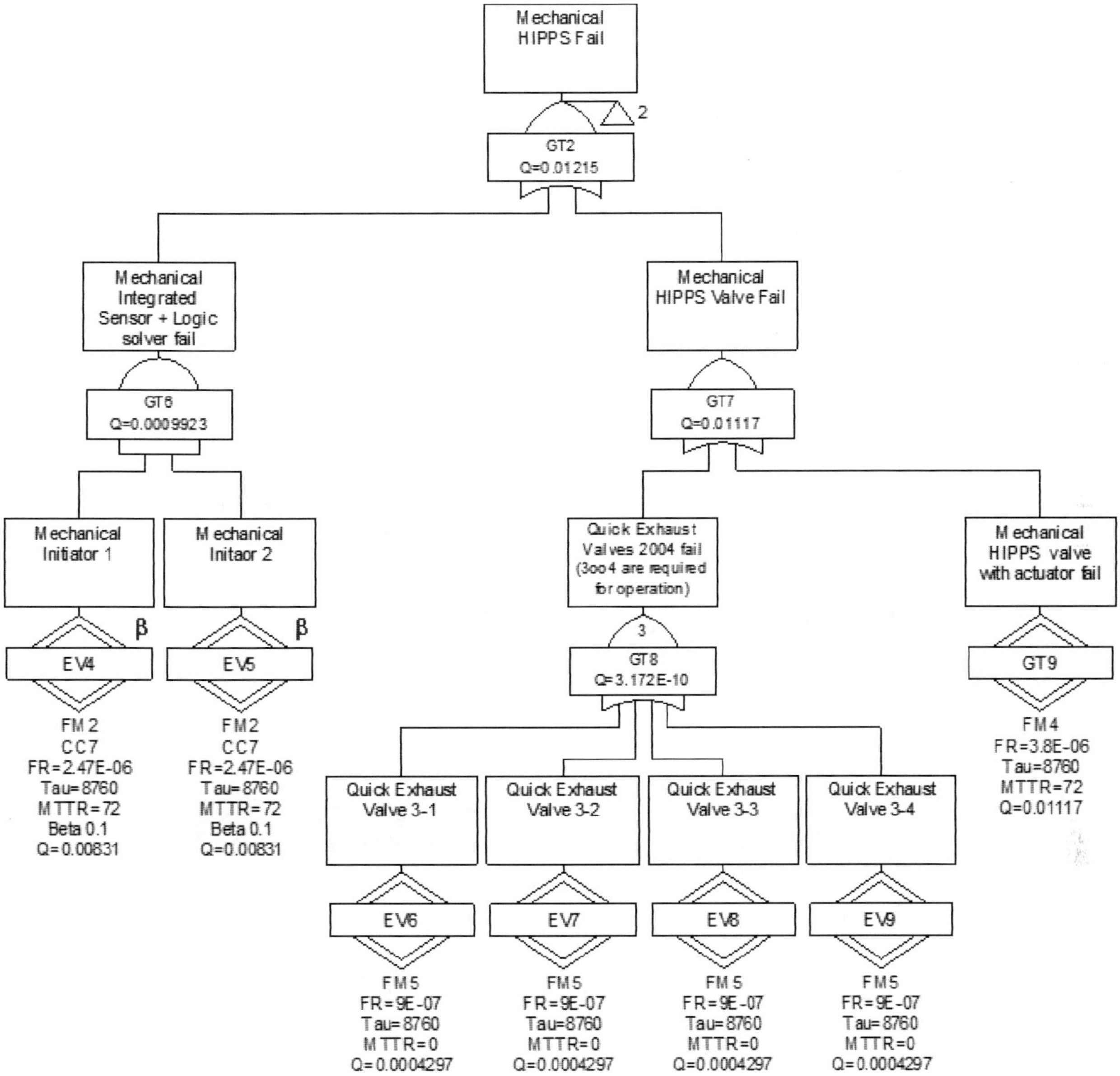

Figure 5. FTA for multiple HIPPS to achieve high risk reduction (contd..)

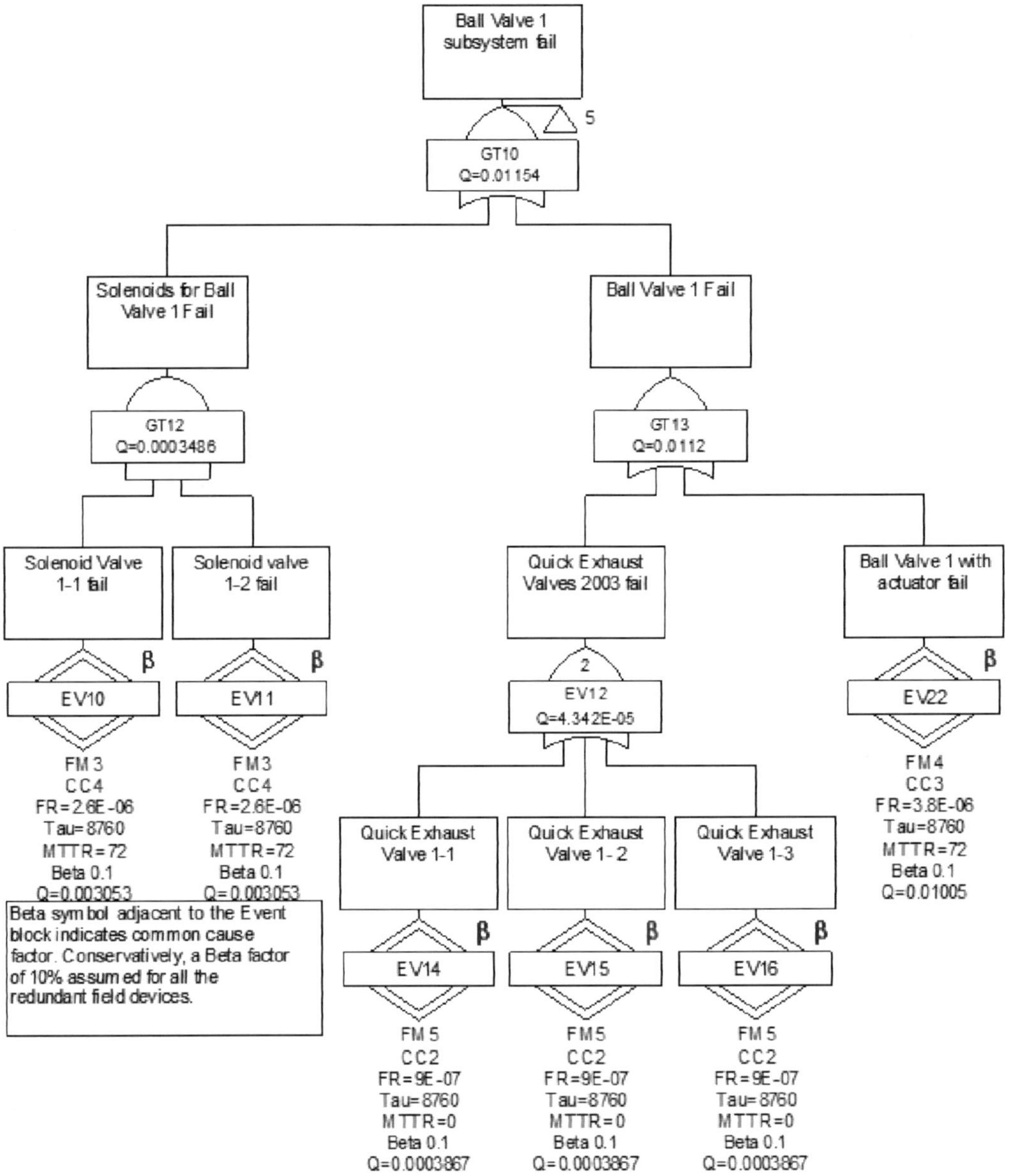

Figure 5. FTA for multiple HIPPS to achieve high risk reduction (contd..)

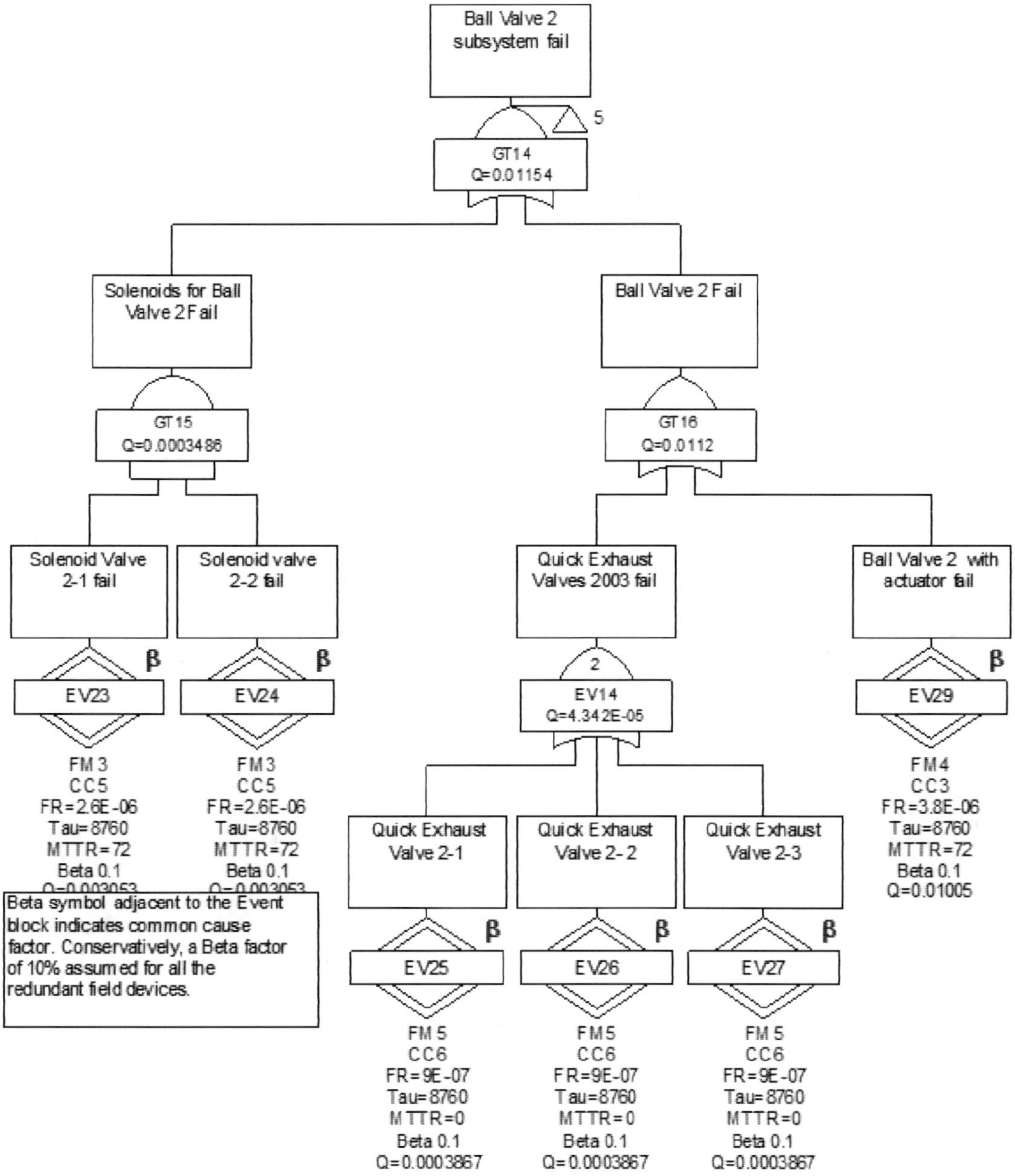

Figure 5. FTA for multiple HIPPS to achieve high risk reduction (contd..)

7 Installation, commissioning, operation and maintenance

Installation and commissioning of the HIPPS must be completed in accordance with the installation drawings, safety manuals, and the SRS.

Operation and maintenance is the longest period in the safety life cycle. The functional safety standards mandate continuous performance monitoring, including demand rates, failure modes and failure rates of various components. The actual performance must be compared with the predicted performance to identify any gaps so that remedial measures can be provided with appropriate safety review.

A robust change management process must be in place to ensure HIPPS integrity throughout the operation and maintenance phase.

8 Conclusion

Assignment of an SIL target and choosing an appropriate technology for a HIPPS is not trivial. In this paper, the significance of PHA and SIL selection for HIPPS has been re-emphasized and a brief account of available options for HIPPS to choose from, their pros and cons, and the selection criteria with examples are provided.

The design, installation, operation, and maintenance of HIPPS shall comply with the requirements of functional safety standards IEC 61508 and IEC 61511. In this article, the challenges in each of the above activities are discussed and appropriate solutions to mitigate the above challenges are provided.

In applications where a high risk reduction greater than 10,000 is required a hybrid HIPPS design approach is proposed. Using FTA demonstrated the modeling of complex hybrid HIPPS to achieve high level risk reduction.

9 References

[1] IEC. IEC 61511 Functional Safety: Safety Instrumented Systems for the Process Industry Sector- Part 1-3, Edition 2, Geneva, 2016.

[2] American Society of Mechanical Engineers (ASME), Boiler and Pressure Vessel Code, Section VIII-Pressure Vessels, New York, 2010.

[3] API, Standard 521, Pressure-relieving and Depressuring Systems, 6th Edition, Washington, DC, 2014.

[4] API, RP17O, Recommended Practice for Subsea High Integrity Pressure Protection Systems (HIPPS), 1 Edition, Washington, DC 2009.

[5] HSE, UK. SPC/TECH/OSD/31, Safety instrumented systems for the overpressure protection of pipeline risers, Available at

http://www.hse.gov.uk/foi/internalops/hid_circs/technical_osd/spc_tech_osd_31.htm, Accessed on Jan 29, 2017.

[6] IEC. IEC 61508 Functional safety of electrical/electronic/programmable electronic safety related systems- Part 1-7, Edition 2, Geneva, 2010.

[7] H.G. Lawley and T.A. Kletz. High-pressure-trip systems for vessel protection, Chemical Engineering, Vol 82, (1975).

[8] Dr. D. J. Smith, Reliability, Maintainability and Risk, 8th Edition, Butterworth-Heinemann, Waltham, MA, 2011.

[9] S. Hauge, S. Habrekke, M.A. Lundeteigen. Reliability Prediction Method for Safety Instrumented Systems- PDS Example Collection, SINTEF Technology and Society, Safety Research, 2010.

[10] Dr. D.J. Smith, and K.G.L. Simpson, The Safety Critical Systems Handbook, 4th Edition, Butterworth-Heinemann, Cambridge, MA, 2016.

475693 Managing Risk Associated w/ Electrical Energy Ignition Sources

Abstract:

In facilities that handle flammable hydrocarbons it is extremely important that all forms of potential ignition sources be carefully managed. This presentation will focus on key points associated with process electrical safety, open flame process heaters, and mechanical forms of ignition sources. OSHA 1910.119 has for many years required that hazardous area classification documentation remain in an evergreen state that requires re-certification at least once every five years. For most process facilities this OSHA mandate presents many problems. In most cases documentation has not been maintained in a current state, the facility perhaps has changed owners if not once many times over the years. The presentation attendee will be exposed to a common sense methodology that focuses primarily on minimizing electrical energy from becoming a source of ignition. A process for updating hazardous area classification documentation will be presented with tips on providing the end-user with an audit proof set of documentation. Once the hazardous area map has been developed then methods of properly documenting and conducting a site compliance assessment will be explained. This compliance assessment is a gap analysis that compares the installed electrical systems against the requirements of the National Electrical Code (NFPA 70 Article 500). Methods of presenting this documentation will be explained so the end-user has a clear understanding of the proper steps for remediation.

The presentation attendee will be exposed to the following:

- Managing risk associated with roadways that are impinged by hazardous classified boundaries
- Open Flame Process heaters encroachment by hazardous classified locations
- Methodology for area classification assessment *(it's determined by the process conditions)*
- Hazard's compliance for instrumentation and electrical system installations in hazardous process locations
- How to interpret documentation in support of the area classification assessments which include:
    - Plan Drawings (maps), Elevations, Site Specific Details. Legends, and notes
    - Assessment report documentation that addresses General Considerations, Identification of Point Sources of Emissions, Physical properties of Materials being Handled, Processed, or Stored, and Gas Dispersion Modeling Graphs and plots
- Methods for mathematical determination of hazard radius thru the concept of commercially available computer assisted gas/vapor dispersion modeling software.
- Application of common standards, codes and installation practices i.e. (NFPA, ISA, API and OSHA)

Detailed handouts as well as a complete audio-visual presentation are used to supplement the presentation.

479320 Review and Analysis of Recent Cyber Security Attacks

Abstract:

Process plants need to ensure that adequate safeguards are in place for a variety of upset conditions, including cyber security attacks, which can result in loss of product, damage to equipment, downtime, and potential safety and environmental disasters. Select cyber security incidents across a variety of industries will be reviewed to determine the modes of cyber security attack and their subsequent impacts. Specific areas of vulnerability will be identified with corresponding mitigation strategies. Possible solutions will be presented for incorporating cyber security into an existing process safety program to ensure a tolerable level of protection for high-consequence events.

478316 Resilience-Based Survival Analysis in the Event of a Process Upset: A Case Study of Utility Plant

Abstract:

Abstract: In a process plant system, safe and reliable operations are highly sensitive to the power or steam failure, loss of cooling water or nitrogen or instrument air. Most of these products and services from utilities plant play an important role as safety barriers. The main focus of this paper is to propose a model for survival evaluation of a process system under an upset condition. This model is based on the holistic resilience-based method that integrates both technical (process parameters variations) and social (policy/regulations, human and organizational) factors. The resilience methodology emphasizes non-linear dynamics, new types of threats, uncertainty, survival through upset conditions and recovery from catastrophic situations [1], [2].

The purpose of this paper is to maximize the performance of safety barriers that will be utilized in case of a process upset condition to prevent the escalation of the event into a catastrophic incident. In this work, a methodology is presented to select the best maintenance strategy [3] while considering the statutory safety regulations for maintenance of equipment like boilers, generators and more. Three different types of maintenance alternatives considered as part of this analysis are: predictive, preventative and condition-based. The key ideas of survival evaluation are presented using a utility plant as the case study.

References

1. Dinh, L. T., Pasman, H., Gao, X., & Mannan, M. S. (2012). Resilience engineering of industrial processes: principles and contributing factors. Journal of Loss Prevention in the Process Industries, 25(2), 233-241.
2. Jackson, S. (2009). Architecting resilient systems: Accident avoidance and survival and recovery from disruptions (Vol. 66). John Wiley & Sons.
3. Vassiliadis, C. G., & Pistikopoulos, E. N. (2001). Maintenance scheduling and process optimization under uncertainty. *Computers & Chemical Engineering*, *25*(2), 217-236.

478816 Model-Based Pressure Relief Design for Reactive Systems

Abstract:

Abstract

Reactive systems behave unpredictably under overpressure conditions. Varying composition impacts the physical properties and response of the mixture to temperature and time. Pressure generation in a reactive system is often a function of both vapor pressure and gas generation rate. Standard methodologies applied in non-reactive pressure relief system design cannot be relied upon to give appropriate sizing for reactive systems. Consistent with DIERS methodologies, model-based design provides accurate, reliable vent sizing. A comprehensive approach to modeling is presented which incorporates vapor-liquid equilibrium, reaction stoichiometry, and reaction rate.

Dynamic Modeling of Inbreathing Requirements for Low-Pressure Storage Tanks

Dona Abou-Chakra
ioMosaic Corporation
93 Stiles Road | Salem, NH 03079
chakra.d.nh@iomosaic.com

Georges A. Melhem, PhD, FAIChE
ioMosaic Corporation
93 Stiles Road | Salem, NH 03079
melhem@iomosaic.com

Ronald J. Willey, PhD
Northeastern University
360 Huntington Avenue | Boston, MA 02115
r.willey@neu.edu

Prepared for Presentation at
American Institute of Chemical Engineers
2017 Spring Meeting and 13th Global Congress on Process Safety
San Antonio, TX
March 26 – 29, 2017

Dynamic Modeling of Inbreathing Requirements for Low-Pressure Storage Tanks

Dona Abou-Chakra
ioMosaic Corporation
93 Stiles Road | Salem, NH 03079
chakra.d.nh@iomosaic.com

Georges A. Melhem, PhD, FAIChE
ioMosaic Corporation
93 Stiles Road | Salem, NH 03079
melhem@iomosaic.com

Ronald J. Willey, PhD
Northeastern University
360 Huntington Avenue | Boston, MA 02115
r.willey@neu.edu

Keywords: Thermal Inbreathing, API 2000, Storage Tanks, Venting.

Abstract

A Process Safety Office™ - SuperChems™ component based model is proposed to study the dynamic behavior of the thermal inbreathing of low-pressure storage tanks upon the occurrence of a sudden cooling rainstorm. The objective here is to establish a model that is capable of properly calculating the design relief capacity of the relief device to be installed on such tanks. The impact of phase-change, particularly vapor condensation, on the thermal inbreathing of the tank assessed. The model was validated against two large-scale experiment datasets and some other published theoretical models. The results were then discussed and compared to the relief requirements obtained using the API 2000 7th Edition method, which is the current standard for venting atmospheric tanks.

1 Introduction

Low-pressure storage tanks are protected against extreme weather changes following the API 2000 (7th edition) standard. The threat of creating a vacuum occurs when the contents of the tank vapor space are suddenly cooled by an intense rainfall causing the vapor space to contract and condense, hence, rapidly decreasing the internal pressure. This pressure drop will pull outside air into the tank's vapor space and acts as the driving force for thermal inbreathing.

The current limitation with the API standard is that it only addresses tanks storing non-condensable gases and have no liquid residue, where in fact most storage tanks contain condensable vapors such as water vapor which will have a big impact on the thermal inbreathing of the tank. This paper assesses the API 2000 relief requirement calculations for low-pressure storage tanks by comparing them to (1) the results of the dynamic simulations from a proposed SuperChems^{TM} based model and (2) to several methods found in the literature (PTB W-22, Salatino, etc.) for both condensable and non-condensable vapor services.

2 API 2000 Methods

The tank contents are affected by the ambient conditions especially when the tank is uninsulated, which is the case for most storage tanks. This method uses air at normal conditions (0°C and 101.325 kPa) as the inbreathed gas. The maximum inbreathing rate, $\dot{V}_{IT}$, is primarily a function of tank volume, insulation, and a C-factor (found in Table 1 below) as shown in Equation (1) below:

$$\dot{V}_{IT} = C\, V_{tk}^{0.7} R_i \tag{1}$$

where $\dot{V}_{IT}$ is the maximum inbreathing volumetric flow rate of air at normal conditions (Nm^3/h), C is the dimensionless factor that depends on vapor pressure, storage temperature and latitude, V_{tk} is the tank volume (m^3), and R_i is the reduction factor for insulation.

Table 1. API 2000 *C*-factor for Thermal Inbreathing

Latitude, φ	**C-factor**			
	Vapor pressure similar or below hexane		Vapor pressure higher than hexane or unknown	
	Average storage temperature (°C)			
	< 25	≥ 25	< 25	≥ 25
φ < 42°	4	6.5	6.5	6.5
42° ≤ φ ≤ 58°	3	5	5	5
φ > 58°	2.5	4	4	4

The assumptions used for the scenario of "tank cooling by rain" are found in Annex E of API 2000 [1] and are listed below:

- Tank is filled with air only.
- Initial vapor space and shell wall temperature are in thermal equilibrium at 55 °C.
- Rain temperature and intensity are constant at 15°C and 225 kg/m^{2}h (This rainstorm has a probability of 1/100 year) respectively; rain angle is at 30° from the horizontal.
- Inside free convection heat-transfer coefficient is 5 W/m^{2}K.
- Outside film cooling heat-transfer coefficient is 5,000 W/m^{2}K.

2.1 API 2000 Annex A Method

The method found in Annex A was used prior to 7th Edition (2008) and is quite different than the current sizing method for venting devices. In order to quantify the thermal inbreathing for storage tanks volumes smaller than 3180 m^3, Annex A [1] assumes that the tank starts cooling from an initial temperature of 48.9°C and at a constant temperature rate of -56 K/h. The thermal inbreathing rate, $\dot{V}_{IT}$ (Nm³/h) is as follows:

$$\dot{V}_{IT} = 0.169\, V_{tk} \quad (2)$$

where V_{tk} is the tank volume (m³). For tank volumes equal or greater than 3180 m³, the thermal inbreathing is assumed to be limited by a constant heat flux, $h\Delta T$, equal to 63 W/m². The thermal inbreathing rate in this case becomes:

$$\dot{V}_{IT} = 0.577\, A_{exp} \quad (3)$$

where A_{exp} is the external tank area exposed to rain (m²). Table 2 below summarizes the thermal normal venting requirements as found in Annex A of API 2000. Volumes greater than 30,000 m³ are outside the scope of this method. However, interpolation is allowed for Table 2 as long as the tank volumes lie between the specified volume limits of 10 m³ to 30,000 m³. [1]

Table 2. API 2000 Annex-A Normal Venting Requirements Due to Thermal Effects [1]

Tank Capacity	Inbreathing	Out-breathing	
Column 1 [a]	Column 2 [b]	Column 3 [c]	Column 4 [d]
		Flash Point ≥ 37.8 °C or Normal Boiling Point ≥ 149 °C	Flash Point < 37.8 °C or Normal Boiling Point < 149 °C
m³	Nm³/h of air	Nm³/h of air	Nm³/h of air
10	1.69	1.01	1.69
20	3.38	2.02	3.38
100	16.9	10.1	16.9
200	33.8	20.3	33.8
300	50.4	30.4	50.4
500	84.5	50.7	84.5
700	118	71.0	118
1,000	169	101	169
1,500	254	152	254
2,000	338	203	338
3,000	507	304	507
3,180	537	322	537
4,000	647	388	647
5,000	787	472	787
6,000	896	538	896
7,000	1,003	602	1,003
8,000	1,077	646	1,077
9,000	1,136	682	1,136
10,000	1,210	726	1,210
12,000	1,345	807	1,345
14,000	1,480	888	1,480
16,000	1,615	969	1,615
18,000	1,750	1,047	1,750
20,000	1,877	1,126	1,877
25,000	2,179	1,307	2,179
30,000	2,495	1,497	2,495

3 Thermal Inbreathing of Non-Condensable Vapors

One of the major assumptions of API 2000 is that the tank experiencing inbreathing is empty (i.e. only filled with dry air) and the vapor space is considered non-condensable. This section will review recent work on inbreathing dynamics for non-condensable dry air services only.

3.1 Salatino's Model

Salatino et al. [2] compare their tank inbreathing model to the older version of API 2000 (1992) that utilized the method of Annex A: a method based on the maximum estimated heat loss rate of 63 W/m^2 upon sudden cooling by a rainstorm of 225 kg/m^{2}h rain density. They recognized that all existing thermal inbreathing models assume a uniform vapor-space temperature, thus his team's contribution was accounting for the temperature non-uniformity inside the tank by dividing the tank into two individual volumes: the shell and the roof.

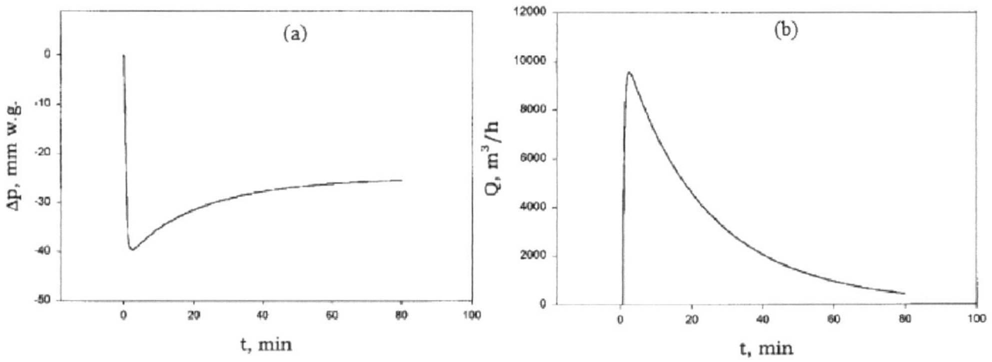

Figure 1. Pressure drop (a) and thermal inbreathing (b) profiles of Salatino's Model [2]

Salatino derived an equation for the maximum thermal inbreathing (Equation 4), stating that the maximum occurs at time t=0 (Figure 1) when the sudden weather change causes (1) vanishing of the radiative flux absorbed by the tank, (2) abrupt drop in ambient temperature and finally (3) rainfall.

$$\dot{Q}_{max} = -\frac{V}{T_G}\frac{dT_G}{dt}\bigg|_{t=0} = \frac{V}{T_{G,1}}\frac{1}{\tau}\,(T_{G,1} - T_{G,2}) \tag{4}$$

$$\tau = \frac{pV}{RT_G}\frac{c}{\sum A_i h_{iG}} \tag{4a}$$

where $\dot{Q}_{max}$ is the maximum thermal inbreathing rate (m^3/h), V is the tank volume (m^3), $T_{G,1}$ and $T_{G,2}$ are the gas temperatures (K) before and after the rainstorm, respectively. τ is the time constant that sums the effects of pressure, volume, heat transfer parameters (Equation 4a).

Salatino et al. further simplified the model by assuming an average value for the heat transfer coefficient of 4 W/m^{2}K and a gas temperature drop (after rainfall) of T_{G1}-$T_{G2} \cong 25$°C claiming that these values will result in a conservative approximation of the inbreathing rate. Replacing these values in Equation (4) yields the simplified maximum inbreathing rate, $\dot{Q}_{max}$, which is a function of the tank volume, *V,* and the height-to-diameter ratio, *f*:

$$\dot{Q}_{max} = 2.6\, V^{2/3} \left(\frac{1 + 2f}{\sqrt[3]{f}} \right) \tag{5}$$

For a 63,000 m^3 cone-roof tank, Salatino predicted a maximum thermal inbreathing value of 9,900 m^3/h, using Equation 5. However, using the API 2000 Annex A method, the maximum inbreathing for the same tank was found to be 4,100 m^3/h. Salatino proved that his method is more conservative than the API 2000 equations and advised that an oversized thermal relief device is considered a safer design in this case.

3.2 Lloyd's Register Energy Americas, Inc. (LR) Model

Lloyd's Register (LR) Energy Americas, Inc. analytically solved the heat balances of a storage tank that is suddenly cooled by rain [3] and compared their results against ISO 28300, which is an equivalent of API 2000 6th edition (2009). The model results are summarized in Figure 2 below.

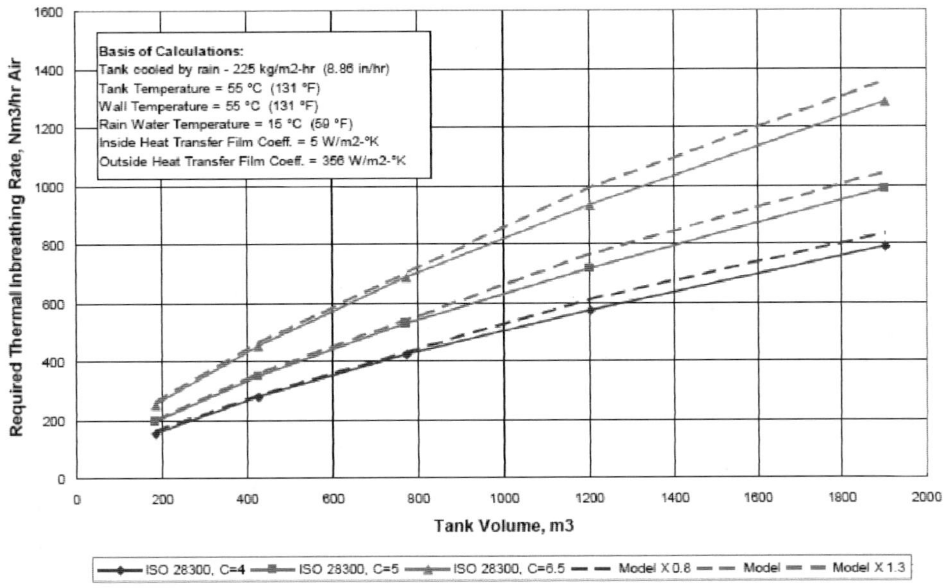

Figure 2. LR Model versus API 2000 6th ed. (ISO 28300)

The LR model was slightly more conservative than the API 2000 thermal inbreathing predictions by factors ranging between 0.5 and 1.3. Note that the C value in Figure 2 refers to the C-factor in the API 2000 thermal inbreathing in Equation (1) of this paper.

The model was then used to study the impact of several variables, some of which had minimal impact on the predicted thermal inbreathing (less than 5% error) such as the tank weight and initial gas temperature. Other variables, like the outside heat-transfer coefficient (Table 3) and rain intensity, had substantial impact on the calculated thermal inbreathing rates.

Table 3. LR Model - Impact of Outside Heat-Transfer Coefficient, h_o [3]

Case No.	Tank Volume m3	Predicted Thermal Inbreathing Rate, Nm3/hr		
		ho = 200 W/m2- K	ho = 300 W/m2- K	ho = 400 W/m2- K
1	186	169.6	191.2	205.1
2	427	308.3	341.4	362.7
3	775	475.1	520.7	548.6
4	1202	678.9	740.5	778.6
5	1903	935.2	1011.6	1058.5

The LR model was used to derive a relationship between the rain intensity (mm/h) and the heat transfer coefficient between the outer tank wall and rain (W/m^{2}K) as shown in Figure 3 below. This linear dependence relationship was implemented in SuperChems^{TM} for this study.

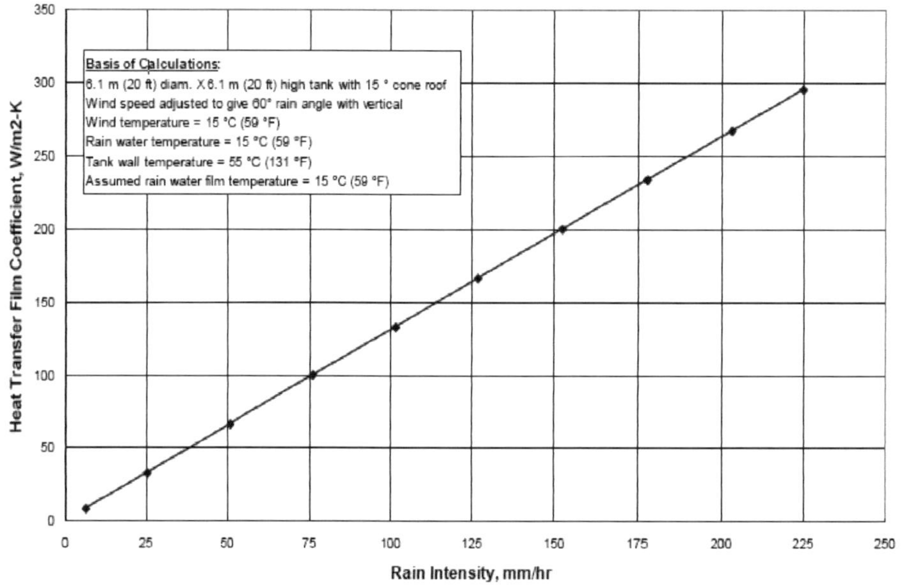

Figure 3. Rain/wall heat transfer coefficient versus of rain intensity

3.3 PROTEGO© Model

Scholz and Weber [4] from PROTEGO© derived the API 2000 7th Edition thermal inbreathing formula (Equation 1 of this paper) and determined that the maximum thermal inbreathing was a function of primarily tank volume with a coefficient that is less than or equal to 9.99:

$$\dot{V}_{max} = Cofficient \cdot V^{2/3} \tag{6}$$

The above equation is applicable to both flat-roof and conical-roof storage tanks. The derivation of Equation (6) was obtained by performing energy balances on the stored gas, the tank walls, and on the rainwater film. The differential equation and its solution are not included in this document, but can be found in Scholz and Weber's publication [4].

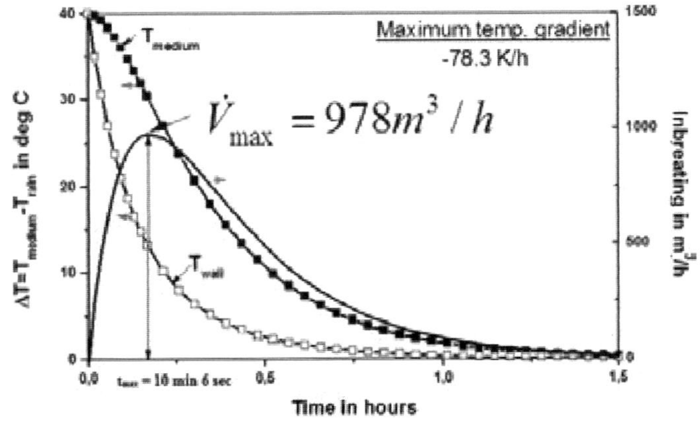

Figure 4. PROTEGO© Model Results

PROTEGO© then compared their results with the Foerster's model [5], whose main assumption was a fully immersed tank, which proved to be in good agreement with the published data. Figure 4 contains the results of a simulation using a conical-roof storage tank of 3,978 m^3 volume and an aspect ratio f = 0.6. Light rain was assumed (75 kg/m^2h) at 15°C. The inside and outside convective heat-transfer coefficients were set to 5 and 37 W/m^2K respectively.

PROTEGO© used their model to study 5 different types of condensable vapors and their effect on the maximum thermal inbreathing rate, however, the results will be discussed in Section 4 of this paper.

4 Thermal Inbreathing of Condensable Vapors

API 2000 neglects the possibility of vapor condensation for the thermal inbreathing/out-breathing relief calculation since only dry air is present in the tank i.e. the tank is empty. One reason to consider the effects of having condensable vapors is the steam cleaning that storage tanks undergo. Hot steam will occupy the tank after the cleaning process is complete and if a rainstorm occurs, the water vapor will condense/contract and pull a vacuum inside the tank.

4.1 Fullarton Model – Water Vapor

Fullarton [6] introduced two dimensionless factors in his model: (1) The reduction factor "η" that accounts for the tank wall thickness by factoring in the metal heat capacity of the walls and (2) the condensation factor "χ" that accounts for the stored-vapor condensation i.e. the additional condensation mass flux and the additional heat of condensation. The formulas of each "η" and "χ" can be found in Fullarton's paper [6].

It is expected that the tank wall temperature drops much faster than the temperature of the vapor stored inside the storage tank because of the metal's higher heat capacity. However, a thicker metal wall acts as a resistance to heat transfer. This means that the tank walls cool at a slower rate than when the wall thickness is neglected. The maximum inbreathing in this case will be lower, hence the reduction factor "η" which "reduces" the maximum thermal inbreathing of a storage tank with thicker walls.

Table 4. Range of χ-factor for Different Condensable Vapors

Condensable Vapor	χ-factor
Water vapor Heptane Hexane	1.5-2
Ethanol	2-2.5
Methanol	3-4

The condensation factor "χ", on the other hand, is primarily a function of the volatility of the condensable vapor being studied. For that reason, Fullarton established a range for the "χ" factor for different vapors as seen in Table 4 above [6].

Therefore, the maximum thermal inbreathing for a condensable vapor space per Fullarton can be obtained by using each of the Naumann approximation, $\dot{V}_N$, the reduction factor, η, and the condensation factor, χ.

$$\dot{V}_{K,max} = \dot{V}_N \eta \chi \tag{7}$$

One must keep in mind that properties of air at 288K were used to derive this equation in addition to the assumption of the initial temperature gradient ($T_{G,0} - T_A$) equal to 40°C. If other values are to be used, the above equation must be re-derived to account for the appropriate conditions.

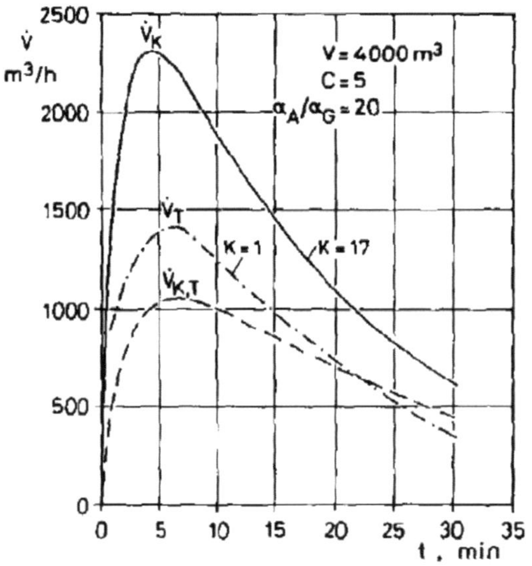

Figure 5. $\dot{V}_{inbreathing}$ **profile – Fullarton Model** [6]

Fullarton performed a dynamic simulation on a 4,000 m³ empty tank intended for the storage of ethanol. The initial gas temperature was 55°C and was suddenly cooled by rain at 15°C. The rain fall was assumed to be an extremely heavy downpour (probability of 1 every 100 years) with $\alpha_A/\alpha_G = 20$ and $\alpha_G = 5$ W/m²K. Figure 5 above shows the resultant inbreathing profile for each of dry air, K=1, and saturated water vapor, K=17. The maximum thermal inbreathing occurs at time t = 6 min where the inbreathing of the condensable vapor space ($\dot{V}_{K=17}$) is *1.6 times larger* than that of dry air ($\dot{V}_{K=1}$).

4.2 PROTEGO© Model – Condensable Vapors

PROTEGO© used their model [4] to obtain the inbreathing curves of five different vapors: methanol, ethanol, saturated water vapor, heptane and decane (Figure 6) and comparing them to the inbreathing curve of dry air in the same tank.

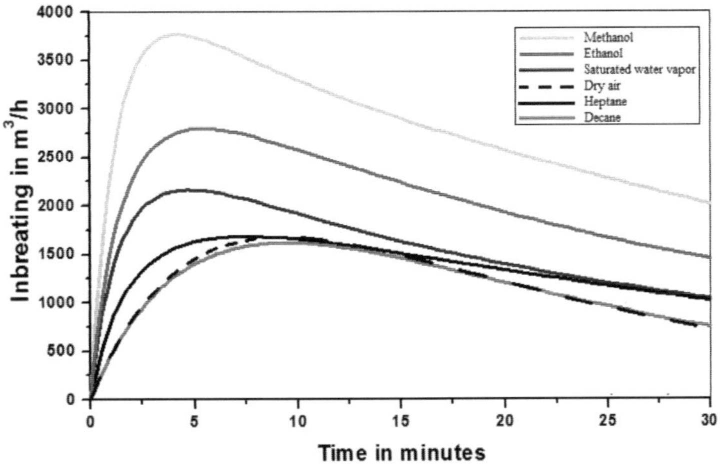

Figure 6. $\dot{V}_{inbreathing}$ profile of different vapors – PROTEGO© Model [4]

It can be seen that the condensation of volatile vapors has a significant impact on the inbreathing rate causing it to have a twofold increase. On the other hand, non-volatiles such as decane do not affect the inbreathing rate and yields the same inbreathing profile as with dry air.

Table 5. Condensation Factor "χ" Results

Vapor	$\chi_{PROTEGO}$	$\chi_{Fullarton}$
Methanol	4.76	3-4
Ethanol	2.23	2-2.5
Heptane	1.61	1.5-2
Water	1.29	1.5-2
Decane	1.01	N/A

PROTEGO© calculated the condensation factor "χ" for each of the simulated vapors and compared their values to that of Fullarton's (Table 4) and summarized the results in Table 5 above. The condensation factor values from this model are all in agreement with the published values [6] except the χ_{methanol} which is greater than the range specified by Fullarton (1986). A possible explanation for the latter is that PROTEGO© may have been more conservative (higher safety margin) when dealing with extremely volatile vapors like methanol.

5 Large-Scale Experimental Data

A thorough literature review yielded two sets of large scale data from actual experiments to study the thermal inbreathing of low-pressure storage tanks. Large scale datasets are preferred for model validation. The first dataset is for non-condensable gases (air) by *Sigel et al.* [7] and the second dataset is for condensable gases by *Holtkoetter et al.* [8].

5.1 Sigel's Experiment

Sigel performed actual data collection for the thermal inbreathing rate from a large tank of 617.5 m^3 capacity exposed to high solar radiation. The tank was empty without any trace of product vapor or liquid residue.

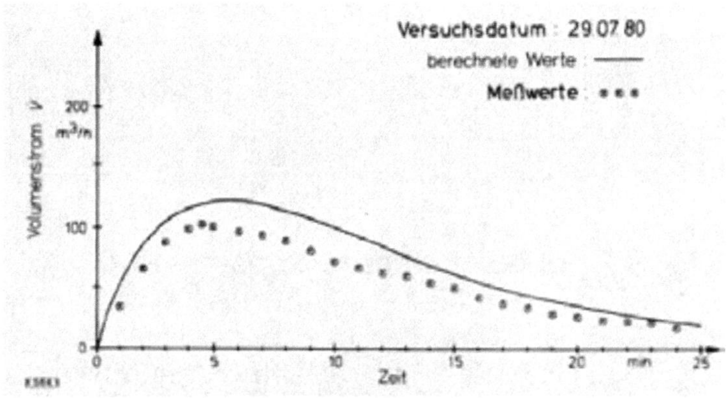

Figure 7. Sigel Experimental Inbreathing Profile

Sigel et al. [7] developed their own thermal inbreathing model to compare it with their experimental results. Figure 7 above displays the measured values of inbreathing (markers on the graph) versus the model (solid line). The inbreathing rates predicted by his model proved to be more conservative than the measured rates.

5.2 Holkoetter's Experiment

Holtkoetter performed an experiment on a 1.18 m^3 tank to measure the actual inbreathing rates and compare the results to his proposed model. This experimental investigation aimed at studying the effects of condensation on thermal inbreathing for three different vapors: water, isopropanol, and methanol.

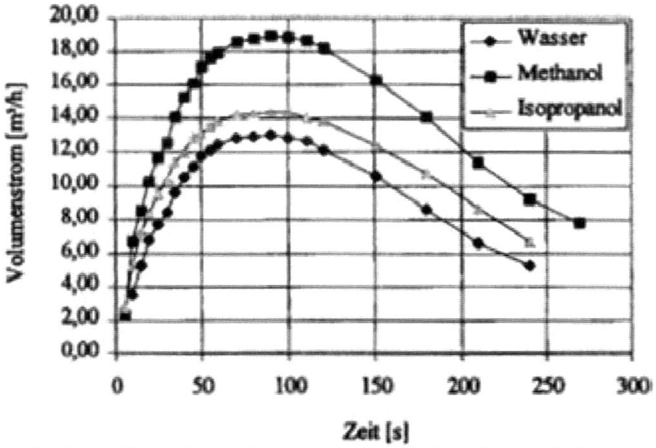

Figure 8. Holtkoetter Experimental Inbreathing Profile

The maximum thermal inbreathing is recorded after almost 95 seconds for all three gases. The maximum thermal inbreathing of methanol $\dot{V}_{max}$ ≈19 m³/h is significantly higher than that for both water ($\dot{V}_{max}$ ≈13 m³/h) and isopropanol ($\dot{V}_{max}$ ≈14 m³/h) due to the fact that methanol is the most volatile vapor.

6 SuperChems™ proposed Model

The inbreathing profiles were generated using SuperChems™ dynamic simulation of storage tanks experiencing a sudden weather change. Dynamic models continually generate results that are used as the initial conditions for the next time interval calculation which perpetually drives the model to completion.

The proposed model first starts by defining a variety of objects to represent components of the system such as the storage tank vessel, the mixture that is stored, the piping segment and relief device, and the water sprays that are used to simulate a rainstorm. A built-in thermodynamics package is selected (modified Peng-Robinson EOS [9]) and scenarios are then created to collect the objects. Once the scenarios are created, single and multi-phase flow dynamics models are associated with the individual scenarios and the dynamic solutions for the inbreathing profiles are produced.

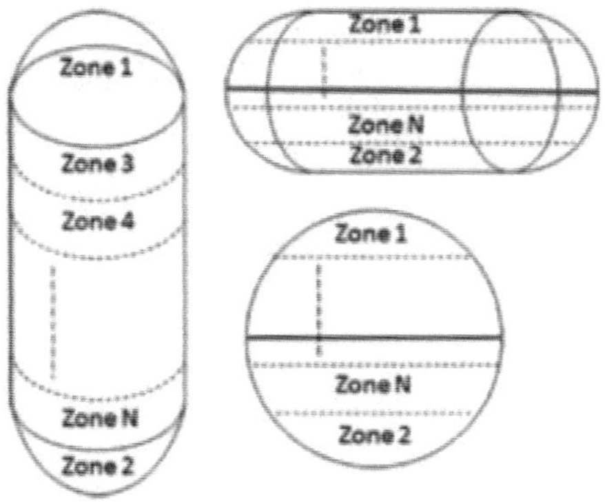

Figure 9. SuperChems™ Segmentation Scheme

An important feature of the software is wall dynamics which segments the vessel into multiple zones (Figure 9) allowing the user to account for heat conduction across the wall and between the zones using the effective conduction length method.

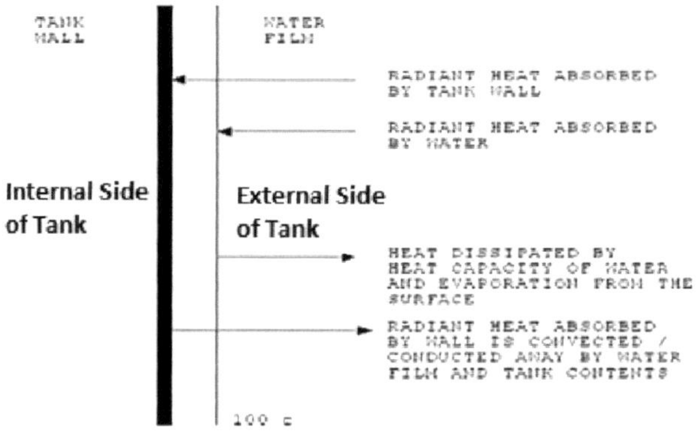

Figure 10: Heat Transfer Mechanisms for a Tank during a Rainstorm

Once segmentation is defined, heat transfer analysis is applied per wall segment (Figure 10) where the impact of rain, solar radiation, wind, and insulation is considered. Inside the tank, the wall-to-fluid heat transfer includes radiation and natural convection for the vapor phase; and nucleate, film and pool boiling for the liquid phase.

It should be noted that many parameters from the existing models and large-scale datasets were missing (vessel, dimensions, wall thickness, mixture composition, heat transfer coefficients, rain intensity, etc.) and had to be scientifically estimated in our dynamic simulations.

6.1 SuperChems™ Model Prediction for Non-Condensable Vapors (Large-scale Data)

The dimensions of the 617.5 m^3 empty tank used in Sigel's experiment [7] were introduced as input parameters to the SuperChems™ vessel object. The rainstorm used in this simulation had an intensity of 0.38 inch/h and a duration of 30 minutes with a temperature of 15°C. The inside heat transfer coefficient value was set to 5 W/m^2K as specified by API 2000 [1].

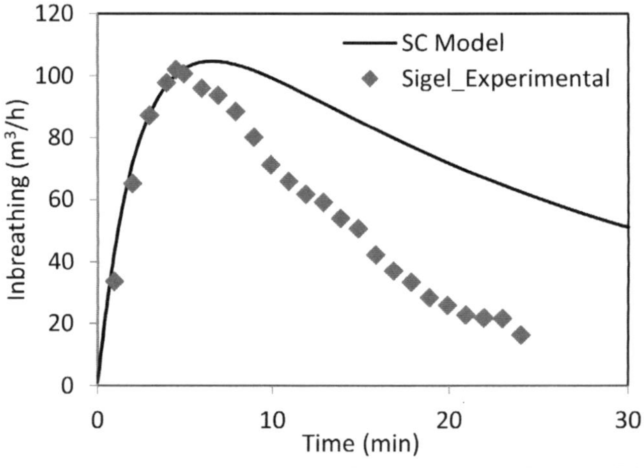

Figure 11. Simulation of Sigel's Tank using SuperChems™

The maximum value for inbreathing predicted by SuperChems was 104 m^3/h at time t = 6.6 min, as seen in Figure 11 above. This value is at a 2% deviation from the actual recorded maximum. The model's slope matches the first part of the inbreathing curve up until the maximum. Although the slopes of the decay of the inbreathing are different, they do not contribute to the design of the relief device because the most important value extracted from such studies is the maximum inbreathing that will act as the design relief capacity of the venting device.

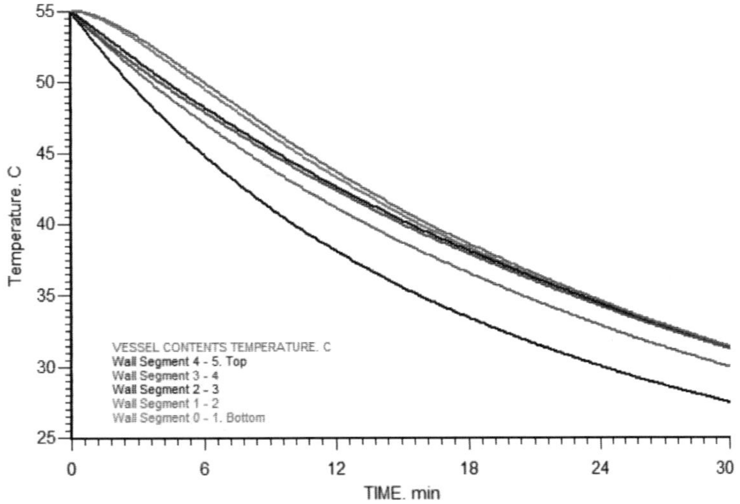

Figure 12: Temperature Profiles of Sigel's Tank using SuperChems™

Figure 12 above shows the temperature profiles of the wall segments and air inside the tank from the SuperChems™ simulation. Since Sigel [7] did not disclose the actual temperature profiles, therefore, we could not compare the computed temperature profile from SuperChems™ to the actual readings from Sigel's experiment.

6.2 SuperChems™ Model Prediction for Condensable Vapors (Large-scale Data)

Holtkoetter's experimental setup [8] consisted of a 1.18m^3 tank with walls that are 1.2 mm wall thick. The three vapors Holtkoetter studied were: methanol, isopropanol and water. For each type of vapor, the tank was filled with a small amount of liquid which was then vaporized using three infrared radiators (2.5-4.2 kW). After vaporization step was complete, water sprays were activated to cool the tank and the thermal inbreathing was recorded (Figure 8).

Using SuperChems™, the vessel was defined, then the rain-wall heat transfer coefficient, h_{rain}, was fixed at 600W/m^2K that corresponds to a rain intensity of 0.1262 kg/m^2s as specified by Holtkoetter. The duration of the rain for this simulation was 300 seconds. The h_{gas} was initially set to the API 2000 value of 5 W/m^2K, but was then adjusted to compensate for the vapor condensation inside the tank.

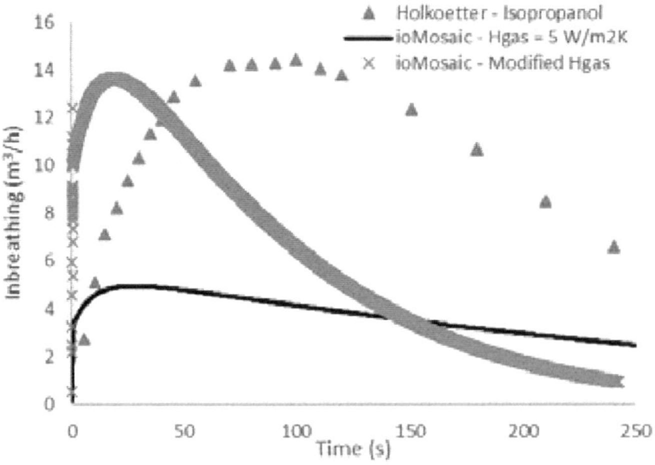

Figure 13. SuperChems™ Simulation of Holtkoetter's Tank with Isopropanol

Figure 13 illustrates the inbreathing profile for two different values of the free-convection heat transfer coefficient of the isopropanol vapor, h_{gas}, inside the tank. It can be clearly seen that using the API 2000 value h_{gas} = 5W/m^{2}K underpredicts the thermal inbreathing by almost 62%. After numerous iterations, a modified value for h_{gas} was found to be 30 W/m^{2}K. This higher value makes physical sense since the condensation of isopropanol occurring on the inner walls of the tank enhances the heat transfer between the vapor and the wall.

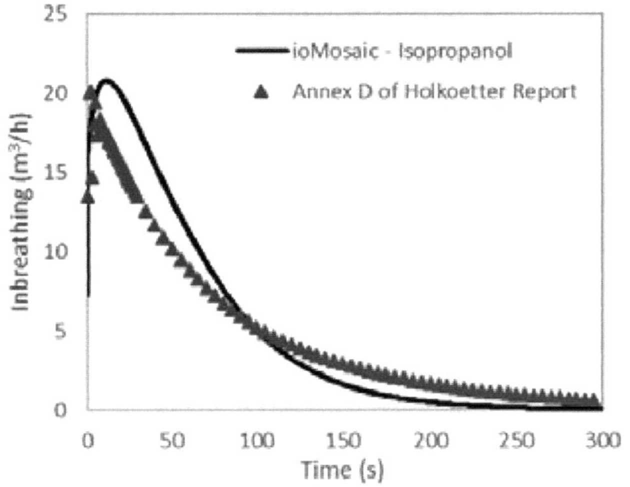

Figure 14. SuperChems™ Reproduction of Holtkoetter's Inbreathing Model (Isopropanol)

Although *Holtkoetter et al.* [8] performed experimental work to measure the maximum inbreathing, they did not, however, compare their experimental results to their proposed model. Hence, the SuperChems™ results were compared to Holtkoetter's model found in "Annex D" of his publication [8].

Figure 14 above proves that the SuperChems™ based model shows good agreement with the Holtkoetter model that considers spontaneous condensation. The predicted value for the $\dot{V}_{max}$ using SuperChems™ was 20.7 m^3/h and is at a +3% error from the value reported by Holtkoetter [8].

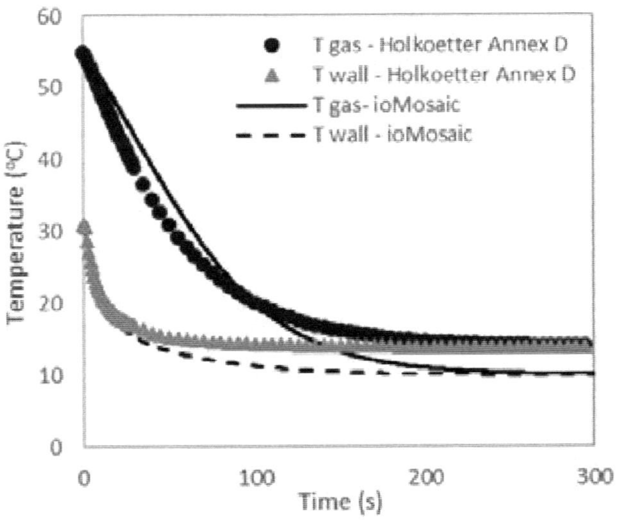

Figure 15. SuperChems™ Reproduction of Holtkoetter's Temperature Model (Isopropanol)

The gas and wall temperature profiles were almost accurately reproduced using the SuperChems™ model and are shown in Figure 15. Holtkoetter attributes the discrepancy between his model and his experimental result to the fact that the experiment was done in the month of December where the available water had a temperature of 8°C instead of 15°C. Colder rain will eventually lead to higher inbreathing rate values, therefore, the inbreathing should be inspected as a function of temperature to account for this 7°C difference in the water temperature. For more information about this method, please refer to the actual publication [8].

6.3 SuperChems™ Model vs. Published Models

In addition to validating our proposed model against large-scale experimental data, the SuperChems™ based model was also compared to the published models listed in Section 3 and 4 of this paper [10].

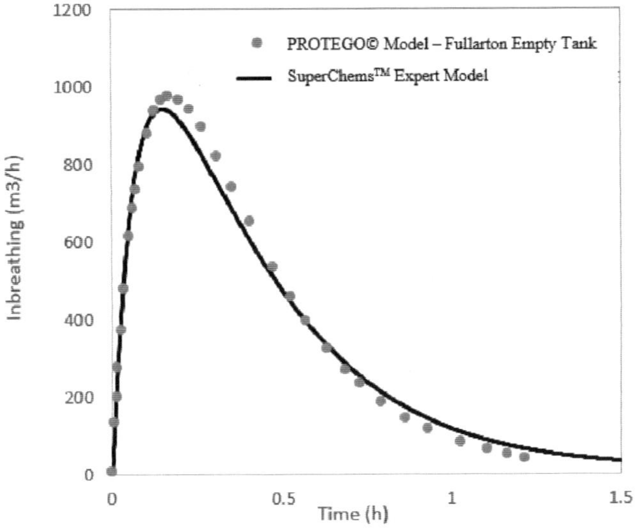

Figure 16. SuperChems™ Inbreathing Profile for Air (Fullarton's tank)

Figure 16 shows the predicted inbreathing profile of dry air using the SuperChems™ model and compared against the PROTEGO© model. The maximum recorded inbreathing from the SuperChems™ Expert simulation is around 940 Nm^3/h which has a -4% deviation from $\dot{V}_{max}$ predicted by PROTEGO©. The temperature and temperature-rate profiles were also calculated and can be found in the actual publication along with the fits for the different models and vapors using SuperChems™ Expert [10].

6.4 Modified Heat Transfer Coefficient

The heat transfer coefficient between the stored vapor and the inner-tank walls, h_{gas}, was modified for each type of volatile condensable vapor [10]. Table 6 below summarizes the values of h_{gas} used in the two-phase dynamic model for a tank-wall thickness of 5 mm. These results show that condensation "enhances" heat transfer thus providing evidence that use of the API 2000 value heat transfer coefficients h_{gas} will underpredict the maximum thermal inbreathing (Figure 13). Also, note that h_{gas} is directly proportional with the vapor's volatility.

Table 6. Modified h_{gas} Values for SuperChems™ Two-Phase Model

h_{gas} (W/m^2K)	Wall Thickness = 5 mm
Methanol	40
Ethanol	30
Water	20
Heptane	5
Decane	7

7 API 2000 Conservativeness

The maximum thermal inbreathing rates, V_{max}, obtained from the SuperChems™ model were compared to those obtained using the API 2000 thermal inbreathing equation, Equation (1), to assess the performance of the API 2000 inbreathing method.

In order to determine the appropriate value for the C-factor in Equation (1), the vapor pressure and storage temperature of the stored gas are needed.

- Since all the investigated vapors (methanol, ethanol, water vapor, etc.) do not have vapor pressures higher than hexane, the vapor pressure was classified as "Similar to Hexane".
- The storage temperature of the gas is 55°C, which is greater than 25°C, so according to Table 1, the C-factor has a *value of 5*.

The storage tanks in this study were all uninsulated, hence, the insulation reduction factor R_i has a value of one. The API 2000 inbreathing equation used for this case now becomes:

$$\dot{V}_{IT} = 5\, V_{tk}^{0.7} \tag{8}$$

To determine whether the API 2000 inbreathing relief requirement is adequate, a deviation percentage of the simulated (or experimental) results the API 2000 $\dot{V}_{max}$ method is computed as follows:

$$\% \, Deviation = \frac{\dot{V}_{max,SC} - \dot{V}_{max,\ API\ 2000}}{\dot{V}_{max,\ API\ 2000}} \times 100 \quad (9)$$

where $\dot{V}_{max,\ API\ 2000}$ is the maximum thermal inbreathing obtained from Equation (8) (m^3/h) and $\dot{V}_{max,SC}$ is the maximum inbreathing from SuperChems™ model simulation (m^3/h).

7.1 API 2000 performance – Large-scale Experiments

Sigel et al. [7] used a tank volume of 600 m^3, which when replaced in Equation (8), results in a maximum inbreathing value of 440 m^3/h. Figure 17 compares $\dot{V}_{max}$ using the API 2000 method to the actual recorded $\dot{V}_{max}$ values during Sigel's experiment. The API 2000 inbreathing prediction is conservative for Sigel's empty tank where the relief device is designed to handle an inbreathing rate 4.2 times larger than the recorded $\dot{V}_{max}$.

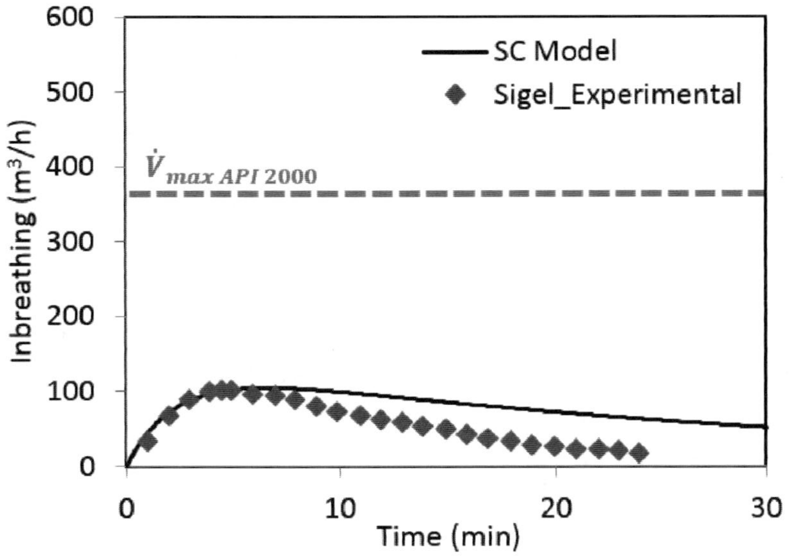

Figure 17. API 2000 Inbreathing Requirements – Sigel's Tank with Dry Air

On the other hand, *Holtkoetter et al.* [8] used a smaller tank with a capacity equal to 1.18 m^3 which results in a maximum inbreathing rate of 5.6 m^3/h according to the API 2000 method – Equation (8). Figure 18 shows that the API 2000 inbreathing relief requirement is insufficient for water, isopropanol, and methanol vapors in Holtkoetter's tank.

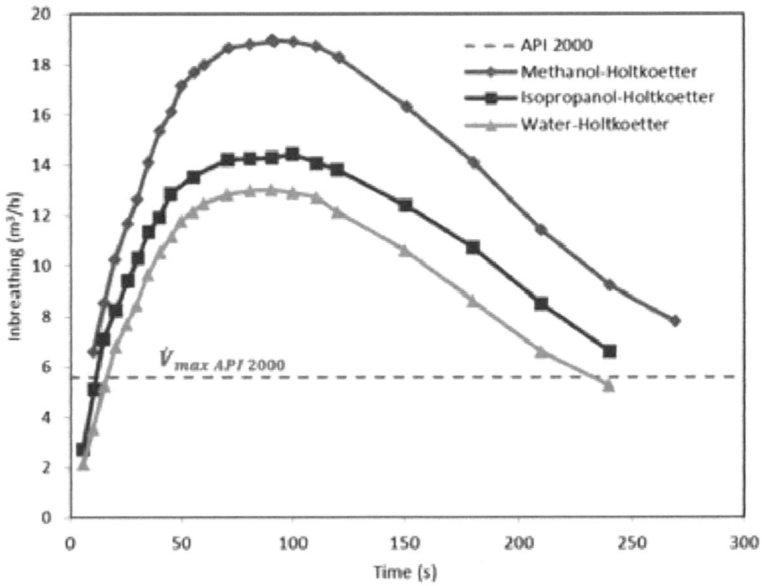

Figure 18. API 2000 Inbreathing Requirements – Holtkoetter's Tank

The recorded inbreathing rates are 2.3 to 3.7 times larger than the rate calculated by API 2000; this leads to an undersized relief device that puts the storage tank at a high risk of failure if a sudden change of weather occurs.

7.2 API 2000 performance – Computer Model

The tank size used in Scholz and Weber's model [4] was 6,283 m^3 to study the condensable vapors. Replacing the tank volumes in Equation (8), results in a maximum inbreathing, $\dot{V}_{max}$ = 2,278 m^3/h. Figure 19 graphically shows the API 2000 $\dot{V}_{max}$ in comparison to the dynamically simulated inbreathing profiles for the different types of vapors, including dry air.

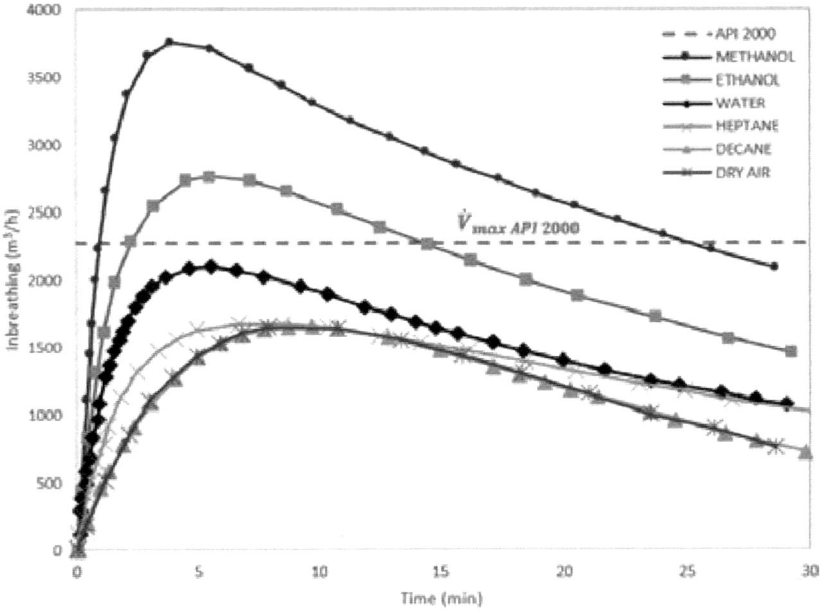

Figure 19. API 2000 Inbreathing Requirements – Different Types of Vapors

It can be seen that API 2000 method has a conservative prediction for the maximum thermal inbreathing of: (1) dry air, (2) heptane, (3) decane, and (4) water services. However, the API 2000 method underpredicts the relief requirement for the thermal inbreathing of a tank saturated with *methanol* and *ethanol*. The relieving capacity, for this case, has to be 1.7 times higher for methanol and 1.2 times higher for ethanol.

8 Conclusion

A dynamic simulation of the thermal inbreathing of low pressure storage tanks experiencing sudden change of weather conditions was performed using a SuperChems™ model. This model was validated by successfully reproducing the thermal inbreathing and temperature profiles of large scale experimental datasets and other available inbreathing models.

The thermodynamic variables specified in the API 2000 7th Edition standard have been tested for adequacy. It was found that the gas/wall free-convection heat transfer coefficient, $h_{gas} = 5W/m^2K$, specified by API 2000 was adequate when the tank was filled with non-condensable vapors like air. However, higher values were suggested for condensing vapors with a range of 10 W/m^2K to 45 W/m^2K depending on the type and volatility of the vapor. As for the rain/wall heat transfer coefficient, the API 2000 value of 5,000 W/m^2K proved to be too conservative, and even unrealistic, therefore leading to an oversized relief device. Suggested values for the rain/wall heat transfer coefficient were found to be in the range of 12 to 600 W/m^2K depending on the intensity and angle of the rainfall.

Finally, the API 2000 7th Edition method for calculating thermal inbreathing was compared with several tanks of different vapor services. The method proved to calculate an adequate relief requirement for tanks storing air, heptane, decane, and even water vapor in one case. However, the application of the API 2000 thermal inbreathing method was not conservative enough when the vapor services became more volatile, like in the case of isopropanol, ethanol, and methanol, that resulted in high risk of failure for these types of storage tanks.

9 References

[1] API Standard 2000, Venting Atmospheric and Low-Pressure Storage Tanks, 7th Edition, March 2014

[2] P. Salatino, G. Volpicelli and P. Volpe. "On the design of thermal breathing devices for liquid storage tanks", Process Safety and Environmental Protection, 77.6, pp. 354-359, 1999.

[3] D. Brooks, "Thermal Inbreathing Requirements of Low Pressure Storage Tanks at Elevated Temperatures". Lloyd's Register Energy Americas, Inc. Spring DIERS Users Group Meeting, Presentation, May 2012.

[4] R. Scholz and R. Weber, "Cooling Down Industrial Tanks". Prepared for PROTEGO Braunschweiger Flammenfilter at API 2000 Meeting Presentation, Fall 2010.

[5] H. Foerster, K. Schampel, and H. Steen, “Atmungsvorgänge infolge von Witterungseinflüssen an Lagertanks für brennbare Flüssigkeiten”, Physikalisch Technische Bundesanstalt (PTB) W-22. Braunschweig Univ. Phys.-Tech. Bundesanst, 1984.

[6] D. Fullarton, J. Evripidis and E.U. Shclünder, "*Influence of Product Vapour Condensation on Venting of Storage Tanks*", Institut für Thermische Verfahrenstechnik, Universität Karlsruhe (TH), 1986.

[7] R. Sigel, B. Kuxdorf, R. Meiss and H. Schwarz, “Praktische und theoretische Untersuchung der Beatmung von Festdachtanks”. Chemie Ingenieur Technik, 55.5, pp. 377-379, 1983.

[8] T. Holtkötter, J. Shang and H.G. Scheckerm “Behälteratmung - Experimentelle Untersuchungen und Entwicklung eines Vorhersagemodells” (Master Thesis), 1995.

[9] G.A. Melhem, R. Saini, and B. M. Goodwin, "A Modified Peng-Robinson Equation of State," *Fluid Phase Equilibria,* 47, pp. 189-237, 1991.

[10] D.D. Abou-Chakra, G.A. Melhem, and R.J. Willey, “Dynamic Modeling of Inbreathing Requirements for Low-Pressure Storage Tanks” (Master’s thesis), 2016. Retrieved from **http://hdl.handle.net/2047/D20209806**

478223 Studying the Relationship Between Inherently Safer Design and Equipment Reliability

Abstract:

Risk associated with a process depends on both likelihood and consequence of an incident. Inherently safer design is based on the use of technologies and chemicals with intrinsic properties that reduce or eliminate hazards. This philosophy is mainly focused on reducing the consequences associated with an incident. However, recent research has revealed that applying principles of inherent safety can have an effect on the likelihood element of risk as well. One of the major drawbacks of implementation of an inherently safer design (ISD) is that it can lead to risk migration, Thus, attempting risk reduction through implementation of ISD by mitigating the consequence element of risk can undesirably lead to an increase in risk by affecting the likelihood element of risk. This likelihood element of risk is the mathematical complement of system reliability. This study mainly focuses on assessing the effect of implementation of ISD principles on the reliability of a chemical process as it progresses along different stages of process design and manipulating this relationship to achieve maximum economic benefits while simultaneously minimizing risk associated with the process.

In this study, safety indices are used for quantification of inherent safety of a process. The choice of safety index depends on the stage of chemical project under consideration. Reliability of an equipment involved in a process is determined through the use of offshore and onshore reliability data (OREDA) and center for chemical process safety (CCPS) reliability databases for process equipment. The reliability of process system is then determined from the reliability of the individual equipment involved in the process. Lastly knowing values of safety indices and reliability of systems for competing designs during different stages of a chemical project (such as process selection stage, conceptual stage etc.), the required relationship between ISD and reliability is established for that particular stage.

Applying this methodology to process selection stage reveals that selecting a process that's inherently safer in terms of the involved reactions, chemicals and equipment might lead to risk migration, since the selected process might have lesser system reliability as compared to other competing process. When this methodology is applied to conceptual stage, where-in the competing processes are similar in terms of reactions and chemicals and vary only in terms of the equipment involved, it is observed that a process that has lesser number of equipment might be more reliable and inherently safer as compared to other competing processes. Lastly, when this methodology is applied for a specific equipment such as a reactor where the reliability is modeled as a function of design parameters of the reactor, it has been observed that within a certain range of design parameters, the reactor becomes sufficiently reliable and inherently safer.

Thus, this study provides a framework for designing of process systems and equipment that are reliable, economically profitable and inherently safer during different stages of project.

479377 Pressure Relief Analysis - It's about More Than Just Compliance

Abstract:

Pressure relief valves and disposal systems are the last layer of defense in petrochemical plants and refineries. Therefore, there are many codes and standards that must be followed to design those systems. Failure to ensure safe plant operations can result in fatalities, injuries, and environmental releases which can result in significant cost and poor public relations.

OSHA's CFR 1910.119 - Process Safety Management of Highly Hazardous Chemicals is designed to manage the hazards associated with facilities which handle highly hazardous chemicals. Many of the requirements in PSM are also valuable for facilities which are not governed by the requirement and application of these principals may improve safety for all facilities.

This paper will present the proposal that Process Hazard Analysis (PHA) be the heart of process for ensuring safe operations to prevent negative impact on companies' image and goodwill with the community.

480249 How to Leverage Limted Design Data Along with Historical Incident Data to Assess Inherent Process Safety During Early Stage of Design?

Abstract:

Inherent safety level quantification remains as one of the challenges for the inherent safety concept to gain industrial acceptance comprehensively. In current practice, most of the time safety assessment is conducted once the process design has been completed. At this stage of design implemented strategies look to control the hazard level. Any design modification necessary as a result of unfavorable safety performance at this stage of design, would be costly when compared to alteration during conceptual design stage [1]. It is beneficial to design industrial processes that are user friendly, inherently safer and have a level of tolerance for the deviations from ideal performance, which will eventually reduce the number of hazardous accidents. Therefore, both industrial experts and researchers are interested in safety indices that can be applied during the early design stage of a process, where the extent of knowledge regarding the process is very limited. Again, the available indices to quantify inherent safety still leave room for enhancement to represent process stream condition more accurately as a mixture rather than pure component(s) and to account for the interaction of all process parameters in determining the level of inherent safety [2].

In this work, a property based methodology to quantify inherent safety level has been proposed by assessing chemical and process unit risk at earlier stage of design. This will assist process designers in producing safer designs by the application of inherent safety principles in a more efficient and cost effective manner. It would offer process/design engineers an alternative to include safety without waiting until detailed design stage, where safety costs increase. Degrees of freedom at this stage would include alternative route selection, equipment sizing, and selection of operating conditions. The additional benefit of safety consideration in the early design stages is that it decreases the level of dependence on safety systems. The proposed safety assessment method here considers the perspective of five main safety factors associated with chemicals such as flammability, explosiveness, corrosiveness, toxicity & reactivity. These factors were linked to the process operating conditions (temperature and pressure). The process unit risk are calculated by classifying them into six main categories (i.e., pump, fired heat exchanger, pressure vessels, etc.) based on API 14C standard and identifying the undesirable events associated with each category. Major incidents database and the probability of unwanted incidents were used to link the undesirable events to operating conditions; this allows for the calculation of unit safety factors using the limited available information. Lastly, an enhancement technique to reflect the contribution of the individual components in the mixture is introduced based on the mixture properties, where the severity score was calculated for the mixture of chemicals in each unit at the actual operating conditions (whenever applicable) and multiplied by the flow rate of the mixture.

This proposed methodology of safety assessment gives a more intuitive, easy-to-understand way to represent the chemical and equipment safety factors posed by each individual unit. The steps in the methodology calculate, compare and prioritize the level of inherent safety of process streams during simulation work that influences the abovementioned safety factors in order to offer the user an insight into the risk associated with alternative design decisions. By prioritizing the streams based on the potential for chemical and equipment safety factors, the design engineers can easily identify the critical streams/units to be considered for improvement in order to avoid or minimize the risk. This methodology of quantifying inherent safety level will provide users another dimension to consider when selecting optimal design with multi objectives (e.g. techno, economic, and environmental).

The assessment of inherent safety level using the proposed method is demonstrated through a case study. HYSYS simulation is used to design two alternative designs for ammonia production.

1. Thiruvenkataswamy, P., et al., *Safety and techno-economic analysis of ethylene technologies.* Journal of Loss Prevention in the Process Industries, 2016. **39**: p. 74-84.

2. Leong, C.T. and A.M. Shariff, *Process route index (PRI) to assess level of explosiveness for inherent safety quantification.* Journal of Loss Prevention in the Process Industries, 2009. **22**(2): p. 216-221.

Inherently Safer Design from Overpressure Protection Viewpoint

Chris Ng
Sachin Kanade
Nancy Faulk
Siemens Energy, Inc.
4615 Southwest Freeway, Suite 900
Houston, TX 77027
chris.ng@siemens.com

Prepared for Presentation at
American Institute of Chemical Engineers
2017 Spring Meeting and 13th Global Congress on Process Safety
San Antonio, TX
March 26 – 29, 2017

Inherently Safer Design from Overpressure Protection Viewpoint

Chris Ng, Sachin Kanade, and Nancy Faulk
Siemens Energy, Inc.
4615 Southwest Freeway, Suite 900
Houston, TX 77027
chris.ng@siemens.com

Keywords: pressure relief, process safety, inherent safety

Abstract

Pressure relief and flare system design is mostly evolving or iterative throughout the various engineering process design stages. This iterative nature stems from the need to capture the impact of engineering design changes on this critical layer of protection, which often is designed as the last line of defense for overpressure events.

The systematic review of pressure relief and flare system design basis at various engineering design and project execution phases will not only help to ensure adequate overpressure protection, but can also highlight opportunities for inherently safer processes and project cost savings. Often, evaluating trade-offs between process safety design improvements against project economics may be challenging.

In this paper, several case studies are presented to demonstrate how inherently safer design options and cost-effective solutions could be implemented across the various engineering design and project execution phases to minimize gaps in terms of overpressure protection between early and as-built designs.

1 Introduction

Figure 1 depicts the stages of a typical engineering project lifecycle, excluding decommissioning. Each stage has associated activities in terms of selection and refinement of design details which contribute to the overall integrity of the facility and economic feasibility of the project. The general approach involving timely review of pressure relief and flare system design could be applied to both large projects involving major Capital Expenditures (CAPEX) as well as smaller projects with limited scope.

Figure 1: Typical CAPEX Project Life Cycle.

During the early phases of design such as conceptual design and Front End Engineering Design (FEED) phases, the scope, objectives, and requirements are first defined, followed by development of a process basis (i.e., heat and material balance), Process Flow Diagrams (PFDs), and preliminary Piping and Instrumentation Diagrams (P&IDs). Equipment, line and instrument lists are then prepared, and preliminary equipment sizing and specifications for process and utility systems are generated. Design options are selected, and design drawings such as equipment layout plans are completed. A budget is developed and costs are estimated.

During later phases such as Detailed Engineering and Procurement, the remaining design information is added, such as vendor data, drains and vents, and piping details. The actual piping layout is verified and updates are incorporated. In addition, electrical, instrumentation, structural and building requirements to support the project are specified. Often Hazard and Operability (HAZOP) reviews are carried out. Typically procurement of the main equipment and auxiliaries takes place followed by construction, installation, commissioning, and Pre-Startup Safety Reviews (PSSRs), in preparation for start-up, operation and maintenance.

However, what is often overlooked during the early phases is input from pressure relief and effluent handling system design engineers. This paper will describe how early involvement by these engineers could help validate relief system design and reduce overall costs and delays later. The case studies described in Section 3 below offer examples of the potential savings.

Relief systems engineering analysis should be performed during various process design and project implemented phases, which allows identification of potential gaps as well as opportunities for improvement. Typical activities include:

- Overpressure scenario analysis
- Determination of relieving loads
- Selection of relief technology
- Sizing of relief systems
- Evaluation of relief device inlet loss and backpressure effects
- Depressuring analysis
- Hydraulic evaluation of common relief headers
- Flare radiation calculations
- Dispersion analysis, including suitability of atmospheric release
- Acoustic induced vibration (AIV) analysis
- Presenting findings and developing mitigation
- Developing comprehensive documentation
- Reviewing impacts of design changes and updating documentation to capture changes are they are implemented.

It is recommended that the reviewer(s) be well-versed in the appropriate industry standards, e.g., American Petroleum Institute (API) Standards 520 Parts I and II, 521, and 2000 [1][2] [3][4]; American Society of Mechanical Engineers Boiler and Pressure Vessel Code (ASME BPVC) Sections I and VIII [5][6]; as well as corporate guidelines and site-specific requirements. In addition, it is recommended that the reviewer establish and utilize a consistent approach for review of all designs, irrespective of the Engineering, Procurement, and Construction (EPC) companies working on the project.

2 Benefits of Early Review of Relief Systems Requirements and Design

2.1 Early Involvement Allows Critical Input

Early involvement of relief system engineers can allow critical input to be incorporated which influences selected design options from an overpressure protection point of view, thus promoting an inherently safer design. By careful consideration of a number of factors, the design team has the potential to reduce overpressure relief and flaring requirements which in turn can reduce sizing of relief devices and associated piping, as well as flare components such as headers, knockout drums, flare stack, and flare tip. Potential reduction of environmental emissions is also realized. The examples below demonstrate some ways inherently safer design options and cost-effective solutions were implemented to minimize gaps in terms of overpressure protection between early and as-built design. In addition, early review can help to reduce or minimize the need for later design changes or mitigating measures, which could add significant costs and delay completion schedule for the project.

2.2 Identify Opportunities to Increase Inherent Safety

2.2.1 Centrifugal Pump Design

Consider a centrifugal pump as a pressure source for downstream equipment, as in Figure 2. As it stands, P-1 is capable of overpressuring the downstream vessel, V-1, in the event of a blocked outlet or an inlet control valve failure.

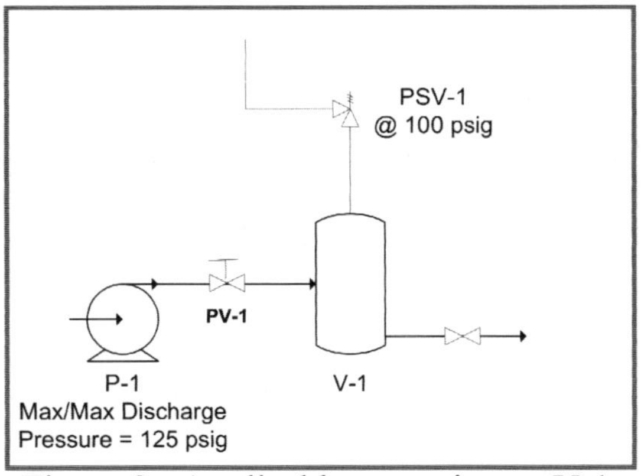

Figure 2: Applicable scenarios on V-1.

By properly sizing this pump during the design phase, however, it may be possible to minimize the relief rates for blocked outlet and control valve failure-open overpressure scenarios of the downstream equipment or even to eliminate the applicability of such scenarios altogether. Working with the process and mechanical engineers as well as rotating equipment designers, the maximum pump impeller size or speed and/ or maximum pump suction pressure could be carefully specified within process operation limits, which in turn limits the maximum pump discharge pressure. By reducing the maximum discharge pressure of the pump to a value less than the downstream equipment Maximum Allowable Working Pressure (MAWP), the above-noted scenarios may not be applicable, as in Figure 3a. An alternative to this, as shown in Figure 3b, is to ensure that the downstream equipment MAWP would be higher than the maximum discharge pressure of the upstream pump. These engineering design choices, which would result in minimizing relief loads or eliminating applicable scenarios, helps make the process inherently safer.

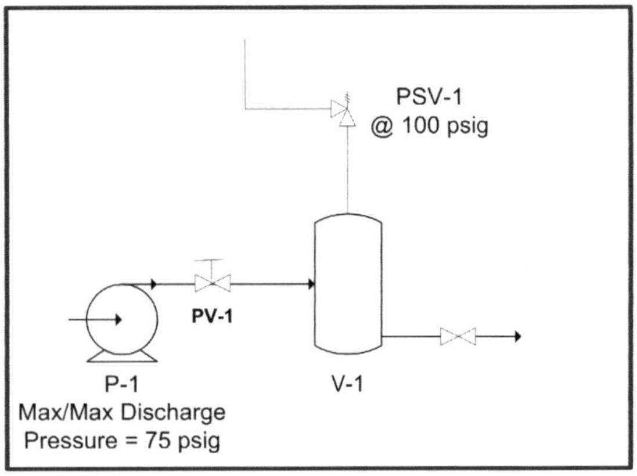

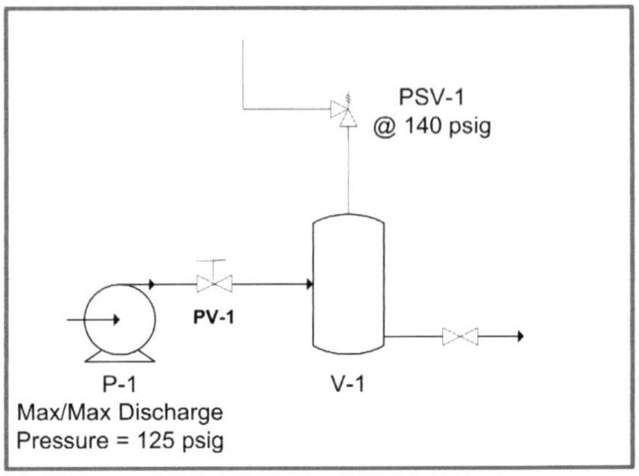

Figure 3: a) Limiting pump discharge pressure. b) Increasing vessel design pressure.

2.2.2 Pump and Compressor Discharge Check Valve Design

A related example involving pumps (and compressors, for that matter) is to ensure appropriate discharge check valve(s) are allocated to mitigate potential for reverse flow, impacting upstream equipment. In some instances, such as in Figure 4, feed could continue into an upstream vessel such as a feed surge drum if the liquid outlet pump is stopped with a discharge check valve stuck open. If this relief scenario is not identified by the relief designer, the relief valve on the feed surge drum may not provide adequate relief capacity.

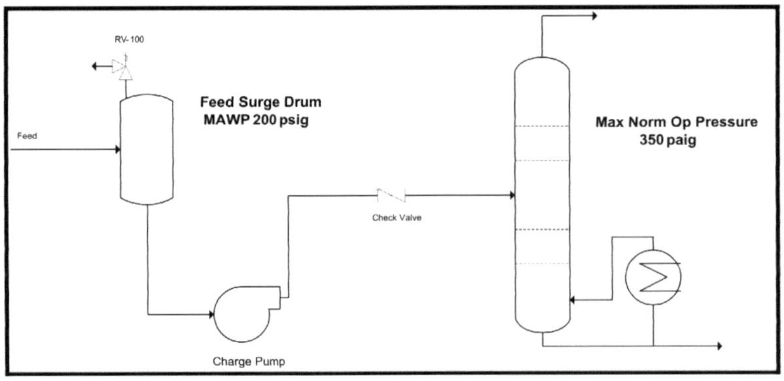

Figure 4: Applicable scenario on Feed Surge Drum

Allocating dual check valves in series of similar (or diverse) design may be required to minimize the backflow rate in the event of a check valve failure; if they meet the requirements set forth in API STD 521, the backflow rate may be limited to check valve leakage instead of check valve failure [7]. Careful specification of the MAWP of the upstream vessel may also help mitigate the overall relief requirement. Consideration of such issues early on is critical to avoid potentially costly design changes later as well as to increase the safety of the process; however, this is commonly missed due to factors such as improper management of change or poor communication between different groups designing upstream and downstream systems.

In a similar fashion, an early design review of several capital projects was performed, focusing on inherently safer design. One recommendation was proposed and accepted to raise the design pressure of the feed surge drum to mitigate the scenario in which a shutdown of a downstream charge pump results in reverse flow to the surge drum, while normal feed to the surge drum continues. By doing so, the feed and reverse flow sources would no longer be capable to overpressure the drum. Early review made the design inherently safer as the scenario was no longer applicable. Without this recommendation, installation of either a safety-instrumented system (SIS) or three 8T10 relief devices would have been necessary.

2.2.3 Control and Bypass Valve Design

Early input is critical in the selection and sizing of a control valve and its bypass valve, since the relief designer must account for the scenario in which *both* the control valve and its bypass valve fail open, as discussed in API STD 521 on "Pressure-relieving and Depressuring Systems". Early engagement of a pressure relief systems expert will help optimize the control valve/bypass and pressure relief requirements to provide adequate overpressure protection within the process operation envelope and requirements, while maintaining industry design and safety best practices. In one instance, a 2" gate valve was specified as the bypass for a 1" control valve, as shown in Figure 5. During a later design stage, it was discovered that this installation resulted in a very undersized relief device for the scenario described above, and thus the need to replace the bypass valve with a different and smaller-capacity type of valve, such as a globe valve. As in this instance, if the nominal size of the installed bypass valve is bigger than the control valve size, it necessitates careful review and possible re-design.

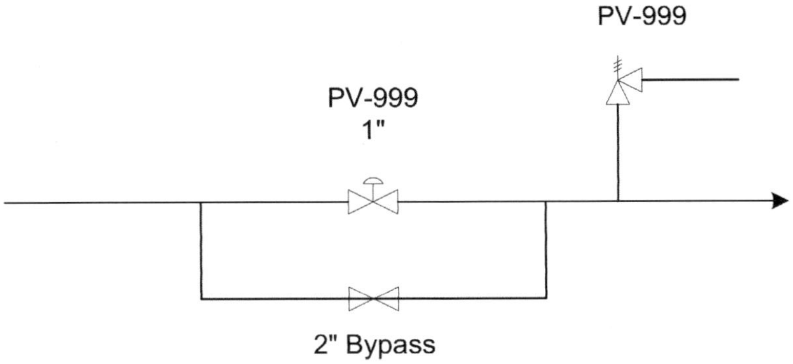

Figure 5: Bypass valve initially specified results in undersized relief device

2.2.4 Static Head

Static head is often overlooked in the preliminary design phases as equipment layouts may not yet be available. Relief designers should encourage vessel designers to specify vessels to be designed to be liquid full at the requested MAWP. Alternatively, relief device set pressures can be reduced to compensate for static head effects provided that the vessels can be filled above the top weld seam (into the top head). In another instance, the effects of static head on the set pressure of multiple relief valves had not been accounted for by an EPC company during detailed engineering. The potential for the internal pressure of affected equipment to exceed MAWP plus allowable accumulation per code was identified, allowing the relief valve set pressures to be adjusted prior to field installation.

2.2.5 Detailed Design Reviews

An independent pressure relief design audit was conducted for a large operating company. Relief design analysis and calculations had been performed by an EPC company as part of an expansion project. A review was conducted during the detailed engineering phase, during which it was found that most of the relief valve and associated piping upgrades which were being proposed by the EPC company were not required. This review included one application for which the EPC company had recommended upgrades to 6 relief valves (including recommendation to install two additional 8T10 relief valves), based on a recirculation flow rate which had been incorrectly included in a blocked outlet scenario relief rate estimate. Upon review, it was concluded that the existing relief valves were adequate and did not need to be upgraded. This reduced the potential for oversized relief capacity.

2.2.6 Design Gaps

In another example of early design review, engineers were able to identify a proposed design gap which would have resulted in costly and unnecessary modifications to a flare system. The proposed design project involved tie-ins of multiple large atmospheric relief valves to the associated flare for a process unit, resulting in inadequate flare capacity and the necessity of making expensive upgrades to this flare system. By performing detailed flare hydraulic modeling, relief engineers were able to show that re-routing the additional relief loads to an alternate adjacent flare having some spare capacity would be more practical.

2.2.7 Design of Feed Surge Drum

During a design review of a feed surge drum, additional examples of recommendations were developed to further increase the inherent safety, such as:

- By limiting the maximum discharge pressure of a feed pump to the proposed design pressure of the downstream equipment, a blocked outlet overpressure scenario was no longer credible for that equipment.
- Rerating the feed/bottom exchangers to account for static head in the event of overfilling in the associated distillation column.
- Car-sealing open (CSO) appropriate valves to ensure that various overpressure scenarios would no longer be considered as design contingencies and relief paths would be available for impacted equipment.

2.2.8 Car-Sealing Strategic Valves

In a final example, several recommendations were developed and adopted prior to start-up for a Coker Vapor Recovery Unit (VRU) revamp project. These recommendations include the following:

- Re-rating equipment
- Car-sealing closed (CSC) the control valve bypass valves
- Reducing the control valve flow coefficient (C_v) to limit maximum flow through the valve in the event of failure open
- Increasing the vessel design pressure to account for static head
- Adding dual check valves to mitigate reverse flow

These changes helped make the overall design inherently safer, so fewer safeguards were required to reduce the risks associated with overpressure scenarios for multiple systems.

2.3 Relief Systems Engineer's Role

Numerous other examples of inadequate design issues as well as missed opportunities to increase the inherent safety of the design are routinely identified by the relief systems engineers after the design has already progressed to a late stage or even after field implementation. Design review can begin early in the process, when preliminary P&IDs are issued for HAZOP or other types of Process Hazard Analysis (PHA). The relief systems engineer should also validate the design once piping isometrics are issued for approval and construction or after other design changes are proposed later. Taking these steps can help to ensure that design shortfalls can be resolved at the earliest opportunity. Making design changes early on is easier and less expensive than during later design stages and even more so than after commissioning and startup, thereby preventing subsequent shutdown, delay or major re-design work.

3 Critical Factors for Success

All pressure relief and effluent handling design documentation and supporting data should be stored in a central repository, for which access can easily be provided to all who have a need for such information. This access can be controlled, so that most viewers will have "read-only" access, and only authorized personnel can make required changes associated with Management of Change (MOC) items, thus keeping overall content up-to-date (evergreen) throughout the life of each facility as MOC activities, major expansion projects, or process modifications are implemented. These activities should trigger design review by relief systems engineers to ensure that relief design continues to be designed to provide adequate relief for all credible overpressure scenarios and that associated documentation is kept up-to-date.

In order for the review process to achieve the desired results, the three parties which are commonly involved (relief systems team, operating company, and EPC company) should work together to understand the goals of each party and how the design inputs from each party should be integrated in the various stages of engineering design and project execution. The following factors have been identified as critical for success:

- Conducting independent relief system design review at the right time in the project scope

- Communication among all parties
- Sharing of critical design data
- Effective management of change
- Discussing and resolving identified issues promptly
- Integrating relief system design review team with project design and execution teams

4 Conclusion

As shown in several examples in this paper, the systematic review of pressure relief and flare system calculations at various process design phases will not only help to ensure adequate overpressure protection, but can also highlight opportunities for inherently safer processes and promote overall project lifecycle cost savings. In addition, such a review process can easily be extended to resolve similar design issues commonly addressed during HAZOP reviews and integrated with the MOC workflow.

5 References

[1] American Petroleum Institute, *API Standard* 520, *Sizing, Selection, and Installation of Pressure-Relieving Devices, Part I-Sizing and Selection*, 9th Edition, 2014.

[2] American Petroleum Institute, *API Standard* 520, *Sizing, Selection, and Installation of Pressure-Relieving Devices, Part II-Installation*, 6th Edition, 2015.

[3] American Petroleum Institute, *API Standard 521, Pressure-relieving and Depressuring Systems,* 6th Edition, 2014.

[4] American Petroleum Institute, *API Standard 2000*, *Venting Atmospheric and Low-pressure Tanks*, 7th Edition, 2014.

[5] American Society of Mechanical Engineers, *ASME BPVC Sec I, Rules for Construction of Power Boilers*, 2015 Edition.

[6] American Society of Mechanical Engineers, *ASME BPVC Sec VIII-1, Rules for Construction of Pressure Vessels*, 2015 Edition.

[7] API STD 521, §4.4.9.3

Disclaimer:

exida

Three Reasons Major Event Risk Management Fails on Capital Projects

Denise Chastain-Knight

P.E., CFSE

exida

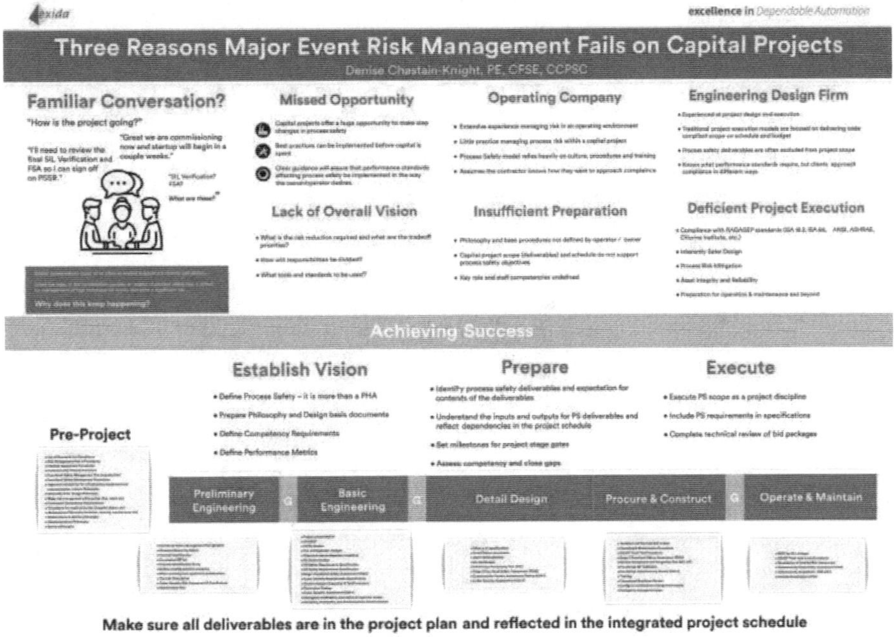

 2

About the Presenter

- **Denise Chastain-Knight, PE, CFSE, CCPSP**
  - BS ChE, Georgia Institute of Technology
  - Senior Functional Safety Engineer
  - More than 31 years experience in engineering design, process improvements, process risk analysis and safety instrumented system implementation
  - Experienced in the chemicals, resins, coatings, monomers, polymers, pharmaceutical, silicate, bio-fuels, petroleum, power...
- **Affiliations**
  - CCPS Technical Steering Committee
  - Past Chair NCEES PE (ChE) exam development committee
  - Fellow AIChE

 3

Familiar Conversation?

How is the project going?

Great we are commissioning now and startup will begin in a couple weeks

I'll need to review the final SIL Verification and FSA so I can sign off on PSSR

SIL Verification? FSA? What are those?

 4

exida

Familiar Conversation?

Similar conversations occur all to often as capital projects are nearing completion.

Often the topic of the conversation involves an aspect of process safety that is critical for management of high consequence events that pose a significant risk.

Why does this keep happening?

 5

exida

Missed Opportunity

- Capital projects offer a huge opportunity to make step changes in process safety
- Best practices can be implemented before capital is spent
- Clear guidance will assure that performance standards effecting process safety be implemented in the way the owner/operator desires.

 6

Operating Company

- Extensive experience managing risk in an operating environment
- Little practice managing process risk within a capital project
- Process Safety model relies heavily on culture, procedures and training
- Assumes the contractor knows how they want to approach complaince

Engineering Design Firm

- Experienced at project design and execution
- Traditional project execution models are focused on delivering code compliant scope on schedule and budget
- Process safety deliverables are often excluded from project scope
- Knows what performance standards require, but clients approach compliance in different ways

exida

PS Activities that Usually Happen

- Hazards Identification and Risk Analysis
- Relief system design
- Compliance with code mandated standards (ASME VIII, NFPA 30, NFPA 70)

exida

PS Activities that Often Fail

- Compliance with RAGAGEP standards (ISA 18.2, ISA 84, ANSI, ASHRAE, Chlorine Institute)
- Inherently Safer Design
- Process Risk Mitigation
- Asset Integrity and Reliability
- Preparation for operation & maintenance and beyond

exida

Why We Fail

- No or inadequate overall vision
- Insufficient preparation
- Deficient project execution

Communication

exida

Lack of Overall Vision

- What is the risk reduction required and what are the tradeoff priorities?
- How will responsibilities be divided?
- What tools and standards to be used?

exida

Insufficient Preparation

- Philosophy and base procedures not defined by operator owner
- Capital project scope (deliverables) and schedule do not support process safety objectives
- Key role and staff competencies undefined

 13

exida

Deficient Project Execution

- Underestimate of the workload required and dependencies with other scope objectives
- Process Safety /Risk Management not treated as a discipline
- Reactive not proactive

 14

How Project work

- Stage gate process
- Scope identified, schedule and budget developed to support scope
- Detail built in layers. Early definition important to minimize rework
- Often involve many contractor teams working on a subset of the scope

 15

Establish Vision

- Define Process Safety – it is more than a PHA
- Prepare Philosophy and Design basis documents
- Define Competency Requirements
- Define Performance Metrics

 16

exida

Prepare

- Identify process safety deliverables and expectation for contents of the deliverables
- Set milestones for project stage gates
- Understand the inputs and outputs for PS deliverables and reflect dependencies in the project schedule
- Assess competency and close gaps

 17

exida

Execute

- Execute PS scope as a project discipline
- Include PS requirements in specifications
- Complete technical review of bid packages

 18

exida

Pre-Project Documents

- List of Standards for Compliance
- Risk Management Plan & Procedures
- Hazards Assessment Procedures
- Cybersecurity Policy & Procedure
- Functional Safety Management Plan (organization)
- Functional Safety Management Procedures
- Approved vendors list for critical/safety equipment and instrumentation
- Alarm Philosophy
- Inherently Safer Design Philosophy
- Major risk management philosophies (fire, relief, etc)
- Contractor Competency Requirements
- Procedures for required studies (dropped object, etc)
- Maintenance Philosophy (isolation, clearing requirements etc)
- Measurement & Metrics philosophy
- Communications Philosophy
- Spares philosophy

 19

exida

Preliminary Engineering

- Functional Safety Management Plan (project)
- Chemical Reactivity Matrix
- Hazards identification
- Conceptual SIF list
- Hazards Identification Study
- Define credible accident scenarios
- Major and long lead equipment specifications
- Controls Description
- Cyber Security Risk Assessment & Gap Analysis
- Maintenance Plan

 20

Basic Engineering

- Project phase HAZOP
- CHAZOP
- HIPPS Studies
- Fire and Explosion Analysis
- Chemical release dispersion modeling
- SIL Determination
- SIS Safety Requirements Specification
- SIF Safety Requirements Specification
- Stage 1 Functional Safety Assessment (FSA1)
- Cyber Security Requirements Specification
- Countermeasure Inspection & Test Procedure
- Penetration Testing
- Cyber Security Assessment (CSA-1)
- Emergency notification, evacuation & response review
- Reliability, Availability, and Maintainability (RAM) analysis

 21

Detailed Design

- Balance of specifications
- Detail Design documents
- Alarm Rationalization
- SIL Verification
- SIS Factory Acceptance Test (FAT)
- Stage 2 Functional Safety Assessment (FSA2)
- Cybersecurity Factory Acceptance Testing (CFAT)
- Cyber Security Assessment (CSA-1)

 22

Procure and Construct

- Technical and financial bid reviews
- Operating & Maintenance Procedures
- SIS/SIF Proof Test Procedures
- Stage 3 Functional Safety Assessment (FSA3)
- SIS Site Acceptance and Integration Test (SAT, SIT)
- Pre-startup SIF Validation
- Pre-startup Cybersecurity Review (CSA 3)
- Training
- Operational Readiness Review
- Configure maintenance management system
- Emergency management plan

 23

Operate & Maintain

- MOC for ALL changes
- SIS/SIF Proof tests & results analysis
- Revalidation of Detailed Risk Assessment
- Cybersecurity Vulnerability Assessment (CVA)
- Cybersecurity Assessment - CSA 4 & 5
- Periodic Revalidation of PHA

 24

exida

exida ®

A Comparative Analysis of Safety Rating System

Linlin Wang[a]

[a]Mechanical and electrical engineering college, China University of Petroleum (East China)

Abstract: With the development of petrochemical enterprises, the scale and petrochemical enterprises frequent accidents, the high risk of petrochemical enterprises has been more and more extensive attention. It's particularly important to establish a sound safety rating system for the timely detection and elimination of risk factors, reduce the probability of accidents appears. However, in China there is less research and application on safety rating system. This paper summarizes the similarities and differences between the various systems and puts forward the development of China's safety rating system.

Keyword: Safety Rating System; Petrochemical enterprises

1 Introduction

Petrochemical enterprises have a lot of flammable and explosive, toxic substances, and the installation of large-scale and continuous production, can easily cause major accidents. In 2013 Sinopec Huangdao Oil pipeline explosion, 62 people were killed, 136 people were injured. With the occurrence of petrochemical enterprises accidents, the safety production of petrochemical enterprises has become one of the issues that government and national focus on. Evaluation and rating of the status quo of enterprise security can promote the enterprises to find their own problems, develop improvement goals and implement continuous improvement. At the same time, it can also improve the government safety supervision departments to find out the status quo of enterprises, determine the key supervision enterprises, and optimize the allocation of resources, enhance the safe production level of enterprises in China, it is also an important means to improve business efficiency. The Chinese government's security supervision department attaches great importance to the assessment of enterprise safety, since 2006, has issued a "Basic norms of enterprise safety production standards", " General specification for safety standardization of hazardous chemicals", "Guidelines for oil industry safety production standardization "etc., The government promotes the implementation of the safety production standardization assessment throughout the country, take the "enterprise self-evaluation, government review" approach to the enterprise rating, and achieved important work results. But in the implementation a lot of problems were also found. Based on the status quo of safety rating of petrochemical enterprises in China, this paper analyzes and compares the implementation process of enterprise safety rating system such as ISRS (International Safety Rating System) ,NOSA five-star comprehensive evaluation system and VPP project implemented by OSHA. It is expected to provide support for the selection and development of our country's petrochemical enterprise safety rating tool.

2 Status Quo of Safety Rating of Petrochemical Enterprises in China

Safety standardization rating is developed, implemented and reviewed by China's relevant laws and regulations, industry standards. It is led by the State Administration of Safety Supervision and implemented on a national scale. It is the most extensive and highest commitment safety rating system in China. In addition, some petrochemical enterprises received the support of government, and independently introduced tools such as ISRS and NOSA. For example, 2014 China Petroleum West Pipeline Company reached the rate 7 of ISRS, and became first enterprise which reached this level in China.

Based on the safety standardization assessment of petrochemical enterprises in China, it is found that there are three main problems in the process of implementing safety standardization assessment in our country:

(1) Safety standardization system and some other implementation systems, such as HSE system, have incompatible problems, resulting in enterprises to repeat engaging in several sets of systems, repeat records, repeat systems. Companies not only feel cumbersome, but also waste human and material resources;

(2) Enterprises have low motivation and a bad attitude to practice safety standardization. Standardization of safety production is a measure imposed by the government to improve the safety level of the enterprise, and enterprises do not really realize its importance. So that the work is easy to become formalize and surface, and did not achieve the purpose of improving the level of safety, but only to let the enterprises to meet the rating and inspection;

(3) China's safety standardization assessment consists of three grades. The grade of the score range is quite large. And the flexibility of the assessment is not sufficient. Besides, in the absence of experimental circumstances, the assessment has a low recognition for some enterprises which take a new process or a new management approach.

3 System Introduction

3.1 ISRS（International Safety Rating System）

ISRS（International Safety Rating System）is a rating system of international security measure, and improving and showing the level of health, safety, environmental and business performance and other aspects of the organization, designed to help organizations manage risk and promote continuous improvement of the evaluation system. Det Norske Veritas (DNV) international safety rating system (ISRS) eighth edition published in 2009 after decades of practicing and improving.

ISRS [7] international safety rating system has 15 elements, each element is divided into several sub-elements of a total of 119 sub-elements, and list the number of points for each sub-element of the assessment, the total number is 661. 15 elements are: leadership, planning and administration, risk assessment, human resources, compliance assurance, project management, training and skills, communication and marketing, risk control, asset management, procurement and contractor management, emergency preparedness , the event learning, risk monitoring, and review the results, in which risk management (including 3 risk assessment,9 risk control, 14 risk monitoring) system for the trunk. During the assessment process, the customer selects the subroutine according to the grade to be evaluated, as follows::

Grade	1	2	3	4	5	6	7	8	9	10
Level range	1-4				5-6		7-8		9-10	
Required number of subroutines	30				60		90		119	
The number of customer chooses	20				15		10		0	
The total number of subroutines	50				75		100		119	
The lowest score of each program%	10	15	20	25	30	35	40	50	60	70
Average minimum score%	20	30	40	60	60	70	70	80	80	90
Live tour lowest score %	65	65	65	65	70	75	80	85	90	95

3.2 NOSA integrated five-star management system

In 1951, the International Occupational Safety Association (NOSA) created a five-star safety management system by conducting a comprehensive hazardous and harmful factors identification and risk assessment to identify all possible hazard identification and develop appropriate management for reduce hazard and risk. NOSA management system regards hazard identification and risk management as the core, Heinrich's "iceberg theory" as the basis and PDCA manner as its operation mode. System consists of five main elements, 72 elements, more than 300 minimum standards, the implementation of 1200 rules and standard instructions (against NOSA Integrated Five Star system revision and reorganization, the system consists of 75 elements). System evaluation content mainly in the following two aspects:

（1） Performance evaluation

Evaluation content of NOSA five-star rating system mainly consists of five areas, particularly in the following table:

NOSA integrated five-star rating system

project	Unit content	The number of elements	score	Share of the share
1	Building and housekeeping (factory) management	11	450	15
2	Mechanical electrical and personal protection	17	600	20
3	Fire and other emergency risk	8	400	13.3

	management			
4	HSE accident recording and investigation	5	400	13.3
5	HSE management organization	31	1150	38.4
total		72	3000	100

Through five major categories of 72 elements of qualitative evaluation, the project almost involved all activities of production and operation. It evaluates each element from the "four levels, three aspects, five grades" to carry out a comprehensive calculation to measure whether the risk of the element has been effectively controlled. "Four levels" means: risk awareness, the file system, compliance, implementation effect. "Three areas" means: safety, health, environment. "Five grades" means: 0, 25, 50, 75, 100 five scoring levels。

（2） Experience assessment

Introducing DIIR (Disabling Injury Incidence Rate) rate of accidents concept of enterprise, and performance assessment combined score, drawn 1 to 5-star level. DIIR shows annual percentage of the total number of employees were injured, according to NOSA claims, statistics should include employees of the enterprise, the tertiary industry and outsourcing construction team personnel. DIIR calculated as follows：

$$\text{DIIR} = \frac{The\ number\ of\ accidents\ caused\ by\ work \times 200000}{Number\ of\ employees\ working\ hours\ per\ year}$$

Where： DIIR——Disabling injury incidence rate；

The number of accidents caused by work - refers to the work caused by or caused by the injury, so that the injured cannot or cannot fully perform the daily designated job work up to one or more working days；

200000-- base by 100 people work for a year (50 weeks), 40 hours a week calculated；

Number of employees working hours per year - Total number of employees × Actual hours per person per year

The final combination of the above two aspects of the results given the final star rating results, the corresponding assessment of the star method as shown in the following table：

NOSA integrated five - star rating system assessment standard table

Star rating	**score（%）**	**DIIR（%）**
NOSCAR	≥95	≤0.8
Five star	≥91	≤1.0
Four star	≥75	≤2.0
Three star	≥61	≤3.0
Two star	≥51	≤4.0
One star	≥40	≤5.0

NOTE: DIIR = the number of industrial accidents × 200000 (base) ÷ annual working hours of employees

3.3 VPP（The Voluntary Protection Programs）

The Voluntary Protection Programs (VPP) is a safety and health management system to prevent workplace accidents and death, by allowing enterprises to implement additional higher than the US Occupational Safety and Health Administration (OSHA) Voluntary standards measures to achieve excellent safety and health management goals. OSHA is through a detailed site assessment and comprehensive site accident rate / illness rate to assess whether the corporate site has a good protection system。

The major elements of the program: management leadership and employee involvement; worksite analysis; hazard prevention and control and safety and health training. To assess something like this, there are three main forms of evaluation methods, including the inspection of documents, on-site inspection and interviews。

Leadership and employee participation	Written safety and health management system, management commitment and leadership, Authority and Line Accountability (authority and responsibility), contractor employees, employees' participation
Workplace analysis	Baseline Hazard Analysis baseline risk analysis, hazard analysis regular activities, regular inspection, hazard reporting systems, accident / incident investigation, trend analysis
Hazard Prevention and Control	Whether there are effective measures to combat control and control, as well as written contingency plans safety and health training

In addition, team members also need to The Total Case Incidence Rate (TCIR)以及The Days Away from Work, Restricted Activity and Job Transfer (DART) Case Incidence Rate evaluated to confirm the company level must be below the national average. After a series of reviews, OSHA decides that the applicant's qualifications are Star, Merit, Demonstration program, if the applicant fails to meet the requirements, return the application, and VPP level is relatively flexible, for those who do not meet the Star designing 1-Year Conditional Star Participants and Star Demonstration Programs.

3.4 Safety standardization assessment

Safety production standardization refers to the establishment of safety production responsibility system, the development of safety management systems and operating procedures, troubleshooting and monitoring of major hazards, the establishment of preventive mechanisms to regulate production behavior, so that the production links in line with relevant safety laws and regulations and standards Standard requirements, people, machines, objects, ring in a good state of production, and continuous improvement, and constantly strengthen the standardization of enterprise safety construction.

Safety standardization evaluation mainly taken independent assessment, external review of the way, self-assessed score of 60 points or more companies can apply for formal standardization rating, the final assessment is divided into one, two, three, one is the highest, from the security Production supervision and management departments to review the classification of supervision and management. During the review process, the following aspects are mainly reviewed: objectives, organization and responsibilities, safety production, laws and regulations and safety management system, education and training, production equipment and facilities, operation safety, hidden trouble investigation and management, major hazard monitoring , Occupational health, emergency rescue, accident learning, performance assessment and continuous improvement of the thirteen major aspects of compliance assessment, the final score to score how much. The enterprises that need to be graded need to carry out self-evaluation on a yearly basis to determine the safety standardization status of the enterprises and to re-order them every three years to urge enterprises to continuously improve the standardization work level.

4 System comparison results

Through the understanding of the information, compared the above systems, found the following the same and different:

1. Purpose

In essence, the above mentioned rating methods are all aimed at checking and enhancing the enterprise to achieve a certain level of security. However, the basic purpose of the rating system is different because of the different backgrounds of the development period. For example, the NOSA five-star rating system was developed in the 20th century, the rapid development of industrial and mining in South Africa and the quality of personnel and other issues caused the high rate of injury accidents, so its basic purpose is to reduce the high incidence of work accidents. In the final analysis, the rating is only a means and security is the purpose.

2. Applicable scope

ISRS is available in 11 languages and has been implemented on thousands of sites worldwide. The industries involved include marine, oil and gas, energy, nuclear, etc., and ISRS is introduced into the oil industry in China, such as China Petroleum West Pipeline Company reached the international security rating of 7.

NOSA five-star rating also can evaluate several kinds of industries including mining, forestry, railways, health care and natural gas, but its vast application in China is applied to the mining industry.

The VPP assessment program is applicable to all companies that use OSHA standards. VPP's participation companies include construction, agriculture, marine, oil and gas, health care and other industries.

Since the implementation of the Safety Production Standardization, under the guidance of the government, Safety Production Standardization has been applied in most enterprises in China, including industry and trade, power industry, transportation, hazardous industries, non-coal mines, construction industry, fireworks, coal industry, Its application covers most of the production enterprises.

3. Evaluators

Safety Production Standardization rating system take the way of "enterprise independent assessment then external assessment " to carry out the rating, Safety Supervision and Management Departments supervises the classification. The external review is conducted by the relevant experts. Similarly, the NOSA five-star rating system is evaluated by the company's team of experts. In the composition of the assessment team, VPP projects and ISRS systems have made innovations.

The VPP site assessment team is composed of experts from all areas of OSHA and SGE (The Special Government Employee), which has performed on-site evaluation. The implementation of the SGE project is intended for enterprises employees to participate in the process of VPP's on-site assessment. The implementation of this innovative project is more conducive to government and corporate information and ideas sharing.

ISRS needs to form an evaluation team consisting of two or more persons prior to evaluation, and to ensure the depth of the HSEQ field in evaluation, the team members contain relevant experts and do not restrict internal / external members, and All evaluation members will receive DNV training courses.

4. Rating method

Whether it is ISRS system, NOSA, VPP or safety standardization rating system are all required to compare the relevant standards and scoring rules with the actual situation of enterprises, through the document inspection, on-site visits and consultation to decide the company's compliance to standards.

However, NOSA and VPP projects need to combine other indicators before the final results. VPP not only requires companies to meet or exceed OSHA standards, but also requires companies to maintain below the average rate of disease released by National Bureau of Labor Statistics. And NOSA combines the evaluation rate of qualified with DIIR ratio to determine the stars should be given.

5. Follow up later

The "PDCA" model has been adequately applied in several rating systems, "plan, implement, check, action", adopt the infinite loop method of " find problems then solve problems, and then find the problem, and then solve the problem" to promote the continuous improvement of enterprise safety management system and continuously improve the level of enterprise security.

But for the rating implementation unit, the follow-up degree is different from others. Among them, the VPP project's follow-up work is very good. VPP participants need to submit an annual self-assessment report, OSHA's site assessment team will confirm whether the enterprise has made a corresponding improvement, and re-examination time interval varies by the enterprise level, such as star business, every 3-5 years conduct a reassessment. For companies with a one-year correction, if they do not meet the requirements after one year, they will withdraw the VPP application.

System comparison table

	ISRS	**NOSA**	**VPP**	**Safety Standardization**
Basic Purpose	get management	Reduce the	prevent	Promote

	system to work more effectively	incidence of industrial accidents	workplace accidents and death	enterprises to achieve standardization and standardization
Applicable scope	ISRS is available in 11 languages. The industries involved include marine, oil and gas, energy, nuclear, etc.,	Including mining, forestry, railways, health care and natural gas, but its vast application in China is applied to the mining industry	all companies that use OSHA standards	Covers most of the production enterprises, such as industry and trade, power industry, transportation, hazardous industries, non-coal mines, construction industry, fireworks, coal industry
Evaluators	Two or more people (DNV training)	Company experts team	OSHA experts in various fields and SGEs who have conducted on-site evaluations	Independent assessment, external review The external review is conducted by the relevant experts
Rating method	Performance evaluation	Performance evaluation and DIIR	companies to meet or exceed OSHA standards, low disease rate	Performance evaluation
Assessment Frequency		Self-evaluati on Twice a year Re-rating depend on clients	Self-evaluation once a year re-examination time interval varies by the enterprise level	Self-evaluation once a year Every three years have re-examination
Grades	10	6	Have Merit、Star、Demonstratio n program etc., grade flexibility is high	3
Voluntary	voluntary	voluntary	voluntary	Forced

5 Conclusion

Combine comparison and thinking above, for the Chinese petrochemical enterprise safety rating system selection, provide the following suggestions:

1. For small private enterprises, Safety Standardization has been fully compatible with security requirements, for other state-owned enterprises, large-scale Sino-foreign joint ventures can choose to introduce foreign security rating system.
2. ISRS system covers the ISO standard and OSHA standard requirements, and also developed a special version of ISRS-Pipe, the rating system also covers the majority of indicators in China oil safety management system, ISRS has been applied in a lot of petrochemical enterprises in China But the ISRS system uses the lowest score of the total score in the first-level elements as the criterion for determining the final grade, so it is not easy to reflect the advantages of the enterprise in the final level judgment.
3. NOSA five-star rating system considers the performance evaluation and DIIR value together in the final evaluation to get the overall level, but when the enterprise finally decided to use the NOSA system, NOSA was originally used for mining assessment should be considered; For the company implementing OSHA standards can apply for the VPP program because VPP determines the enterprise level by checking whether the company is performing a higher-than-OSHA standard.

Acknowledgement

This work is supported by the National Natural Science Foundation of China (21406271) and the China Scholarship Council "Innovative Talents International Cooperation Program" (Liu Jin Mei [2016] 7988). Thank you very much.

Reference

1. https://www.dnvgl.com/services/isrs-for-the-health-of-your-business-2458
2. https://www.nosa.co.za/

478942 The Difficulty and Prospect of the Development of Quantitative Assessment Technology in the Field of Chemical Safety in China

Abstract:

Abstract: With the increasingly high demand for safety of chemical industry quantitative assessment technology has been widely recognized in chemical industry safety field because of its more accurate assessment results. Nowadays many existing quantitative assessment techniques,(such as quantitative risk assessment (QRA), reliability assessment, safety integrity level assessment (SIL) .etc)have been gradually introduced into the filed of China chemical safety to provide more assistance for the risk management of chinese chemical enterprises. However, some basic data of the quantitative assessment on chemical industry in China are lost due to the lack of systematic statistical analysis of data(such as equipment failure data, accident event data .etc), which leads to the situation that our country often adopt foreign related data when conducting a quantitative assessment. In result, results of quantitative assessment are out of their authenticity. In addition, because there is no more specific and more operational assessment guidelines for the part of quantitative assessment techniques issued in our country, it also affects the accuracy and scientific of the results. According to analyzing some misunderstandings and difficulties in chinese chemical safety field when carrying out the quantitative assessment technology, the paper puts forward some suggestions on the change and improvement of related industries to make quantitative assessment technology to provide greater help for the development of chemical safety field in China.

PROCESS
SAFETY MATTERS

Recipe for a *Sizzling* HAZOP

Poster Presentation to 13th GCPS

Angus Keddie, Process Safety Matters

PROCESS
SAFETY MATTERS

Benefits of Running an Effective HAZOP

- Commercial: Meeting schedule; Risk optimisation of facility
- Safety: Pick up all significant potential hazards
- Psychological: Limiting stress on participants
- Chairman: High probability that the chairman will be invited back

Recipe Success? – Key Components

- Recipe Disasters……
- Provenance
- Preparation
- Supporting Team
- Kitchen Knives
- Ingredients
- Spice and Flavour
- Oven Conditions
- ….and How to Avoid Them
- Ta Dah!

Recipe Disasters

Recipe Disasters

- Behaviour to avoid if Effective HAZOP to be achieved:
  - Falling asleep
  - Fighting with team members
  - Allowing fighting between team members
  - Silence of participants
  - Permitting interminable discussions on minor issues
  - Scribing yourself
  - Working well in excess of 7 hours per day

HAZOP Chair Provenance

To paraphrase the Bard "Some or born to HAZOPness, some achieve HAZOPness and some have HAZOPness thrust upon them"

There are around 5000 HAZOP chairmen worldwide all of them being equal, but some are more equal than others.

In order to optimise process safety for existing and new facilities worldwide, excellent PHSERs need to be carried out. A key ingredient of this is the performance of the facilitator. What personal characteristics combine to create this performance?

HAZOP Chair Provenance

- Undergraduate degree in Chemical Engineering
- Professional exposure to relevant aspects of industry:
  - Process Design
  - Operations/Commissioning
  - Team Leadership/Management
  - Process Safety
- Top up with Masters in Process Safety
- Modal age range: 40 to 60
- CPD of IChemE Training Course
- Observation of Other respected HAZOP chairs to evolve best practice
- But like all great recipes, a little variety doesn't do any harm and may enhance the final product.

HAZOP Chair Characteristics

Desirable attributes for the leader include:

- wide experience of all the stages of process hazard studies, including quantitative risk assessment;
- extensive experience as a HAZOP study member and, preferably, some as a HAZOP study scribe;
- training in the leadership of HAZOP studies;
- technical competence and the ability to quickly understand the system and its operation;
- meticulous attention to relevant detail;
- good analytical thinking;
- motivational skills including the encouragement of creativity and open speaking;
- independence from the project itself with no other direct responsibility to the project manager other than completing the scope of the HAZOP study.

Fail to Prepare……

- Hold a Pre-meeting with key project stakeholders to establish significant draft tenets:
  - Project Background and Scope
  - Stakeholder Drivers
  - Optimal Methodology
  - Key Hazards
  - Putative Assumptions
  - Principal Personalities
- Outcomes and Terms of Reference inform Intro presentation and initial approach

The Scribe - Sous chef or trusty side kick?

LinkedIn HAZOP Discussion:

- **Does a PHA scribe need to have a technical background? – Ron Brachten, Nov 2014**
- **AK contribution: scribal attributes -**
  - Fast and accurate typing skills
  - Native or excellent English language skills
  - Sufficient *confidence* to capture the consensus and to overlook the inevitable spelling and grammar gremlins
  - *Stamina*
  - Good humour
  - Interest in the process
  - Threshold technical knowledge

  Technical knowledge is handy but....

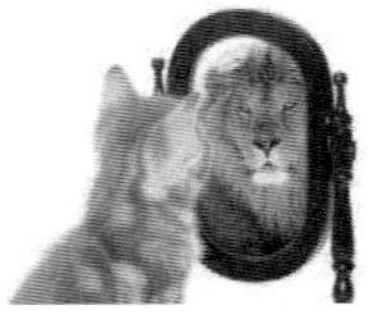

HAZOP Tools – Sharpen those knives

- Climate controlled room, good natural light
- A1 PIDs on walls or second screen
- Availability of Flip charts or White boards
- Wifi/LAN access to upload project tech info and wider references
- Good food during breaks and lunch
- Effective HAZOP recording software

Ingredients

- Meat and 2 Veg:
  - Process Description
  - Noded P&IDs
  - Intro Presentatiion
- Side Dishes
  - PFDs
  - Cause and Effect Diagram
  - Report from any previous HAZOP(s), HAZID
  - HAZOP Terms of Reference
  - Facility Design Standards
  - Plot Plan
  - Line List
  - Piping Classification Overview
  - Toxicity/flammability data for materials being handled
  - Relevant National/International Codes/Laws/Standards

The texture and flavour – the local element!

The experience of HAZOP can be vary significantly depending on technical, cultural and stakeholder perspectives:

Stakeholders

- Operators focus on operability and consistency of design. **(gooey and sweet)**
- The EPC contractor's focus is on schedule and no changes (hard and dry)
- The operator's safety team focus on strict code compliance **(tart and bitter)**

Technical

- Upstream – no steady state; little standardisation
- Downstream – Large through puts/inventories; replication

Cultural

- "Many nations separated by a common language – HAZOP!"

Oven Conditions – HAZOP Ambience

The team works better if relaxed and motivated – try to make this happen early on:

- Introduce yourself to team members
- Get everyone to introduce themselves round the table, including what contribution they expect to make
- Directly engage all team members in the first 30 mins
- Allocate responsibilities based on member skills and attributes
- If team is flagging, call a break
- It two team members become adversarial, take a time out
- Smile

PROCESS SAFETY MATTERS

Recipe Disasters – and their remedies

- Behaviour to avoid if Effective HAZOP to be achieved:
  - Falling asleep – stand up and walk around when facilitating
  - Fighting with team members – develop excellent rapport with team from the off;
  - Allowing fighting between team members – call time out; verbal warnings; firing culprit(s)
  - Silence of participants – engage in meetings (and in breaks).
  - Permitting interminable discussions on minor issues – set time limits for discussion; "is there anything we need to record at this point?"
  - Self Scribing – refuse!
  - Working well in excess of 7 hours per day – planning; negotiate schedule; identify key risks up front; prioritise nodes (Shell approach – 6 working hours per day, 4 days per week, 4 weeks out of 5)

PROCESS SAFETY MATTERS

PROCESS SAFETY MATTERS

Ta Dah!

Angus Keddie

Director

Process Safety Matters

+44 (0) 7515897869

+44 (0) 1344 455800

info@processsafetymatters.com

477894 Bridging the Divide - OHS and Process Safety

Abstract:

The occupational health and safety field is well established, with a number of professional bodies, undergraduate programs and a comprehensive body of knowledge. Conversely, the process safety field has evolved largely over the past 40 years as an engineering discipline, with a focus on engineering solutions. Both of these fields utilise a risk management basis for decisions and actions. But both fields still largely operate independently of each other. There is much to be learned from sharing knowledge and responsible collaboration to deliver safer plants and operations from an OHS and a process safety perspective. This paper will explore how we can bridge the divide between OHS and process safety to achieve better safety outcomes. This includes better education and information flow between both professions as well as leveraging off existing systems to advance process safety.

20 Years Of Personal Process Safety Lessons

Steve Streblow
ACS Engineering
10914 Milano Ct.
Richmond, TX 77406
streblowsr@yahoo.com

Prepared for Presentation at
American Institute of Chemical Engineers
2017 Spring Meeting and 13th Global Congress on Process Safety
San Antonio, TX
March 26 – 29, 2017

20 Years of Personal Process Safety Lessons

Steve Streblow
ACS Engineering
10914 Milano Ct.
Richmond, TX 77406
streblowsr@yahoo.com

Keywords:

Amine tower; carbon dioxide; communication and restriction; competency; compliance; culture; documentation; failure analysis; intimidation; investigation; lessons learned; material of construction; mechanical integrity; pressure; stress; risk assessment; vessel repair.

Abstract

This paper presents in detail one of the many process safety issues encountered by the author in his 20 years of industrial experience. This issue deals with an amine tower, but touches upon the following topics: carbon dioxide; communication and restriction; competency; compliance; culture; documentation; failure analysis; intimidation; investigation; lessons learned; material of construction; mechanical integrity; pressure; stress, risk assessment and vessel repair.

It describes better ways to address the situation and applicable RAGAGEP. It is hoped that sharing this information will enable others to learn from the past and prevent similar issues in the future.

1 Introduction

This paper presents in detail one of the many process safety issues encountered by the author in his 20 years of industrial experience. This issue deals with an amine tower, but touches upon the following topics: carbon dioxide; communication and restriction; competency; compliance; culture; documentation; drawings; failure analysis; intimidation; investigation; lessons learned; material of construction; mechanical integrity; pressure; stress, risk assessment and vessel repair.

It describes better ways to address the situation and applicable RAGAGEP. It is hoped that sharing this information will enable others to learn from the past and prevent similar issues in the future.

2 Amine Tower Leak

In one of the operating companies I worked for, I was supposed to be the technical support for many gas processing facilities, which included amine systems for removal of carbon dioxide. I was verbally informed of a small leak in one of the facilities amine regenerator towers which was stopped by placing a screw in the side of the tower, as shown in the photo (taken later) immediately below.

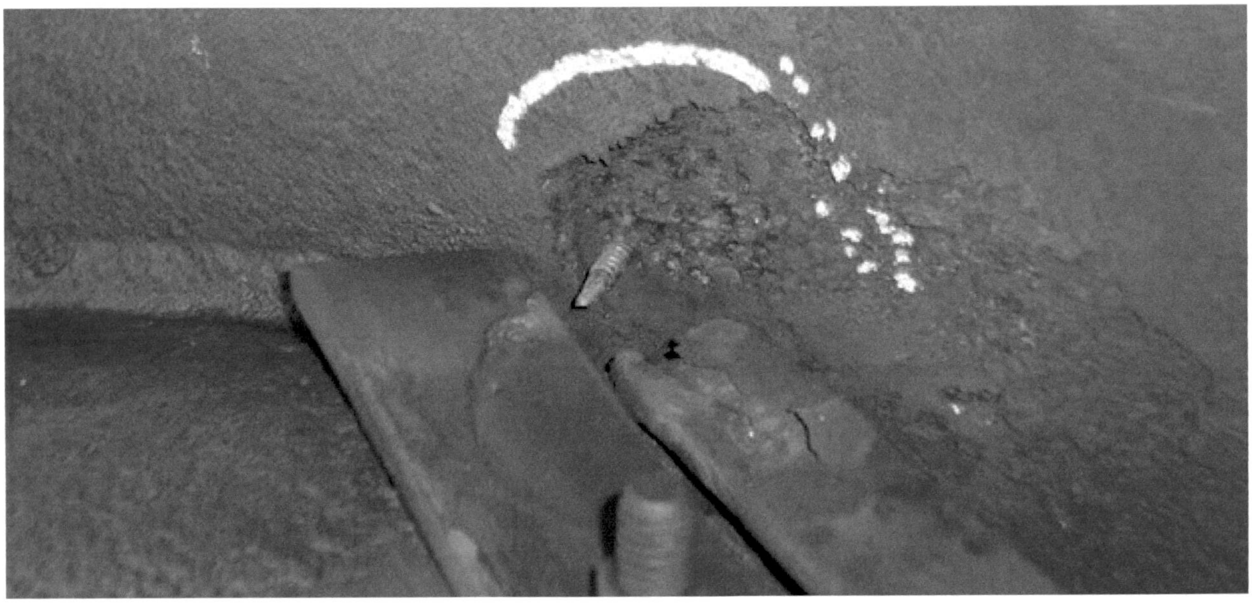

I was shocked for several reasons. First, I was not consulted for my technical opinion on how to proceed. Second, because after I communicated with the plant, they stated they were told by a service provider that they normally use that this type of repair was proper.

So much of this situation seemed wrong to me, so I provided the plant suggestions to restrict access to the area nearby the tower, while I determined how to proceed. The plant did put this area restriction in place.

I started to research what to do. First, I verified the pressure vessel law, as one can do by checking NB-370 at:
https://www.nationalboard.org/PrintAllSynopsis.aspx?Jurisdiction=Select
The location was in the state of Texas, which did not have a pressure vessel law. I next tried to find RAGAGEP for this. I called some pressure vessel manufacturers and had some brief conversations, which helped me focus my search. I found out that to keep the ASME code status, the repair should have been performed through a company which has an R stamp and filled out the appropriate paperwork. The R stamp status of a company can be confirmed at:
http://www.nationalboard.org/ManufacturerDirectory.aspx
The RAGAGEP to be used to maintain ASME code compliance is NBBI NB 23 (NBIC), National Board Inspection Code, Parts 1-3, reference 1.

If the vessel is a non-code vessel or if ASME code status does not need to be maintained, then one can refer to RAGAGEP API 510, Pressure Vessel Inspection Code: In-service Inspection, Rating, Repair, and Alteration, Tenth Edition, May 2014, reference 2.

After seeing that there was much technical knowledge needed for this activity that I did not have, I checked with management to see if there was someone better qualified than I inside the company, and I was disappointed to learn that there was not. Management did not seem concerned about the lack of expertise or the fact that I was being asked to work outside my area of expertise. It would be good to have a dedicated mechanical integrity expert in the company, since there is a great breadth and depth of technical knowledge needed.

I then tried to get the results of the previous mechanical integrity inspections for the tower, which I was told existed, but I was not provided. This withholding of information without any explanation seemed suspicious to me. This is a weakness in the culture. So, I tried to determine what the inspections should be for this item.

I tried to get the paperwork on the tower screw “repair”, but none was provided, again showing a cultural problem.

I informed the plant that the company they were using did not have an R stamp and the screw action was not a proper repair.

I suggested contacting an inspection company to perform some external vessel wall thickness inspections, to see if additional thin spots were present. The external inspection was performed, which showed some additional thin areas.

I was verbally informed of a second small leak in the subject amine regenerator tower which was stopped by placing a screw in the side of the tower, as shown in the photo (taken later) immediately below.

It was disappointing that the facility took the same actions as last time, when they had been advised otherwise. I notified my management, and they did not seem concerned. I again reminded the plant that the company they were using did not have an R stamp and the screw action was not a proper repair.

I was verbally informed of a third small leak in the subject amine regenerator tower which was stopped by placing a screw in the side of the tower, as shown in the photo (taken later) immediately below.

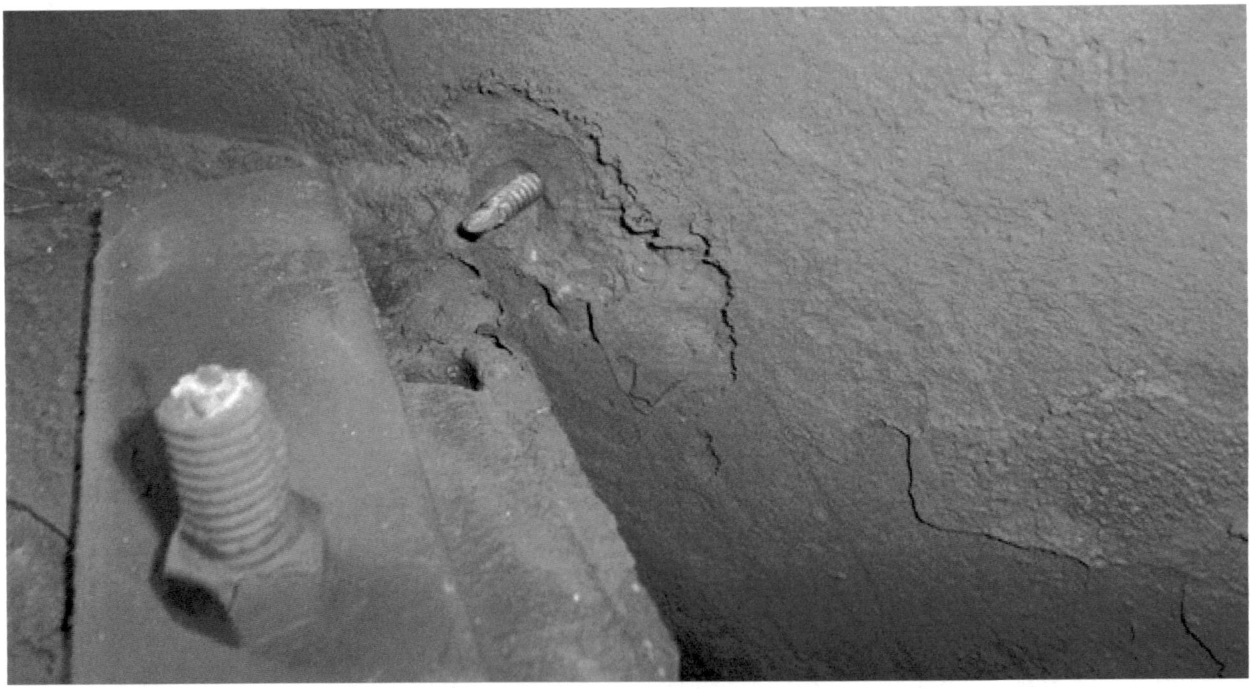

It was disappointing that the facility took the same actions as the last two times, when they had been advised otherwise. I again reminded the plant that the company they were using did not have an R stamp and the screw action was not a proper repair.

This time when I notified my management, they may have over reacted, since they now changed their direction drastically in stating that the right thing to do was shut down the facility, again without trying to get input from me or the plant manager. I was concerned about this drastic change in direction. When I spoke with management about this, they said this was a great success, since the facility was shut down in response to the latest leak. I informed management that I did not view it as a success to take this action in response to the third instance, but would have viewed it more positively if more action was taken on the first issue. A consequence analysis could have been conducted to determine the extent of the consequences. Then the likelihood could have been determined. Combining the consequence and the likelihood results in the risk, which could be used to help make a quantitative risk based determination, rather than a subjective decision.

I then asked management if we could share the learnings from this one site with the other 30+ sites in the company, and they said no, they would not do so until the incident happened at 3 or more facilities. It greatly disturbed me that management would wait for more issues to happen rather than sharing knowledge to prevent future incidents. This shows another cultural weakness.

Since the facility was now shutdown, it presented an opportunity to do some internal inspection, which was done. This effort resulted in many pictures, including the ones in this paper, as well as many wall thickness readings. The wall thickness readings were as thin as 0.110".

The original vessel wall thickness was supposed to be 0.375", with 0.125" corrosion allowance, which shows that the end of life wall thickness should be 0.250", well above the 0.110" measured. The vessel had a 50 psig design pressure, was 78" ID, with an 18 Ft skirt, 64 feet shell height and was made from SA-516-70.

I put together a spreadsheet, based on ASME Boiler and Pressure Vessel Code, Section VIII, Division I, 1980, reference 9, to evaluate what wall thickness was needed to handle different scenarios. Joint efficiencies were taken from reference 9, UW-12. Allowable stresses and overstress for combined loads were from reference 9, UG-23.

Seismic and wind speed were considered by calculating the moment at the bottom head to shell weld, and then converting to metal thickness.

I got the original calculations for the tower, and found that the original seismic calculations were in error. Seismic tension resulted in a required wall thickness of 0.048", which is less than the 0.110" measured wall thickness, so the wall thickness was acceptable. Seismic compression resulted in a required wall thickness of 0.113", which is just above the 0.110" measured, meaning the wall is not thick enough. A useful reference for seismic design is Guidelines for Seismic Evaluation and Design of Petrochemical Facilities, Second Edition, ASCE, Reston, Virginia, 2011, reference 10.

For the wind loading, the tower was divided into 4 different zones. Wind tension resulted in a required wall thickness of 0.072", which is less than the 0.110" measured wall thickness, so the wall thickness is acceptable. Seismic compression resulted in a required wall thickness of 0.169", which is above the 0.110 measured, meaning the wall is not thick enough. A useful reference for wind loading is ASCE 7-05, Chapter 6, Figure 1A Basic Wind Speed – Western Gulf of Mexico Hurricane Coastline, reference 11. A second useful reference is Wind Loads for Petrochemical Facilities and Other Industrial Facilities, ASCE, Reston, Virginia, 2011, reference 12.

Reference 9, UG-27 was used to determine the minimum thickness for internal pressure and liquid height. The thickness needed for circumferential stress is 0.176", which is above the 0.110 measured, meaning the wall is not thick enough. The thickness needed for longitudinal stress is 0.052", which is less than the 0.110" measured wall thickness, so the wall thickness is acceptable. The thickness needed for compressive stress is 0.085", which is less than the 0.110" measured wall thickness, so the wall thickness is acceptable.

For the combined load temperature, pressure, level, wind and seismic, the thickness needed for longitudinal stress is 0.144", which is above the 0.110 measured, meaning the wall is not thick enough. The thickness needed for compressive stress is 0.306", which is above the 0.110 measured, meaning the wall is not thick enough.

Management tried to direct me to just solve the issue by reducing design pressure until the wall was thick enough. However, this was not a reasonable path forward, for several reasons, such as that if the same corrosion allowance was used, the remaining life (0.110" measurement minus 0.125" corrosion allowance) was 0. Also, there was no mention of the other aspects that would need to be considered, such as increased relief valve sizing at lower pressure. Also, the wall was not thick enough for several other cases listed above, such as seismic and wind, which are independent of the design pressure. Reducing the design pressure from 50 to 15 psig, meant that the wall thickness was still not adequate for the combined loading.

I also did some process simulation to determine if there could be changes made to the process and/or equipment to reduce the corrosion, such as:

- Changing type of amine
- Changing flowrate of amine
- Changing tower diameter
- Changing tower materials of construction
- Changing heat input to the reboiler

I was told to stop on this effort, which meant no better design would be in place when the next tower has problems. Production again won out over process safety when I was told to get a new tower the same as the last one.

A fitness for service assessment could have been conducted, however, production again won out over process safety when I was told to get a new tower the same as the last one.

A risk based mechanical integrity program could have been developed and put into place, but it wasn't. The emphasis was on production over process safety.

A root cause analysis could have been conducted, but the emphasis was on production and appearance over process safety. This is demonstrated by the operations direction to rapidly take the tower offsite to keep the plant appearance attractive. There was not an effort to engage in dialogue to discuss the issue. A better approach is to engage in dialogue with employees and seek their input and comments on suggestions. When management eventually realized the value of the investigation process, the tower was already destroyed. This was unfortunate, because the same type of failures starting appearing at other sites.

I told management that they needed to institute a mechanical integrity group. They eventually did, however, when I was informed of the head of the mechanical integrity group and the recent hire for our area, I could see that they were not qualified. I informed management of this, and I was told that was not my concern. This again shows a poor process safety culture. No attempt was made to resolve my valid concern, just to dismiss it.

I finally put many items in writing to my management, and was provided an individual visit by HR from headquarters. In this closed-door meeting, I was told many unacceptable things, including that I should not write about process safety, but management is glad to discuss orally. This shows a poor process safety culture. I resorted to writing when it was clear that oral was not effective. I did not agree with HR. I was then sent to an individual training session with the head of compliance, which was titled "Write Now, Pay Later." It is clear that the company tried to restrict internal written communication on safety and health, when it is an employee right to write written complaints to external agency, OSHA. This does not show a safety culture.

3 Additional RAGAGEP

One good resource is Guidelines for Mechanical Integrity Systems, AIChE CCPS, 2006, Wiley, reference 3. For pressure vessels, a useful item is "Table 9-1 RAGAGEPs for Pressure Vessels".

Another useful item in reference 3 is "Table 9-13 Mechanical Integrity Activities for Pressure Vessels". It has example activities and typical frequencies for:

- New Equipment Design, Fabrication, and Installation
- Inspection and Testing
- Preventive Maintenance
- Repair

Reference 3 also has Technical Basis for Activity and Frequency, Sources of Acceptance Criteria, Typical Failures of Interest, Personnel Qualifications, Procedure Requirements, Documentation Requirements for the following categories:

- New Equipment Design, Fabrication, and Installation
- Inspection and Testing
- Preventive Maintenance
- Repair

Reference 3, Chapter 11 contains an introduction to a risk based approach for mechanical integrity.

Reference 3, Section "12.4 Equipment Failure and Root Cause Analyses" provides useful information such as:

- "Step 1 - Evaluate Conditions at the Site."
- "Step 2 - Perform Preliminary Component Assessment."
- "Step 3 - Preserve "Fragile" Data Sources."
- "Step 4 - Perform Macroscopic Examination."
- "Step 5 - Perform Microscopic Examination (if warranted)."
- "Step 6 - Perform More Detailed Testing (if warranted)."
- "Step 7 - Test Under Simulated Conditions (if warranted)."
- "Step 8 - Make the Failure Mechanism Determination and Develop Corrective Actions."

Listed below are several additional API documents that contain excellent information related to mechanical integrity, with specific applicable sections listed.

API Recommended Practice 571, Damage Mechanisms Affecting Fixed Equipment in the Refining Industry, First Edition, December 2003, is reference 4. See sections:

- "4.2.14 Erosion / Erosion-Corrosion"
- "4.3.6 CO2 Corrosion"
- "5.1.1.1 Amine Corrosion"
- "5.1.2.2 Amine Stress Corrosion Cracking"

API Recommended Practice 572, Inspection of Pressure Vessels (Towers, Drums, Reactors, Heat Exchangers, and Condensers), Second Edition, February 2001, is reference 5. See sections:

- 7 for reasons for inspection
- 8 for causes of deterioration
- 9 for frequency and time of inspection
- 10 for inspection methods and limitations
- 11 for methods of repair

API Recommended Practice 579, Fitness for Service, First Edition, March 2000 is reference 6. Reference 6 states "The ASME and API design codes and standards for pressurized equipment provide rules for the design, fabrication, inspection and testing of new pressure vessels, piping systems, and storage tanks. These codes do not address the fact that equipment degrades while in-service and that deficiencies due to degradation or from original fabrication may be found during subsequent inspections. Fitness-For-Service (FFS) assessments are quantitative engineering evaluations which are performed to demonstrate the structural integrity of an in-service component containing a flaw or damage. This Recommended Practice provides guidance for conducting FFS assessments using methodologies specifically prepared for equipment in the refining and petrochemical industry. The guidelines provided in this recommended practice can be used to make run-repair-replace decisions to help ensure that pressurized equipment containing flaws which have been identified by inspection can continue to operate safely."

API Recommended Practice 580, Risk-based Inspection, First Edition, May 2002 is reference 7. Reference 7 states "1.1 PURPOSE -The purpose of this document is to provide users with the basic elements for developing and implementing a risk-based inspection (RBI) program. The methodology is presented in a step-by-step manner to the maximum extent practicable."

API Recommended Practice 581, Risk Based Inspection Technology, Second Edition, September 2008 is reference 8. In reference 8, "PART 1- INSPECTION PLANNING USING API RBI TECHNOLOGY", see sections:

- 4 for RBI Concepts
- 5 for Pressure Vessels

In reference 8, "PART 2 - DETERMINATION OF PROBABILITY OF FAILURE IN AN API RBI ASSESSMENT", see section 8 for "Damage Factor – Amine Cracking".

Reference 8, "PART 3 - CONSEQUENCE ANALYSIS IN AN API RBI ASSESSMENT" covers hole size, release rate calculation, release inventory, impact of detection systems, flammable and explosive consequences, toxic consequences, non-flammable non-toxic consequences, component damage and personnel injury consequence areas, financial consequences.

4 Conclusions

This paper presented in detail one of the many process safety issues encountered by the author in his 20 years of industrial experience. This issue deals with an amine tower, but touches upon the following topics: carbon dioxide; communication and restriction; competency; compliance; culture; documentation; failure analysis; intimidation; investigation; lessons learned; material of construction; mechanical integrity; pressure; stress, risk assessment and vessel repair.

It describes better ways to address the situation and applicable RAGAGEP. It is hoped that sharing this information will enable others to learn from the past and prevent similar issues in the future.

5 References

[1] NBBI, NB 23 (NBIC), National Board Inspection Code, Parts 1-3.

[2] API 510, Pressure Vessel Inspection Code: In-service Inspection, Rating, Repair, and Alteration, Tenth Edition, May 2014.

[3] Guidelines for Mechanical Integrity Systems, AIChE CCPS, 2006, Wiley.

[4] API Recommended Practice 571, Damage Mechanisms Affecting Fixed Equipment in the Refining Industry, First Edition, December 2003.

[5] API Recommended Practice 572, Inspection of Pressure Vessels (Towers, Drums, Reactors, Heat Exchangers, and Condensers), Second Edition, February 2001.

[6] API Recommended Practice 579, Fitness for Service, First Edition, March 2000.

[7] API Recommended Practice 580, Risk-based Inspection, First Edition, May 2002.

[8] API Recommended Practice 581, Risk Based Inspection Technology, Second Edition, September 2008.

[9] ASME Boiler and Pressure Vessel Code, Section VIII, Division I, 1980.

[10] Guidelines for Seismic Evaluation and Design of Petrochemical Facilities, Second Edition, ASCE, Reston, Virginia, 2011.

[11] ASCE 7-05, Chapter 6, Figure 1A Basic Wind Speed – Western Gulf of Mexico Hurricane Coastline.

[12] Wind Loads for Petrochemical Facilities and Other Industrial Facilities, ASCE, Reston, Virginia, 2011.

Identifying Opportunities of Enhancing Safeguard Stewardship through IPL Rationalization

SreeRaj R Nair
Technical Safety, Chevron U.S.A Inc.
Houston, USA
snair@chevron.com

Bernadette Thornton
Safety Systems, Chevron
ETC, Houston, USA
bernadette.thornton@chevron.com

[Rev2]

Prepared for Presentation at
American Institute of Chemical Engineers
2017 Spring Meeting and 13th Global Congress on Process Safety
San Antonio, TX
March 26 – 29, 2017

Identifying Opportunities of Enhancing Safeguard Stewardship through IPL Rationalization

SreeRaj R Nair
Technical Safety, Chevron U.S.A Inc.
Houston, USA
snair@chevron.com

Bernadette Thornton
Safety Systems, Chevron
bernadette.thornton@chevron.com

Keywords: Process Hazard Analysis, Layer of Protection Analysis, Independent Protection Layers, Rationalization

Abstract

Ensuring effective safeguard stewardship is an integral part of the journey to address the challenge of moving from 'I think" to "I know" that a major accident hazard is managed. One of the steps in achieving effectiveness is to ensure that the process design is robust and that the safeguards provide the necessary protection to reduce the process risk.

Process Hazard Analysis and Layer of Protection Analysis are carried out to identify hazards, determine the potential consequence and evaluate the adequacy of the safeguards. Opportunities for further risk reduction are identified and risks are eliminated or reduced through established risk management programs. Though the study methodologies have been used for decades, there are opportunities for improvement.

This paper specifically addresses the opportunities in the evaluation of safeguards during the design of major capital projects. It explores some of the key concerns including lack of independence and variations in modifier application that could lead to over or under engineering; this could in turn may result in increased risk exposure. Additionally, this paper explains how the use of a defined methodology for the rationalization of Independent Protection Layers can assist in overcoming these challenges. The significance of performing IPL rationalization at appropriate stage of engineering progress is key in ensuring effective safeguards through design. A comparison of the advantages of rationalization exercise performed at various project phases is also discussed in the paper.

1 Introduction

A critical starting point in process safety management lies with the identification of facilities with potential for high impact (Safety and Environment) events and an understanding of the high consequence scenarios. For the high impact facilities, a detailed Process Hazard Analysis is conducted to determine the scenarios of concern. The potential impact from realizing those scenarios and the likelihood considering the safeguards (prevention and mitigation) is evaluated. A process safety management system is established with a system of "controls" (including critical safeguards verifications, and validation) to prevent the occurrence of these incidents or reduce their severity.

Layers of Protection Analysis (LOPA) was introduced as a detailed frequency evaluation method compared to Hazard and Operability Studies and as a simpler version compared to Quantitative Risk Assessment. This method is used to assess the adequacy of independent protection layers (IPL) and to ensure that required risk reduction is achievable (i.e., there will be no gaps post implementation of the risk reduction plan). LOPA is a systematic approach for applying IPL which follows guided principles and rulesets as prescribed in the industrial standards (AIChE 2001). CCPS publication (AIChE, 2007) provides guidance on choosing and evaluating initiating event frequency and independent layers of protection.

An independent protection layer (IPL) is a device, system, or action that is capable of preventing a scenario from proceeding to its undesired consequence independent of the initiating event or the action of any other layer of protection associated with the scenario (AIChE, 2015).

An effective IPL should be independent of the initiating causes, survive the initial event, and available to perform at the expected level for the intended duration. In order to determine the effectiveness, each safeguard which has been identified as a potential IPL needs to have the entire core attributes; the attributes are independence, specificity, integrity, reliability, auditability, access security, management of change. IPL can be either 'Active' or 'Passive' with Active IPL being those with the ability to detect (sense a condition or problem), decide (whether an action or intervention is necessary or not) and deflect (action capable of preventing consequence in a timely manner) and Passive IPL being those which primarily deflect to reduce the impact.

The following sections review some of the challenges in determining the effectiveness of active IPL and how to overcome the challenges through a structured approach known as IPL rationalization. It is followed by discussion on the inherently safer approach and its significance in improving the performance of the LOPA results. The potential economic benefits of IPL rationalization (Capital and Operational) are also evaluated with relevant project case studies.

2 LOPA Process Concerns

After evaluation of multiple projects, we have determined that the major LOPA concerns tend to fall into three categories: lack of independence, variations in modifier application, and appropriate documentation. The challenges in IPL related documentation can be addressed effectively by implementation of Process Safety Information management plan.

2.1 Lack of Independence

The first issue, lack of independence, relatively an easy one to eliminate, yet it continues to appear in reviews. Most of this is due to a failure to understand packaged equipment controls and a failure to identify the critical actions required by a safety function. Packaged equipment controls typically are not well defined in early Process Hazard Analysis (PHA) / Hazard and Operability (HAZOP) studies. Assumptions concerning the independence of systems and poorly defined functionality tend to lead to IPL credit being given to these systems in error. For example, vibration monitoring systems tend to send requests to other systems, such as the Safety Instrumented System (SIS), to complete critical actions such as the closure of valves or motor stops. If a Safety Instrumented Function (SIF) exists with the same critical actions, then credit cannot be taken for both. The critical actions of each would have to be defined to remove the conflict and the system common cause would need to be reviewed to ensure that the combination will provide the appropriate layers of protection. Failure to validate packaged equipment independence as the project progresses leads to discovery late in the project that the required number of IPL is unavailable.

2.2 Modifier Application

Differences in the team make-up and the facilitation of the LOPA studies leading to differences in the application of enabling events and conditional modifiers and therefore poor application of IPLs. Enabling events and conditional modifiers are applied to change the likelihood of a consequence or severity of a consequence, so these take into account factors such as the likelihood of personnel being in the area and the phase of operation. Failing to apply enabling events and conditional modifiers properly can lead to major operational issues, including alarm floods and excessive instrumentation. For example, if one facilitator applies a conditional modifier to a scenario and another facilitator does not, the required risk reduction is greater for the second facilitator and will require the application of additional IPL. This inconsistency in the application of enabling events and conditional modifiers leads to under-design (with gaps) and over-design (most-often) of instrumented systems and, potentially requiring additional maintenance and testing requirements.

Effective IPL rationalization can reduce the concerns of lack of independence and modifier application by providing an opportunity to look at the application of IPL on a facility basis rather than a scenario by scenario basis.

3 IPL Implementation Challenges

Poor application of IPL in the processes could result in facilities operated at elevated risk levels. The key challenges in the implementation of LOPA findings include the following:

- Risk reduction during design primarily relies on engineered systems, leaning to the route of 'Add-on' safety;
- Lack of credits for non-engineered / soft barriers as additional layers of protection;
- Operational inefficiency due to overly engineered protection system (multiple interlocks, trips for dedicated systems designed without considering overall facility reliability requirements);

- Regulatory requirements – prescriptive versus performance based approach;

Kletz and Amyotte [Ref. 3] have said, "Intensification, substitution, attenuation, and limitation of effects produce inherently safer design because they avoid hazards instead of controlling them by adding protective equipment. The term 'inherently safer' implies that the process is safer because of its very nature and not because equipment has been added to make it safer". However, this very point is often overlooked during design and often adding additional layers (mainly instrumented and hardware) are considered as improvement to safety. Industries' traditional solution to increased levels of hazard has been through more controls. For example, an offshore facility designed to be operated as 'normally unmanned' could become 'normally manned' just because of the manning distribution required to ensure the effectiveness (through inspection and maintenance) of the 'added on' engineered safeguards.

Projects have become more complex and personnel involved in projects rarely see the effect of their designs, so design complexity is less of a consideration and the more cost effective solution, typically an instrumented one, is implemented to ensure enough safeguards are present to manage the risk. Unfortunately, this leads to the massive problem of facilities having to manage large numbers of safeguards and IPL.

While industry is moving towards performance based approaches to define risk reduction, prescriptive regulatory environments still exist. Prescriptive environments ensure a minimal level of risk reduction is implemented, but they also can add challenges to the implementation of a performance based approach. Prescribed testing and testing procedures may provide some reduction in the overall risk of a facility, the tests rarely are sufficient to provide assurance of the effectiveness of the prescribed functions as IPL; therefore, additional testing must be performed on top of the regulatory required testing potentially requiring more personnel to complete ensure the functions meet their risk reduction target.

4 How to Overcome the Challenge

Eliminating hazards using the inherently safer design route is the most effective way to overcome the challenge of safeguard stewardship simply because there will be only fewer safeguards as IPL (Kletz and Amyotte, 2010). Time and effort can be spent on managing the asset integrity of pipe, vessels, and major equipment rather than managing hundreds of safeguards. In such cases, the projects compare options for managing risk. If elimination is not a viable option considering safety and operability factors, decisions are made to proceed with acceptable levels of risk using adequate safeguards/IPL. The following words of wisdom shall be considered as part of ISD approach (Nair 2009, Nair 2017, Stuart, et. al 2003, Kletz and Anyotte, 2010):

- More is not always safer
- Identify all hazards (you can't eliminate or manage a hazard which have not been identified)
- Design against accidental loads
- Design plants for easy and safe maintenance (*reliable equipment which requires less maintenance*)
- Simplify the process and minimizing the requirement to rely on human intervention

While companies acknowledge that management of IPL is necessary, the number of IPL being identified prioritizes finding ways to reduce the load on maintenance and testing personnel. One way to reduce the load is by using IPL rationalization. IPL rationalization can be defined as a process of defining the most effective set of IPL to provide the appropriate risk reduction while also reducing the number of management activities and operator tasks and the maintenance and testing load a facility must endure. So the question of how to define the most effective set of IPL must be answered.

IPL rationalization uses the results of a LOPA to identify IPL which are unnecessary to provide the facility with the risk reduction required. The process requires a functional safety Subject Matter Expert (SME), a process engineer and an operations representative to review the application of enabling events and conditional modifiers, identify SIF with multiple Safety Integrity Level (SIL) and ensure that SIF, critical alarms and basic process control system (BPCS) loops identified as IPL are, in fact, effective. During a LOPA, many of these items may be addressed, but as projects require more and more facilitators to make large HAZOP more manageable, more and more issues creep into the design. The smaller group allows for a more in-depth review of independence, consistency in identification of the SIF and its actions, and a way to compare the results of one study with another study to ensure IPL chosen between similar units and similar scenarios provide the appropriate risk reduction while applying the same enabling events and conditional modifiers. Most importantly, IPL rationalization provides a way for the project to review the IPL from a bird's eye view as well as scenario by scenario to identify whether IPL are appropriate to mitigate the risk for the overall facility as well as the individual scenarios. The most surprising element of the process is that it requires very little time to complete, especially with the correct pre-work completed.

The process of IPL rationalization is simple. First, the team must first identify the order of preference for implementation of IPL. This preference will take into account the facility's maintenance staff size, the competency of the staff and the operating basis of the facility. For example, some owners may prefer to take more SIFs as they have a good management program for them and no management program for other engineering IPL. Other owner may prefer other instrumented IPL over SIF as they do not expect to have the competency they would prefer to manage the IPL.

Secondly, the team reviews the application of enabling events and conditional modifiers to guarantee their effectiveness or applicability. Next, the IPL are reviewed in order of preference and where different IPL have been provided for scenarios with the same consequence, the use of different IPL is either rationalized or the IPL are revised to provide the most effective set. Unused IPL are either removed or removed from the IPL list.

Finally, the team reviews any recommendations to identify if there are LOPA gaps that may be addressed using existing IPL or whether new IPL must be created.

Experience has demonstrated that the process can be effectively used for brownfield facilities as well. However, it need to be noted that there could be differences in the IPL preferences and that unused IPL are less commonly found as facilities choose to use existing IPL rather than add new ones, even if the final result may be less efficient.

5 A Representation of the Benefits

The benefits of completing IPL rationalization depends on three things:

1. The objective of the project
2. The timing of the study
3. The regulatory environment of the facility

The objective of the project may affect IPL rationalization results as looking for ways to eliminate or reduce the numbers of a particular IPL type or may require the addition of IPL in other categories. This activity may not be a reduction of IPL, but rather a reallocation of IPL. Alternatively, the project may be looking to reduce the number of IPL overall by ensuring the application of IPL is streamlined. Illustrations of IPL rationalization benefits in terms of IPL and potential economic saving are given in Figure 1 and Figure 2. A recent project looking to reduce the number of alarms was able to achieve a 90% reduction in alarms with little effect to the number in other IPL categories due to the proper application of conditional modifiers. If one assumes that maintenance of these alarms requires approximately \$100 per year and that the number of alarms was reduced from 1000 to 100, then it can be found that the facility saves approximately \$90000 per year in alarm maintenance alone (*representative figures only*).

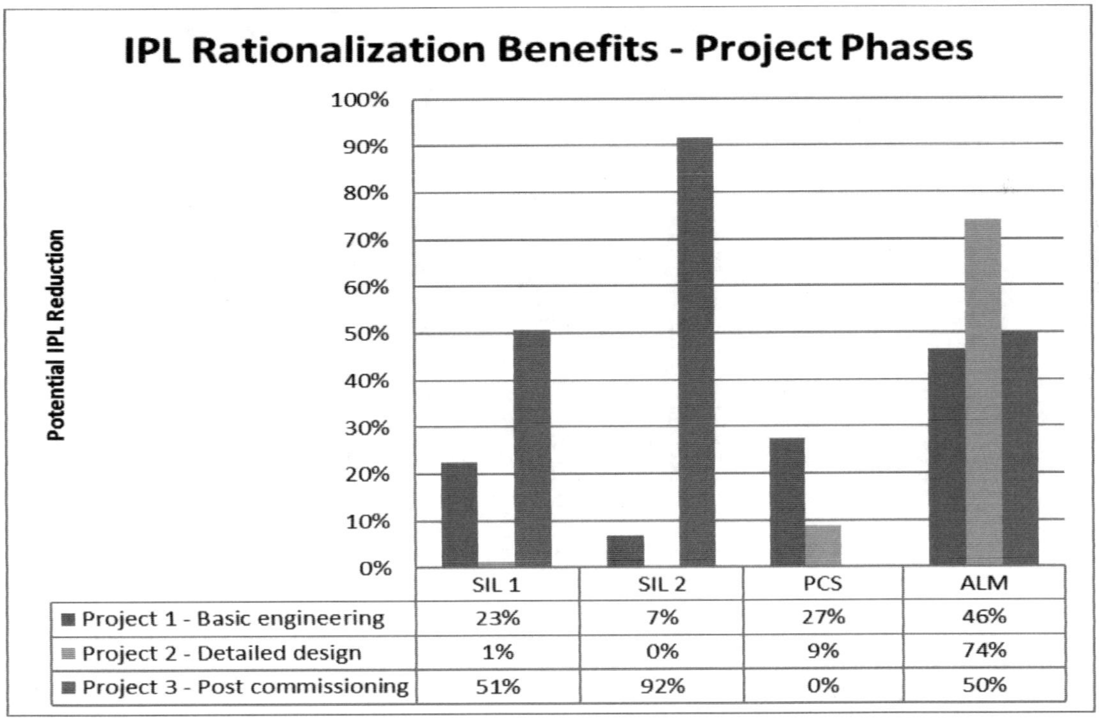

	SIL 1	SIL 2	PCS	ALM
Project 1 - Basic engineering	23%	7%	27%	46%
Project 2 - Detailed design	1%	0%	9%	74%
Project 3 - Post commissioning	51%	92%	0%	50%

Figure 1: IPL Rationalization Benefits – At Different Phases of Project

The timing of the study has a significant effect on the outcome of the IPL rationalization. Performing rationalization before the design is complete (i.e., before construction) provides opportunity to eliminate IPL, such as SIF, that are found to be unnecessary (Figure 1). A project may choose to keep the functions or alarms for operational or reasonable risk reduction justification/demonstration; but take the opportunity to reduce the testing or maintenance associated with these. In addition, the team may be able to identify designs which could be modified to provide a more inherently safer facility. A study performed late in the project will not achieve the same benefits. They may be able to make reductions in some areas, but if the design or installation is almost complete, there is only limited opportunity to remove unnecessary IPL (Figure 1). Alarm priorities may be reduced and maintenance and testing requirements may be relaxed, but the facility will likely be forced to retain most IPL. Performance of IPL rationalization post-commissioning provides even less opportunity for savings. IPL have not only been designed and installed, but they have already been commissioned and, in many cases, changing maintenance and testing requirements will require changes to maintenance managements systems that likely have already been completed, potentially adding cost short-term. A review of projects showed potential IPL reduction of 25-30% for projects in Basic Design / Front End Engineering Design (FEED) and 10% reduction for those in Detailed Design or later.

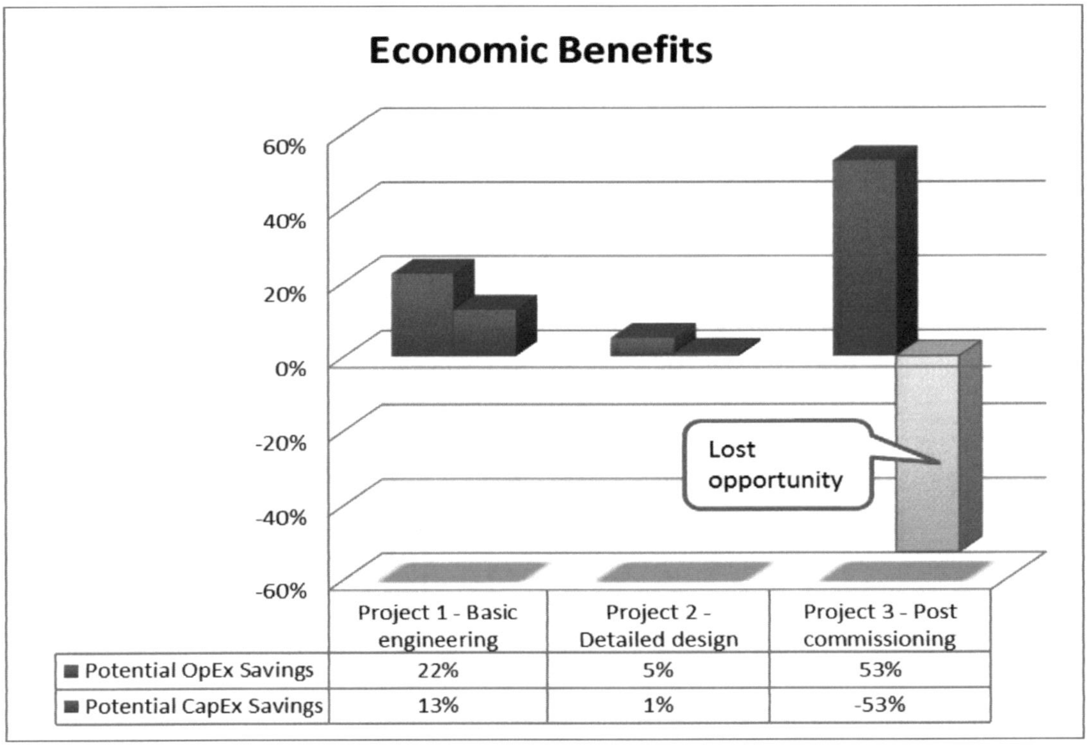

Figure 2: IPL Rationalization and Potential Economic Savings

When taking into account both Capital Expenditure (engineering design, construction, installation) and Operational Expenditure (inspection, testing and maintenance), the savings to a facility can be significant if the appropriate evaluation is carried out in early project stages; even in later stages of the project, the suggested approach can reduce IPL requirement and remove some of the maintenance and testing cost of the so-called "gold plating" that occurs on projects (Figure 2).

Finally, the regulatory environment can have an impact on the study. Where regulators follow a prescriptive regimen for safety instrumented systems (e.g., U.S. Gulf of Mexico), then the benefits of IPL rationalization are realized through the removal of extensive testing and maintenance related to IPL, but not through removal of IPL or through the removal of regulatory required testing. The potential to gain savings still exists though, so the exercise provides benefit by removing the "gold plating" of the facility.

6 Conclusion

Critical safeguards identified, performance requirements established and an evergreen process to ensure safeguards remain in place are key to keep high consequence scenarios from occurring. Through safety studies and by following industrial codes and standards, multiple layers of protection are factored in to prevent or mitigate potential events. However, it is experienced that often the designs and facility become complicated due to 'add-on' safety. Such facilities could experience operational inefficiency, high cost of maintenance and potentially higher risk exposure to personnel.

IPL rationalization should be part of the design engineering process to overcome the challenges and a systematic approach will help in:

- establishing 'effective' safeguard
- optimizing IPLs
- lifecycle benefits

The paper reviewed some of the challenges in determining the effectiveness of IPL and how to overcome the challenges through a IPL rationalization exercise. The paper discussed the inherently safer approach and its significance in improving the performance of the LOPA results. The timing of the exercise has a significant effect on the outcome of the IPL rationalization. The potential economic benefits (Capital and Operational) are also evaluated with relevant project case studies.

7 References

1. CCPS. "Guidelines for Independent Protection Layers and Initiating Events", American Institute of Chemical Engineers, New York, NY, 2007
2. CCPS, "Guidelines for Initiating Events and Independent Protection Layers in Layer of Protection Analysis", American Institute of Chemical Engineers, New York, NY, 2015
3. CCPS, "Layer of Protection Analysis – Simplified Process Risk Assessment", American Institute of Chemical Engineers, New York, NY, 2001
4. Nair SreeRaj, "Determining the criteria for evaluation of toxic hazards, Journal of HSE & Fire Engineering", ASFE, Issue 2 March 2009
5. Nair SreeRaj, "'Inherently Safer' – the Concept and its Application in Oil & Gas Sector," International Conference on Safety and Fire Engineering, SAFE-'17, 2017
6. Stuart Chia, Kieran Walshe and Ed Corpuz, "Application of inherent safety challenge to an offshore platform design for a new gas field development – approaches and experiences," Symposium Series No. 149, IChemE, 2003
7. Trevor Kletz and Paul Amyotte, "Process Plants: A Handbook for Inherently Safer Design," Second Edition, 2010

Process Safety Site Assessment Program (PSSAP) – 5 years of Assessments

Andrew P. Broadbent
American Petroleum Institute (API)
1800 West Loop South, Suite 475
Houston, TX 77027
broadbenta@api.org

Prepared for Presentation at
American Institute of Chemical Engineers
2017 Spring Meeting and 13th Global Congress on Process Safety
San Antonio, TX
March 26 – 29, 2017

Process Safety Site Assessment Program (PSSAP) – 5 years of Assessments

Andrew P. Broadbent
American Petroleum Institute (API)
1800 West Loop South, Suite 475
Houston, TX 77027
broadbenta@api.org

Keywords: Process safety, assessments, benchmarking, protocols, scoring, rating, weighting

Abstract

The API Process Safety Site Assessment Program was developed in 2011 as part of the AFPM and API Advancing Process Safety Programs. PSSAP began conducting assessments in 2012. By the end of 2016, the API Site Assessment program will have conducted 52 General (7 protocols) and 27 HF Alkylation/RP 751 assessments. These assessments have been conducted 59 different refineries and petrochemical facilities. The areas assessed are: Process Safety Leadership, Management of Change, Mechanical Integrity, Safe Work Practices, Operating Practices, Facility Siting, Process Hazards Analysis, and HF Alkylation/RP 751.

This presentation will provide a brief overview of the program and present some the learnings, trends and benchmarking developed from the assessments based on the five-year history of the program.

In 2010, the American Petroleum Institute (API), American Fuel & Petrochemicals Manufacturers (AFPM) and industry joined together to address improving process safety performance with the goal of preventing process safety events at refineries and petrochemical plants. API, AFPM and the industry wanted this effort, known as Advancing Process Safety, to be:

- Focused on higher risk process safety activities;
- Refining & petrochemical focused;
- Voluntary; and
- An avenue to communicate and share experiences and knowledge.

Advancing Process Safety (APS) consists of six different programs which are:

- Process Safety Metrics and Analysis (RP 754)
- Process Safety Hazard Identification
- Event Sharing
- Focused Improvement
- Regional Meetings
- Process Safety Site Assessments Program (PSSAP).

The intent of all the programs is to raise the level of process safety performance at the sites to prevent process safety events.

The management of these programs is split between API and AFPM. API manages the Process Safety Site Assessment Program, Process Safety Metrics and Analysis, and the Focused Improvements effort. AFPM manages Process Safety Hazard Identification, Event Sharing and Regional Networks.

Benefits/Opportunities

The Process Safety Site Assessment Program is perhaps the most hands-on of the six programs since it involves a team of assessors being onsite at a refinery or petrochemical plant for an entire week. The program uses credible, third-party assessors who have specific expertise in one or more of the eight areas being assessed. The team utilizes a set of protocols that are applicable to both refineries and chemical plants. These protocols were developed through a collaborative effort of subject matter experts from a wide range of companies.

Through these site assessments, the program provides the sites with feedback regarding what they are doing well and what improvements they can make to their process safety programs. Additionally, the program provides the sites with benchmarking data so the sites can compare how they are doing versus their industry peers.

The assessments are not process safety management compliance audits, and so we refer to them as assessments and not audits. We speak to process safety rather than process safety management so people will understand that these assessments are not the three-year process safety management (i.e., 29 CFR 1910.119) compliance audits, rather they are assessments of how sites are performing relative to 'what good looks like" based on the experience of industry and the expert assessors. As you will see, the assessments only cover seven areas of process safety (compared with the 14 areas in OSHA 1910.119) as well as HF alkylation/API RP 751.

Protocols

The eight protocols address:

- Process Safety Leadership
- Operating Practices
  - Procedures
  - Shift Handover
  - Operator Rounds
  - Alarm Management, and
  - Workplace Conditions
- Mechanical Integrity – Fixed Equipment
- Safe Work Practices
  - Procedures
  - Work Permit Procedures
  - Utilities, etc.
- Management Of Change
- Process Hazards Analysis
- Facility Siting
  - Permanent Buildings
  - Portable Buildings, and
  - Tents
- HF Alkylation/API Recommended Practice (RP) 751

The first seven elements are areas that could prevent a process safety event from occurring or reduce the severity of an event (i.e., Facility Siting). The eighth area – HF alkylation – deals with process safety in this specialized operation. Assessments involving the first 7 protocols are referred to as "General" Assessments and the HF Alkylation/RP 751 ones are known as "HF In-Depth RP 751" Assessments.

Note: Since the HF Alkylation protocol specifically follows API RP 751, the HF-In-Depth RP 751 Assessment does meet the RP 751 requirement to conduct an audit once every three years.

Sites may also choose to do just some of the protocols (for example, MOC and Facility Siting) rather than all seven at a time. These are known as "Partial" assessments.

These protocols were developed by industry for industry utilizing 54 subject matter experts from a wide range of companies. The protocols were developed, balloted using the API standards development consensus process, and then piloted at three sites. The protocols were then reviewed and modified based on learnings from the pilot assessments, including input from the assessors.

Keys to the Program

There are three very important components the PSSAP program, which are:

- Benchmarking
- Mentorship
- Expertise of the Assessors

Benchmarking

Refineries and petrochemical plants always like to see how they compare to their peers in the industry. In the PSSAP program this is done by providing a score for each protocol and each protocol question.

The 7 General Assessment protocols have a total of 577 questions and the HF In-Depth protocol has 394 questions.

When developed, each protocol question was given a weighting indicating the question's importance to preventing a process safety event from occurring. The question's weighting is a fixed number between 1 and 4 in half increments (i.e., 0, 0.5, 1, 1.5, etc.) with 4 indicating the most importance. During the assessment the assessor will rate each question based on the information and evidence the site provides. This is done on a scale of 0 to 4, also in half increments. The score is a simple calculation of the weighting times the rating.

Keeping the scoring objective is done through ensuring the protocols are well written so that different assessors will understand and interpret the question the same way. Consistency is also addressed through the training provided and calibration sessions with the assessors at the start of the assessments. Additionally, the PSSAP Program Manager provides oversight so that a consistent approach is used by the assessors.

Mentorship

One of the most important pieces of the program is the information exchange the sites get with the assessors. As mentioned, the assessors have considerable experience and expertise based on their years of service in the industry and through performing PSSAP assessments.

We inform the sites (we typically visit the site six months prior to the assessment to set up all the logistics), to get the most value out of the assessment, that their employees should be interviewing the assessors as much as the assessor is interviewing the employee.

As an example, on one assessment the assessor sat down with three site employees for an hour and discussed various ways to manage car seals.

We encourage sites to take advantage of the time with the assessors throughout the week, including the daily debriefs, lunch time discussions, and the closing conference of Friday.

Assessors

As mentioned, the assessors are major piece of the PSSAP program. They are handpicked based on their expertise and industry background/knowledge. Typically they are retired from industry companies, though some work for consulting firms (or have their own firm) but worked in the industry at some point in time.

An assessor candidate with appropriate qualifications and references is selected to attend a 1-1/2 day training course which explains the API Process Safety Site Assessment Program to them. The candidate must undergo an interview with a three-person panel – the PSSAP Manager and two industry representatives before being selected to be an assessor. Not all candidates that attend the training get selected. Also, assessors are picked for specific protocols (i.e., we do not look for Mechanical Integrity assessors to be Operating Practices experts and vice versa). Currently, over 70 people have inquired about being an assessor and we have around 20 in our assessor pool.

To date, a typical assessment team is averaging 39 years of industry experience. Typically the team has 7 assessors.

An in-depth HF alkylation unit assessment follows the API RP 751 Recommended Practice in detail. This assessment can be done at the same time as a "basic" assessment or completely separate and consists of 5 assessors. As noted, an In-Depth RP 751 assessment can be used by the site as their 3-year RP 751 audit.

Assessments

An assessment gets started when the site contacts the PSSAP Program Manager to schedule a week for the assessment. It is important that the site assessment date, be selected well in advance – at least 6 months – so appropriate time is available for planning, including scheduling the assessors.

A pre-assessment visit is then scheduled and conducted at the site by the PSSAP Program Manager. This meeting serves two purposes. First, it allows the site management team to learn more about the program and ask any questions. The second purpose is to meet with the site contact (the person who will work with the PSSAP Program Manager to ensure all arrangements are made prior to the start of the assessment) to go over the logistical needs for the assessment - including conference rooms, scheduling interviews, documents needed prior to the assessment, etc.

Conducting the Assessment

On the week of the assessment the team arrives on Sunday afternoon/evening ready to go. The assessors will have already spent about 8 hours reviewing site documents, provided by the site through an FTP file share site, so they will be well versed on the site's programs relative to the protocols they are assessing.

Monday begins with an opening conference to review the ground rules and usually a quick tour of the site. After the tour, each assessor interviews key people to learn more about the site's management systems, programs and procedures. These interviews have been pre-scheduled by the site using a template provided to them by the PSSAP Program Manager.

The assessors also work to spend as much time in the field as possible. The intent is to not only review whether the site has good management systems, but to provide them with insight on how well their systems are being implemented in the field.

On Tuesday, Wednesday, and Thursday, the assessment team meets with site staff, spends time in the field, and holds a debrief with the site personnel, typically around 4:00 pm. These daily debriefs have been extremely beneficial as they are excellent opportunities for site personnel to continue to interact with the assessors. Often when the debrief has been completed, there are follow-up conversations with the assessors so site staff can gain more insights.

Thursday night and Friday morning is when the assessors compile all the information and the protocol ratings. The assessors enter observations and ratings for each line item in each protocol, thus the facility is provided with a considerable amount of information.

On Friday, an hour to an hour and a half closing conference is held with the site personnel, including the site management team. The site receives a PowerPoint overview, the completed protocol spreadsheets, and some preliminary benchmarking data. The draft report is provided to the site the following week. Distribution of the reports is as designated by the site.

When the assessment team leaves the site on Friday, the assessors return all hard copy documents and delete any electronic documents that were provided to them by the site.

Handling Report & Data

The site will typically be given two to three weeks to comment on the draft final report. This review period is not a discussion of the scores, but whether the wording of the Observations and Good Practices is clear. A final report is then issued by the program manager to the site.

Once the final report is issued, the data and information including the good practices and observations, with the permission of the site, are blinded and entered into the database for trending and benchmarking purposes. API does not ask for corrective action plans or updates.

Program Status, Experience & Keys

As of year-end 2016, we have conducted 53 General site assessments, 27 HF alkylation/API RP 751 in-depth assessments and 5 Partial assessments. The program has been to 28 different companies at 60 different locations (note: several companies have had a site assessment at more than one site.)

Item	Total	Average per Assessment
Protocol Questions	> 41,200	577 Basic 394 HF
Interviews	> 8,000	>100
Observations	> 2,800	~ 35
Good Practices	> 2,400	~ 30

Below are a couple of comments received from sites that have been assessed:

- "You cannot fool these Assessors — they have

done *your* job for 30 years"

- *"In one week the assessment team found issues that our internal teams could not find in 3+ weeks, and it was stuff we needed to know and address. Well focused, deep dive process."*

Benchmarking Data

The benchmarking data is provided in several formats. For example, the data is analyzed and presented by:

- Protocol
- Sub-protocol (some protocols have multiple areas of review – Shift Handover, Operator Rounds, etc.)
- Specific protocol sections (groupings of questions for specific areas)
- Management Systems vs. Implementation (having the procedure vs. how is it being done in the field)
- Size of site
- Size of company
- Quartile

Benchmarking data updates are provided generally every six months to sites that have had an assessment.

Data Review – Good Practices and Observations

Below are some examples of the types of Good Practices and Observations that the assessors have been identifying through the site assessments. If you note the last bullet on both slides this is an example of exactly what we are after. At some sites it's an item that needs to be addressed while at other sites we have seen them doing it well.

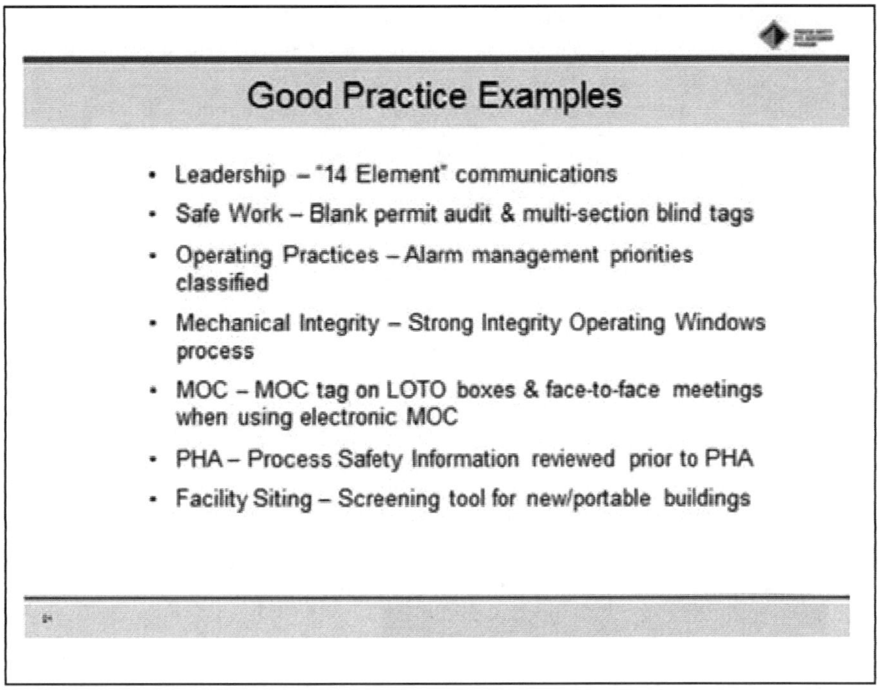

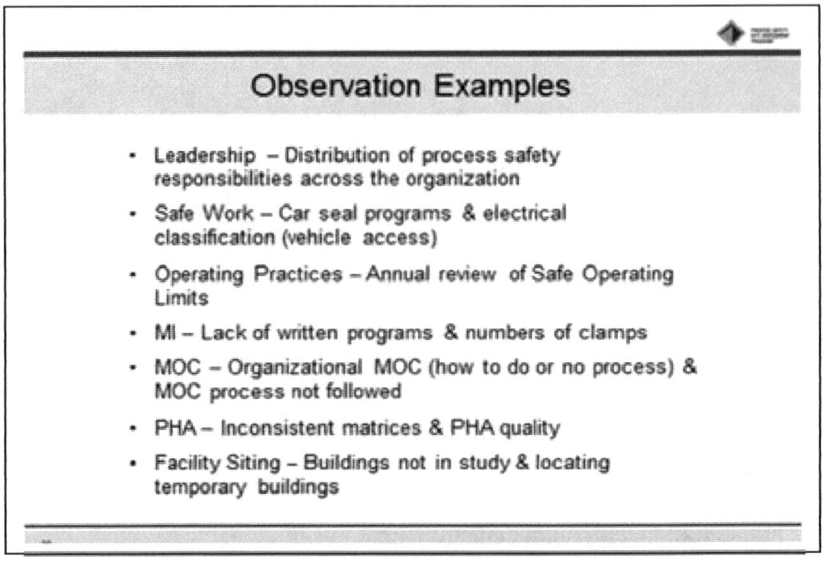

This points to a key element of the site assessment program - sharing the information from sites doing it well with others that are working to put systems/programs/procedures in place at their site. A variety of methods are used for sharing the practices learned through the site assessment program including:

- The Advancing Process Safety Program
    - Regional networks
    - Hazard Identification program
- Industry Conferences and Seminars
- API standards

The data, information, and learnings from the site assessments have been used to focus resources to develop informational and guidance documents regarding areas identified as improvement opportunities. For example, an informational document for mechanical integrity regarding small bore piping was developed based on observations during PSSAP site assessments and data from the 754 metrics and Event Sharing database..

Conclusion

In conclusion, the "boots on the ground" performance of site assessments began April of 2012. Feedback from these assessments has told us that the process of preparing for an assessment (going through the protocols prior to the assessment) and the mentorship/interaction with the assessors has been very beneficial to the sites. And now that we have done enough sites to allow the sharing of benchmarking data the sites are getting even more value from the program.

Alarm Management Meets SIS

SIS Alarm Requirements

How many alarms must be configured per the SIS Standard (ISA-84 Part 1) for an SIF?
Zero

How many alarms must be configured per the AM Standard (ISA-18.2) for an SIF?
Zero

How many alarms are necessary for an SIF to correctly perform its function?
Zero

Pre-alarm – What makes sense?

Pre-alarm before an SIS impending trip?
True – This gives the operator an opportunity to prevent the trip, keeping the plant running and reducing the demand frequency on the SIF.

Pre-alarm should always be on the BPCS control or indicator tag rather than the SIS transmitters?
False – Choice of where to put the pre-alarm is up to the facility, and it's a good idea to document this in the Alarm Philosophy.

Each transmitter in a voting arrangement should alarm individually?
False – Are we trying to keep our plants operating safely or do we just throw multiple alarms at the operator?

SIS and BPCS transmitter disagreements among redundant SIS transmitters should be alarmed?
True – With this alarm in place, there is no need for a pre-alarm on more than one transmitter.

Introduction

Most plants focus on meeting the alarm management requirements laid out in ISA 18.2. Even if the proper approach is taken in configuring the BPCS, there is one area that is sorely neglected, the SIS system. SIS is addressed based on the requirements laid out in ISA 84. Unfortunately, this standard does not address alarming related to SIS events. ProSys contends that a successfully implemented alarm management philosophy is only effective when taken as a holistic view of the BPCS and SIS systems.

Unified Alarming Scheme

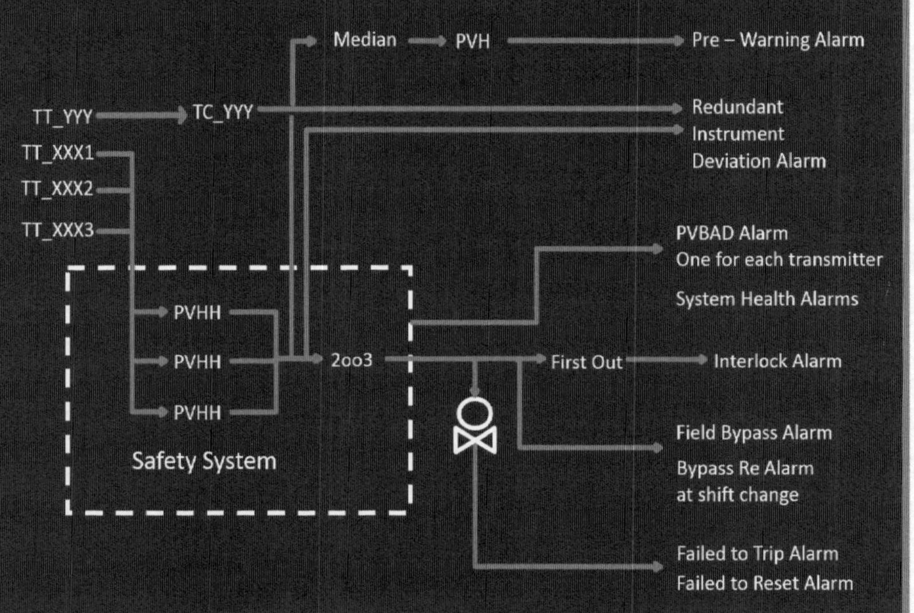

Trip – What makes sense?

First out alarm indicating the reason for a trip?
True – There is no better way to announce to the operator that a trip has been initiated.

Alarm SIS valve moving to trip position?
False – Going to the trip position is the expected response to an SIS trip – better, provide a failed to trip alarm, which makes noise if the valve doesn't work as expected.

Bypass – What makes sense?

Alarm board operator initiated SIS bypass?
False – Although a common practice, this alarm adds little or nothing to the integrity of the system: "Hey Joe/Mary/Sam, you just pressed a button!"

Alarm field operator initiated SIS bypass?
True – This may come as a surprise to the console operator, so it's not a bad thing to annunciate.

If a bypass is in place, trip point alarm on the SIS transmitter?
False – Why give the operator an alarm that isn't reliable or won't be believed? Find an alternative monitoring plan.

Re-alarm on SIS Bypass?
Neutral – Bypass and shift changeover procedures exist, but do we follow them 100%? This is a facility decision. If used, re-alarm shortly after shift change.

PROSYS® Maximize Your Plant™ www.ProSys.com

ENHANCING PROCESS SAFETY MANAGEMENT INTEGRATION TO IMPROVE PROCESS SAFETY AND PROFITABILITY IN THE FUELS AND PETROCHEMICALS INDUSTRY

Thomas D. Watson, P.E.
TDWatson Consulting LLC
Pittsburgh, PA USA
twatson@tdwatson.com

Ronald D. Fuchs, P.E.
RD Fuchs Engineering LLC
Lake Jackson, TX USA
rdfuchs13@aol.com

Harold Phillippi, P.E.
Independent Process Safety Consultant
Charleston, SC USA
butch.phillippi@gmail.com

Keywords: PSM, Process Safety Management, Relief Systems, Incident Investigation, Mechanical Integrity.

Abstract

Process Safety Management (PSM) in the Fuels and Petrochemicals industry has the paradox of being well established and yet constantly challenged to improve its effectiveness and impact on process safety and profitability. Such a paradox is a natural consequence of an industry exhibiting a positive process safety culture that maintains a sense of vulnerability and sees the imperative to improve process safety. This paper addresses the frequently asked question, "How do we continue to improve our organization's process safety culture?" and presents opportunities and examples for organizations to enhance their existing PSM systems and culture by further integrating PSM elements both within an organization's established PSM systems and its other systems, processes, and procedures. The paper illustrates integration and enhancement opportunities using Relief System Design Basis (RSDB) documentation, a component of the PSM element - Process Safety Information (PSI). Illustrations show RSDB integration with the PSM elements of Mechanical Integrity (MI), Process Hazard Analysis (PHA), and Incident Investigation (II). The paper then illustrates opportunities to increase profitability by highlighting RSDB integration with partial downtime throughput opportunities and improved project definition. The illustrations challenge organizations that truly "own" a process safety culture to look for additional opportunities to further enhance and integrate established PSM systems into their organization to further improve safety and profitability.

1. Introduction

The Fuels and Petrochemicals Industry in partnership with governing regulatory agencies have established effective Process Safety Management (PSM) systems to handle hazardous materials and energies over the past several decades. Examples include the U.S. Occupational Health and Safety Administration (OSHA) PSM Standard and the European Commission Seveso II Directive. The Center for Chemical Process Safety (CCPS) offers an extensive reference list of regulatory/industry/system requirements.[1] The catalyst for these industry systems have often been the tragic circumstances around industry incidents and catastrophes. Today, PSM in the Fuels and Petrochemicals Industry has the paradox of being well established and yet constantly challenged to improve its effectiveness and impact on process safety and profitability. Such a paradox is a natural consequence of an industry exhibiting a positive process safety culture that maintains a sense of vulnerability and sees the imperative to improve process safety.[2] The vulnerability and sense of urgency are well founded as the industry continues to experience preventable incidents. The Marsh Report states the 100 largest property damage losses in the hydrocarbon industry from 1974 to 2013 had a combined property damage cost of approximately US$34 billion.[3] Organizations demonstrating a committed process safety culture continually challenge and address the question, "How do we improve our organization's process safety culture?"

Organizations that have successfully managed the PSM implementation and sustainment phases have already experienced and moved past the common misconception that process safety and profitability have conflicting objectives. These organizations further realize they must move beyond a "separate but equal" PSM system established among their current business management systems, and fully integrate PSM into their systems to further improve their organization's impact on process safety and profitability. Recent efforts in the Fuels and Petrochemicals Industry to improve process safety through management system integration include industry efforts such as:

- The American Fuel and Petrochemical Manufacturers (AFPM) "Advancing Process Safety Programs," which promotes sharing PSM learnings across member organizations of AFPM
- The CCPS publication containing guidelines for integrating management systems and metrics to improve process safety performance.[4]

This paper identifies and presents specific examples and opportunities for organizations to continue to enhance their existing PSM systems and process safety culture by further integrating PSM elements both within an organization's established PSM systems and its other business systems, processes, and procedures. The paper illustrates integration and enhancement opportunities using Relief System Design Basis (RSDB) documentation, a component of the PSM Process Safety Information (PSI) element. Specifically, the examples illustrate RSDB integration with the PSM elements of:

- Mechanical Integrity (MI)
- Process Hazard Analysis (PHA)
- Incident Investigation (II)

RSDB integration with other business processes also present opportunities to directly improve profitability. Two examples presented here include RSDB integration with:

- Production planning (partial downtime throughput opportunities)
- Improved project definition

The following sections provide a general introduction to RSDB and illustrate the five integration opportunities in more detail.

2. An Introduction to RSDB and its Relationship to PSM

Within the Fuels and Petrochemicals Industry, many processes involve highly hazardous chemicals. To adequately assess and address a process' hazards an organization must adequately define the Process Safety Information of the process. Process Safety Information (PSI), includes information regarding the process' hazards, technology, and equipment.[5] Examples of hazard information include toxicity and permissible exposure limits. Examples of technology information include a process' chemistry and operating limits. Finally, examples of equipment information include materials of construction and the Relief Systems Design Basis (RSDB). There are many references for what constitutes RSDB and how to effectively manage RSDB documentation.[6] The American Petroleum Institute states RSDB documentation should demonstrate that all pressure-containing equipment has been identified and the relief requirements adequately analyzed. Documentation should include specific engineering guidance, methodology, analysis, and reference information that collectively reflects Recognized and Generally Accepted Good Engineering Practices (RAGAGEP). [7]

RSDB is a key component of PSI, but RSDB is also an important component in maintaining other Process Safety Management (PSM) elements such as Mechanical Integrity (MI), Process Hazard Analysis (PHA) and an Incident Investigation (II). Furthermore, RSDB can be a significant aspect of process production constraints and project opportunities associated with processes. The next sections discuss integrating RSDB into the aforementioned PSM elements, production planning, and project development processes.

3. RSDB Integration with Mechanical Integrity (MI)

Integrating Relief System Design Basis (RSDB) analysis with Mechanical Integrity's (MI's) Risk Based Inspection (RBI) through a combined risk assessment can identify acute process equipment vulnerabilities. Mechanical Integrity (MI) activities ensure the proper design, installation, operation, and maintenance of process equipment. Risk Based Inspection (RBI) is one component of an effective MI program that links risk assessment and mitigation activities with appropriate inspection and process control activities. RSDB analysis, on the other hand, identifies the applicable relieving scenarios and required relieving rates affecting pressure-containing process equipment to ensure the relieving scenarios do not expose the equipment to operating conditions outside of design limits. The American Petroleum Institute's Recommended Practice (RP) 581 identifies the interdependency of the Pressure Relief Device (PRD) with the fixed equipment the PRD is protecting.[8] RP 581 evaluates PRD criticality to set mechanical integrity activities such as inspection frequency. The RBI methodology combines the PRD's anticipated demand rate with the probability of failure on demand of the PRD to determine an overall risk for the PRD and the process equipment the PRD protects. While the API RP 581 methodology is sound, when the RSDB documentation is not available or the RSDB analysis has identified deficiencies in the PRD design, the RBI methodology develops shortcomings that should be addressed.

As an example, consider a process vessel containing a flammable hydrocarbon liquid. An RBI of the vessel defines a risk associated with the vessel, the vessel's PRD and the anticipated overpressure scenarios. The RBI analysis is predicated in part, however, on the PRD adequately protecting the vessel in the anticipated overpressure scenarios should the PRD open. The RSDB documentation identifies fire as the only applicable overpressure scenario but concludes the PRD is undersized. The RSDB can provide additional insight into the level of overpressure that would occur (and the anticipated consequence). The additional insights would significantly impact the RBI risk assessment and potentially warrant a reassessment of inspection and mitigation actions for the PRD and the protected vessel.

The example illustrates how integrating RSDB with MI's RBI can identify acute equipment vulnerabilities and aid in further prioritizing MI efforts and RSDB deficiency resolution, including risk mitigation efforts. Consequently, integrating both RSDB and MI can further enhance PSM integration and improve process safety and profitability.

4. RSDB Integration with Process Hazard Analysis (PHA)

Integrating RSDB with PHA can reduce costs and improve the effectiveness of both the PHA and the RSDB. As most readers are aware, a Process Hazard Analysis (PHA) in the United States and many other countries, is a requirement of a Process Safety Management (PSM) program, (for example, 29 CFR 1910.119 requires PHAs and periodic revalidations). No matter which methodology a PHA employs [Hazard and Operability study (HAZOP), "what-if," Failure Modes and Effects Analysis (FMEA), etc.], capable teams must perform the analysis. The teams typically range from 4 to 8 (or more) persons, consisting of owner's personnel and often outside consultants/specialists. No matter who is present, the participants represent a necessary and significant staff resourcing cost. Furthermore, each PHA finding should be well documented, especially those requiring additional evaluation and stewardship before the findings are declared either a confirmed issue or not. The PHA finding documentation and corresponding follow-up/stewardship also represent a significant staff resourcing cost. Finally, actions required to address the confirmed issues, whether facilities changes, work process changes, or administrative changes represent significant costs.

With the PHA costs sufficiently identified, now consider how integrating RSDB with PHA reduces costs and improves PHA and RSDB effectiveness. PHAs always identify questions and concerns about the adequacy of the over (or under) pressure protection systems as well as items such as Maximum Allowable Working Pressure (MAWP), hydrotest pressure, materials of construction, etc. Furthermore, PHAs require the most recent drawings, heat and material balances, and other Process Safety Information (PSI). Similarly, the most recent RSDB should have many (if not all) of these items assembled and documented, with any changes since the most recent RSDB tracked and provided as a part of the facility's Management of Change (MOC) program. The overlap of PSI and resources between RSDB and PHA is substantial. Consequently, completing RSDB analysis prior to and integrating it with the PHA process and follow-up activities introduces cost savings. Common savings occur relating to over (or under) pressure findings. If properly completed and documented, the RSDB can provide the PHA participants with immediate feedback regarding the adequacy of pressure protection devices, which in most cases are code items and must be addressed. A benefit of having a quality RSDB done prior to the PHA is that the over or under pressure concerns can be quickly resolved. On many occasions, no action item follow-up work is necessary. This reduces costs both during the PHA and the PHA follow-up. Conversely, the PHA serves as an audit check of the RSDB as a capable team reviews the RSDB basis as needed.

As a quick example, a PHA team questioned the adequacy of an exchanger shell's design pressure during a blocked-in condition. The maximum pressure generated from a blocked-in condition from a manual valve closure was greater than the design pressure by approximately 10 percent. The PHA team referenced the RSDB. The RSDB previously identified the same issue and investigated the situation in more detail. The RSDB identified the exchanger's MAWP to be approximately 20 percent more than the noted design pressure, and the hydrotest pressure was based on 1.5 times the MAWP. The U-1 form that was part of the RSDB package clearly indicated this. The RSDB package also noted the exchanger shell was adequate based on ASME VIII, Appendix M, M-5.7, in this situation, because the pressure was below 116% of MAWP and it would be immediately noted by operators, the valve closure was not identified as an issue and the PHA follow-up item was closed. Because the PHA utilized the RSDB, the PHA saved costs on discussion and documentation of an action item and the necessary follow-up and tracking of a code related item. Conversely, the PHA team validated the RSDB, enhancing PSM integration and improving process safety and profitability.

5. RSDB Integration with Incident Investigation (II)

Incident Investigation (II) is also a mainstay of any effective personnel or Process Safety Management (PSM) program. II is an opportunity to learn from mistakes, system failures, equipment failures, and/or overlooked vulnerabilities. II is also an opportunity to evaluate the effectiveness of preventive and control measures installed to prevent or mitigate an incident. In the CCPS Vision 20/20 Characteristics of Companies with Great Process Safety Performance the authors write: "To reduce incidents, employers and employees must enthusiastically support a culture that is driven to learn from many sources including benchmarking, near misses and incidents, and jobs done well."[9] For incidents involving release from a relief device and/or a flaring event, II intersects Relief System Design Basis (RSDB) and together they provide a learning opportunity that enhances PSM and improves process safety.

Yogi Berra said; "When you come to a fork in the road, take it." When an incident occurs investigate it. Definitions may vary for what constitutes an incident. Indeed, there are legal requirements for investigation of safety and environmental incidents. These form the baseline for what organizations must do but do not limit organizations from doing more. For instance, some organizations with very effective PSM programs investigate all flaring events. The organization calling the event an incident, a near miss, or flaring event is least important, while the organization's learning opportunity from the event is most important.

By integrating RSDB with II organizations can do real case study evaluation of the RSDB's assumptions and technical predictions. Integration can help answer such questions as:

- Did the scenarios of the RSDB represent the incident that occurred?
- Did the relief valve effectively handle a release rate lower than the controlling case without issues such as chattering?
- Were the release rate, pressure limitation, and system temperature change what the calculations predicted they would be?
- Did the flare handle the relieving event without an environmental release or safety issue? For example: flare smoking, toxic release, or high temperature exposure of personnel and adjacent equipment.

RSDB integration with II may facilitate getting to the root cause of the incident. The RSDB contains valuable information about the relief system response and may help identify the original process design intent. If the relief system response was not as intended, the relief system may be the cause of the incident. Additionally, if there are layers of protection designed to make flaring or release valve opening an infrequent event but flaring is occurring often, the preventing layers of protection may need further evaluation.

The RSDB and II integration may also help identify additional inspections and corrective actions. In this case II also integrates Mechanical Integrity (MI). For example, damage to the relief system may have occurred from too-high or too-low temperature or pressure. Mechanical damage resulting from corrosive chemicals or relief valve chattering may have occurred. Additionally, the RSDB may shed light on damage that the incident caused to other process equipment. Finally, if the RSBD does not accurately predict what will occur during such an incident there should be a corrective action to change the RSDB and possibly modify the relief system. Consequently, the II serves to further validate the RSDB and the RSDB enhances the II. An important point to note: incorporating RSDB into II requires someone knowledgeable in RSDB be on the II team.

In conclusion, a relief device release or flaring event II that does not consider the RSDB leaves a valuable tool unused. Integrating RSDB with II enhances PSM integration and improves process safety and profitability by gaining additional benefit from both RSDB and II.

6. RSDB Integration with Production Planning

Opportunities often present themselves in complex facilities to ensure process safety and increase profitability by integrating Relief System Design Basis (RSDB) with production planning. Many facilities integrate processing units in a complex manner. Such facilities often have multiple production constraints preventing individual processing units from running maximum capacities simultaneously. Production constraints may include downstream processing capacity, feed qualities, processing unit downtimes, permitting limits, or even facility RSDB constraints. When a facility RSDB constraint limits production, understanding the relationship between other production constraints and the facility RSDB constraint becomes critical to ensure process safety and optimal profitability. Hence the need and opportunity to integrate RSDB with production planning.

Before presenting RSDB and production planning integration examples, there are a few points to note. First, RSDB analysis for an individual processing unit typically establishes the unit's safe operating window at a maximum capacity. A facility RSDB, however, reflects multiple processing units operating simultaneously. Second, due to the myriad of production constraints a complex facility may have, the facility RSDB often reflects a limited set of facility production constraints (often with a very conservative basis). Third, in some instances, the facility RSDB may constrain various processing units to a fraction of their maximum production capacity. Finally, when production constraints setting the facility RSDB change, production planning needs to evaluate alternative production options that the facility RSDB does not adequately reflect.

In one example, a complex facility relief system constraint of excessive relief system backpressure for low set point relief devices exists. The excessive backpressures result from a potential utility failure scenario impacting multiple processing units. This facility RSDB constraint limits the facility's production. When the processing unit with the low set point relief devices is down for maintenance, however, the facility RSDB constraint relaxes. Thus, opportunities present themselves for other processing units to increase production. In another example, a complex facility schedules a portion of the facility relief system maintenance downtime with processing units having a substantial impact on the facility relief system constraint. During the downtime, however, the startup timing for the relief system is delayed due to discovery work and opportunities arise to increase profitability by minimizing production cuts within a newly defined facility RSDB constraint.

7. RSDB Integration with Project Development

Integrating Relief System Design Basis (RSDB) with projects presents opportunities to improve project definition, increase project profitability (through project cost savings), and enhance Process Safety Management (PSM). Having a preliminary well documented RSDB at the on-set of a project — often called the Front End Engineering Design (FEED) stage — will identify the various scenarios that should be a part of the sizing of not only the individual relief devices, but also the relief device collection system functional requirements. The potential costs of the project's impact on the relief devices and systems can then be reviewed and optimized in coordination with the project's functional equipment. The following examples best illustrate these points.

In a larger project a liquid feed drum was required. The feed drum's initial design was 50 psig, and the feed sources ranged from 35 psig to 100 psig. Because the feeds to the drum were greater than the drum's proposed design pressure, the overpressure device and corresponding system had to be designed for liquid overfill. All the sources that were greater than 50 psig represented an overfill potential of around 70,000 barrels a day, and consequently the relief device and the corresponding collection system size and costs were unattractive. Upon further investigation and review of the options, the project team increased the design pressure of the feed drum to 110 psig, which had minimal cost differential when compared to the necessary relief system design. Thus, the 110 psig design was chosen and the relief device was then sized for fire and some potential back-flow on stoppage of the feed pumps. Cost savings were in the millions of dollars.

A second project involved design and installation of equipment that had to be tied into the flare system. A preliminary RSDB was done early in the project. Upon analysis of some typical global scenarios such as loss of power, loss of cooling water, etc., the new project's primary relief device(s) would require increasing the collection system's piping and flare size. Furthermore, the sizing of the relief device and piping alone had a large estimated cost.

The relief device's size was driven by the main tower's selected design pressure and steam reboiler pressure available. The project team, in coordination with the RSDB personnel, were tasked with optimizing the reboiler and the relief system. During the optimization phase, the team identified 2 key items: 1) the reboiler steam had only one pressure available, but it could be let down to a lower pressure, and 2) at a slightly lower pressure the steam could easily provide the available heat necessary for the process.

Additional investigation by the team revealed that the tower material and thickness had not initially taken advantage of the potential MAWP. The material thickness needed was ~0.82 inches, and 0.875 inches was to be selected. This allowed for a MAWP higher than design. However, because the steam pressure was reduced, good engineering practice would require that the design consider failure of the steam pressure control valve.

Failure of the steam pressure control valve would provide sufficient heat to overpressure the tower based on the new design pressure and the relief device to the flare was still needed. To address this failure, the steam system (downstream of the pressure let-down valve) was provided with its own pressure relief valve which could be vented to atmosphere.

With the team's recommended changes noted above, the only scenario for the main tower relief device was fire, which had a lower cost of installation and did not require an update of the flare system. Having the RSDB done early in the project allowed the designers and project team to make some easy (paper) cost effective changes, and did not require any modifications to the existing collection and flare systems.

The examples above demonstrate organizations integrating RSDB during project development to improve project definition, reduce project cost, and even make processes inherently safer. Organizations realize further PSM enhancements by integrating a process unit's existing RSDB documentation with project development activities. In addition to the benefits illustrated above, organizations incorporate project impacts on a process unit's RSDB with minimal duplication of effort (minimal cost), and enhance the process unit's RSDB integrity. From an overall PSM perspective, the project development integration enhances both the PSM and project development system.

8. Conclusion

This paper presents opportunities for organizations to enhance their PSM systems and process safety culture by integrating Relief System Design Basis (RSDB) with the following PSM elements and business processes:

- Mechanical Integrity (MI)
- Process Hazard Analysis (PHA)
- Incident Investigation (II)
- Production planning
- Project development.

Organizations that truly "own" a process safety culture should use each successful opportunity implementation as a spring board to further enhance PSM integration and improve process safety and profitability.

9. References

1. CCPS, "Guidelines for Integrating Management Systems and Metrics to Improve Process Safety Performance", Appendix A, Center for Chemical Process Safety, American Institute of Chemical Engineers, New York, NY 2016.

2. CCPS, "Vision 20/20 Booklet – Committed Culture", Center for Chemical Process Safety, American Institute of Chemical Engineers, New York, NY 2016.

3. Marsh and McLennan Companies, "The 100 Largest Losses", 23rd edition (2014).

4. CCPS, "Guidelines for Integrating Management Systems and Metrics to Improve Process Safety Performance", Center for Chemical Process Safety, American Institute of Chemical Engineers, New York, NY 2016.

5. Occupational Health and Safety (OSHA) "Process Safety management of Highly Hazards Chemicals" 29CFR1910.119(d).

6. Thomas Watson, "AM-13-24 Considerations for Effectively Managing Emergency Relief Systems Design," American Fuels & Petrochemicals Manufacturers Annual Meeting, 2013.

7. American Petroleum Institute, "API Standard 521 – Pressure-relieving and Depressuring Systems", section 4.7.3, sixth edition (2014).

8. American Petroleum Institute, "API RP 581 – Risk-Based Inspection Technology, Part 1", Section 7, second edition (2008).

9. Center for Chemical Process Safety's Vision 20/20: "Characteristics of Companies with Great Process Safety Performance" (2013, McCavit, Berger, and Nara).

479098 Reactive Hazards Management: Lessons Learnt from Major Incidents

Abstract:

Chemical process industry has long suffered from major incidents due to lack of proper understanding of reactive hazards. The methyl-isocyanate leak in Bhopal, India, has been by far the worst industrial incident in history which led to the development of many process safety regulations. However, the formation and implementation of regulations has not prevented the occurrence of these incidents. Many incidents such as Concept Sciences Inc. where a hydroxylamine manufacturing unit was destroyed, the explosion at the Sterigenics International Inc., ethylene oxide sterilization facility, Bayer CropScience Pesticide Chemical Runaway Reaction Explosion and West Fertilizer Explosion have highlighted the importance of having a better understanding of the reactive hazards. The focus of this presentation is to identify different types of reactive hazards through examples of chemicals incidents and analyze how a better understanding of these hazards could have avoided these incidents.

479245 The Study of Safety Evaluation Technique for High Temperature Process

Abstract:

There are many accidents of a firing furnace and kneading machine by overheating and the abnormal injection of the product. When the gas which suddenly expanded by thermal decomposition destroys plant equipment and enters the explosion concentration range, a gas explosion is caused and damage spreads. The material which it is easy to decompose is invested by mistake by a firing furnace and has the trouble that the pressure of the furnace abnormally rises to in our factory. Also, troubles such as an explosion of the plumbing by the thermal decomposition of the polymer by the overheat, vent explosion by the thermal decomposition by the false injection of an additive having low heat stability has broken out with kneading machine

These troubles are caused by the thermal decomposition by a sudden temperature rise. The temperature program of the analytical instrument which is generally used by a thermal analysis is 10-20 /min. The conventional thermal analysis equipment cannot evaluate the temperature rise rate at the time of the trouble exactly to become than 20 /min. It becomes the problem that the slow temperature rise may be different from rapid heating up in a decomposing reaction.

We produced a device capable of quickly raising temperature to solve these problems.

This device is comprised of the vertical plumbing which it sealed and heating furnace, and it is very simple structure. The sample holder is fixed to the upper part in the plumbing by a magnet. And the holder drops in the heating zone by keeping a magnet away from the plumbing and is heated rapidly. The materials of the plumbing are SUS316L, and, as for the use upper limit temperature, 500 , the use upper limit pressure are 1MPa. Because the deterioration of SUS materials by the heat is intense when it is raised the temperature of the device to 750 , the plumbing is disposable.

The inside of the plumbing is replaced with inert gases such as nitrogen, argon. The maximum temperature rising rate is 150 /min (furnace temperature: 750).

We introduce an example of the trouble analysis of a firing furnace and kneading machine using this device.

[Case example 1] Study of firing furnace trouble

After burning with a new additive at 720 , abnormal pressure rise happened in a furnace, and a plant stopped. After measuring quantity of outbreak gas using the rapid heating device, it was recognized that gas of 7 times was generated and a gasification rate tripled in comparison with a conventional additive.

[Case example 2] Study of kneading machine trouble by overheating

The additive which had low heat stability was spent mistakenly in large quantities to a kneading machine, and a vent board exploded. After performing a rapid temperature rise examination under a kneading temperature condition, as for the pressure raising rate when an additive disintegrated, 1.8MPa/min, the quantity of gasification became 190L/kg. The additive concentration became (mixing ratio with the resin) more than at least 75%, and it was recognized that a large amount of additives were invested mistakenly.

Fortunately, this example does not become the big accident with resulting in injury or death. In this explosion example, a vent board penetrated a checker plate. It might become the casualty when there was a person near a kneading machine.

In comparison with the plant of the petrochemical and fine chemistry that heat risk analytical technique establishes, the firing furnace and kneading machine is often run without detailed risk assessment. The rapid heating device becomes the powerful tool to do the heat risk evaluation of these high temperature processes.

In future, it is going to be improved the rapid heating device changes the plumbing materials, and to use it to the temperature ranges more than 1,000 .

Handling uncertainty in LOPA and SIF design

Akshat Khirwal
IRESC
Unit 2801, Enterprise Square 3
39 Wang Chiu Road
Kowloon Bay
Hong Kong
Akshat.khirwal@irescglobal.com

Arjan Abeynaike, Venkatesh Sourirajan
IRESC
Arjan.abeynaike@irescglobal.com, Venkatesh.s@irescglobal.com

Keywords: Uncertainties, LOPA, SIL, SIF design, reliability

Abstract

Layer of protection analysis (LOPA) is a widely used semi-quantitative method for determining the reliability requirement for safety instrumented functions (SIFs). An advantage of LOPA is that it specifies a numerical reliability target - the average probability of failure on demand (PFD_{avg}). However, uncertainties in the input parameters and simplifications within the LOPA method lead to a margin of error in the target PFD_{avg} that is often not accounted for in the design of the SIF. Further uncertainties are introduced when carrying out a reliability analysis of the SIF design to ensure that the target PFD_{avg} is met. The result of overlooking such uncertainties is that the target PFD_{avg} may be set low and the calculation of the actual PFD_{avg} achieved by the design may be optimistic, leading to a SIF that is under-designed. The need to account for uncertainty is emphasized in the latest version of IEC 61511, the international standard for SIF design.

This paper investigates and highlights the sources of such uncertainties in LOPA and SIF reliability analysis. In particular, the quality of the data used as input parameters to LOPA and reliability models has a large effect on the accuracy of the results. These parameters include failure rates, conditional modifiers, beta factors for common cause failures, and proof test coverage factors. Available data for these parameters can be sparse and may not be representative of the conditions in which the SIF is applied. Where the confidence intervals of the data are available, these can be incorporated into the calculations to reach a more conservative result. Alternatively, a safety factor can be applied to try and capture all the uncertainties. In this approach, the target PFD_{avg} obtained from LOPA is reduced by the safety factor to give a more stringent target PFD_{avg} for the SIF design to meet. The methods for handling uncertainty are reviewed in this paper.

1 Introduction

SIL Determination is the process of assigning a Safety Integrity Level (SIL) to Safety Instrumented Functions (SIFs) following any of the methods i.e. Risk Graph, Risk Matrix or Layer of Protection Analysis (LOPA); as laid down in IEC 61511 and IEC 61508 [1] [2]. Of these methods, the semi-quantitative LOPA approach is increasingly being adopted for SIL Determination as compared to other qualitative approaches such as Risk Graph or Risk Matrix. An advantage of LOPA is that it specifies a numerical reliability target for a safety instrumented functions (SIF), which is expressed in terms of the target average probability of failure on demand (target PFD_{avg}), or target risk reduction factor (RRF, where RRF = 1/ PFD_{avg}). This target PFD_{avg} then forms the basis for design of the SIF wherein a reliability analysis is carried out to ensure that the target PFD_{avg} is met. However, uncertainties in the input parameters and simplifications within the LOPA method lead to a margin of error in the target PFD_{avg} that is often not accounted for in the design of the SIF. Further uncertainties are introduced when carrying out a reliability analysis of the SIF design. Without accounting for these uncertainties in the assessment, the SIF may be under-designed since LOPA may result in under prediction of the target PFD_{avg} or the reliability analysis may show that the target PFD_{avg} has been achieved although this may not be the case. This paper has reviewed the sources of uncertainties in the LOPA method and design of the SIF (SIF reliability analysis) and provides a way forward for handling such uncertainties.

2 Uncertainties in LOPA

Initiating event frequency, consequence severity, and likelihood of failure of Independent Protection Layers (IPLs) are the key parameters used in LOPA to approximate the risk of a scenario in case the SIF fails to act when the demand arises. In addition to the above, conditional modifiers [3] may also be used to factor in specific parameters pertaining to the demand on the SIF. The simplicity of LOPA allows it to be easily understood by engineers of all disciplines, leading to its widespread use. The drawback of this simplicity is the greater uncertainty in the results, as described in the following sub-sections. The governing equations and the LOPA worksheet used for assessing the scenarios discussed in this paper are referenced from IEC 61511-3 [1] and CCPS, Simplified Process Risk Assessment [4].

2.1 Initiating event frequency

The demand on any SIF may arise due to one or more events, each with a frequency in the range of once per year to once in plant life (for example, 30 years). There is some approximation involved in the selected initiating event frequency values especially when using generic databases. For instance, the source data in the generic database may not be appropriate for application to the facility under review or there may not be enough confidence in the dataset. Power failure or failure of other utilities such as cooling water or steam is a good example. The frequency of these failures from generic databases is typically 1x 10^{-1} per year (once in 10 years) [4]. However, in the author's experience, based on feedback from many operators, these values are likely to be somewhere between once per year to once in ten years. Similarly, the frequency of pump trip based on generic databases is 1x 10^{-1} per year, however operational experience suggests that a value in the range of once in 2 years to once in 3 years may be more appropriate. An analysis of the OREDA [5] data also provides a value close to once in 3 years. Additionally, when there are multiple demand scenarios for a particular SIF, failure to consider the cumulative frequency of all the demand scenarios can also lead to under-estimation of the overall initiating frequency. A simple approach is to consider a suitable approximation for the cumulative demand frequency [6] [7]. In addition, common cause events specific to the facility is also a contributor which needs to be identified. Without such considerations, the required target PFD_{avg} may be under-predicted. All of the above introduce uncertainties in the LOPA assessment noting that the initiating event frequency is the starting point of a LOPA assessment. Therefore, care must be taken in selecting an appropriate value for this parameter.

To better understand the dependence on the initiating event frequency, consider a typical distillation column as shown in Figure 1. There are many scenarios such as overhead condenser failure, reflux pump failure, reboiler control failure, etc. which may lead to a demand on the high high pressure SIF provided on the column. For the sake of brevity, only the case of Reflux Pump trip is used further in this example.

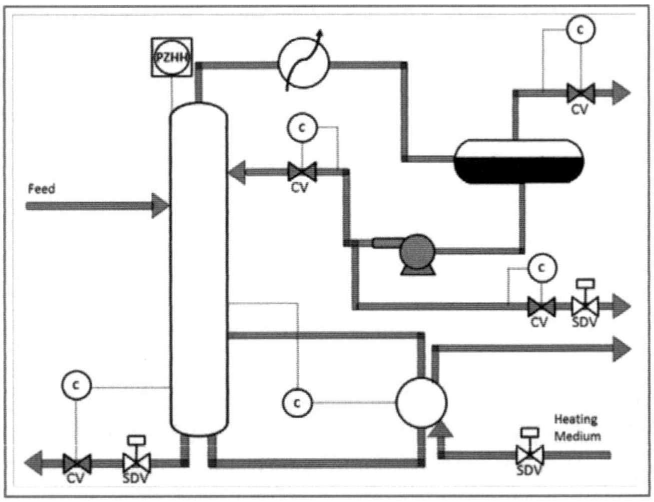

Figure 1. Typical Distillation Column

Assuming an initiating event frequency of 1 x 10^{-1} per year for pump trip and considering a Target Mitigated Event Likelihood (TMEL) of 1 x 10^{-5} per year for multiple fatalities along with an ignition probability of 0.1 and taking credit for pressure safety valve (PSV) on the column (PFD_{avg} of 1 x 10^{-2}), this would result in a SIL 1 with Risk Reduction Factor (RRF) 10 (target PFD_{avg} of 1 x 10^{-1}). However, if the initiating event frequency for pump trip is changed to 3.33 x 10^{-1} per year (once in 3 years), this translates to SIL 1 with RRF of 33 (target PFD_{avg} of 3 x 10^{-2}). This example demonstrates the dependence of initiating event frequency selection on the target PFD_{avg}.

2.2 Consequence Severity and Conditional modifiers

The TMEL assigned to the SIF is directly associated with the degree or the severity of the consequence assumed (e.g. single fatality: 1 x 10^{-4} per year, multiple fatalities: 1 x 10^{-5} per year). This in turn has a direct impact on the SIL assigned to the SIF as well as its target PFD_{avg}. Therefore, careful consideration must be given to the selection of the consequence severity during a LOPA review. For instance, the impact area or hazard zone due to a loss of containment event can be determined using consequence modelling [7], the result from which can be used to assign the severity (e.g. single fatality vs multiple fatalities). Additionally, this can be used to determine the relevant conditional modifiers such as occupancy factor within the impact zone. This approach is more reliable than making coarse assumptions which may introduce uncertainties in the analysis. Such refinement can also be applied to other modifiers such as ignition probability where using a more specific value representative of the hazardous fluid and/or the facility features may be appropriate as compared to use of generic values.

If we revisit the example of typically distillation column presented in Section 2.1; considering an initiating event frequency of 3.33 x 10^{-1} per year and assuming a TMEL of 1 x 10^{-4} per year (single fatality), an ignition probability of 0.1 and taking credit for PSV on the column (PFD_{avg}

of 1 x 10^{-2}), this would result in a RRF of 3 (target PFD_{avg} of 3 x 10^{-1}), which is below SIL 1 requirement. However, if a higher severity (TMEL: 1 x 10^{-5} per year corresponding to multiple fatalities) is selected based on consequence modelling results and an ignition probability of 0.3 is selected based on refinement, the assigned SIL will increase significantly to SIL 2 with RRF 100 (target PFD_{avg} of 1 x 10^{-2}). Although this example shows an increase in the assigned SIL, the application of detailed assessment for determining severity could also show a reduction.

2.3 Independent Protection Layers

Independent Protection Layers (IPLs) are devices, systems or actions that prevent the realization of an undesired consequence (e.g. the consequence the SIF is safeguarding against). Each IPL has an associated PFD_{avg} which is assigned and used in the LOPA methodology for SIL Determination. For example, process deviation alarm with sufficient time for operator intervention has a PFD_{avg} of 1 x 10^{-1}, an independent BPCS control has a PFD_{avg} of 1 x 10^{-1} while a PSV has a PFD_{avg} of 1 x 10^{-2}.

2.3.1 Effect of Common Cause Failures

One of the fundamental assumptions in the LOPA approach is that these IPLs are completely independent and do not share any common cause failures (CCF). In reality, multiple protection layers may be impacted or their performance may be affected due to similar environmental conditions or deviations from desired process parameters. Corrosion or deposition leading to simultaneous degradation of a pressure safety valve and a shutdown valve is one such example. Other examples of CCF include scenarios where multiple PSVs are required to meet total relief requirements, or where an alarm and control, although independent safeguards, both reside on the BPCS. Additionally, dual check valves for preventing backflow could also be impacted by CCF. Although such scenarios may take a period of time to develop and the likelihood may be low, such CCF can have a direct impact on the PFD_{avg}. For example, consider a case where the safety system comprises a SIF and an IPL and the target PFD_{avg} for this system is 1 x 10^{-3}. The PFD_{avg} of the IPL is 5 x 10^{-2}. If the SIF and IPL fail independently only, then the target PFD_{avg} required of the SIF is (1 x 10^{-3}) / (5 x 10^{-2}) = 2 x 10^{-2}. However on inclusion of CCF, even if a low probability of CCF is assumed (1% of the PFD_{avg} of the IPL), the target PFD_{avg} for the SIF then becomes 1 x 10^{-2}.

The authors do acknowledge that for the purpose of time-effective LOPA review, it may not be practical to dwell deeper into CCF for IPLs as it may require an in-depth analysis (for which sufficient details may also not be available at the time of LOPA review). However, it is important to be aware of the inherent uncertainties introduced in the analysis by not accounting for the CCF associated with IPLs.

2.3.2 Effect of Multiple Protection Layers

There is another inherent uncertainty associated with IPLs apart from CCF. The PFD_{avg} for any protection layer (whether mechanical protection, alarm with operator response, BPCS or SIF) is not static but increases over time until a required test or maintenance activity is conducted to check for faults [4]. This relationship between PFD and time is shown graphically in Figure 2. In LOPA, the PFD_{avg} values of the IPLs are multiplied together to estimate the PFD_{avg} of the overall safety system [4]. This calculation, while simple to perform, may tend to underestimate the overall PFD_{avg}. In a more quantitative analysis of reliability, the time-dependent PFDs of each protection layer are multiplied together to yield a result similar to Figure 3, which shows the PFD of the safety system. The PFD_{avg} for the system can then be calculated.

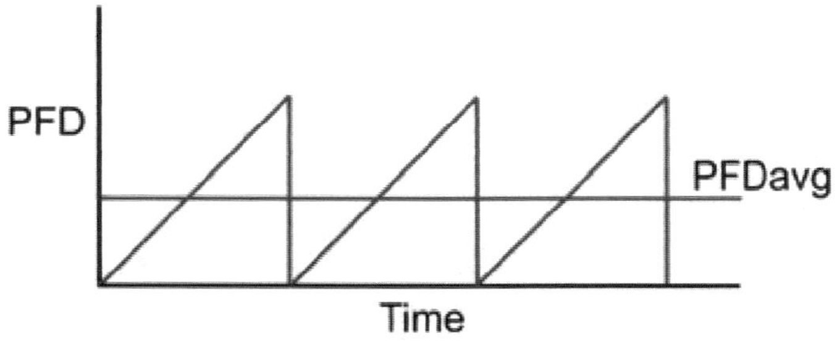

Figure 2. PFD as a function of time for a single protection layer. Periodic testing / maintenance was assumed to detect all faults in this example.

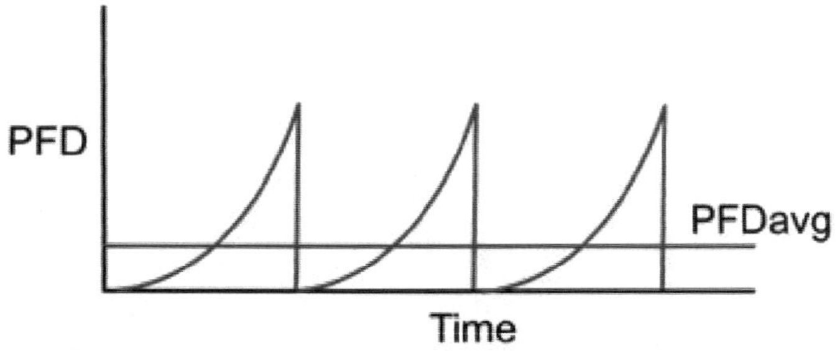

Figure 3. PFD as a function of time for a safety system with two protection layers (one SIF and one other IPL). Proof testing was assumed to detect all faults in this example.

To illustrate the difference in the results provided by LOPA and a more rigorous analysis, the same example from Section 2.3.1 is used. LOPA estimates the PFD_{avg} of the SIF to be 2×10^{-2} (excluding CCF), considering one IPL with a PFD_{avg} of 5×10^{-2} per year and a target overall PFD for the system of 1×10^{-3} (see calculation in Section 2.3.1). Using the PDS method of reliability analysis [8] gives a SIF PFD_{avg} of 1.5×10^{-2}, i.e. LOPA overestimates the target PFD_{avg} by 33%. As the number of protection layers increases, this error becomes more significant, as shown in Table 1. For two or more IPLs with a SIF, the error in the target PFD_{avg} for the SIF is at least 100% using the PDS method; i.e. the LOPA value for target PFD_{avg} is double that given by the PDS method. This gross overestimation in the target PFD_{avg} for the SIF suggests that the LOPA method may not be sufficiently accurate where two or more IPLs are considered. In such cases a more rigorous analysis may be conducted outside of the LOPA session (refer Section 3). Such rigorous approach, if adopted, may also help reduce the error for cases where only one IPL is applicable.

Table 1: Effect of the number of IPLs on the error in PFD_{avg} for a safety system as calculated by LOPA

Number of IPLs	Error in target PFD_{avg} for SIF
0	0%
1	33%
2	100%
3	220%
n	$\frac{2^{n+1}}{n+2} - 1$

2.4 Assessing the cumulative uncertainty for multiple parameters in LOPA

The above sections have discussed how the uncertainty of each individual parameter affects the target PFD_{avg} calculation. In practice, however, the cumulative effect of the uncertainties for multiple parameters also needs to be considered. As a simple illustration and keeping in line with the discussion above, the same example of the distillation column high high pressure SIF is used. As shown in the table below, the changes in various parameters results in a significant difference in the target PFD_{avg} requirements. One additional parameter presented in the table below is the use of PFD_{avg} of 3 x 10^{-2} for IPLs considering multiple PSVs are required to meet relief requirements. As expected, the required RRF is much more stringent when the uncertainties are modelled. However, as highlighted in earlier sections, depending on the SIF under review, in some cases, the requirements may also be less stringent upon further refinement.

Table 2: Effect of cumulative uncertainty of multiple parameters in LOPA

Description*	Initiating Event Frequency	IPL	Ign. Prob.	Total Mitigated Event Frequency	TMEL	Target PFD_{avg}	RRF
Case 1	1.00E-01	1.00E-02	1.00E-01	1.00E-04	1.00E-04	1.00E+00	1
Case 2	3.33E-01	1.00E-02	3.00E-01	1.00E-03	1.00E-04	1.00E-01	10
Case 3	1.00E-01	1.00E-02	1.00E-01	1.00E-04	1.00E-05	1.00E-01	10
Case 4	3.33E-01	1.00E-02	3.00E-01	1.00E-03	1.00E-05	1.00E-02	100
Case 5	1.00E-01	3.00E-02	1.00E-01	3.00E-04	1.00E-04	3.33E-01	3
Case 6	3.33E-01	3.00E-02	3.00E-01	9.00E-04	1.00E-04	1.11E-01	9

3 Approaches for handling uncertainty in LOPA

Based on the various modes of introduction of uncertainties in the LOPA analysis as discussed in the preceding sections, the question that arises is whether the direct use of target PFD_{avg} from LOPA as input to SIF design is optimal or is it desirable to investigate possible ways to refine the target PFD_{avg} from LOPA to ensure the SIF is not under-designed. Some ways to refine the target PFD_{avg} are discussed in the following paragraphs.

One proposed approach for dealing with uncertainty without compromising the simplicity that makes LOPA an attractive method is to conduct a more rigorous risk assessment outside of the LOPA session for high consequence / high risk scenarios to confirm that the SIF reliability target PFD_{avg} obtained from LOPA is sufficient. Such an assessment may include a quantitative reliability analysis such as a Fault Tree Analysis or Markov Model to obtain a more accurate estimate of the required target PFD_{avg}. These methods can also take into account common cause failures and the time dependency of PFD. In fact the most recent version of IEC 61511 [1] requires that a quantitative methodology be used for cases where the required risk reduction is greater than 10,000 (target PFD_{avg} from LOPA is equal to or lower than 1×10^{-4} per year). By using LOPA as a screening tool, this strategy can ensure that high consequence / high risk scenarios receive the more rigorous assessment they deserve, while for lower risk scenarios the more time-efficient LOPA can be considered sufficient.

Alternatively, a number of different cases can be considered for each LOPA scenario, such as in Table 2. Each case considers different values for the various parameters used in the LOPA calculation, resulting in a range of RRFs, from the more conservative to the more liberal. Taking Table 2 as an example, the results range from an RRF of 1 to 100. A final RRF at some point in between these two extreme cases may be selected for the SIF and carried forward into the SIF design. In this way, a level of uncertainty can be accounted for without being overly conservative.

Another approach is to introduce a "Safety Factor" to the reliability target set during LOPA. In this approach, it is recommended that the target Risk Reduction Factor obtained from LOPA should be increased by a safety factor to give a more stringent target for the SIF design to meet. For instance, if the target based on results from LOPA is RRF 100, the SIF design can conservatively consider a target PFD_{avg} associated with RRF 300. Similarly, if the target is RRF 500, then SIF design may consider RRF 800. However, when the RRF is at or close to borderline values, the SIF design may consider the next interval, e.g. for RRF 800 and above, the SIF design may consider RRF 1300 or alternately the RRF used can be justified by conducting a more rigorous quantitative analysis as mentioned above. Some engineering judgement should be applied when selecting this safety factor depending on the level of conservatism adopted in the initial LOPA assessment. As can be inferred from Table 2 above, when uncertainties are accounted for, the target PFD_{avg} can actually vary by one order of magnitude. Applying a safety factor to the results will therefore account for the uncertainties.

4 What PFD_{avg} to assume when Risk Graph Method is adopted for SIL Determination?

Risk Graph Method [1] is the alternate method commonly used for SIL Determination. Unlike the semi-quantitative LOPA approach, which specifies a numerical reliability target (PFD_{avg}) for the SIF in addition to the assigned SIL, the Risk Graph approach relies on the qualitative definitions for the parameters involved to deduce the assigned SIL without setting an actual numerical target (PFD_{avg}). Furthermore, this target PFD_{avg} value from Risk Graph is provided as a range. Hence, where Risk Graph method is used, it is not an uncommon practice to use the upper bound of the PFD_{avg} associated with the assigned SIL. For example, the LOPA method may provide an RRF of 40 (PFD_{avg} of 2.5×10^{-2} per year) for a SIL 1 function, while the same SIL 1 function following the Risk Graph method may adopt the upper bound RRF of 10 (PFD_{avg} of 1×10^{-1} per year).

Consider the same example of a distillation column as discussed in Section 2. If the column high high pressure SIF is assessed using the Risk Graph approach assuming that a trip of Reflux Pump could be expected once a year, taking credit for the pressure safety valve on the column as an IPL (credit of 100) and taking no credit for occupancy and avoidance; the resulting analysis show that Risk Graph method also determines a SIL 2 which is similar to the results from LOPA are accounting for uncertainties. The key difference being the overall PFD_{avg} of 3.33×10^{-3} per year (RRF 300) obtained from the LOPA which will be more stringent than the PFD_{avg} of 1×10^{-2} per year (RRF 100) in case the upper bound value of SIL 2 is used.

The above example shows that using the output from the Risk Graph directly may lead to under-prediction of the target PFD_{avg} and thereby leading to under design of the SIF. The concept of "Safety Factor" as discussed in Section 3 may therefore also be applied where Risk Graph is used. Based on this, it is recommended that the SIF design can use the mid-point of the upper bound and lower bound values based on PFD_{avg} range applicable to the SIF corresponding to the assigned SIL. The safety factor may also be selected on a case by case basis depending on the severity of the associated consequences (for example, using a value above the mid-point of upper and lower bound may be justified for the most severe consequences).

5 Uncertainties in SIL reliability analysis

The uncertainties in terms of obtaining an appropriate target PFD_{avg} for a SIF using LOPA methodology are discussed in Section 2. It is however, worth noting that potential for introduction of uncertainties into the analysis is not limited to this step alone. Moving on in the safety life cycle, the next step after LOPA involves a reliability analysis to ensure that the target PFD_{avg} is met for each SIF. This activity is commonly termed as "SIL Verification" in the industry, but strictly with reference to IEC 61511, it is considered within the domain of "SIS Design and Engineering". The target PFD_{avg} that is considered for the reliability analysis should be the same as the target PFD_{avg} that was calculated in LOPA (a safety factor may be applied as described in Section 3). Where Risk Graph methodology is applied, the target PFD_{avg} can be a mid-point value in the assigned SIL band as suggested in Section 4.

This reliability analysis also involves a number of input parameters, some of which may introduce uncertainties into the analysis. It is important to understand and evaluate the nature of these uncertainties associated with input parameters such as failure rate data for SIF components, beta factor for common causes and proof test coverage factors in SIF reliability analysis as discussed in the subsequent sub-sections. In this section, PFD_{avg} results that were calculated using exSILentia software are presented to illustrate the effects of the input parameters on the PFD_{avg}. The numerical PFD_{avg} results are only applicable for the specific systems and assumptions used in the calculations.

5.1 Failure Rate Data

Failure rate data for the individual SIF components is the basis of the reliability analysis. It is therefore of utmost importance that the data used is based on sufficient sample size and is representative of the conditions in which the SIF is applied. The quality of the data used as input parameters to reliability models has a large effect on the accuracy of the results.

Ideally, an appropriate method for determining the failure rate is to collect field data from a large, randomized sample of units. The operational time until failure should be recorded for each unit. Such data mining however; is a long term project, requiring the co-operation of multiple operators, ideally from a wide range of different facilities and operating environments. For this reason, such studies are rare and normally "generic data", i.e. data applicable for a broad type of component (such as pressure transmitter or shut-down valve) is used rather than data for a specific model or a specific environment. The OREDA database [5] is one example, which provides such generic data. Alternately, some vendors may commission a field study to collect failure data specific to their components or build their own failure data based on reported failures. However, given the smaller sample size as well as the gaps in ensuring that each failure is reported appropriately by the operating facilities, there will be uncertainties in the reported data and these values may tend to be on the optimistic side. For example, while the failure rate for a ball valve from a generic data source is 9.8×10^{-7} per hour [9], several vendors of ball valves report failure rates of less than 1.0×10^{-7} per hour. Another issue is that such data may only be sourced from a limited number of facilities operating in a specific environment and hence, this data may not be applied to a range of facility types or operating environments.

One approach to iron out these uncertainties is to conduct a Failure Mode and Effect Analysis which is a workshop activity bringing together a group of experienced personnel (operations, maintenance, instrumentation, design, etc.) for brainstorming the possible failure modes associated with each component and its associated failure rate using operational experience and field data. The quality of such an exercise is highly dependent on the experience of the participants and their engineering judgement. Hence, even though conducting such an exercise will be fruitful, some margin of error is still expected.

5.2 Beta Factor

Typically, the overall reliability of a SIF is estimated based on both independent and dependent failures. For functions with redundant components, multiple components may fail simultaneously due to a common failure mode (e.g. stress, maintenance errors, etc.). These dependent failures are the percentage of the failures attributed to the common causes and are denoted as Beta Factor in the reliability analysis.

Taking note of the uncertainties in failure data for components as highlighted in earlier sections, it is worth highlighting that the chances of availability of detailed field data for beta factors covering a large sample size is even lower, thereby leading to a greater degree of uncertainty. IEC 61508 [2] advocates a method for estimation of the beta factor using a questionnaire and a scoring system. Based on the scoring system, typically the beta factor value ranges from 1 - 10%. This is a somewhat subjective approach. In reality, beta factors higher than 10% have also been reported in the industry [10]. This may however represent an extreme operating environment pertaining to a particular facility. Therefore, some engineering judgement is required.

It is to be noted that beta factor has a direct impact on the PFD_{avg} achieved for systems with redundant components. Since common cause failures are often the dominating contributor to PFD_{avg}, an increase in beta factor leads to a significant increase in PFD_{avg}. This relationship can be seen in the PFD_{avg} results given in Figure 4, for SIFs with various configurations (1oo2, 2oo3, etc.). Hence, selection of a low beta factor may lead to underestimating the PFD_{avg} and potentially also under design of a SIF. For configurations without redundancy (e.g. 2oo2), the beta factor has little impact on the PFD_{avg}.

In the author's experience, some references and guidelines from operating companies provide recommended beta factor values. However, taking note of the findings presented in this section, it should be ensured that such recommended values have a sound basis and are as accurate as possible. Over-estimating beta factor value may lead to over-design of the SIF which in turn also has an associated cost impact. Conversely, under-estimating the beta factor may lead to SIF being under-designed.

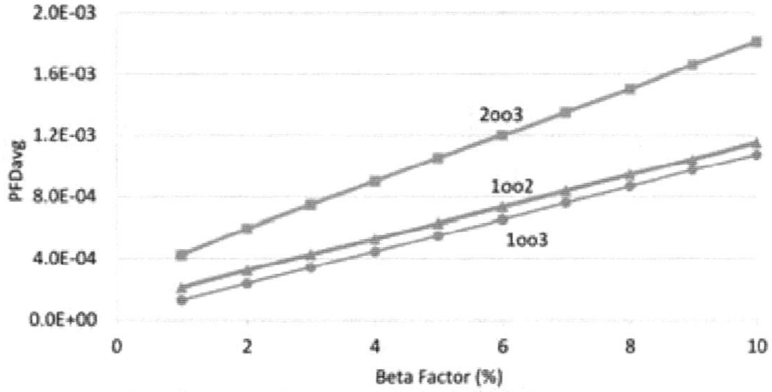

Note: Calculation performed using exSILentia software with the following parameters: λ = 2675 FIT; MTTR = 8 hrs; PTI = 1 yr; PTC = 100%; Mission Time = 10 yrs.

Figure 4. Impact of Beta Factor on PFD_{avg} calculation

5.3 Proof Test Interval

A proof test is performed periodically to detect failures and restore the system / component to an "as good as new" condition. The actual proof test interval may be governed by factors associated with SIF design and operating philosophy specific to a facility. However, the selected interval forms the basis of SIF design since the reliability of the SIF can be compromised with regard to the desired target PFD_{avg} in case the proof test interval used for SIF design is not followed. The uncertainty in this regard lies in whether the recommended proof test interval is rigidly followed or is only conducted on as-need basis in the actual operational phase. Operators may tend to postpone a shutdown for testing purposes if it may result in a loss of production, thereby compromising the SIF reliability. The emphasis to rigidly adopt the recommended proof test intervals therefore rests with the Operating Company and the kinds of safety culture they advocate or follow.

5.4 Proof Test Coverage

One of the important parameters associated with proof tests is the coverage factor which represents the fraction of the potential failures that will be identified during a full proof test. A perfect proof test (i.e. coverage factor of 100%) is unlikely. As far as the authors are aware, there is also no field data available to determine the effectiveness of proof tests. Vendors may recommend a coverage factor based on a detailed FMEA exercise for the associated components. Such an FMEA can identify which failure modes will be successfully captured by a proof test, and thus estimate the coverage factor. Proof test coverage factors can sometimes be lower for components with good diagnostic coverage because the remaining failure modes that are undetected by diagnostics are more difficult for a proof test to detect. In the author's experience, typical coverage factors range from 80 - 99% for sensors and logic solvers and 70 - 80% for final elements (e.g. shutdown valves). While this is currently the best approach for determining the coverage factor, the uncertainty in the result is usually not estimated.

The proof test coverage factor has a significant impact on the calculated PFD_{avg} of a SIF element, as can be seen in Figure 5. For example, for an element with a 1oo1 configuration, decreasing the coverage factor from 90% to 80% resulted in a 46% increase in the PFD_{avg}. Given this effect, caution should be exercised when selecting coverage factors for use in SIF design since they will impact the risk reduction that can be achieved. The actual test performance should ideally be verified to match the assumptions (this becomes especially critical when higher factors e.g. 90% or above are used).

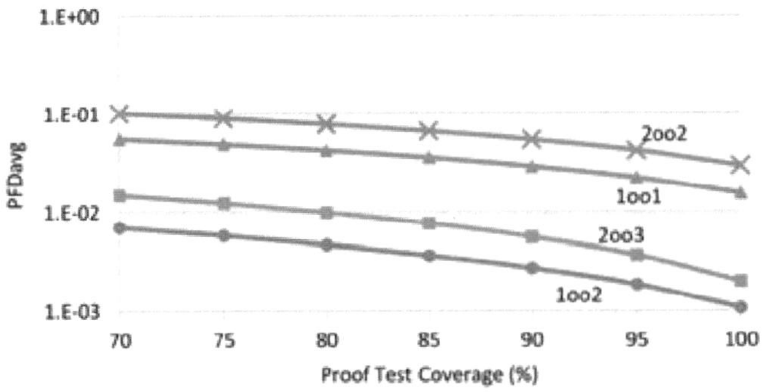

Note: Calculation performed using exSILentia software with the following parameters: λ = 1810 FIT; MTTR = 8 hrs; PTI = 2 yrs; Mission Time = 20 yrs; beta factor = 5%

Figure 5. Effect of Proof Test Coverage Factor on PFD_{avg} for various voting configurations

An alternative method for considering failures that are not detected by a proof test is to include a "probability of test independent failures", or P_{TIF}. This approach, which is advocated in the

PDS method [8], differs from the proof test coverage approach in that P_{TIF} is a constant contributor to the PFD_{avg}, whereas the effect of proof test coverage on the PFD_{avg} increases over time. Even when using P_{TIF}, however, there remains uncertainty in the value used for this parameter, with implications on the accuracy of the calculated PFD_{avg}.

5.5 SIF Operational Life

Operational life of a SIF refers to the life time of the various components of the SIF, sometimes also denoted as mission time. This also serves as a key input in SIF design. In the author's experience, typically a mission time in the range of 20 to 30 years is assumed. Lower mission times such as 5 or 10 years may be impractical in terms of actual implementation on site since this implies that the corresponding components will be replaced or completely overhauled after a 5 or 10 year period, which also has a significant cost impact. This assumption on operational life of the SIF has a direct impact on the risk reduction achieved by a SIF, provided that all other parameters are kept constant. Figures 6 and 7 show that a higher mission time is likely to result in a higher PFD_{avg} (or in other terms, a lower risk reduction). Figure 7 shows that the relationship between mission time and PFD_{avg} is almost linear, with a steeper gradient at a lower proof test coverage. This shows that the mission time has a greater impact on PFD_{avg} when the proof test coverage is lower. If the proof test coverage is taken as 100%, the mission time has no effect on PFD_{avg}. However, as discussed in Section 5.4, 100% proof test coverage is not likely to be realistic.

While operators may typically consider the cost of replacing equipment when deciding upon mission time, consideration should also be given to the effect that mission time will have on the reliability of SIFs. A shorter mission time can help SIFs to meet the target PFD_{avg} and may even allow a lower level of redundancy to be used for the SIF components. This is particularly relevant for SIL 3 rated SIFs, which may require a high level of redundancy as well as high reliability of components, resulting in a high capital cost. Reducing the mission time may help optimize the capital cost of the SIF. A cost benefit analysis may be conducted to justify optimization of the mission time. However, some engineering judgement must be exercised in order to keep the mission time within practical limits. In this regard, 15 years may be considered as a practical lower limit for mission time but this may vary on a case by case basis also depending on the owner's operating philosophy. In such cases, it is important to be aware that in case the capital replacement is not performed after 15 years, this will potentially leave the system exposed to higher probability of failures.

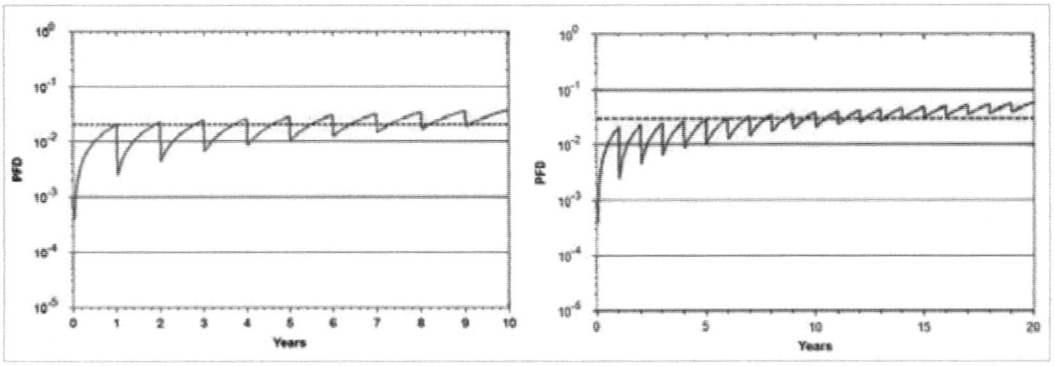

Figure 6. PFD_{avg} vs SIF Operational Life / Mission Time (Years)

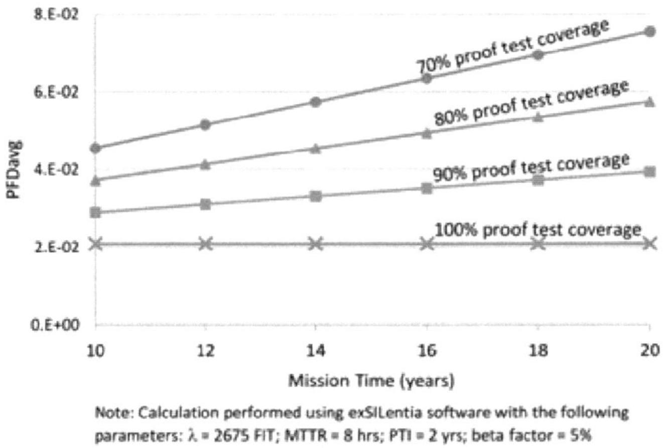

Figure 7. Effect of Mission Time on PFD_{avg} for a SIF with 1oo1 configuration

5.6 Assessing the cumulative uncertainty for multiple parameters in SIF Reliability Analysis

The above sections discuss how the uncertainty of each individual parameter affects the PFD_{avg} calculation. In practice, however, the cumulative effect of the uncertainties for multiple parameters also needs to be considered. As a simple illustration, Table 3 presents PFD_{avg} values calculated for a high pressure SIF with 1oo2 component configuration, using three different combinations of beta factor and proof test coverage, representing a best case, base case and worst case scenario. The beta factor was varied between 1% and 10%, while the proof test coverage was varied between 70% and 90%, which are ranges that can be reasonably expected for these parameters. The PFD_{avg} for the worst case scenario was more than double that of the base case and more than five times that of the best case. In these scenarios, beta factor was found to be the main contributor to the variation in PFD_{avg}, as the uncertainty in the beta factor may be +100% / -60%, whereas the proof test coverage may typically have an uncertainty of only ±13%.

Table 3: PFD_{avg} for a high pressure SIF in three scenarios

Scenario	Beta Factor	Proof Test Coverage	PFD_{avg}	Variation from base case
Best Case	1%	90%	2.93E-04	-70%
Base Case	5%	80%	9.88E-04	0
Worst Case	10%	70%	2.10E-03	+113%

Note: Calculation performed using exSILentia software with the following parameters: λ = 600 FIT; MTTR = 8 hrs; PTI =3 yrs; Mission Time = 15 yrs.

6 Conclusion

In this paper, the authors have highlighted multiple modes through which uncertainties are introduced into the LOPA review and subsequently in the reliability analysis for SIF design. Through working examples, this paper demonstrated that without accounting for such uncertainties, the SIF can easily be under-designed or over-engineered. The emphasis on accounting for such uncertainties is also highlighted in the most recent edition of IEC 61511. This paper investigated further into the nature of uncertainties associated with various input parameters associated with LOPA and SIF reliability analysis including approaches using which such uncertainties can be accounted appropriately.

7 Acknowledgement

The authors wish to acknowledge the support received from IRESC including some of the information included in this paper, as well as the contribution from Ms. Nicole Yeung.

8 References

[1] The International Electrotechnical Commission (IEC), Functional Safety – Safety Instrumented Systems for the Process Industry Sector, IEC-61511, Second Edition

[2] The International Electrotechnical Commission (IEC), Functional Safety of Electrical/ Electronic/ Programmable Electronic Safety Related Systems, IEC-61508, First Edition

[3] Guidelines for Enabling Conditions and Conditional Modifiers in Layer of Protection Analysis, CCPS

[4] Layer of Protection Analysis, Simplified Process Risk Assessment, CCPS

[5] Offshore and Onshore Reliability Data (OREDA), Volume 1 – Topside Equipment, 6th Edition, 2015

[6] Akshat Khirwal, Venkatesh S., Consideration of Cumulative Frequency of Demand Scenarios in SIL/LOPA Review, Hazards Asia Pacific, Kuala Lumpur, Malaysia, 2015

[7] Akshat Khirwal, Venkatesh S., Inherently Safer Design & Robust Protection Layers for High Consequence Hazards, Mary Kay O'Connor Process Safety Center, 19th Annual International Symposium, College Station, Texas, USA, 2016

[8] Reliability Prediction Method for Safety Instrumented Systems, PDS Method Handbook, SINTEF, 2010 edition

[9] Safety Equipment Reliability Handbook, Exida, Third Edition, 2007

[10] S. Hauge et al., Common cause failures in safety-instrumented systems: Using field experience from the petroleum industry. Reliab. Eng. Sys. Saf. 151 (2016), 34-45.

479264 Fundamental Research at Purdue Process Safety Research Center

Abstract:

Fundamental Research at Purdue Process Safety Research Center

Significant industrial incidents continue to occur such as fires, explosions and gas releases resulting in serious injuries, loss of life, and damage to the environment and facilities. Such incidents are not solely germane to the oil & gas industry, but occur in a wide variety of industries including agriculture, pharmaceuticals, and manufacturing. Two years ago the 'Purdue Process Safety and Assurance Center' (P2SAC) was established within the School of Chemical Engineering to conduct fundamental research to prevent such incidents. The presentation will address the four research projects the Center is currently pursuing with funding from its initial Sponsors: BP, Chevron, ExxonMobil, Honeywell, Eli Lilly and Shell.

The initial projects include:

- **optimal placement of gas detectors to prevent fires and explosions**: the research uses CFD-based dispersion simulations and rigorous mathematics to perform optimal placement over hundreds of thousands of leak scenarios and potential detector locations.

- **design & synthesis of catalysts to prevent undesired hazardous byproducts**: zeolite catalysts are widely used in the refining industry and certain structures can promote undesired byproducts leading to potential over-pressures and explosions. Zeolite synthesis and characterization methods are being developed.

- **real time process optimization to reduce risk**: Novel continuous manufacturing processes can reduce the risk related to powder handling. Continuous manufacturing, as with pharmaceuticals, requires efficient process monitoring and control.

- **application of inherently safer separation technology** to significantly reduce processing volume, solvent use, and extent of unit operations. Advanced mixed solvent extraction & simulated moving bed techniques are used to recover polycarbonates and flame retardants from plastic waste.

Indirect benefits from the Center are the depth of the core Chemical Process Safety course required of all undergraduates, as well as undergraduate process safety research initiatives in such as areas as dusts, reactive chemicals and nano-materials.

Current focus is on growing the Center and encouraging participation by a wide variety of industries such as pharmaceuticals, agriculture and manufacturing, in addition to oil & gas, and chemicals.

10-13-16

479277 Bringing New Life to Process Safety

Abstract:

<>Abstract:

Whilst risk cannot be eliminated, it can be better managed. The world of Process Safety Management continues to evolve as organizations find better ways of managing risks. In the simplest form, managing the risk is to reducing inherent process risk to the tolerable level.

First step in managing risks is to identify source of Hazards and quantifying risk associated to Hazards. Traditionally organizations heavily rely on expert judgment and brainstorming methodologies like HAZOP for finding operational and design risks. In addition, quality of such analysis are directly related to the precision of assumptions and information from operation and maintenance of the equipment under control.

In this session we will explore how organizations are using new techniques such as dynamic risk visualization, cumulative risk techniques, operational risk assessments and cyber Hazop's to advance the boundaries of process safety management.

This paper will explore how existing systems and data sources, risk visualization and intelligence techniques are being used to bring static process safety data to life. Flat "1 dimensional" information such as lists, forms and databases are now being transformed into "2 dimensional" dynamic information such as plot plans, barrier models, risk matrix, bow ties etc. that provide greater context of relationships between information.

This information will resonate with engineers and technician who rather working with meaningful information of operation and maintenance systems instead of some abstract numbers and concepts that need extensive expertise in the reliability engineering.

We also will look at how advances in Operational Risk Assessments and Cumulative Risk allow operators to determine in real time if the overall cumulative risk is tolerable before actions are taken, devices are removed from service, maintenance activities are deferred etc. In other words, leading indicators and metrics will be in disposal of operators to proactively facilitate decision making for managing risk. This could avoid incidents that developing in the operation due to degradation of control and safety systems or deviations from initial operation and maintenance procedures and may lead to severe incidents.

And finally, since new technologies and architectures for instrumentation, control and safety systems introduced new risks of cyber intrusions and attacks that need to be considered in the risk management and the world of security standards (ISA 62443-3-2 Security Risk Assessment and System Design) and safety standards (ISA-TR84.00.09-2013 Security countermeasures related to Safety Instrumented Systems and Associated IACS) advanced in the last few years, we will look at how Cyber Process Hazard Analysis (PHA) techniques are being applied to provide the same systematic approach to security as traditionally applied to safety.

"Good process safety management does not happen by chance and require constant active engagement", in this paper we suggested novel approaches that can revive and improve the process safety.

479367 Developing an Effective System for Action Item Resolution

Abstract:

PSM action items are are regularly identified through many different means in a typical chemical processing facility. Knowing what to do with these action items and how best to proceed is no easy task. Not all action items are created equally, or even with the same goal in mind. An action item identified in a PHA may present a direct hazard to an employee, while an action item generated from a pressure relief systems study may present a documentation concern, a deviation from current practices, a direct hazard to an employee, or any combination thereof. Many facilities have a system in place for shepherding action items to completion, but these systems typically assume all action items are created equally. In this paper, we will demonstrate the importance of having a system that handles each action item in the most effective way; from evaluating the nature of the concern, considering the risks associated with the action item, developing the action plan, to completing the action plan. We will also cover whose input should be considered at each stage in the process to help ensure the best solutions possible are developed.

Where Does Management of Change (MOC) Begin?

Juliana Schmitz
Praxair, Inc.
Av. Pastor Martin Luther King Jr., 126, bloco 10/ala A,
Corporate Shopping Nova América, Rio de Janeiro, RJ, Brazil
juliana_schmitz@praxair.com

Patti Jones
Praxair, Inc.
175 E Park Dr, Tonawanda, NY, 14150, USA
patti_jones@praxair.com

Gabriel Drummond
Praxair, Inc.
Av. Pastor Martin Luther King Jr., 126, bloco 10/ala A,
Corporate Shopping Nova América, Rio de Janeiro, RJ, Brazil
gabriel_drummond@praxair.com

Prepared for Presentation at
American Institute of Chemical Engineers
2017 Spring Meeting and 13th Global Congress on Process Safety
San Antonio, TX
March 26 – 29, 2017

Where Does Management of Change (MOC) Begin?

Juliana Schmitz
Praxair, Inc.
Av. Pastor Martin Luther King Jr., 126, bloco 10/ala A, Corporate Shopping Nova América, Rio de Janeiro, RJ, Brazil
juliana_schmitz@praxair.com

Patti Jones
Praxair, Inc.
175 E Park Dr, Tonawanda, NY, 14150, USA
patti_jones@praxair.com

Gabriel Drummond
Praxair, Inc.
Av. Pastor Martin Luther King Jr., 126, bloco 10/ala A, Corporate Shopping Nova América, Rio de Janeiro, RJ, Brazil
gabriel_drummond@praxair.com

Keywords: MOC, engineering, process safety

Abstract

Management of Change is often affiliated with an operating facility. However, several projects have the common challenge of being developed within the constraints of an aggressive schedule, considering the time from project kick-off until unit start-up. In order to comply with the deadlines, early releases of engineering design documentation become subject to changes. During project development, these changes may come as scope changes, supplier updates, etc. that may affect more than one engineering discipline and any on-going construction/assembly activities. Another change source may be non-conformances identified during the construction/assembly phase. In this case, considering project milestones and costs, an option taken into account is to leave the system or portion the way it is instead of re-working to match the original design. For all of the given cases, a robust Engineering Management of Change (EMOC) process must be in place throughout the project life cycle to guarantee modifications are properly identified, communicated, and evaluated by technical experts for possible effects on process and design safety prior to implementation, or at a minimum, prior to startup of the system/subsystem. In addition, an EMOC process gives the opportunity to analyze the reasons for the modifications in order to avoid increased costs and design errors on future projects, leading to improved quality control. This paper presents some previous accidents related to management of changes, and addresses EMOC implementation, work process and elements of a successful EMOC program.

1 Introduction

Preliminary design activities begin early in the process to define scope, cost and product delivery requirements. Changes are a natural part of these efforts.

As the project progresses through the engineering detail design phase, a process hazard analysis (PHA) is conducted. At this moment, a multidisciplinary team evaluates if the process is safe and if any modifications are required. The project team is informed of the revised documentation and proceeds with each skill activity.

Subsequent changes may occur as, for example: updated supplier documentation, the team that will effectively operate the plant in the future may raise a new need for the project, some may be identified during detail design requiring changes and change requests during the construction phase. Any modification after the PHA may not be properly communicated to the adequate professionals and analyzed, resulting in a possible unsafe condition.

This paper focuses on project execution phases until plant startup and turnover to operations where the Management of Change (MOC) work process begins. It also explores the challenges identified when developing an Engineering Management of Change (EMOC) work process and shares an approach taken within Praxair, Inc.

2 Why is the Engineering Management of Change Important?

Process Safety is a relatively new engineering area, being developed from the early 60's and mostly based on catastrophic incidents, which negatively affected the industry. Kerin [5] confirms this point of view and lists a series of landmark events, their key learnings and influences in process safety. These events had different causes, mostly related to poor management of various safety areas, such as hazard identification, training, maintenance, management of change (MOC) and others.

The numerous and notable MOC related accidents that happened in the industry over the last century showed the world the need of a well-developed and maintained MOC procedure.

2.1 Titanic, 1912

The sinking of the Titanic, in 1912, is a well-known catastrophic incident. "The design of the ship included 44 water tight compartments giving rise to the perception that the ship was unsinkable and in the event of an accident would be its own lifeboat." [2] An accelerated construction schedule was the catalyst in making changes that proved to be fatal. A robust management of change process and a cultivated safety culture may have prevented the incident.

2.2 Flixborough, 1974

Venart [3] describes the Flixborough accident, which occurred on June 1st, 1974, causing 28 fatalities and several property damages near the Nypro plant in North Lincolnshire, UK. In this plant, there was a process area responsible to produce caprolactam, an intermediary for the Nylon production. Caprolactam production was conducted through the oxidation of the cyclohexane, a highly flammable chemical. There were six reactors in series where this oxidation reaction was performed.

Two months before the explosion, a crack was observed in the 5th reactor of the chain. With the objective of reducing the impact in production, the responsible plant personnel decided to connect directly the 4th and 6th reactors while the 5th was removed for maintenance. To proceed with the change, a "dog-leg" pipe was used to connect the reactors 4 and 6, while the original installation considered a straight pipe between these reactors. This is a modification that clearly changed the characteristics of fluid flow between reactors 4 and 6. The dog-leg line support was a temporary installation, made of scaffolding bars. These changes were never evaluated in detail and during start-up. On June 1st, the bellows connected to the reactors flanges failed due to an overstress and tons of cyclohexane was released into the atmosphere. With the vapor dispersion, an ignition source was found in the plant, and a vapor cloud explosion occurred resulting in fatalities, as well as significant property damage. [4]

2.3 CONDEA Vista Company, 1998

U.S. Chemical Safety and Hazard Investigation Board [5] promotes a discussion about the reaction vessel explosion in CONDEA Vista Company, on October 13th, 1998. In this unit, linear alkyl benzene was produced by mixing aluminum chloride and hydrocarbons. Around 3 months before the incident the process was modified, involving a mixing of powdered aluminum with hydrogen chloride to form the aluminum chloride. After this change was implemented, the bottom nozzle of the reactor got plugged due to the formation of a sludge residue, which the operators were not able to reach and remove during maintenance. Considering this scenario, the chemist of the site and the process engineer decided to use water to dissolve the sludge and to absorb the heat generated due to an exothermic reaction between the water and the sludge. This process was started and once it was apparently under control, the process engineer and chemist left the plant.

The shift supervisor decided to make a new change and inject steam at the nozzle. The process supervisor who was not in the plant agreed with the new procedure. At the shift turnover, the day shift supervisor talked with the night shift supervisor about the steam injection. The night shift supervisor took responsibility to inject steam and oriented an operator to execute this procedure.

After a few minutes of continuous steam injection, the reactor exploded, leaving three workers injured and property damage estimated in $13 million USD. Among other shortcomings related to the catalyst change that resulted in sludge formation, U.S. Chemical Safety and Hazard Investigation Board **[5]** affirms that this accident highlights the value of having an authorization or approval step as part of an MOC system for abnormal situations. If such a procedure had been in place, a technical manager would have reviewed the proposed procedure involving steam and may have detected its deficiencies.

2.4 Prevention

Among others, these accidents demonstrate the importance of Management of Change in process safety. This was emphasized when MOC was included as one of the main pillars of the Process Safety Management (PSM) standard, issued by the U.S. Occupational Safety and Health

Administration and of the Risk Management Plan of the Environmental Protection Agency (EPA).

However, not only changes performed after the plant startup will result in safety accidents. Engineering Management of Change (EMOC) is also a key area to prevent accidents and should be addressed in a proper manner by industrial companies.

One example of a poor EMOC related event can be an updated drawing sent by an equipment supplier during the project execution phase indicating a decrease of a vessel Maximum Allowable Working Pressure (MAWP). If this modification is not communicated to the proper professionals, the pressure relief device set point designed for the equipment may not be adjusted as required, which can lead to lack of vessel overpressure protection and possible vessel rupture.

Another example can be a change of a pipe routing during project phase, without revision of the flexibility study. This could result in pipe over stress, leading to a line rupture and energy release.

Several other examples of poor EMOC related accidents could be discussed showing proper management of changes during engineering is indispensable for the safety of the plant, workers and environment. Figure 1 indicates some sources of modifications that could impact the safe startup/operation of the system.

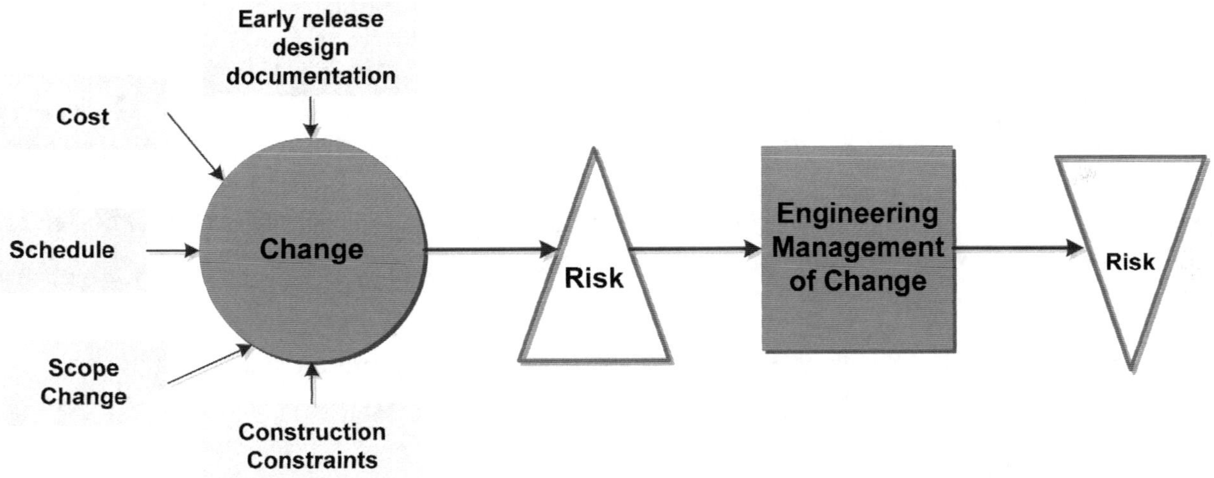

Figure 1 Integrated management of change system

Unfortunately, EMOC use is not as widespread as MOC, which still gives space for industrial accidents to happen. The main purpose of defining a work process for managing changes, even during the project development phases, is to guarantee all changes are adequately communicated, reviewed, new hazards are identified and analyzed and a mitigation plan is defined to ensure that a safe installation and operation is delivered to the customer.

3 EMOC Implementation

A change is known as a deviation from the original design. Modifications may come as scope changes, supplier updates, or other sources, and they may affect more than one engineering discipline and any on-going construction/assembly activities.

"Changes in construction projects are very common and likely to occur from different sources, by various causes, at any stage of a project, and may have considerable negative impacts on item such as costs and schedule delays." [6] If there is no formalized change management process, impacts not only on cost and schedule may be observed, but also process safety events, as mentioned on the previous section of this paper.

According to CCPS [7], the level of detail covered by the management of change system should be based on

- The hazards or risks of the process
- The existing process safety culture at the location

The company should develop a system for the control of modifications implemented on the project under design, construction, commissioning or startup phases. This work process should include approval of the change by a qualified professional or team, depending on the complexity of the hazards associated.

Based on the hazards involved in the specific process, the type of changes that need evaluation by an engineering expert from the appropriate discipline, as well as changes that do not, shall be defined when designing the specific EMOC work process for a certain company. For example; changes that can affect safety systems, critical operating parameters, shutdown set points, relief valve settings, materials of construction, etc. should undergo EMOC approval. Minor corrections to documentation or clarifications may not.

The management of change work process may not be the same in every industry and this does not mean one is better than another. The culture of the company and the location may require different applications of the MOC activities. [7]

3.1 Critical Phase

Changes during construction inherently introduce unexpected risk, especially under schedule and cost pressure. Some changes that can affect process safety may not be obvious to construction personnel. A simple yet important example is relocating a support on the discharge of a relief valve due to interference. Another may be substituting flange bolt material. Regardless of the origin of the change, evaluation and approval should take place prior to starting up the system/subsystem. Identification of all sources which can generate an engineering change must be considered as a natural part of the overall work process.

4 Proposed EMOC Work Process

Typically, during the engineering phase, several design safety meetings are conducted, such as P&ID Reviews, Layout Reviews, PHAs and FSSs. These reviews shall be conducted with the participation of all engineering disciplines related to the plant design. After these project landmark meetings, several changes may occur and their evaluation is a critical part of risk management throughout the project life cycle. At this moment, after the main safety reviews are performed, the EMOC work process is expected to begin.

As shown on Figure 2, all changes related to the project design are identified and communicated by the professional who initiated the modification, to a single point of contact within the project team.

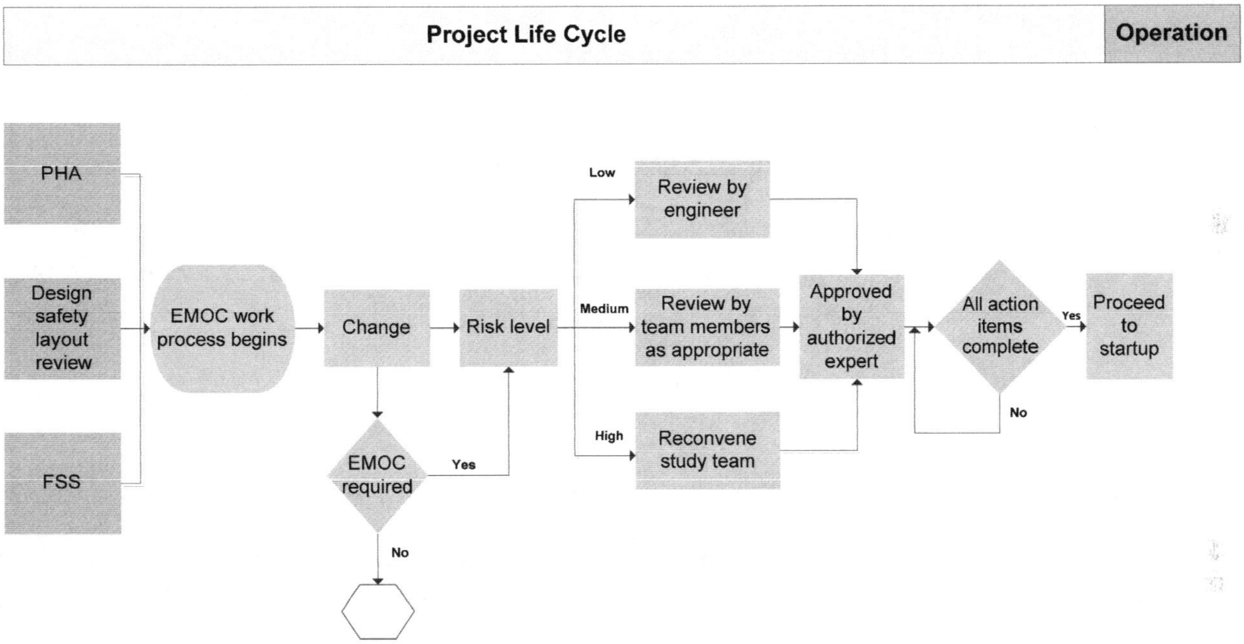

Figure 2 High-level EMOC Workflow Diagram

In cases where a modification is only a correction, clarification or replacement in a kind, there is no need to go through the full EMOC work process. The change can be recorded in a simple spreadsheet containing the type of change, a brief explanation and a mitigation plan to make sure nothing has been missed.

The modification should then be classified as low, medium or high risk per the EMOC owner and the level of review is defined based on the associated risk. The risk ranking criteria shall be described on the company's EMOC program; e.g.:

- Low: addition of a local pressure indicator on the design for maintenance purposes only.
- Medium: manual isolation valve removal, while valve not indicated at PHA as part of a hazard scenario.

- High: customer addition of a tie in point near the pipeline battery limit, which could lead to misdirected or reverse flow, with potential for process safety impact.

After reviewing the change, an approval should be required by a Subject Matter Expert (SME) before proceeding to change implementation and mitigation plan. The plant will be allowed to start-up only after all EMOC items were reviewed and approved.

It will be responsibility of the company implementing EMOC to determine which kind of changes will require a complete review, the risk categories that will be available to be assigned, both qualitatively and/or quantitatively and the required level of review according to these risks. In the example presented in Figure 2, a low risk results in a review by an engineer, a medium risk results in a review by the team members and a high-risk change will require the engineering team involved in the HazOp, Design Safety Layout review or the FSS evaluations to reconvene.

Several different approaches are possible, as long as they are simple, easy to understand by those responsible for implementation and provide a robust EMOC work process.

5 Elements of a Successful Engineering Management of Change Program

Successful EMOC programs define the key elements applicable to their industry. Some will be common such as commitment from leadership, roles and responsibilities, etc. Included below are some elements to take into consideration.

- Commitment from Leadership;
- Roles and responsibilities as they apply to specific skill sets (i.e. process engineering, controls system engineering, manufacturing, construction management, etc.);
- Clear definition of changes that must undergo engineering evaluation, including examples;
- Definition of changes that are not considered EMOC related;
- Risk level associated with the change;
- Approvers limited to subject matter experts;
- A streamlined work process focused on function over form, aligned with the various phases of the project life cycle;
- Identification of change sources;
- In process verification at predetermined milestones;
- Robust training that includes internal and external examples accompanied by a test with feasible scenarios;
- Associated PSI updated accordingly;

- Auditable;
- Evergreen
    - Periodic calibration of the work process;
    - Input from those responsible to implement the process;
    - Captures internal data to enhance understanding of what constitutes a change that requires EMOC review.
- Lessons learned captured and implemented on future projects;
- Leverage key learnings from external events in where MOC was identified as the cause or a contributing factor.

6 Conclusion

An aggressive schedule may lead to early release of design documentation that is still subject to change, which could result in the execution of PHA studies earlier than the optimal period. Although the early study may provide an opportunity to identify key risks earlier in the process, a significant number of action items and changes will probably occur, which may affect the integrity of the PHA.

A well-defined and implemented EMOC program is an essential part of risk management. The methodology applied allows systematic control of project changes which helps to ensure process, personnel and environmental safety. An additional benefit is facilitating a proactive means to analyze the underlying cause for unplanned changes. These lessons learned can assist on future projects to manage cost and improved quality control.

The main purpose of the management of change is to avoid loss of primary containment (LOPC), especially of hazardous chemicals, which could lead to catastrophic consequence. Once the EMOC process is successfully implemented, the reduction of quality issues will become a natural part of the process, resulting in a safer and higher quality facility/assembly/operation. An effective Engineering Management of Change work process is critical to the safe commissioning and startup of a chemical plant.

7 References

[1] T. Kerin. The evolution of process safety standards and legislation following landmark events – What have we Learnt?, Process Safety Progress 35 (2016), 165-170.

[2] CCPS. *Incidents that Define Process Safety*, Center for Chemical Process Safety, American Institute of Chemical Engineers, New Jersey, 2008.

[3] J.E.S Venart, Flixborough: The explosion and its aftermath. Process Safety and Environmental Protection 82 (B2): 105 – 127 (2004).

[4] C. Sadee, D.E. Samuels and T.P. O'Brien, The Characteristics of the explosion of cyclohexane at the Nypro (UK) Flixborough plant on June 1974, Journal of Occupational Accidents, 1 (1976/1977) 203 – 205.

[5] U.S. Chemical Safety and Hazard Investigation Board, Management of Change (No. 2001-04-SB) (2001).

[6] Q. Hao et all, Change Management in Construction Process, International Conference on Information Technology in Construction, Santiago, Chile, 2008

[7] CCPS. *Guidelines for Management of Change*, Center for Chemical Process Safety, American Institute of Chemical Engineers, New Jersey, 2008.

479400 Effectively Utilizing Calarp/RMP/PSM Contractors to Assist in Optimizing Process Safety Culture, Compliance with Standards, and Workforce Involvement

Abstract:

Since the inception of Occupational Safety and Health Administration (OSHA) Process Safety Management (PSM) in 1992, facilities have frequently contracted process safety consultants to satisfy regulatory needs, as well as assist in complying with industry best practices.

The primary objective of this paper is to provide tangible suggestions for effectively utilizing process safety contractors for assisting in optimizing process safety culture, compliance with regulatory requirements and industry best practices, as well as tips to improve workforce involvement and coordination. The core of this paper will detail various management styles of the prevention program and an evaluation of influences on process safety culture based on varying degrees of involvement from the facility and usage of contractors. This paper aims to provide an in-depth analysis of the pros and cons of contractor involvement and its efficacy relative to the scale of the facility and intricacy of corporate structure based on anecdotal evidence and industry experience.

Key topics include:

- Overview of different types of process safety culture in relation to prevention program management and contractor usage,
- The Challenge - Striking an optimal balance between contractor use and facility involvement whilst maintaining economic viability,
- The Challenge for Multi-Facility Corporations - Ensuring changes in corporate standards are successfully communicated to all children facilities ranging from documentation updates and organizational changes,
- Suggestions and Improvements - Implementation of effective prevention program management systems to ensure workforce involvement and accountability, as well as practical allocation of contractor resources to accomplish critical tasks whilst balancing facility resource commitment.

A Streamlined Approach for Full Compliance with SIF Implementation Standards

William G. Bridges, President
PROCESS IMPROVEMENT INSTITUTE, INC. (PII)
1321 Waterside Lane, Knoxville, TN 37922
Phone: (865) 675-3458
Fax: (865) 622-6800
e-mail: wbridges@piii.com

Art Dowell III, Principal Engineer
PROCESS IMPROVEMENT INSTITUTE, INC. (PII)
2437 Bay Area Blvd PMB 260
Houston, TX 77058
Phone: 713-865-6135
e-mail: adowell@piii.com

Prepared for Presentation at
13th Global Congress on Process Safety
San Antonio, TX
March 27-29, 2017

UNPUBLISHED

A Streamlined Approach for Full Compliance with SIL Implementation Standards

William G. Bridges, President
PROCESS IMPROVEMENT INSTITUTE, INC. (PII)
e-mail: wbridges@piii.com

Art Dowell III, Principal Engineer
PROCESS IMPROVEMENT INSTITUTE, INC. (PII)
e-mail: adowell@piii.com

Keywords: Safety Instrumented System, SIS, Safety Integrity Level, SIL, LOPA, Layer of Protection Analysis, Independent Protection Layers, IPLs, human factors

Abstract

Many companies put FAR too much redundant effort into determining what SIL (safety integrity level) is needed and then verifying the SIF (safety instrumented function) design will give the SIL targeted. This paper shows how to apply the qualitative definition of independent protection layers (IPLs) within the setting of a process hazard analysis (PHA) to get most of the gain from Layer of Protection Analysis (*LOPA)* without doing a LOPA (without using numerical values). We show how we use a PHA team to identify when a SIF is needed and to select the proper target SIL. This portion of the SIL evaluation and the identification and labeling of the IPLs during the PHA/HAZOP does not take any longer than a normal PHA/HAZOP, once the right habits are established. Note that this approach eliminates the need for a separate SIL Evaluation Study to identify the SIFs and select the target SIL. Then, this paper describes how to perform the SIL Verification and Safety Requirements Specification (SRS) remotely, again without the need for a redundant team meeting. This approach has been used at many sites and for thousands of SIFs.

Background

Typically, hazardous consequences, their causes, and safeguards are identified in a PHA/HAZOP (process hazard analysis / hazard and operability analysis)[1]. The PHA may determine if additional protection layers are required or if additional analysis is required.[2]

In many organizations, a separate LOPA (layer of protection analysis)[3] meeting is held to determine which HAZOP safeguards are independent protection layers (IPLs) and to determine if additional IPLs are needed to meet the risk tolerance criteria. If SIFs are needed, the required risk reduction (PFD, probability of failure on demand) is determined in LOPA, which then sets the SIL.

Some organizations may even have a separate SIL evaluation study, which is certainly redundant to a well-run PHA and is wasteful if the PHA is properly staffed and managed and/or if any subsequent LOPA are performed well.

Fundamentals of SIL Assessment

The lifecycle requirements for safety instrumented functions are specified in industry standards IEC 61511 (ANSI/ISA 84.00.01-2004, Part 1)[4,5]. The required steps to specify the SIS (safety instrumented system) are:

1. Safety Instrumented Function (SIF) identification.
2. Determining the Safety Integrity Level (SIL) for each SIF.
3. Designing the SIF to meet the required SIL.
   SIL Verification Calculation (actually, this calculation is iterative with step 3, but the end calculation is a deliverable that proves 3 is correct). The SIL standards allow the following methods to use to determine the required SIL (ANSI/ISA 84.00.01-2004, Part 3, Section 3.8)[4]:
   - Fault Tree Analysis (FTA), Full Quantitative Risk Analysis (QRA)
   - LOPA – simplified quantitative analysis
   - Qualitatively within PHA (HAZOP, etc.)

> "A qualitative method may be used as a first pass to determine the required SIL of all SIFs. Those which are assigned a SIL 3 or 4 by this method should then be considered in greater detail using a quantitative method to gain a more rigorous understanding of their required safety integrity." – *ANSI/ISA 84.00.01-2004, Part 3, Section 3.8*[4]

Table 1 compares risk analysis approaches for Steps 1 and 2.

FTA and QRA are very detailed and labor-intensive and would be required for complicated systems, particularly where there is not complete independence between safeguards.

LOPA is a simplified analysis tool that requires fewer resources, but is heavily dependent on the requirement that all the safeguards be completely independent.

However, for many processes, there are a number of scenarios (cause leading to a consequence) where the PHA team can readily determine sufficient independent protection layers are already in place.

Table 1: Comparison of Risk Analysis Approaches

Approach	Assessment Methods	Risk Judgment Method	Risk Judgment Method	Estimated Range of the Results
Qualitative Only	HAZOP, FEMA	Expert Voting, focusing on site data	Capable >95% of time	Plus or minus 1/2 order of magnitude
Simplified Quantitative	LOPA, Risk Graph	Multiplication of statistical averages of general failure rate data; with broad assumptions on management systems	Needed on about 5% of the scenarios	Plus or minus 1 order of magnitude
Full Quantitative	FTA, ETA, QRA, HRA		Needed for less than 0.01% of the scenarios	Plus or minus 1 order of magnitude

Figure 1 contrasts the typical approach to SIL assessments with the optimized streamlined approach.

Typical SIL Determination versus Streamlined Approach

In the typical approach,

- **Step 1**: The PHA/HAZOP is performed using a multidisciplinary team of engineers, operators, instrumentation staff, maintenance staff, and perhaps vendors. Then in a separate SIL study, a multidisciplinary team of engineers, instrumentation staff, maintenance staff, and perhaps vendors brainstorm the hazard scenarios (again) and decide which are candidates for SIFs. A semi quantitative approach, such as RiskGraph, or layer of protection analysis, determines if an SIF is required.
- **Step 2:** The RiskGraph dictates the required SIL, or the LOPA compares the mitigated risk to the risk tolerance criteria and determines the SIL from the order-of-magnitude risk reduction needed.

In contrast, in the streamlined approach,

- **Step 1:** The PHA/HAZOP is performed using a multidisciplinary team of engineers, operators, instrumentation staff, maintenance staff, and perhaps vendors (to this point the same as the typical approach). Within the PHA/HAZOP, the team estimates the risk of each scenario using qualitative judgment (based on experience, knowledge, and memory of site-specific data). The PHA/HAZOP team then qualitatively determines if the best risk reduction method is an SIF.
- **Step 2:** The PHA/HAZOP team qualitatively determines the SIL needed (typically by judging the level of redundancy needed based on the value they place on each IPL and the new or modified SIF needed)

Figure 1: Comparison of the Typical and the Optimized Streamlined Approach from PII for SIL Assessments (Steps 1 and 2)

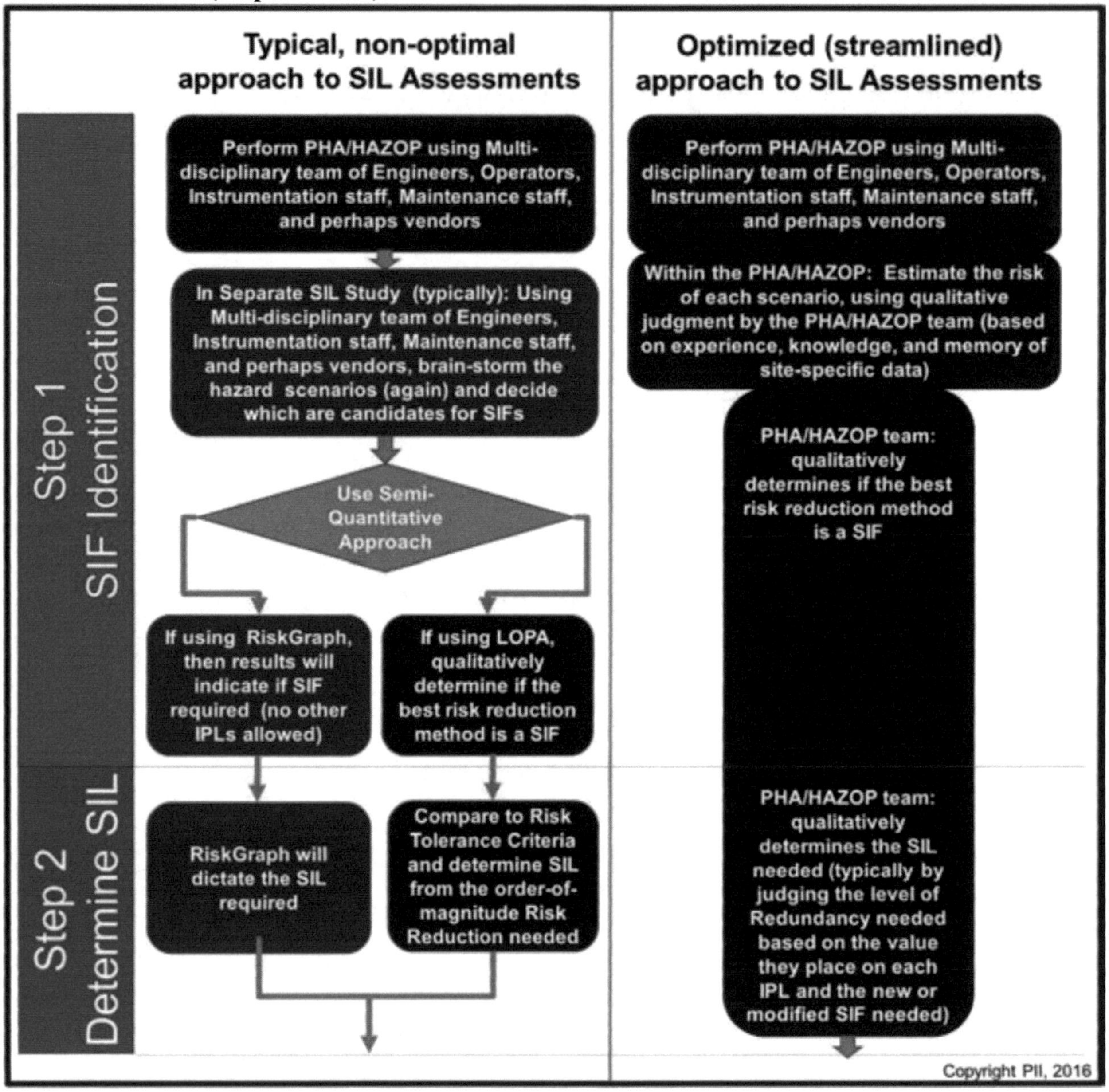

Figure 2 also contrasts the typical approach to SIF design and verification with the streamlined approach. Steps 3 and 4 are iterative.

In the typical approach,

- **Steps 3 and 4**:
  - Decide on basic architecture of SIF, from experience with other SIFs.
  - Decide on vendors of components
  - Provide full design of SIFs from experience

- Perform trial SIL verification calculations using software such as exSILentia™ using actual component data
- If the target SIL is achieved, issue the SIL verification calculation report and the safety requirement specification (SRS)
- If the target SIL is not achieved, revise the design and recalculate.

Figure 2: Comparison of the Typical and the Optimized Streamlined Approach from PII for SIL Assessments (Steps 3 and 4)

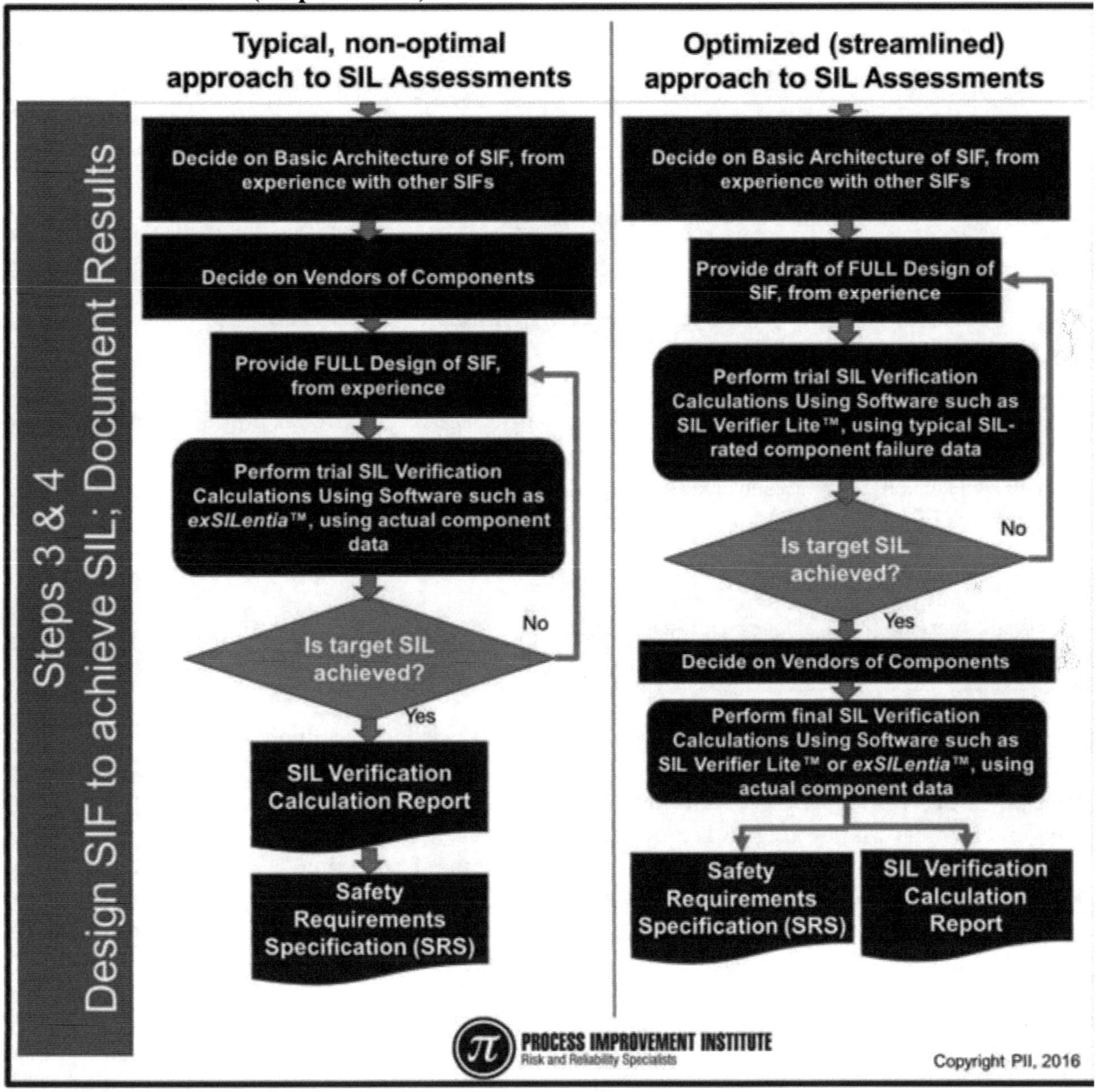

In the streamlined approach,

- **Steps 3 and 4**:
  - Decide on basic architecture of SIF, from experience with other SIFs.

- Provide draft of full design of SIF, from experience.
- Perform trial SIL verification calculations using software such as SIL Verifier Lite™ using typical SIL-rated component failure data.
- If the target SIL is not achieved, revise the design and recalculate.
- If the target SIL is achieved,
    - Issue the SIL verification calculation report and the safety requirement specification (SRS).
    - Decide on vendors of components.
    - Perform final SIL Verification Calculations Using Software such as *SIL Verifier Lite*™ (from PII) or *exSILentia*™ (from exida) using actual component data.
    - Issue the SIL verification calculation report and the safety requirement specification (SRS).

The advantage of the streamlined approach is determination of the need for an SIF and of its SIL quicker and with fewer resources than the typical approach.

Requirements for the Qualitative Streamlined Approach for SIL Determination

For the qualitative streamlined approach to be used effectively, there are important requirements for the PHA/HAZOP team.

- PHA/HAZOP team leader is knowledgeable and proficient in LOPA.
- PHA/HAZOP team members have the following information in memory:
    - Consequence severity.
    - Initiating cause frequency.
    - Independent protection layer (IPL capability).
    - Risk tolerance criteria.

An experienced PHA team, who is knowledgeable in the process technology, can effectively and efficiently analyze a scenario and mentally perform a layer of protection analysis. The team determines the consequence severity and the initiating cause frequency. A highly competent team understands which safeguards are truly independent protection layers and can judge the risk reduction provided by each. Then the team can determine if additional IPLs are required to meet the risk tolerance criteria. For some scenarios, an SIF may be required and the team can determine the required probability of failure on demand for the SIF and can thus determine the required SIL.

A best practice is that the PHA team recommends the functional description and the SIL for the proposed SIF. The functional description does not specify the details of the SIF design, rather it specifies what process variables should be considered and what action should be taken to detect the upset and to prevent the consequence. For example, the functional description may say "monitor the level of a vessel and stop the flow into the vessel to prevent overflow".

If the PHA team is uncertain, or if the scenario is complex, then the PHA can recommend a detailed LOPA be done for the scenario.

These conclusions are based on more than 8000 PHA/HAZOPs, 3000 LOPAs, and about 100 QRAs and HRAs. We have observed PHA participants becoming knowledgeable about LOPA and IPL requirements. The participants then began to apply the principles of LOPA spontaneously in the PHA. With good leadership, the PHA team could then easily move to the full streamlined approach.

The experience and competency of the PHA leader and the PHA team is extremely important. If the PHA team is not comfortable with the principles of LOPA, and especially the principles of IPLs, the streamlined approach is not appropriate. Likewise, the PHA team must have the required information and understanding in its memory. Otherwise, the qualitative approach can degenerate into the "arm wrestling" arguments that PHA teams used qualitatively to determine SIL requirements before the introduction of LOPA.

Suggestions for the Streamlined Approach for SIF Design

Likewise, there is a streamlined approach for the SIF design. Organizations have developed a recipe book of typical SIF designs for SIL 1, SIL 2, and SIL 3. Depending on the required SIL, the designer can choose a design from the recipe book (as described in Figure 2, "from experience"). The recipe book provides a starting point for the detailed SIF design. Figure 3 shows a small sample of SIF designs that could be included in a recipe book.

Figure 3: Typical SIF Architectures for Information

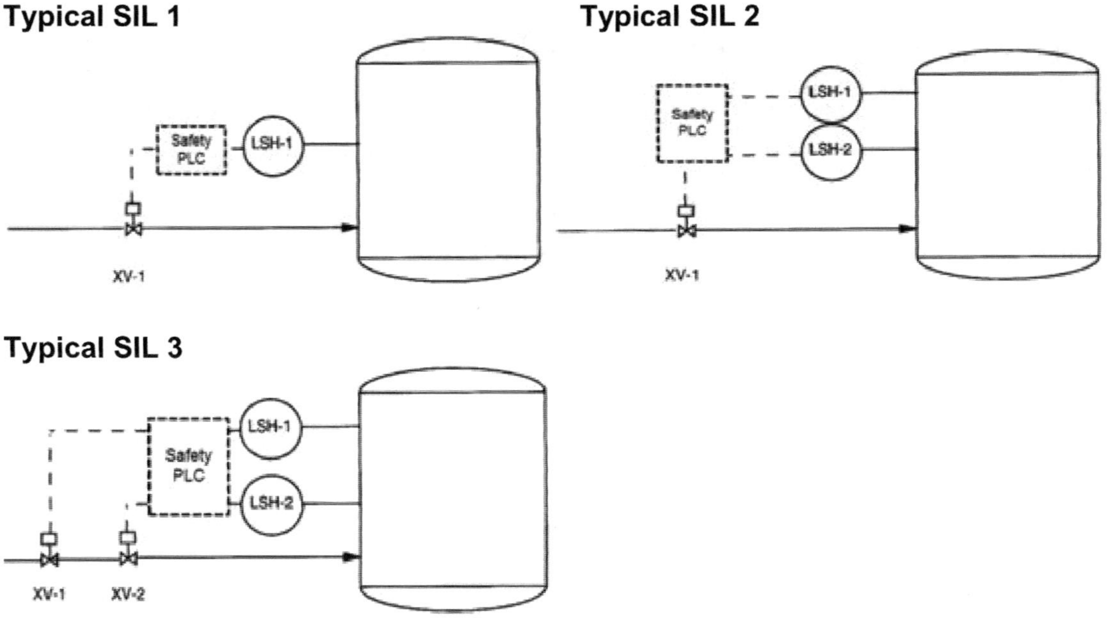

Case Study 1

An example of the streamlined approach is summarized in Table 2 for a facility with storage spheres (in China; the PHA and other analyses were performed in 2014). The table shows two deviations from the PHA. IPLs are identified and existing SIFs are assigned an SIL, based on the architecture. (Note that the achieved SIL for each of the SIFs (existing or recommended) must be confirmed by a SIL verification calculation that includes the failure rate of the components, the voting architecture, and the proof test interval.)

For the high-level deviation, the high-level SIF was assigned SIL 1. The overflow through the pressure equalization line to other spheres was determined to be an IPL.

For the low level deviation, the low level indication and low level alarm were determined not to be an IPL because they are part of the initiating cause. The PHA team concluded that feeding from two spheres at all times is an IPL for this facility because of the unlikelihood for both spheres to have low level at the same time. The PHA team also thought that an additional SIF level alarm with operator response to switch tanks within 60 minutes could be a possible IPL, if the action of the operator is quick enough. The team wrote a recommendation to ensure that the operator response could be an IPL.

In this case study, the PHA team was able to determine in the PHA meeting which safeguards were IPLs. They also determined the nominal SIL for existing SIFs (to be confirmed in a later SIL verification). And the team determined that operator response could be an IPL if there was a documented procedure and an annual drill. For these two deviations, the PHA team did the work that would be done in a later LOPA meeting in the typical approach.

Next, the SIL Verification was performed for the SIL 1 SIF (and all other SIFs in the unit). This SIL Verification was accomplished using exSILentia™ software and the analysis and documentation took less than 2 hours.

Thus, the entire requirements for a SIL identification, SIL target level determination, and SIL Verification were accomplished very quickly compared to typical approaches.

Table 2: Storage Sphere PHA

No.: 2 XXXX storage spheres xxx-T-XX A/B/C/D/E/F/G/H/I/J/K/L (1 of 12)					
#	**Dev.**	**Causes**	**Consequences**	**Safeguards**	**Recommendations**
2.1	High level	Too much flow to one sphere from XX Plant (through their pump; about 40 bar MDH)	High pressure (see 2.5)	High level SIF with level sensors voted 2oo2, to close inlet valve - **SIL 1** Overflow thru pressure equalization line to other spheres (through normally open [NO] valve) - **IPL**	
2.2	Low level	Failing to switch from the sphere with low level in time (based on level indication)	Low/no flow - Liquid from spheres through high pressure product pumps to the vaporizer (see 4.2) Low/no flow - Unqualified liquid from spheres back to Plant (see 6.2)	Level indication and low level alarm, inspected each year, per government regulation **(not IPL; part of the cause)** 9 other spheres with possibly enough level to switch to Feeding from two spheres at all times, so unlikely for BOTH spheres to have low level at the same time - **IPL** Two level indication from SIS level transmitter, with low level alarm, with more than 60 min available to switch tanks (SIF driven alarm and response) - **possible IPL, if action of the operator is quick enough**	Rec 4. Make sure the Human IPL of response to low level in all spheres and tanks is described in a trouble-shooting guide (like an SOP) and practiced once per year per unit operator. This will make this response a valid IPL.

Case Study 2

In 2014 and 2015, an identical approach to that of Case Study 1 (the streamlined approach of this paper) was performed for a large gas producer in South America. In that work, the PHAs for more than 500 nodes plus all procedures of all operating modes was completed in about 10 weeks of PHA meetings. The PHA team identified more than 800 SIFs, along with their target SILs, using the approach in Steps 1 and 2.

Next, staff within PII but separate from the PHA team performed about 60 LOPA to clarify certain issues, some related to SIFs. But less than 5% of the SIFs were established with LOPA; 95% were established with the PHA alone.

Finally, staff separate from the PHA performed SIL Verification of the 800+ SIFs, using exSILentia™. The verification was for a combination of SIL 1 and SIL 2 SIFs; a couple of SIL 3 SIFs were also analyzed and debated. The relatively few LOPAs took about 2-3 hours each (due to the complexity of the 5% of scenarios that went to LOPA) and each SIL Verification took about 2-3 hours to perform and document.

Caution – the issue of human factors for SIL 2 and SIL 3 SIFs[6]

There is an important aspect of important aspect of SIFs and the possible achieved SIL that is not addressed in this paper. Relying on a high integrity SIF, without accounting for human error during interventions, is also a waste of resources. You may believe that you have reduced the risk by a factor of 100 for a SIL 2 SIF, or a factor of 1000 for a SIL 3 SIF, when in fact the human factors during commissioning and proof testing may degrade the overall risk reduction factor to about 10. If the human factor is accounted for in the SIF design phase and the SIL verification phase, then the human error can also be prevented to allow the higher reduction in risk that is targeted.

Conclusions

Based on experience in a multitude of PHAs, LOPAs, QRAs, and HRAs, we have seen that experienced, competent PHA teams with an experienced, competent leader can effectively apply the principles of LOPA to many scenarios during the PHA meeting. The team can determine the approximate SIL of existing SIFs, the team can determine which safeguards are truly IPLs, and the team can recommend additional IPLs, including SIFs with specific SILs. If the scenario is too complex, or the PHA team is uncertain, the team can recommend a detailed LOPA. The streamlined qualitative makes better use of resources and time than the typical approach with separate meetings for PHA and LOPA.

It is critical that the PHA leader is experienced and competent in LOPA. It is critical that the PHA team has in its memory the data for consequence severity, initiating cause frequency, IPL requirements, and the risk tolerance criteria.

Acronyms Used

AIChE– American Institute of Chemical Engineers
CCPS – Center for Chemical Process Safety (an AIChE technology alliance)
HAZOP – Hazard and Operability Analysis
HRA – Human Reliability Analysis
IEC – International Electrotechnical Commission
IEF – Initiating Event Frequency
ETA – Event Tree Analysis
FMEA – Failure Mode Effect Analysis
FTA – Fault Tree Analysis
HRA – Human Reliability Analysis
IPL - Independent Protection Layer
LOPA – Layer of Protection Analysis
PHA – Process Hazard Analysis
PII – Process Improvement Institute, Inc.
QRA – Quantitative Risk Analysis
SIF – Safety Instrumented Function
SIL – Safety Integrity Level
SIS – Safety Instrumented System
SRS – Safety Requirements Specification

References

1. *Guidelines for Hazard Evaluation Procedures*, 2008, CCPS/AIChE.
2. Bridges, W., & Dowell, A., "Identify SIF and Specify Necessary SIL, and other IPLs, as part of PHA/HAZOP," *12th Global Congress on Process Safety*, AIChE, 2015.
3. *Layer of Protection Analysis: Simplified Process Risk Assessment,* 2001, CCPS/AIChE.
4. *Functional Safety: Safety Instrumented Systems for the Process Industry Sector - Part 3: Guidance for the Determination of the Required Safety Integrity Levels*; ANSI/ISA 84.00.01 Part 3, 2004.
5. IEC 61511, *Functional Safety: Safety Instrumented Systems for the Process Industry Sector - Part 1: Framework, Definitions, System, Hardware and Software Requirements*, International Electrotechnical Commission
6. Bridges, W., and Thomas, H., "Accounting for Human Error Probability in SIL Verification Calculations," *8th Global Congress on Process Safety,* Houston, AIChE, April 2012.

479509 The Art of the Audit: Being Audit Ready, Everyday

Abstract:

Have you ever made an appointment for an annual physical and suddenly join a gym? We all know we should be exercising every day, but it becomes more important before going in for that checkup. In reality, not exercising has consequences to your health and not your doctors, so why do we do this? To impress them, instead of doing it to take care of ourselves? This phenomena of doing things you do not normally do in preparation for an event is unfortunately how too many of us treat audits. We hear an auditor is coming and end up running around trying to organize paperwork, convince the operators to clean up the plant, and everyone in management becomes much more attentive. Does this behavior really set the proper tone to develop a safety culture? Auditors are there to check up on our compliance of rules that benefit only us to follow, particularly in the field of safety.

A successful operation will be 'Audit Ready' every day. 'Audit ready' means that operations are in compliance with rules, personnel are safe and exhibit safe behaviors, and all of the team members understand why they perform the functions of their job. Compliance tools, personnel awareness and routine self-audits are three key topics in becoming ready for an audit every day.

Personnel awareness is among the most vital areas to ensuring safe and effective operations, as well as successful audits. Audits involve people from across almost all departments, and as such mirror the requirements of multifunctional teams in everyday operations.

Compliance tools are important to track work that is being completed, and create convenient locations to store data which proves compliance. This can include tracking the progress of MOCs through weekly notifications, using databases to track when procedures need to be reviewed, and more. Visibility of what needs to be accomplished is important.

Routine self-audits help us to find gaps in compliance, and reduce complacency. These can include a simple spot check of one area of compliance, or be more comprehensive, and there are pros and cons to both. A spot check keeps everyone on their toes and takes a short enough time such that they can be done more frequently.

Ensuring sustainable records and checking everyday habits will go a long way to creating a safe operating environment, and with that a workforce that is audit ready, every day.

479546 Economic Justification for Better Maintenance in a Safety System

Abstract:

It is not difficult to think of numerous ways in which human factors might impact functional safety. For example, imperfect repair, improper calibration, faulty installation, etc. can all contribute to decreases in functional safety performance. However, if actions are to be taken to mitigate these impacts then it is essential to have some quantitative measure of how this impacts the PFDavg, so that the actions' effectiveness can be assessed. We will show how a simple measure system called the Site Safety Index^{TM} (SSI) is used to adjust safety metrics, including the PFDavg and the MTTFS, on a site specific basis. The lifecycle economic impact of these metrics is then used to show how improved operations and maintenance can be justified. Economic justification for better maintenance can be easily verified through this approach being further mindful of the safety culture.

479547 How Important Is Realistic Failure Rates for Safety Instrumented Functions?

Abstract:

A given Safety Instrumented Function design must be verified to meet the required SIL level by calculating probability of failure on demand. However, this does require realistic failure rate data for each device. Fortunately, many device manufacturers are getting their products certified per IEC 61508 and many of the certificates now have failure rate data. Other manufacturers supply analysis reports with failure rate data. But how good is this data supplied by the manufacturers? All such data must be checked to make certain it is applicable for the process industries as some data may have been derived based on machine safety or automotive applications.

This paper will show some examples of manufacturer provided failure rate data and a methodology created to validate device type failure rate data for the process industries.

479917 Enabling Sustainable Relief System Management Programs through Independent Reviews

Abstract:

Overpressure protection and flare systems usually represent the last line of protection for personnel and the environment in industrial facilities operating under high temperatures and pressures. During the design phase of either Greenfield or Brownfield projects, assessing the basis for major relief systems and making clear recommendations to reduce the risks associated with each system has helped companies realize substantial safety and economic benefits. Independent and safety focused reviews, with an emphasis on value added solutions has enabled organizations to set the right foundation for a sustainable relief system management program. This, in turn, has helped enable the facilities in meeting the ever-changing regulatory needs in a timely manner.

This paper will give a typical breakdown of the workflow process for each of the entities involved in the project - the client, the contractor, and the independent reviewer. Four distinct work activities are typically identified for such a review - development of preliminary overpressure protection, engineering design of relief systems, validation of the design, and validation of as-built documentation. After each stage, the output from the independent analysis is compared with the contractor's work and updates are performed before continuing to the next stage. This paper will also give an overview of these stages and discuss the various steps and value added for each stage of the project.

Jack Rabbit II: Experimental Large-Scale Chlorine Release Trials

Mark Whitmire[1], Shannon Fox[2], Thomas Spicer[3], Leo Stockham[4], Steven Hanna[5], Joseph Chang[6], Michael D. Sohn[7], Thomas Mazzola[8], Ronald Meris[9], Damon Nicholson[10], Andy Byrnes[11]

[1]Noblis, Falls Church, VA, USA; [2]DHS S&T Chemical Security and Analysis Center, Aberdeen Proving Ground, MD, USA; [3]University of Arkansas, Fayetteville, AR, USA; [4] Leidos, Albuquerque, NM, USA; [5]Hanna Consultants, Kennebunkport, ME, USA; [6]ANSER, Falls Church VA, USA; [7]Lawrence Berkeley National Laboratory, Berkeley CA, USA; [8] Engility, Lorton, VA, USA; [9]DTRA, Fort Belvoir, VA, USA; [10]Dugway Proving Ground, UT, USA; [11]Utah Valley University, Provo, UT USA

INTRODUCTION

Chlorine is a potent toxic inhalation hazard (TIH) chemical, and each year millions of tons of it are transported in bulk via road, water, and rail. Consequently, an accidental or deliberate chlorine release near a populated area may cause a considerable number of fatalities, and would create other health, safety, and environmental problems. Hazard prediction models are inconsistent with the evidence and observations from chlorine disasters, and the understanding of such releases is otherwise insufficient to guide and inform an effective emergency response. So beginning in 2010, DHS and its Health, Safety and Environment (HSE) partners and stakeholders planned the Jack Rabbit (JR) Program. The program was led by the DHS Chemical Security Analysis Center (CSAC) located at Aberdeen Proving Grounds-Edgewood Area MD.

OVERVIEW

The Jack Rabbit (JR) program is comprised of a series of projects that includes JRI (2010), JRII Phase 1 (2015), and JRII Phase 2 (2016). The JRI (Fox and Storwold, 2011) project was completed to develop, test, and evaluate a mechanism for the controlled, rapid release of liquefied, pressurized gases (specifically, ammonia and chlorine), to observe the transport and dispersion of a disseminated vapor and aerosol cloud, and to investigate chemical deposition and reactions with water and soil. The project also evaluated instruments, test methods, and strategies that may be used to study more industrial scale tests. Using either chlorine or ammonia, two 'pilot' tests and eight 'record' tests of either 1 or 2 tons were completed. For chlorine, among other findings, the results demonstrated that downwind transport and turbulent mixing are initially resisted by a dense persistent gas and aerosol cloud, that source phenomena were non-linear with increasing release volumes, and that the reactivity with soil containing water and organic matter is an important removal mechanism.

JACK RABBIT II

JRII leveraged and extended the work of JRI with controlled chlorine field experiments on a scale corresponding to railcars and tank-trucks; that is, 5 to 20 tons each. The goals were to collect data on the release source, measure chemical reactions with the environment, observe and evaluate exposure effects on equipment and infrastructure, and assess urban impacts using a Mock Urban Environment (MUE) or an unobstructed dissemination and downwind dispersion environment, both located on the Urban Test Grid (UTG) at Dugway Proving Grounds UT. Data was collected in real time via ground based and unmanned aerial vehicle systems, video cameras, and spectroscopic instruments, in situ point detectors, path integrated detectors, and concentration and dosage determination instruments. The data was thoroughly and qualitatively reviewed and shared with all project participants in prescribed and common formats. The generated data and findings will improve the understanding of and operational hazard prediction modelling for TIH releases, and lead to more effective emergency response and training, national preparedness, and mitigation strategies.

URBAN TEST GRID (UTG)

The UTG (Nicholson, Hedrick, and Lian, N; 2017) was the designation for the JRII chlorine release site surveyed and built specifically for the JRII tests; it was at Dugway Proving Grounds west of Granite Mountain on Goodyear Road. It included the playa and adjacent support areas, concrete pad, gravel pad, and surveyed grid extending 11 km downrange and aligned roughly south to north.

The UTG included a circular, 0.5 ft. deep, 82 ft. diameter, rebar reinforced concrete disseminator vessel pad with 3 embedded temperature sensor stations, and a 1 inch steel lip to contain chlorine liquid. The center of the circular pad was the center of the UTG site, Northing 4445633.945, Easting 288109.182 (Latitude, 40.134347; Longitude, -113.487076). The concrete pad rested on a 24 inch deep, 400 ft. × 600 ft. gravel pad level with the concrete pad and symmetric about it.

Figure 1. Chlorine release from the disseminator on the concrete pad, with CONEX container array on the gravel pad of MUE, looking upwind (south) from atop the 2 x 3 CONEX container stack. The dense cool cloud remains low as it moves downwind and laterally around and through the Mock Urban Environment made from CONEX containers.

Winds predominate from approximately 165° from June to September, so the UTG was oriented about a centerline from that direction. The UTG included the playa within a 90° angle bisected by the UTG centerline with vertex at the disseminator vessel location, and extending to the 11 km arc. The "Near Field" was the UTG region from the disseminator up to and including the 2 km arc and inside the 90°

angle; the "Extended Field" was the UTG region past the 2 km arc to the 11 km arc and inside the 90° angle.

The Mock Urban Environment (MUE) was constructed on the gravel pad and surrounding the concrete disseminator pad using a staggered row array of 86 CONEX containers (CONtainer EXpress shipping containers, either 18 or 40 ft. long) to observe and study the transport and dispersion of dense gas around and above urban structures. The array included a two wide by three high (2 x 3) stack of six 20 ft.-CONEX containers located directly downwind of the source, and two 2-axle trailers nearby with insulation, dry wall, and trailer skirting for infiltration studies. Aside from the CONEX containers and trailers, up to seven vehicles were included in the array. Three fire trucks, one ambulance, and three cars were located on the north edge of the gravel pad during Trials 1-3. During Trial 4, four vehicles were placed on the playa past the 250 m arc; during trial 5, three vehicles were moved between the rows of CONEX containers but adjacent to the concrete disseminator pad, and three vehicles remained on the playa past the 250 m arc.

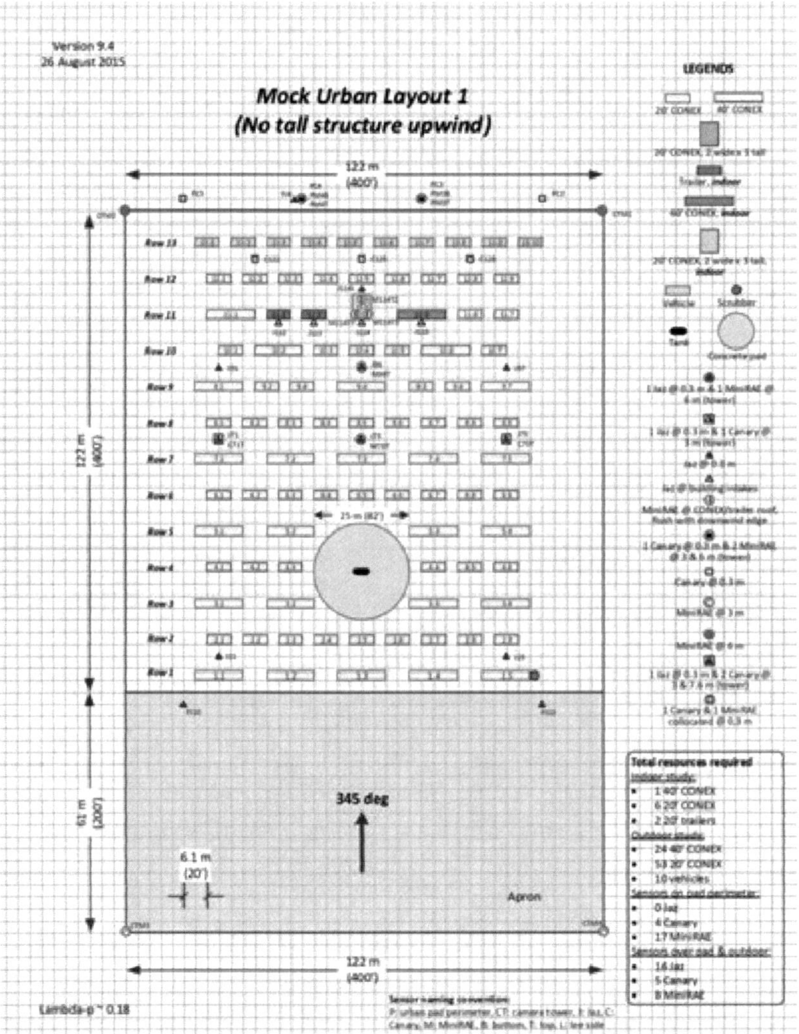

Figure 2 Mock Urban Environment (MUE) schematic showing gravel pad around circular pad for Chlorine disseminator; CONEX containers and instrument locations are also shown.

Instrument locations for the UTG "Near Field" and "Extended Field" were surveyed, recorded, and reported. Point sensors included about 170 MiniRAE 3000s equipped with volatile organic compound

(VOC) photoionization detectors; about 60 ToxiRAE Pro units; a Gasmet™ DX-4000 (FTIR) gas analyzer; 28 Ultraviolet (UV) Canary units for UTG monitoring and indoor studies; and 16 Ocean Optics Jaz instruments for UTG monitoring and indoor studies. Standoff Detectors included two Cerex Monitoring Solutions UV Sentry chlorine-specific integrated path detector, three Ultraviolet (UV) Light Detection and Ranging (LiDAR) systems for monitoring the moving chlorine gas cloud, and elastic backscatter LiDAR systems for monitoring the moving aerosol cloud. A GRIMM Aerosol portable laser aerosol spectrometer and dust monitor was also used for particle sizing model using laser scattering techniques. Other equipment included remotely operated high-definition (HD), high-speed (HS), and FLIR® IR Camera SC660 micro-bolometer cameras positioned throughout the test grid to visualize and record the chlorine release and resulting plume.

Meteorology equipment included 34 Portable Weather Information Display Systems (PWIDS) collecting data at 2 m; three 32 m towers each equipped with sonic anemometers collecting data at heights between 2 and 32 m; and 31 3D ultrasonic anemometer to gather wind and turbulence statistics, and fluxes of heat and momentum. Energy Balance Station (EBS) provided in situ measurements of the surface vertical fluxes of sensible heat, latent heat, net radiation, and soil surface heat flux. The MiniSODAR™ high-frequency Doppler acoustic sounding system designed to measure the atmospheric vertical wind profile from 15 to 200 m. Numerical gridded forecast meteorology using embedded observations and real-time meteorological data from the test grid were used for HPAC modelling leading up to each trial and to determine "GO/NO GO" test status.

DOWNWIND DISPERSION AND TRANSPORT ENVIRONMENT

For tests in 2016, and other than two vehicles, a CONEX container, and a modular office trailer used for follow-up infiltration experiments, the UTG was cleared of the simulated buildings and obstructions used in 2015. The project goals were to monitor and collect the cloud concentration data from the initial phase of a very dense, two-phase cloud near the source to dispersion of the vapor cloud farther downwind and up to 11 km away using point detectors and stand-off spectroscopic instruments.

DISSEMINATION

The disseminator vessel for use as the chlorine release source during the tests was located at the center of the concrete pad with its major axis perpendicular to the UTG centerline. The 2000 gal. tank was mounted in a 3-dimensional dynamic mass measurement load cell system. Other vessel mounted instrumentation included a redundant system for determining the remaining mass of chlorine in the vessel, differential pressure sensors on the vertical axis at various depths in and below the vapor space, guided wave radar liquid level measurement, and instrumentation for absolute pressure and direct contact temperature profiles inside the vessel. At 80-90% operating capacity, the vessel could contain from 8.5 to 10 tons chlorine.

Each release was accomplished using a blind flange with exploding bolts rated to safely hold the pressure of a vessel loaded with liquid chlorine at ambient temperature. To blow the flange and release the chlorine, the bolts were remotely detonated; when the bolts broke, the blind flange was blown off with the aid of the chlorine weight and head pressure, and chlorine flowed through the vessel's circular opening. A quick release valve was not used to simulate the release because of the likelihood of internal flashing flow in the piping and valves; that is, the mass release rate would be less than what would be observed from a breach of containment. In 2015, every release was directly down (180°) through a six inch hole. In 2016, each release was either directly down (Test 6); at 135° and downwind (Test 7), or directly up (Test 8). After Tests 7 and 8 any liquid chlorine remaining in the vessel was discharged via a remotely operated valve in the bottom of the tank. Before inert gas purging and chlorine refill operations, the ejected blind flange was replaced using standard bolts. The last release was from a tanker truck that had been removed from service, and containing up 19 t chlorine. The tanker truck's skin was breached with a remotely activated explosive shape charge.

Figure 3. Test Release 6; 10 Tons chlorine on August 31, 2016. A moment after the release the liquid chlorine may be seen splashing onto the concrete pad. As the chlorine aerosolizes and vaporizes it cools the surroundings and condenses water vapor.

OBSERVATIONS AND FINDINGS

From the JRII trials the program observed that:

- Upon breaching the tank, liquid chlorine exits the tank and rapidly flashes to a two phase aerosol and vapor cloud with a liquid layer deposited on the pad, ultimately all vaporizing and transporting downwind.
- The escaping chlorine was found to auto-refrigerate to its -34 degree Celsius boiling point and lower due to further evaporative cooling outside the tank.
- The force of the escaping chlorine jet generated strong initial turbulence and momentum to drive the expanding cloud over nearby Conex structures.
- Evacuation of the tank during 5-ton releases took approximately 35 seconds.
- Evacuation of the tank during 9-ton releases took approximately 50 seconds.
- Structures were observed to channel the escaping and expanding cloud laterally.
- Significant upwind movement of the chlorine clouds was observed due to the momentum of the escaping gas, expansion of the gas, and gravity-slumping.
- Visible clouds were observed to persist in UTG for approximately 20-30 minutes depending upon release amount and winds.
- Long visible clouds observed to stretch from UTG to approximately 2 km as downwind dispersion progressed.

- Data sets from Jack Rabbit II have been quality-assured, finalized, and recently released, enabling in-depth studies to begin on cloud concentrations, movement, and behavior to be quantified.

EMERGENCY RESPONSE EXPERIMENTS

For each test, researchers from Utah Valley University placed two "Witness Boards" on surveyed arcs adjacent to the disseminator in the urban environment to observe the effect of liquid chlorine and high concentrations of chlorine gas (approaching 25%) on common building components, urban surfaces, and fire equipment. Go-Pro® and HD color surveillance cameras captured the exposure on video. The selection of the test materials emphasized high organic content (creosote treated telephone pole and railroad tie, road asphalt, shingles, rubber, and bare wood), but copper, aluminum, galvanized steel, and carbon steel coupons were included. The copper and carbon steel surfaces began to be discolored 1-2 hours after the exposure, and the chlorine sometimes dissolved the petroleum based materials.

INDOOR INFILTRATION STUDIES

The Defense Threat Reduction Agency (DTRA) is evaluating various hazardous airborne contaminant applications for transition to DOD CBRN consequence models. Their current indoor application models presume that any contaminant is inert and neutrally buoyant, thus the models may not be useful for predicting the indoor effects of a dense, reactive gas like chlorine. Lawrence Berkeley National Laboratories (LBNL) proposed a series of experiments to DTRA and the JR program to assess the applicability of the models for predicting the indoor concentration of chlorine in the event of an outdoor chlorine release.

LBNL successfully completed a series of indoor infiltration experiments in 2015 in support of the DTRA effort. The LBNL experiments were completed to understand and characterize the dominant infiltration pathways for reactive dense gases into buildings, the uptake of such gases into structures, and the re-emission of them from the structures after the peak of the plume had passed. Three single-story CONEX structures, a single three-story CONEX structure, and 2 modular office trailers placed on the gravel pad were used to represent indoor office, school, or home environments. Follow-on studies were completed in 2016 using a single-story CONEX structure at the same, but unobstructed site. The CONEX structures were configured for controlled experiments of airflow and transport in single- or multi-story structures, and modified to include insulation and temperature-control. Chlorine penetration into and through the structure, and transfer of it across the building envelope was monitored via controlled artificial “leaks” and exhaust fans. The model offices were furnished with realistic insulation, surface materials, windows, and HVAC system characteristics, thus the simulated modular offices helped to relate the data from the controlled CONEX structures to real buildings. The data may be used for calibration, verification and/or validation of the indoor model, and to support DOD concepts of operation for sheltering in the event of an outdoor CBRN release.

PROGRAM STATUS

The program is entering a new phase after completing field testing. In 2017, the program will be processing and reporting data from 2015 and 2016, and transitioning it for permanent data storage, and to stakeholders. CSAC will also be presenting a JR overview, narrative, and plans in presentations to interested groups. The program will be proposing, designing, managing, and performing small and large scale experiments during chamber testing, and reducing and analyzing data to demonstrate or elucidate the reaction of chlorine with various substrates to suggest large-scale toxic inhalation hazard (TIH) chemical release mitigation strategies, inform emergency response strategies that may reduce the impact to affected populations and infrastructure, calculate the value of modeling parameters, and otherwise support stakeholders interests.

ACKNOWLEDGEMENTS
This research has been supported by the US Department of Homeland Security (DHS) and the US Defence Threat Reduction Agency (DTRA). Additional support (funding, personnel and instruments) is coming from numerous partners, with the list growing with time. We appreciate guidance from many scientists who are members of the planning groups and/or have provided comments on the scientific approach.

REFERENCES

1. Nicholson, D., Hedrick, A., and Lian, N., Final Data Report for Jack Rabbit (JRII), 2017

2. Fox, S. and D. Storwold, 2011: Project Jack Rabbit: Field Tests. Chemical Security Analysis Center, Science and Technology Directorate, U.S. Department of Homeland Security. CSAC 11-006, 162 pp.

479394 Functional Safety - Pitfalls and Good Practices

Abstract:

The implementation of effective functional safety is valuable in ensuring that safety critical control and protective systems are appropriately specified, designed, installed and maintained. Through the International Electro-technical Commission (IEC) 61508 and 61511 suite of functional safety standards, the concepts of the Safety Lifecycle and Safety Integrity Levels (SIL) are now well known. However, DNV GL's experience from numerous projects across different industry sectors suggests that there remain challenges to be overcome.

DNV GL has conducted a number of Functional Safety Assessments (FSAs) on various electrical/ electronic and programmable electronic safety systems in the process sector. The FSAs are a mandatory element of IEC 61511 which provide an independent judgment on the integrity of the safety system and should be performed by competent independent Functional Safety Practitioners. At least one FSA must be performed on a Safety Instrumented System and should be completed prior to the introduction of hazards. Through DNV GL's experiences in functional safety assessments this paper will discuss;

- How to avoid common mistakes in the management and implementation of functional safety;
- Examples of good practice in the application of functional safety;
- How good practice can help optimize test regimes, increase plant availability and reduce maintenance workload whilst ensuring the system continues to provide the required risk reduction.

General Method for Uncertainty Evaluation of Safety Integrity Level (SIL) Calculations – Part 2 Analytical Methods

Raymond "Randy" Freeman
S&PP Consulting

For Presentation at the American Institute of Chemical Engineers
2017 Spring Meeting in San Antonio, Texas -- March 26-30, 2017

Abstract

The IEC 61511 standard requires a verification calculation that a proposed design for a safety instrumented function (SIF) achieves the desired safety integrity level (SIL). The evaluation of the safety integrity level of a new or existing safety instrumented system requires detailed calculations based on the failure rates of the device and the planned maintenance/testing cycle for the system. The failure rates of the devices are often taken from standard failure rate tabulations of equipment. The maintenance and testing plans are developed based on plant experience. All of the data used in the SIL calculations are uncertain. This paper develops a general method for uncertainty analysis of the SIL calculations. The general method is based on the application of probability theory - variance contribution analysis (VCA) – to the equations presented in ISA TR 84.00.02-2115. An example is worked to demonstrate the methodology.

Background

The calculation of the probability of failure on demand (PFD) is a common engineering task when designing an interlock or safety system that is to be in compliance with IEC 61511 [Ref. 1]. The calculation of the PFD is often done using approximate equations defined in the ISA TR84.00.02 technical report [Ref. 2]. The simplified equation method in the ISA report is commonly used and is based on the use of reliability block diagrams where the field sensors, Safety Instrumented System (SIS) logic solver and final control elements are considered independent of each other in the sense of not sharing common devices or systems. The PFD is then calculated using the failure rates of the devices, planned test intervals, vendor supplied estimates on diagnostic coverage of the devices and an allowance for the potential for common cause failures. Almost all of these parameters are uncertain. The failure rate data is often taken from generic data sources which show wide ranges in the observed values.

Because of the uncertainty in the parameters, the design engineer makes allowances in the design by the use of safety factors or rules of thumb to improve the chances that the final interlock installation will work as intended. Since each engineer has a different set of safety factors and rules of thumb, two designs may differ significantly in the way a hazard is controlled.

A more formal method for handling the underlying uncertainty in the calculation of the PFD of an interlock is needed. Previously, Freeman and Summers [Ref. 3] published an uncertainty analysis of the PFD of an interlock. Two different methods were used in this analysis:

- Monte Carlo Simulation
- Variance Contribution Analysis (VCA)

Monte Carlo Simulation requires that the engineer build a model of the interlock using specialized computer software. The use of VCA requires that the sensitivity of the interlock model be determined either by numerical methods or by direct analytical calculations. The Freeman paper demonstrates that VCA can be used for the uncertainty analysis. However, the paper does not present a complete analysis method that can be applied to any system defined in the ISA technical report TR84.00.02. The goal of this paper is to develop a general set of analytical equations that will allow VCA to be used in uncertainty analysis of any interlock developed per the IEC 61511 standard.

Review of Interlock Design

The IEC standard looks at an interlock as a series of three major components (see Figure 1). First is the sensor set which sends an indication of an abnormal event to a logic solver. The logic solver determines is the signal from the sensor meets the conditions required to activate (trip) the interlock. If the interlock is to be activated, a signal is sent to the final control elements to take action (stop flow, close valve, etc) to prevent a process safety event from occurring.

Based on the devices selected the design engineer then evaluates whether the proposed interlock design meets the design criterion for probability of failure on demand. Typically, the design engineer is given the target PFD as a statement such as “provide a SIL-2 interlock to stop flow on high level in a vessel.” A SIL-2 interlock requires a PFD of at most 0.01 or a risk reduction factor of at least 100. The needed risk reduction is often determined in a Layer of Protection Analysis (LOPA) [Ref. 4] and represents the managements decision on how risk is to be managed in a system. The resulting SIL-2 interlock design must have a high likelihood of achieving the target risk reduction as the LOPA team may be relying on it to be part of an overall risk reduction plan to prevent a process safety incident.

The ISA technical report (TR 84.00.02) [Ref. 2] provides shortcut methods for the evaluation of the PFD of an interlock design. Tables 1 and 2 are taken from the ISA technical report [Ref. 2]. Typical parameters need to compute the PFD include:

- DC is the diagnostic coverage;
- DI is the diagnostic interval;
- TI is the proof test interval,
- λ^D is the dangerous failure rate;
- MTTR is the mean time to restore the system to operation
- β is the common cause failure parameter that is always is between 0 and 1

In practice most if not all of these parameters are uncertain. Using parameter values obtained from plant records, vendor data or generic data bases the design engineer proceeds to develop the model for the interlock.

What is needed is a means to rapidly evaluate the impact of parameter uncertainty on the interlock PFD. The remainder of this paper applies the methods of variance contribution analysis (VCA) to determine the mean (expected value) and the standard deviation of the interlock PFD for a given design.

Table 1. Simplified PFD_{avg} formulas for Non-Repairable System without considering CCF, Diagnostics or MTTR.			
Configuration	**Function**	**PFD based on "Average before" failure rate**	**Appendix A Equation Number**
1oo1	$F_1 =$	$\left[\lambda^D \times \frac{TI}{2}\right]$	A-1
1oo2	$F_2 =$	$\left[(\lambda^D)^2 \times \frac{TI^2}{4}\right]$	A- 2
1oo3	$F_3 =$	$\left[(\lambda^D)^3 \times \frac{TI^3}{8}\right]$	A-3
2oo2	$F_4 =$	$\left[\lambda^D \times TI\right]$	A-4
2oo3	F5 =	$\frac{3}{4} \times \left[(\lambda^D)^2 \times TI^2\right]$	A-5
3oo3	F6 =	$3 \times \left[\lambda^D \times \frac{TI}{2}\right]$	A-6

Where TI is the proof test interval, λ^D is the dangerous failure rate.

Table 2. Simplified PFD_{avg} Formulas for Repairable System considering CCF, Diagnostics and MTTR			
Configuration	**Function**	**PFD based on "Average before" failure rate**	**Appendix A Equation Number**
1oo1	$F_7 =$	$\left[\frac{(1-DC)\times\lambda^D\times TI}{2}+\frac{DC\times\lambda^D\times DI}{2}+\lambda^D\times MTTR\right]$	A-24
1oo2	$F_8 =$	$\left[\frac{(1-DC)\times(1-\beta)\times\lambda^D\times TI}{2}+\frac{DC\times(1-\beta)\times\lambda^D\times DI}{2}+(1-\beta)\times\lambda^D\times MTTR\right]^2+\left[\frac{(1-DC)\times\beta\times\lambda^D\times TI}{2}+\frac{DC\times\beta\times\lambda^D\times DI}{2}+\beta\times\lambda^D\times MTTR\right]$	A- 25
1oo3	$F_9 =$	$\left[\frac{(1-DC)\times(1-\beta)\times\lambda^D\times TI}{2}+\frac{DC\times(1-\beta)\times\lambda^D\times DI}{2}+(1-\beta)\times\lambda^D\times MTTR\right]^3+\left[\frac{(1-DC)\times\beta\times\lambda^D\times TI}{2}+\frac{DC\times\beta\times\lambda^D\times DI}{2}+\beta\times\lambda^D\times MTTR\right]$	A-26
2oo2	$F_{10} =$	$2\times\left[\frac{(1-DC)\times\lambda^D\times TI}{2}+\frac{DC\times\lambda^D\times DI}{2}+\lambda^D\times MTTR\right]$	A-27

Table 2. Simplified PFD_{avg} Formulas for Repairable System considering CCF, Diagnostics and MTTR			
Configuration	**Function**	**PFD based on "Average before" failure rate**	**Appendix A Equation Number**
2oo3	F_{11} =	$3\times\left[\frac{(1-DC)\times(1-\beta)\times\lambda^{D}\times TI}{2}+\frac{DC\times(1-\beta)\times\lambda^{D}\times DI}{2}+(1-\beta)\times\lambda^{D}\times MTTR\right]^{2}+\left[\frac{(1-DC)\times\beta\times\lambda^{D}\times TI}{2}+\frac{DC\times\beta\times\lambda^{D}\times DI}{2}+\beta\times\lambda^{D}\times MTTR\right]$	A-28
3oo3	F_{12} =	$3\times\left[\frac{(1-DC)\times\lambda^{D}\times TI}{2}+\frac{DC\times\lambda^{D}\times DI}{2}+\lambda^{D}\times MTTR\right]$	A-29

Review of Variance Contribution Analysis (VCA) Methodology

The mean and variance of a function of random variables can be approximated using the method described by Haugen [Ref. 5] and applied by Freeman [Ref. 3, 6, 7]. Define an arbitrary function of a set of random variables, x_i, as:

Let

$$Y = F(x_i) \qquad \text{(Eq 1)}$$

The mean of Y can be estimated using the following approximation:

$$E(Y) = F[\ E(x_i)] \qquad \text{(Eq 2)}$$

Where:
$E(Y)$ = expected value of random variable Y = mean of Y
$E(x_i)$ = expected value of random variable x_i = mean of x_i

The variance of Y can likewise be estimated as:

$$V(Y) = \sum_{i=1}^{n} \left[\frac{\partial Y}{\partial x_i}\right]^2 V(x_i) \qquad \text{(Eq 3)}$$

Where:

$V(Y)$ = variance of random variable Y as defined above in Equation 1

$V(x_i)$ = variance of random variable x_i as defined above in Equation 1

Note that the variance is simply the square of the standard deviation. Using the variance will simplify the mathematics that is described below. The contribution of each independent random variable to the overall variance in the function is:

$$V(Y \text{ from } x_i) = \left[\frac{\partial Y}{\partial x_i}\right]^2 V(x_i) \qquad \text{(Eq 4)}$$

The relative contribution of each term to the overall variance V(Y) is a measure of the importance in the uncertainty in the particular random variable, x_i. In effect, this is a sensitivity analysis combined with a uncertainty evaluation. The variance contribution combines the sensitivity in the answer to changes in the uncertain random variable, x_i, with a measure of the uncertainty in the random variable, x_i. The overall variance in Y is found by summing the sensitivity weighted variances from each random variable.

Recommended Interlock PFD Uncertainty Analysis Method

Previously, Freeman and Summers [Ref. 3] suggested a framework for the inclusion of uncertainty analysis in the calculations completed for a new interlock per the ISA TR [Ref. 2]. The following uncertainty analysis method has been expanded to incorporate critical decisions that the process management must make in the design process.

1. Complete the interlock design using the methods outlined in IEC 61511 [Ref. 1].

2. Review with the process system management and determine if the proposed interlock should be considered repairable or non-repairable. A simple flow chart for this decision making is presented in Figure 2. The basic question is: "can the interlock be repaired safely while the process operates." This is a management question and management should be the one that decides the answer to this important question.

3. Create interlock performance equation as the mathematical model for the combination of sensor, logic solver and final control elements using the methods outlined in ISA technical report (TR84.00.02) [Ref. 2]. Equations needed to evaluate the uncertainty of the models specified in the ISS report are developed in Appendix A of this paper. For non-repairable systems, Table 3 can be used. For more complex systems such as non-repairable redundant systems with the potential for common cause failures, use the repairable equations of Table 4 and set DI, DC, and MTTR all equal to zero. For systems where common cause failures (CCF), Diagnostics and repair are to be considered, use the recommendations of Table 4. There are design situation where the equations developed in Appendix A will not exactly apply. Specifically, designs where the system contains redundancy subject to common cause failures (CCF) that should be considered non-repairable during operation. The failure models and the equations needed to evaluate the uncertainty in these systems are developed in Appendix B to this paper.

Table 3. Roadmap for Mean and Variance of Non-Repairable System Without Considering CCF, Diagnostics and MTTR			
Configuration	**Function**	**Mean PFD Using Appendix A Equation**	**PFD Variance Using Appendix A Equation**
1oo1	F_1 =	A-9	A-18
1oo2	F_2 =	A- 10	A- 19
1oo3	F_3 =	A-11	A-20
2oo2	F_4 =	A-12	A-21
2oo3	F_5 =	A-13	A-22
3oo3	F_6 =	A-14	A-23

Table 4. Roadmap for Mean and Variance of Repairable System Considering CCF, Diagnostics and MTTR			
Configuration	**Function**	**Mean PFD Using Appendix A Equation**	**PFD Variance Using Appendix A Equation**
1oo1	F_7 =	A-30	A-44
1oo2	F_8 =	A- 31	A- 86
1oo3	F_9 =	A-32	A-94
2oo2	F_{10} =	A-33	A-48
2oo3	F_{11} =	A-34	A-102
3oo3	F_{12} =	A-35	A-50

Note that in all of the calculations indicated by the equations referenced in Table 4, the expected value of any the random variables is used to complete the calculations.

4. Define the uncertainty in the parameters and variables of the interlock model specified in step 2. The uncertainty can be given as the upper and lower range of the possible values (uniform probability distribution), as the upper, lower and recommended values (triangular distribution), or as a mean and standard deviation (normal distribution). See the example above for guidance in the evaluation of safety instrumented system interlocks.

5. Compute the expected value of each variable in the interlock performance equation. The equations for the mean and variance for the uniform, triangular and normal probability distributions are presented for various probability distributions in Vose [Ref. 8]. The example interlock calculations used a triangular distribution to represent the uncertainty in the parameters.

6. Compute the expected value or mean of the interlock PFD using the mean value of each of the variables in Step 5. See Tables 4 and 5 for recommended equations.

7. Compute the sensitivity of the result from the interlock performance equation by use of the partial derivative of the basic interlock performance equation with respect to each of the variables as presented in Appendix A.

8. Compute the variance of the interlock performance equation PFD by use of the variance contribution using equations from Tables 3 or 4. This entails multiplying the variance of each of the uncertain variables in the basic interlock performance equation by the square of its sensitivity (obtained in step 7), as evaluated at the variable mean. Sum the resulting terms to obtain the overall variance of the PFD in the interlock performance equation. See Tables 4 and 5 for recommended equations.

9. Determine the level of risk that the owner/operator wishes to take that the final interlock will not work. Note that this is a management decision not an instrument engineer decision! In this paper, the 95% level of risk reduction has been used:
   - 5% chance of failure or a 95% chance of the interlock achieving the desired risk reduction

10. Assuming that the interlock owner operator wishes to take a low risk (5%) of the interlock failing to achieve its design target PFD, compute the 95% upper confidence limit on the computed PFD by use of the standard normal factor, Z, [Ref. 9] as:

$$Z = \left[\frac{xi - E(x)}{\sigma}\right] \qquad \text{(Eq 5)}$$

Where:

σ = standard deviation of the PFD of the interlock of interest from the interlock performance equation obtained from step 7. Note that the variance of a random variable is the square of the standard deviation of the random variable.

E(x) = the expected value of the PFD of the interlock of interest from the interlock performance equation obtained from step 5

For the 95% upper limit, Z = 1.645. Rearranging Eq. 5 allows for the direct calculation of the corresponding value of the 95% upper confidence limit on the PFD as:

$$X95\% = 1.645\,\sigma + E(x) \qquad \text{(Eq 6)}$$

Where:

X95% = the upper 95% limit on the computed PFD of the interlock of interest from the interlock performance equation.

Compare the 95% upper confidence limit on the PFD of the interlock of concern with that established as the desired PFD for risk reduction. If the 95% confidence of the RRF is greater than the desired RRF, the design is complete. If not, revise the design or change inspection test intervals to achieve the desired RRF. If it is not possible to achieve the desired target RRF economically, revisit the LOPA study accordingly to incorporate better information obtained in the uncertainty analysis. Improve the integrity of the LOPA IPLs or identify additional IPLs to drive the risk to a tolerable level. Continue this process until the computed RRFs are greater than the desired RRFs for risk reduction and risk management.

Example Interlock Problem

Previously, Freeman and Summers [Ref. 3] published an example in the use of VCA to determine the likelihood that the interlock would provide the risk reduction desired by the management and the LOPA team. The process system is shown in Figure 3. The process uses a compressor to increase the pressure of a process stream prior to additional processing. The gas being compressed is toxic and flammable. Of concern is a slug of liquid being sent to the compressor. If a liquid slug is sent to the compressor, significant damage to the compressor is expected with probable seal damage and a subsequent release of flammable and toxic material into the work area. A large fire and/or explosion could result if the release were to be ignited. If the release is not ignited, the nearby workers could be exposed to the toxic gas resulting in death or injury. A Layer of Protection Analysis (LOPA) review of this system has been completed. Among several recommendations, the LOPA team recommended the installation of a SIL-2 high level interlock in the Compressor Knock Out Drum to stop the compressor prior to liquids entering the system.

Figure 1 shows a simplified diagram for the interlock. Three level sensors are provided in the Compressor Knock Out Drum. The SIS logic solver will use 2oo3 voting to detect high level in the Compressor Knock Out Drum. The SIS logic solver will also monitor the difference in level signal from each of the three level sensors and will activate an alarm if the deviation is excessive. Two independent methods of stopping the compressor are provided. The SIS will directly signal the motor controller on the compressor to stop. In addition, the SIS logic solver will also signal two additional relays to open the power supply to the compressor motor. These two different shutdown will both be activated in the event of high level in the Compressor Knock Out Drum. Either one of the two shutdowns is capable of stopping the compressor by turning off the electric power supply to the motor.

Models

The first step in the calculation of the "goodness" of an interlock is to establish the model to be used in the calculations. Note that the sensors are 2oo3 voting and the final control elements are each 1oo1. Appendix A presents the equations for various models that can be used for describing this system.

The overall probability of failure on demand (PFD) of the interlock is given as:

PFD = PFDs + PFDsis + PFDfce (Eq 7)

Where:

PFD = Probability of failure on demand of the interlock as a whole

PFDs = Probability of failure on demand of the sensors (voting as 2oo3)

PFDsis = Probability of failure on demand of the SIS logic solver

PFDfce = Probability of failure on demand of the final control elements.

Since there are two final control elements arranged in series, the PFDfce = sum of the PFDs of the final control elements (Relay PFD and MCC PFD).

PFDce = PFDr + PFDmcc (Eq 8)

Where:

PFDr = Probability of failure on demand of the two relays voting as 1oo2 to shutdown gas compressor.

PFDmcc = Probability of failure on demand of the MCC to shutdown the gas compressor

For this example, the following selections are made to model the performance of the interlock.

Model for Sensors (2oo3)

The sensors are considered repairable as one sensor can be replaced while the other two function and provide continuous functionality of the interlock. There are three sensors that will be used in a 2oo3 voting system. From Appendix A, using Equation A-7, the model for the sensors becomes:

$$\text{PFDsavg} = 3\times\left[\frac{(1-DCs)\times(1-\beta s)\times\lambda^{Ds}\times TIs}{2}+\frac{DCs\times(1-\beta s)\times\lambda^{Ds}\times DIs}{2}+(1-\beta s)\times\lambda^{Ds}\times MTTRs\right]^2+$$

$$\left[\frac{(1-DCs)\times\beta s\times\lambda^{Ds}\times TIs}{2}+\frac{DCs\times\beta s\times\lambda^{D}\times DIs}{2}+\beta s\times\lambda^{Ds}\times MTTRs\right] \quad \text{(Eq 9)}$$

Where:

PFD_{savg} is the average probability of failure on demand of the sensors

DC_s is the diagnostic coverage for sensor failure

DI_s is the diagnostic interval for the sensors

$MTTR_s$ is the mean time to restore the sensors to functionality given a sensor failure

TI_s is the test interval for the sensors

β_s is the common cause failure parameter

λ^{Ds} is the failure rate to a dangerous condition for the sensors

MTTR is the mean time to restore the system from the time that failure occurs

Model for SIS Logic Solver

Use a fixed probability of failure on demand as specified by the vendor. Unless detailed design information is provided by the SIS logic solver vendor, this will be the normal default condition for most studies. For this problem a typical PFD of 1.30×10^{-4} was selected to represent the SIS Logic Solver.

Model for Final Control Elements

There are two separate paths to shutdown the gas compressor. First is by the SIS logic solver commanding the MCC to shutdown power to the the gas compressor motor. The second is for the SIS logic solver to issue a shutdown command to two interposing relays (R1 and R2) which will cause the power to the gas compressor motor to stop. Two different models are needed to the final control elements. These two subsystems are considered non-repairable in this analysis,

Model for Interposing Relays (1oo2)

There are two interposing relays (R1 and R2 in the interlock). From Appendix A, use Equation A-25 setting parameters DC=DI=MTTR=0 for the non-repairable interposing relays, the model becomes:

$$PFDravg = \left[\frac{(1-\beta r)\times\lambda^{Dr}\times TIr}{2}+\right]^2+\left[\frac{\beta r\times\lambda^{Dr}\times TIr}{2}+\right] \quad \text{(Eq 10)}$$

Where:

PFD_{ravg} is the average probability of failure on demand of the relays voting 1oo2 to shutoff the gas compressor.

TI_r is the test interval for the relays

β_r is the common cause failure parameter

λ^{Dr} is the failure rate to a dangerous condition for the relays

Model for MCC (1oo1)

From Appendix A and using equation A-1 for the non-repairable MCC, the model becomes:

$$\boldsymbol{PFDmccavg} = \frac{\lambda^{Dmcc} * \mathbf{TImcc}}{2} \quad \text{(Eq 11)}$$

Where:

PFD_{mccavg} is the average probability of failure on demand of the MCC to shutoff the gas compressor.

TI_{mcc} is the test interval for the MCC

$\lambda^{Dmcc}$is the failure rate to a dangerous condition for the MCC

Data

The calculation of the PFD of the interlock requires a set of data to be used to represent the system. Tables 5, 6, and 7 are taken from the Freeman-Summers paper [Ref. 3] and presents the data used to represent the interlock system. Note that these data were originally taken from generic data sources and do not represent any particular device or system.

Table 5. Uncertainty Data for Level Sensors Used in Example Interlock

(All variable probability distributions assumed to be triangularly distributed)

Variable	Min	Mode	Max	Mean	Variance
λ^{Ds}- Fail Dangerous Rate	2.84×10^{-3} Yr^{-1}	5×10^{-3} Yr^{-1}	8.5×10^{-3} Yr^{-1}	5.45×10^{-3} Yr^{-1}	1.36×10^{-6} Yr^{-1}
DC_s – Diagnostic Coverage *	0.8	0.9	0.99	0.897	1.51×10^{-3}
DI_s – Diagnostic Interval **	5.71×10^{-5}	5.71×10^{-5}	5.71×10^{-5}	5.71×10^{-5}	0
TI_s – Test Interval	4 Yr	5 Yr	6 Yr	5 Yr	5.56×10^{-2} Yr^2
β_s- Common Cause Failure Fraction *	0	0.02	0.1	0.04	4.67×10^{-4}
$MTTR_s$ - Mean Time to Restore ***	1.37×10^{-3} Yr	8.22×10^{-3} Yr	1.92×10^{-2} Yr	9.59×10^{-3} Yr	1.34×10^{-5} Yr^2

* Unit less

** 0.5 hours, assumed to be deterministic

***Min of 12 hr, Mode of 72 hours, Max of 168 hrs

Table 6. Uncertainty Data for Relays Used in Example Interlock (All variable probability distributions assumed to be triangularly distributed)					
Variable	**Min**	**Mode**	**Max**	**Mean**	**Variance**
λ^{Dr}- Fail Dangerous Rate	8.76×10^{-9} Yr^{-1}	2.00×10^{-3} Yr^{-1}	4.73×10^{-2} Yr^{-1}	1.64×10^{-2} Yr^{-1}	1.19×10^{-4} Yr^{-2}
DC_r – Diagnostic Coverage *	NA	NA	NA	NA	NA
DI_r – Diagnostic Interval*	NA	NA	NA	NA	NA
TI_r – Test Interval	1 Yr	1 Yr	2 Yr	1.33 Yr	0.056 Yr^2
β_r- Common Cause Failure Fraction	0	0.02	0.1	0.04	4.67×10^{-4}
$MTTR_r$ - Mean Time to Restore*	NA	NA	NA	NA	NA

* Not used in relay model

Table 7. Uncertainty Data for MCC Used in Example Interlock (All variable probability distributions assumed to be triangularly distributed)					
Variable	**Min**	**Mode**	**Max**	**Mean**	**Variance**
λ^{Dr}- Fail Dangerous Rate	1.74×10^{-4} Yr^{-1}	1.31×10^{-3} Yr^{-1}	3.00×10^{-2} Yr^{-1}	1.05×10^{-2} Yr^{-1}	4.76×10^{-5} Yr^{-2}
DC_r – Diagnostic Coverage*	NA	NA	NA	NA	NA
DI_r – Diagnostic Interval*	NA	NA	NA	NA	NA
TI_r – Test Interval	1 Yr	1 Yr	2 Yr	1.33 Yr	5.6×10^{-2} Yr^2
β_r- Common Cause Failure Fraction*	NA	NA	NA	NA	NA
$MTTR_r$ - Mean Time to Restore*	NA	NA	NA	NA	NA

* Not used in MCC model

Uncertainty Analysis Results

Using the VCA method to estimate the PFD of the interlock design yields the results shown in Table 8.

Table 8. Results of VCA on Example Interlock Design

Subsystem	Mean PFD	Variance of PFD
Sensors	6.41E-5	1.7657E-9
Logic Solver	1.34E-4	0
Relays	5.46E-4	2.0333E-7
MCC	6.98E-3	4.52E-5
Totals	**7.73E-3**	**4.54E-5**

Note: Calculations are carried to an adequate number of places to allow for the resulting PFD to be determined. Rounding to significant number of figures is done at the end.

The mean PFD of 0.00773 implies a mean risk reduction factor of:

RRF = 1/PFD = 1/0.00773 = 129 (Eq 11)

The variance of 4.54E-5 implies a standard deviation in the PFD of

Std Dev = $(\text{Variance})^{1/2}$ = 0.006737 (Eq 12)

The corresponding 95% level of confidence in the PFD:

X95% = 1.645 std Dev + Mean PFD (Eq 13)

X95% = 1.645 (0.006737) + 0.00773= 0.01881 (Eq 14)

The RRF for 95% certain PFD is

RRF95% = 1/0.01881 => 53 (Eq 15)

This is essentially the same result previously reported by Freeman and Summers [Ref. 3] using either Monte Carlo Simulation or numerical approximation methods for the VCA sensitivities. The calculated PFD at the 95% level of 0.01881 indicates that there is only a 5% chance that the interlock will provide an RRF worse that 53. Since this level is only SIL-1 capable and not
SIL-2 capable as desired by management, a revised design will be needed to achieve the desired risk reduction or a different protective measure will be needed to control risk.

Acknowledgement

This paper is an expanded version of a paper entitled ''General Method for Uncertainty Evaluation of Safety Integrity Level (SIL) Calculations'' presented at the Mary Kay O'Connor 2016 Process Safety Symposium, October 26, 2016, College Station, TX. The author wishes to thank those individuals who gave comments and suggestions for improving this article.

References

1. International Society of Automation, Functional Safety: Safety Instrumented Systems for the Process Sector—Parts 1, 2, and 3, ANSI/ISA 84.00.01, 2004 [USA implementation of IEC 61511].
2. International Society of Automation, *Safety Integrity Level (SIL) Verification of Safety Instrumented Functions*, ISA TR84.00.02, Research Triangle Park, NC. (2015).
3. Freeman, R. And Summers, A, "Evaluation of Uncertainty in Safety Integrity Level Calculations," *Process Safety Progress*, Vol 35, No 4, pp 341-348, December 2016. Published Wiley Online Library (wileyonlinelibrary.com). DOI 10.1002/prs.11805, 23 January 2016
4. Center for Chemical Process Safety, *Layer of Protection Analysis – Simplified Process Risk Assessment*, American Institute of Chemical Engineers, New York, 2001
5. Edward B. Haugen, *Probabilistic Approaches to Design*, John Wiley, New York, 1968
6. R.A. Freeman, "Quantifying LOPA Uncertainty," *Process Safety Progress*, Vol 31, No. 3, pp 240-247, 2012
7. R. A. Freeman, "Simplified Uncertainty Analysis of Layer of Protection Analysis Results," *Process Safety Progress*, Vol. 32, No 4, pp 351-360, December 2013, Published online in Wiley Online Library (wileyonlinelibrary.com) DOI 10.1002/ prs11585.pdf, 3 June 2013
8. David Vose, *Risk Analysis – A Quantitative Guide*, 3rd Edition, John Wiley, 2008
9. Paul L. Meyer, *Introductory Probability and Statistical Applications*, John Wiley, New York, 1972

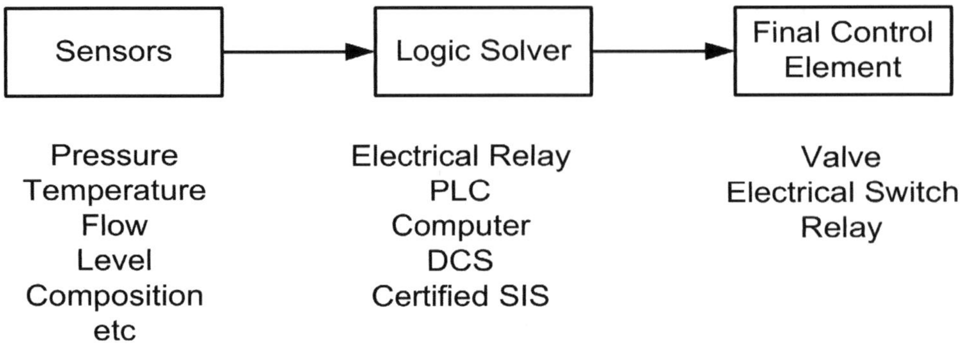

Figure 1. Typical Components of an SIS Interlock

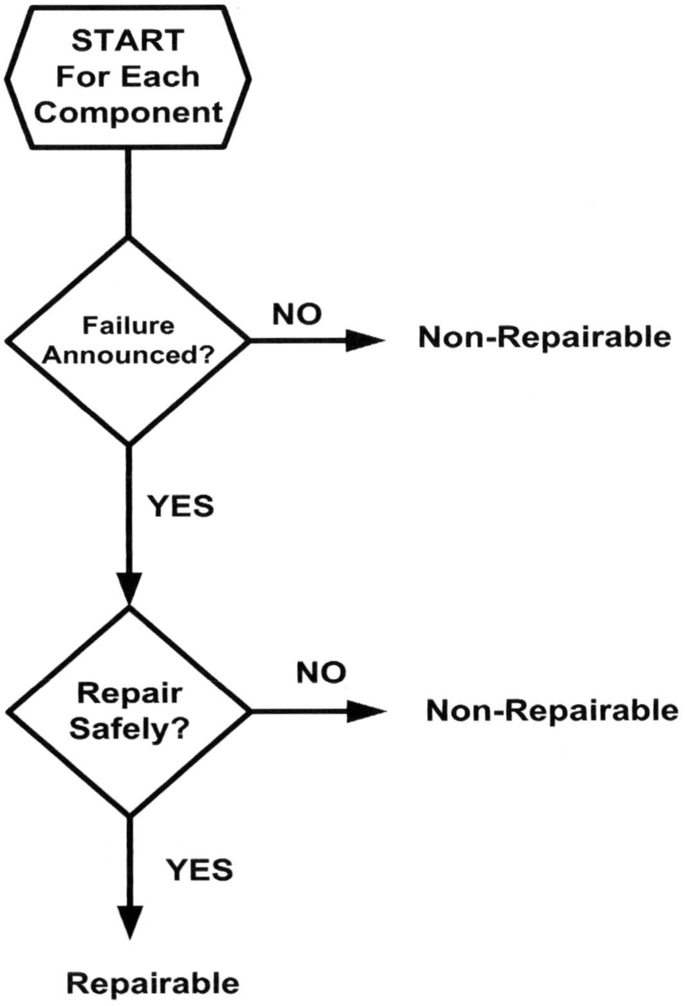

Figure 2. Decision Tree for Determination of Repairable or Non- Repairable

Figure 3. P&ID Sketch for Example Interlock

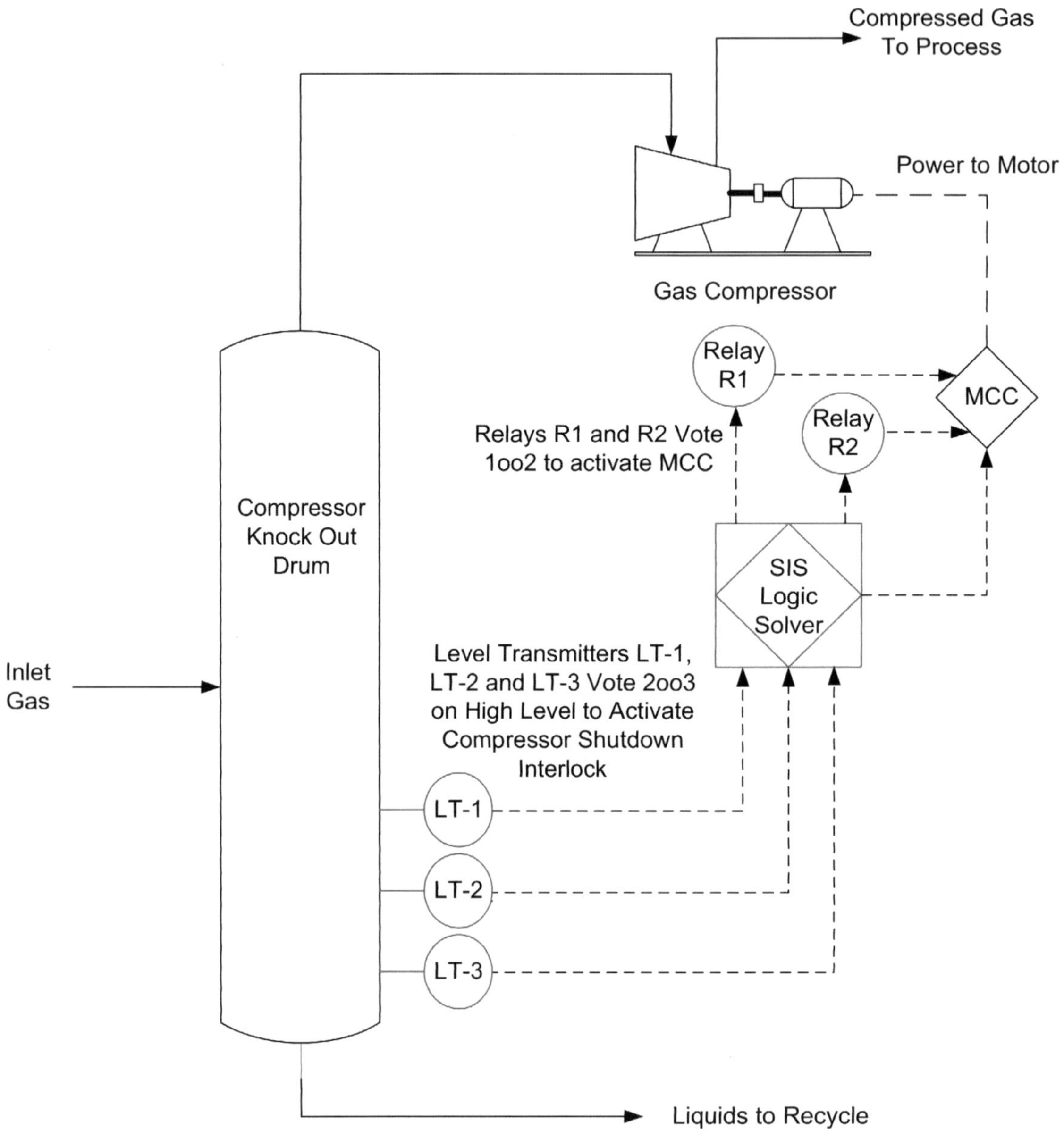

APPENDIX A -- DEVELOPMENT OF UNCERTAINTY EQUATIONS

This appendix develops the equations needed to evaluate the uncertainty in the SIL PFD as computed using the methods outlined in the ISA Technical Report [Ref. 2]. This appendix is divided into two sections:

- Non-Repairable Systems
- Repairable Systems

The uncertainty equations for other configurations not covered in the ISA report are developed in Appendix B.

NON-REPAIRABLE SYSTEMS

For systems that are considered non-repairable, the simplified equations of Table A.1 (taken from the ISA Technical Report [Ref. 2]) are used for the analysis.

Table A.1 Simplified PFD_{avg} formulas for Non-Repairable System without considering CCF, Diagnostics or MTTR.

Configuration	Function	PFD based on "Average before" failure rate	Equation Number
1oo1	$F_1 =$	$\left[\lambda^D \times \frac{TI}{2}\right]$	A-1
1oo2	$F_2 =$	$\left[(\lambda^D)^2 \times \frac{TI^2}{4}\right]$	A- 2
1oo3	$F_3 =$	$\left[(\lambda^D)^3 \times \frac{TI^3}{8}\right]$	A-3
2oo2	$F_4 =$	$\left[\lambda^D \times TI\right]$	A-4
2oo3	F5 =	$\frac{3}{4} \times \left[(\lambda^D)^2 \times TI^2\right]$	A-5
3oo3	F6 =	$3 \times \left[\lambda^D \times \frac{TI}{2}\right]$	A-6

Where TI is the proof test interval, λ^D is the dangerous failure rate.

The expected value of the probability of failure on demand is found by calculating the PFD using the expected value (mean) of the random variables.

Let y a function of some random variables, x_i, as:

$$y = F(x_i) \quad \text{(Eq A-7)}$$

The mean of y can be estimated using the following approximation:

$$E(y) = F[E(x_i)] \quad \text{(Eq A-8)}$$

Where:

E(y) = expected value of random variable y = mean of y

$E(x_i)$ = expected value of random variable x_i = mean of x_i

For the configurations defined in Table A.1, the corresponding expected PFD are presented in Table A.2.

Table A.2 Expected Value of PFD_{avg}formulas for Non-Repairable System without considering CCF, Diagnostics or MTTR.			
Configuration	**Function**	**Expected PFD based on "Average before" failure rate**	**Equation Number**
1oo1	$E(F_1) =$	$\left[E(\lambda^D)\times\frac{E(TI)}{2}\right]$	A-9
1oo2	$E(F_2) =$	$\left[(E(\lambda^D))^2\times\frac{E(TI)^2}{4}\right]$	A- 10
1oo3	$E(F_3) =$	$\left[(E(\lambda^D))^3\times\frac{E(TI)^3}{8}\right]$	A-11
2oo2	$E(F_4) =$	$\left[E(\lambda^D)\times E(TI)\right]$	A-12
2oo3	$E(F_5) =$	$\frac{3}{4}\times\left[(E(\lambda^D))^2\times E(TI)^2\right]$	A-13
3oo3	$E(F_6) =$	$3\times\left[E(\lambda^D)\times\frac{E(TI)}{2}\right]$	A-14

Where E(x) is the expected value of random variable x.

The variance of the PFD is found by taking the weighted sum of the variance of the random variables. The weighting function is the sensitivity of the PFD with respect to random variable. In mathematical terms the variance of a function may be found in terms of the variance of the random variables as:

Once again let

$$y = F(x_i) \qquad \text{(Eq A-15)}$$

The variance of Y can likewise be estimated as:

$$V(y) = \sum_{i=1}^{n} \left[\frac{\partial y}{\partial x_i}\right]^2 V(x_i) \qquad \text{(Eq A-16)}$$

Where:

V(y) = variance of random variable y as defined above

$V(x_i)$ = variance of random variable x_i as defined above

The sensitivity of y with respect to a random variable x_i is:

$$\text{Sensitivity of y with respect to } x_i = \left[\frac{\partial y}{\partial x_i}\right] \qquad \text{(Eq A-17)}$$

For the configurations defined in Table A.1, the corresponding variance of the PFD are presented in Table A.3.

Table A.3 Variance of PFD_{avg} formulas for Non-Repairable System without considering CCF, Diagnostics or MTTR.			
Configuration	**Function**	**Variance in Function F_i based on "Average before" failure rate**	**Equation Number**
1oo1	$V(F_1) =$	$\left[\frac{E(TI)}{2}\right]^2 V(\lambda^D) + \left[\frac{E(\lambda^D)}{2}\right]^2 V(TI)$	A-18
1oo2	$V(F_2) =$	$\left[E(\lambda^D)^2 \frac{E(TI)^2}{2}\right]^2 V(\lambda^D) +$ $\left[E(\lambda^D)^2 \frac{E(TI)}{2}\right]^2 V(TI)$	A- 19
1oo3	$V(F_3) =$	$\left[3\times E(\lambda^D)^2 \frac{E(TI)^3}{2}\right]^2 V(\lambda^D) +$ $\left[\frac{3}{8} E(\lambda^D)^3 E(TI)^2\right]^2 V(TI)$	A-20
2oo2	$V(F_4) =$	$E(TI)^2 V(\lambda^D) + E(\lambda^D)^2 \times V(TI)$	A-21
2oo3	$V(F_5) =$	$\left[\frac{3}{2}\times E(\lambda^D)\times E(TI)^2\right]^2 V(\lambda^D) +$ $\left[\frac{3}{2}\times E(\lambda^D)^2 \times E(TI)\right]^2 V(TI)$	A-22
3oo3	$V(F_6) =$	$\left[\frac{3}{2}\times E(TI)\right]^2 V(\lambda^D) +$ $\left[\frac{3}{2}\times E(\lambda^D)\right]^2 V(TI)$	A-23

Note that for the calculation of the variance functions, V(Fi), the random variables are evaluated at the expected value (mean).

REPAIRABLE SYSTEMS

For systems that are considered repairable, the simplified equations of Table A.4 are used for the analysis. The expected value of the probability of failure on demand (PFD) is found by substituting the expected value of each of the random variables into the corresponding equation. The expected value of the PFD for each of the system configurations is presented in Table A.5

Table A.4 Simplified PFD_{avg} Formulas for Repairable System considering CCF, Diagnostics and MTTR

Configuration	Function	PFD based on "Average before" failure rate	Equation Number
1oo1	$F_7 =$	$\left[\frac{(1-DC)\times\lambda^D\times TI}{2}+\frac{DC\times\lambda^D\times DI}{2}+\lambda^D\times MTTR\right]$	A-24
1oo2	$F_8 =$	$\left[\frac{(1-DC)\times(1-\beta)\times\lambda^D\times TI}{2}+\frac{DC\times(1-\beta)\times\lambda^D\times DI}{2}+(1-\beta)\times\lambda^D\times MTTR\right]^2+\left[\frac{(1-DC)\times\beta\times\lambda^D\times TI}{2}+\frac{DC\times\beta\times\lambda^D\times DI}{2}+\beta\times\lambda^D\times MTTR\right]$	A- 25
1oo3	$F_9 =$	$\left[\frac{(1-DC)\times(1-\beta)\times\lambda^D\times TI}{2}+\frac{DC\times(1-\beta)\times\lambda^D\times DI}{2}+(1-\beta)\times\lambda^D\times MTTR\right]^3+\left[\frac{(1-DC)\times\beta\times\lambda^D\times TI}{2}+\frac{DC\times\beta\times\lambda^D\times DI}{2}+\beta\times\lambda^D\times MTTR\right]$	A-26
2oo2	$F_{10} =$	$2\times\left[\frac{(1-DC)\times\lambda^D\times TI}{2}+\frac{DC\times\lambda^D\times DI}{2}+\lambda^D\times MTTR\right]$	A-27

Table A.4 Simplified PFD_{avg} Formulas for Repairable System considering CCF, Diagnostics and MTTR			
Configuration	**Function**	**PFD based on "Average before" failure rate**	**Equation Number**
2oo3	$F_{11} =$	$3\times\left[\frac{(1-DC)\times(1-\beta)\times\lambda^D\times TI}{2}+\frac{DC\times(1-\beta)\times\lambda^D\times DI}{2}+(1-\beta)\times\lambda^D\times MTTR\right]^2+\left[\frac{(1-DC)\times\beta\times\lambda^D\times TI}{2}+\frac{DC\times\beta\times\lambda^D\times DI}{2}+\beta\times\lambda^D\times MTTR\right]$	A-28
3oo3	$F_{12} =$	$3\times\left[\frac{(1-DC)\times\lambda^D\times TI}{2}+\frac{DC\times\lambda^D\times DI}{2}+\lambda^D\times MTTR\right]$	A-29

Where:

DC is the diagnostic coverage;

DI is the diagnostic interval;

TI is the proof test interval,

λ^D is the dangerous failure rate;

MTTR is the mean time to restore the system to operation

β is the common cause failure parameter that is always is between 0 and 1

Table A.5 Expected PFD Formulas for Repairable System considering CCF, Diagnostics and MTTR			
Configuration	**Function**	**PFD based on "Average before" failure rate**	**Equation Number**
1oo1	$E(F_7) =$	$\left[\frac{(1-E(DC))\times E(\lambda^D)\times E(TI))}{2}+\frac{E(DC)\times E(\lambda^D)\times E(DI)}{2} + E(\lambda^D)\times E(MTTR)\right]$	A-30
1oo2	$E(F_8) =$	$\left[\frac{(1-E(DC))\times(1-E(\beta))\times E(\lambda^D)\times E(TI)}{2}+\frac{E(DC)\times(1-E(\beta))\times E(\lambda^D)\times E(DI)}{2}+(1-E(\beta))\times E(\lambda^D)\times E(MTTR)\right]^2 + \left[\frac{(1-E(DC))\times E(\beta)\times E(\lambda^D)\times E(TI)}{2}+\frac{E(DC)\times E(\beta)\times E(\lambda^D)\times E(DI}{2})+E(\beta)\times E(\lambda^D)\times E(MTTR)\right]$	A- 31
1oo3	$E(F_9) =$	$\left[\frac{(1-E(DC))\times(1-E(\beta))\times E(\lambda^D)\times E(TI)}{2}+\frac{E(DC)\times(1-E(\beta))\times E(\lambda^D)\times E(DI)}{2}+(1-E(\beta))\times E(\lambda^D)\times E(MTTR)\right]^3 + \left[\frac{(1-E(DC))\times E(\beta)\times E(\lambda^D)\times E(TI)}{2}+\frac{E(DC)\times E(\beta)\times E(\lambda^D)\times E(DI)}{2}+E(\beta)\times E(\lambda^D)\times E(MTTR)\right]$	A-32

Table A.5 Expected PFD Formulas for Repairable System considering CCF, Diagnostics and MTTR			
Configuration	**Function**	**PFD based on "Average before" failure rate**	**Equation Number**
2oo2	E(F$_{10}$) =	$2\times\left[\begin{array}{l}\frac{(1-E(DC))\times E(\lambda^D)\times E(TI))}{2}+\\ \frac{E(DC)\times E(\lambda^D)\times E(DI)}{2}\\ +E(\lambda^D)\times E(MTTR)\end{array}\right]$	A-33
2oo3	E(F$_{11}$) =	$3\times\left[\begin{array}{l}\frac{(1-E(DC))\times(1-E(\beta))\times E(\lambda^D)\times E(TI)}{2}+\\ \frac{E(DC)\times(1-E(\beta))\times E(\lambda^D)\times E(DI)}{2}+\\ (1-E(\beta))\times E(\lambda^D)\times E(MTTR)\end{array}\right]^2+$ $\left[\begin{array}{l}\frac{(1-E(DC))\times E(\beta)\times E(\lambda^D)\times E(TI)}{2}+\\ \frac{E(DC)\times E(\beta)\times E(\lambda^D)\times E(DI)}{2}+\\ E(\beta)\times E(\lambda^D)\times E(MTTR)\end{array}\right]$	A-34
3oo3	E(F$_{12}$) =	$3\times\left[\begin{array}{l}\frac{(1-E(DC))\times E(\lambda^D)\times E(TI))}{2}+\\ \frac{E(DC)\times E(\lambda^D)\times E(DI)}{2}\\ +E(\lambda^D)\times E(MTTR)\end{array}\right]$	A-35

The variance of the PFD of the repairable systems is found in the same manner as that in the non-repairable cases. Once again let:

$$y = F(x_i) \quad \text{(Eq A-36)}$$

The variance of Y can likewise be estimated as:

$$V(y) = \sum_{i=1}^{n} \left[\frac{\partial y}{\partial x_i}\right]^2 V(x_i) \quad \text{(Eq A-37)}$$

Where:

V(y) = variance of random variable y as defined above

$V(x_i)$ = variance of random variable x_i as defined above

The sensitivity of y with respect to a random variable x_i is:

In the case of the repairable systems, there are several more potentially uncertain or random variables:

- DC is the diagnostic coverage;
- DI is the diagnostic interval;
- TI is the proof test interval,
- λ^D is the dangerous failure rate;
- MTTR is the mean time to restore the system to operation
- β is the common cause failure parameter that is always is between 0 and 1

We must evaluate the partial derivative of the PFD function with respect to each and then combine them using equation A-37. The partial derivatives of function F7 with respect to each random variable are found as follows

$$\frac{\partial F_7}{\partial DC} = \frac{-\lambda^D \times TI}{2} + \frac{\lambda^D \times DI}{2} \quad \text{(Eq A-38)}$$

$$\frac{\partial F_7}{\partial \lambda^D} = \frac{(1-DC)\times TI}{2} + \frac{DC \times DI}{2} + MTTR \quad \text{(Eq A-39)}$$

$$\frac{\partial F_7}{\partial TI} = \frac{(1-DC)\times \lambda^D}{2} \quad \text{(Eq A-40)}$$

$$\frac{\partial F_7}{\partial DI} = \frac{DC \times \lambda^D}{2} \quad \text{(Eq A-41)}$$

$$\frac{\partial F_7}{\partial MTTR} = \lambda^D \qquad \text{(Eq A-42)}$$

We can now compute the variance of function F7 as:

$$V\left(F_7\right) = \sum_{i=1}^{n} \left[\frac{\partial F_7}{\partial x_i}\right]^2 V(x_i) \qquad \text{(Eq A-43)}$$

$$V\left(F_7\right) = \left[\frac{-\lambda^D \times TI}{2} + \frac{\lambda^D \times DI}{2}\right]^2 \text{V(DC)} + \left[\frac{(1-DC)\times TI}{2} + \frac{DC\times DI}{2} + MTTR\right]^2 \text{V(}\lambda^D\text{)+}$$

$$\left[\frac{(1-DC)\times\lambda^D}{2}\right]^2 \text{V(TI)} + \left[\frac{DC\times\lambda^D}{2}\right]^2 \text{(DI)} + \left[\lambda^D\right]^2 \text{V(MTTR)} \qquad \text{(Eq A-44)}$$

Equation 44 represents the variance of the probability of failure on demand of the 1001 configuration when the system is repairable. We note the following relationships between the 1001, 2oo2 and 3oo3 configurations:

$F_{10} = 2\ F_7$ (Eq A-45)

$F_{12} = 3\ F_7$ (Eq A-46)

We can directly determine the variance of functions F_{10} and F_{12} from the properties of the variance operator, V(X).

$V(F_{10}) = V(2\ F_7)$ (Eq A-47)

$V(F_{10}) = 4\ V(F_7)$ (Eq A-48)

$V(F_{12}) = V(3\ F_7)$ (Eq A-49)

$V(F_{12}) = 9\ V(F_7)$ (Eq A-50)

The determination of the variance of functions F_8, F_{10}, and F_{11} is done in a similar manner. To simplify the presentation of the derivation of variance of functions F_8, F_{10}, and F_{11}, we introduce two functions H and Q as:

$$H = \left[\frac{(1-DC)\times(1-\beta)\times\lambda^D\times TI}{2} + \frac{DC\times(1-\beta)\times\lambda^D\times DI}{2} + (1-\beta)\times\lambda^D\times MTTR\right] \qquad \text{(Eq A-51)}$$

$$Q = \left[\frac{(1-DC)\times\beta\times\lambda^D\times TI}{2} + \frac{DC\times\beta\times\lambda^D\times DI}{2} + \beta\times\lambda^D\times MTTR\right] \qquad \text{(Eq A-52)}$$

We note the following relationships:

$F_8 = H^2 + Q$ (Eq A-53)

$F_9 = H^3 + Q$ (Eq A-54)

$F_{11} = 3\,H^2 + Q$ (Eq A-55)

The variance of F_8, F_9 and F_{11} are found as:

$V(F_8) = V(H^2 + Q)$ (Eq A-56)

$V(F_9) = V(H^3 + Q)$ (Eq A-57)

$V(F_{11}) = V(3\,H^2 + Q)$ (Eq A-58)

We will need the sensitivity of F_8, F_9 and F_{11} to each of the potential random variables.

$$\frac{\partial F_8}{\partial x} = \frac{\partial(H^2 + Q)}{\partial x} = \frac{\partial(H^2)}{\partial x} + \frac{\partial(Q)}{\partial x} \quad \text{(Eq A-59)}$$

Likewise

$$\frac{\partial F_9}{\partial x} = \frac{\partial(H^3 + Q)}{\partial x} = \frac{\partial(H^3)}{\partial x} + \frac{\partial(Q)}{\partial x} \quad \text{(Eq A-60)}$$

$$\frac{\partial F_{11}}{\partial x} = \frac{\partial 3(H^2 + Q)}{\partial x} = 3\frac{\partial(H^2)}{\partial x} + 3\frac{\partial(Q)}{\partial x} \quad \text{(Eq A-61)}$$

Using the chain rule for partial derivatives, we now write:

$$\frac{\partial F_8}{\partial x} = 2H\frac{\partial(H)}{\partial x} + \frac{\partial(Q)}{\partial x} \quad \text{(Eq A-62)}$$

$$\frac{\partial F_9}{\partial x} = 3H^2\,\frac{\partial(H)}{\partial x} + \frac{\partial(Q)}{\partial x} \quad \text{(Eq A-63)}$$

$$\frac{\partial F_{11}}{\partial x} = 6H\frac{\partial(H)}{\partial x} + \frac{\partial(Q)}{\partial x} \quad \text{(Eq A-64)}$$

We now evaluate the derivatives of function H with respect to each potential random variables:

- DC is the diagnostic coverage;
- DI is the diagnostic interval;

- TI is the proof test interval,
- λ^D is the dangerous failure rate;
- MTTR is the mean time to restore the system to operation
- β is the common cause failure parameter that is always is between 0 and 1

$$\frac{\partial(\mathrm{H})}{\partial DC} = \frac{-(1-\beta)\times\lambda^D\times TI}{2} + \frac{(1-\beta)\times\lambda^D\times DI}{2} \qquad \text{(Eq A-65)}$$

$$\frac{\partial(\mathrm{H})}{\partial TI} = \frac{(1-DC)\times(1-\beta)\times\lambda^D}{2} \qquad \text{(Eq A-66)}$$

$$\frac{\partial(\mathrm{H})}{\partial\lambda^D} = \left[\begin{array}{l}\dfrac{(1-DC)\times(1-\beta)\times TI}{2} + \\ \dfrac{DC\times(1-\beta)\times DI}{2} + (1-\beta)\times MTTR\end{array}\right] \qquad \text{(Eq A-67)}$$

$$\frac{\partial(\mathrm{H})}{\partial MTTR} = (1-\beta)\times\lambda^D \qquad \text{(Eq A-68)}$$

$$\frac{\partial(\mathrm{H})}{\partial\beta} = \left[\begin{array}{l}\dfrac{(1-DC)\times(-1)\times\lambda^D\times TI}{2} + \\ \dfrac{DC\times(-1)\times\lambda^D\times DI}{2} + (-1)\times\lambda^D\times MTTR\end{array}\right] \qquad \text{(Eq A-69)}$$

$$\frac{\partial(\mathrm{H})}{\partial DI} = \frac{DC\times(1-\beta)\times\lambda^D}{2} \qquad \text{(Eq A-70)}$$

In a similar manner we evaluate the partial derivatives of function Q with respect to each potential random variables.

$$Q = \left[\frac{(1-DC)\times\beta\times\lambda^D\times TI}{2} + \frac{DC\times\beta\times\lambda^D\times DI}{2} + \beta\times\lambda^D\times MTTR \right] \quad \text{(Eq A-71)}$$

- DC is the diagnostic coverage;
- DI is the diagnostic interval;
- TI is the proof test interval,
- λ^D is the dangerous failure rate;
- MTTR is the mean time to restore the system to operation
- β is the common cause failure parameter that is always is between 0 and 1

$$\frac{\partial(Q)}{\partial DC} = \frac{(-1)\times\beta\times\lambda^D\times TI}{2} + \frac{\beta\times\lambda^D\times DI}{2} \quad \text{(Eq A-72)}$$

$$\frac{\partial(Q)}{\partial TI} = \frac{(1-DC)\times\beta\times\lambda^D}{2} \quad \text{(Eq A-73)}$$

$$\frac{\partial(Q)}{\partial\lambda^D} = \left[\frac{(1-DC)\times\beta\times TI}{2} + \frac{DC\times\beta\times DI}{2} + \beta\times MTTR \right] \quad \text{(Eq A-74)}$$

$$\frac{\partial(Q)}{\partial MTTR} = \beta\times\lambda^D\times MTTR \quad \text{(Eq A-75)}$$

$$\frac{\partial(Q)}{\partial\beta} = \left[\frac{(1-DC)\times\lambda^D\times TI}{2} + \frac{DC\times\lambda^D\times DI}{2} + \lambda^D\times MTTR \right] \quad \text{(Eq A-76)}$$

$$\frac{\partial(Q)}{\partial DI} = \frac{DC\times\beta\times\lambda^D}{2} \quad \text{(Eq A-77)}$$

We can now determine the sensitivity of the functions F8, F9 , F11 to the random variables of interest. Restating equation A-62 for sensitivity of F8 with respect to the random variables.

$$\frac{\partial F_8}{\partial x} = 2H\frac{\partial(H)}{\partial x} + \frac{\partial(Q)}{\partial x} \quad \text{(Eq A-78)}$$

For F_8 we find

$$\frac{\partial F_8}{\partial x} = 2\mathrm{H}\frac{\partial(\mathrm{H})}{\partial x} + \frac{\partial(\mathrm{Q})}{\partial x} \qquad \text{(Eq A-80)}$$

$$\frac{\partial F_8}{\partial DC} = 2\begin{bmatrix} \frac{(1-DC)\times(1-\beta)\times\lambda^D\times TI}{2} + \\ \frac{DC\times(1-\beta)\times\lambda^D\times DI}{2} + (1-\beta)\times\lambda^D\times MTTR \end{bmatrix} \times$$

$$\left[\frac{-(1-\beta)\times\lambda^D\times TI}{2} + \frac{(1-\beta)\times\lambda^D\times DI}{2}\right] +$$

$$\frac{(-1)\times\beta\times\lambda^D\times TI}{2} + \frac{\beta\times\lambda^D\times DI}{2} \qquad \text{(Eq A-79)}$$

$$\frac{\partial F_8}{\partial TI} = 2\begin{bmatrix} \frac{(1-DC)\times(1-\beta)\times\lambda^D\times TI}{2} + \\ \frac{DC\times(1-\beta)\times\lambda^D\times DI}{2} + (1-\beta)\times\lambda^D\times MTTR \end{bmatrix} \times \left[\frac{(1-DC)\times(1-\beta)\times\lambda^D}{2}\right] +$$

$$\frac{(1-DC)\times\beta\times\lambda^D}{2} \qquad \text{(Eq A-81)}$$

$$\frac{\partial F_8}{\partial MTTR} = 2\begin{bmatrix} \frac{(1-DC)\times(1-\beta)\times\lambda^D\times TI}{2} + \\ \frac{DC\times(1-\beta)\times\lambda^D\times DI}{2} + (1-\beta)\times\lambda^D\times MTTR \end{bmatrix} \times$$

$$\left[(1-\beta)\times\lambda^D\right] + \beta\times\lambda^D\times MTTR \qquad \text{(Eq A-82)}$$

$$\frac{\partial F_8}{\partial \beta} = 2\left[\begin{array}{l}\frac{(1-DC)\times(1-\beta)\times\lambda^D\times TI}{2}+\\ \frac{DC\times(1-\beta)\times\lambda^D\times DI}{2}+(1-\beta)\times\lambda^D\times MTTR\end{array}\right]\times$$

$$\left[\left[\begin{array}{l}\frac{(1-DC)\times(-1)\times\lambda^D\times TI}{2}+\\ \frac{DC\times(-1)\times\lambda^D\times DI}{2}+(-1)\times\lambda^D\times MTTR\end{array}\right]\right]+$$

$$\left[\frac{(1-DC)\times\lambda^D\times TI}{2}+\frac{DC\times\lambda^D\times DI}{2}+\lambda^D\times MTTR\right] \tag{Eq A-83}$$

$$\frac{\partial F_8}{\partial DI} = 2\left[\begin{array}{l}\frac{(1-DC)\times(1-\beta)\times\lambda^D\times TI}{2}+\\ \frac{DC\times(1-\beta)\times\lambda^D\times DI}{2}+(1-\beta)\times\lambda^D\times MTTR\end{array}\right]\times\left[\frac{DC\times(1-\beta)\times\lambda^D}{2}\right]+$$

$$\frac{DC\times\beta\times\lambda^D}{2} \tag{Eq A-84}$$

$$\left[\frac{\partial F_8}{\partial \lambda^D}\right] = 2\left[\begin{array}{l}\frac{(1-DC)\times(1-\beta)\times\lambda^D\times TI}{2}+\\ \frac{DC\times(1-\beta)\times\lambda^D\times DI}{2}+(1-\beta)\times\lambda^D\times MTTR\end{array}\right]\times\left[\begin{array}{l}\frac{(1-DC)\times(1-\beta)\times TI}{2}+\\ \frac{DC\times(1-\beta)\times DI}{2}+(1-\beta)\times MTTR\end{array}\right]+$$

$$\left[\frac{(1-DC)\times\beta\times TI}{2}+\frac{DC\times\beta\times DI}{2}+\beta\times MTTR\right] \tag{Eq A-85}$$

Using the sensitivities for F8 calculated above, the variance in F8 due to uncertain or random variables is found as:

$$V(F_8) = \left[\frac{\partial F_8}{\partial DC}\right]^2 V(DC) + \left[\frac{\partial F_8}{\partial TI}\right]^2 V(TI) + \left[\frac{\partial F_8}{\partial \lambda^D}\right]^2 V(\lambda^D) +$$

$$\left[\frac{\partial F_8}{\partial MTTR}\right]^2 V(MTTR) + \left[\frac{\partial F_8}{\partial \beta}\right]^2 V(\beta) + \left[\frac{\partial F_8}{\partial DI}\right]^2 V(DI) \tag{Eq A-86}$$

Using the equation A-63, we now determine the variance of F_9.

$$\frac{\partial F_9}{\partial x} = 3\mathrm{H}^2 \frac{\partial(\mathrm{H})}{\partial x} + \frac{\partial(\mathrm{Q})}{\partial x} \qquad \text{(Eq A-87)}$$

$$\frac{\partial F_9}{\partial DC} = 3\left[\frac{(1-DC)\times(1-\beta)\times\lambda^D\times TI}{2} + \frac{DC\times(1-\beta)\times\lambda^D\times DI}{2} + (1-\beta)\times\lambda^D\times MTTR\right]^2 \times$$
$$\left[\frac{-(1-\beta)\times\lambda^D\times TI}{2} + \frac{(1-\beta)\times\lambda^D\times DI}{2}\right] +$$
$$\frac{(-1)\times\beta\times\lambda^D\times TI}{2} + \frac{\beta\times\lambda^D\times DI}{2} \qquad \text{(Eq A-88)}$$

$$\frac{\partial F_9}{\partial TI} = 3\begin{bmatrix} \frac{(1-DC)\times(1-\beta)\times\lambda^D\times TI}{2} + \\ \frac{DC\times(1-\beta)\times\lambda^D\times DI}{2} + (1-\beta)\times\lambda^D\times MTTR \end{bmatrix}^2 \times \begin{bmatrix} \frac{(1-DC)\times(1-\beta)\times\lambda^D}{2} \end{bmatrix} +$$

$$\frac{(1-DC)\times\beta\times\lambda^D}{2} \qquad \text{(Eq A-89)}$$

$$\frac{\partial F_9}{\partial MTTR} = 3\begin{bmatrix} \frac{(1-DC)\times(1-\beta)\times\lambda^D\times TI}{2} + \\ \frac{DC\times(1-\beta)\times\lambda^D\times DI}{2} + (1-\beta)\times\lambda^D\times MTTR \end{bmatrix}^2 \times \begin{bmatrix} (1-\beta)\times\lambda^D \end{bmatrix} +$$

$$\beta\times\lambda^D\times MTTR \qquad \text{(Eq A-90)}$$

$$\frac{\partial F_9}{\partial \beta} = 3\begin{bmatrix} \frac{(1-DC)\times(1-\beta)\times\lambda^D\times TI}{2} + \\ \frac{DC\times(1-\beta)\times\lambda^D\times DI}{2} + (1-\beta)\times\lambda^D\times MTTR \end{bmatrix}^2 \times$$

$$\begin{bmatrix} \begin{bmatrix} \frac{(1-DC)\times(-1)\times\lambda^D\times TI}{2} + \\ \frac{DC\times(-1)\times\lambda^D\times DI}{2} + (-1)\times\lambda^D\times MTTR \end{bmatrix} \end{bmatrix} +$$

$$\begin{bmatrix} \frac{(1-DC)\times\lambda^D\times TI}{2} + \frac{DC\times\lambda^D\times DI}{2} + \lambda^D\times MTTR \end{bmatrix} \qquad \text{(Eq A-91)}$$

$$\frac{\partial F_9}{\partial DI} = 3\left[\frac{(1-DC)\times(1-\beta)\times\lambda^D\times TI}{2} + \frac{DC\times(1-\beta)\times\lambda^D\times DI}{2} + (1-\beta)\times\lambda^D\times MTTR\right]^2 \times \left[\frac{DC\times(1-\beta)\times\lambda^D}{2}\right] + \frac{DC\times\beta\times\lambda^D}{2} \quad \text{(Eq A-92)}$$

$$\left[\frac{\partial F_9}{\partial \lambda^D}\right] = 3 \times \left[\frac{(1-DC)\times(1-\beta)\times\lambda^D\times TI}{2} + \frac{DC\times(1-\beta)\times\lambda^D\times DI}{2} + (1-\beta)\times\lambda^D\times MTTR\right]^2 \times \left[\frac{(1-DC)\times(1-\beta)\times TI}{2} + \frac{DC\times(1-\beta)\times DI}{2} + (1-\beta)\times MTTR\right] + \left[\frac{(1-DC)\times\beta\times TI}{2} + \frac{DC\times\beta\times DI}{2} + \beta\times MTTR\right] \quad \text{(Eq A-93)}$$

Using the sensitivities for F_9 calculated above, the variance in F_9 due to uncertain or random variables is found as:

$$V(F_9) = \left[\frac{\partial F_9}{\partial DC}\right]^2 V(DC) + \left[\frac{\partial F_9}{\partial TI}\right]^2 V(TI) + \left[\frac{\partial F_9}{\partial \lambda^D}\right]^2 V(\lambda^D) + \left[\frac{\partial F_9}{\partial MTTR}\right]^2 V(MTTR) + \left[\frac{\partial F_9}{\partial \beta}\right]^2 V(\beta) + \left[\frac{\partial F_9}{\partial DI}\right]^2 V(DI) \quad \text{(Eq A-94)}$$

Finally, in a similar manner the variance in F_{11} is computed as:

$$\frac{\partial F_{11}}{\partial x} = 6\mathrm{H}\frac{\partial(\mathrm{H})}{\partial x} + \frac{\partial(\mathrm{Q})}{\partial x} \qquad \text{(Eq A-95)}$$

$$\frac{\partial F_{11}}{\partial DC} = 6\left[\begin{array}{l}\frac{(1-DC)\times(1-\beta)\times\lambda^{D}\times TI}{2}+\\ \frac{DC\times(1-\beta)\times\lambda^{D}\times DI}{2}+(1-\beta)\times\lambda^{D}\times MTTR\end{array}\right]\times$$

$$\left[\frac{-(1-\beta)\times\lambda^{D}\times TI}{2}+\frac{(1-\beta)\times\lambda^{D}\times DI}{2}\right]+$$

$$\frac{(-1)\times\beta\times\lambda^{D}\times TI}{2}+\frac{\beta\times\lambda^{D}\times DI}{2} \qquad \text{(Eq A-96)}$$

$$\frac{\partial F_{11}}{\partial TI} = 6\left[\begin{array}{l}\frac{(1-DC)\times(1-\beta)\times\lambda^{D}\times TI}{2}+\\ \frac{DC\times(1-\beta)\times\lambda^{D}\times DI}{2}+(1-\beta)\times\lambda^{D}\times MTTR\end{array}\right]\times\left[\frac{(1-DC)\times(1-\beta)\times\lambda^{D}}{2}\right]+$$

$$\frac{(1-DC)\times\beta\times\lambda^{D}}{2} \qquad \text{(Eq A-97)}$$

$$\frac{\partial F_{11}}{\partial MTTR} = 6\left[\begin{array}{l}\frac{(1-DC)\times(1-\beta)\times\lambda^{D}\times TI}{2}+\\ \frac{DC\times(1-\beta)\times\lambda^{D}\times DI}{2}+(1-\beta)\times\lambda^{D}\times MTTR\end{array}\right]\times\left[(1-\beta)\times\lambda^{D}\right]+$$

$$\beta\times\lambda^{D}\times MTTR \qquad \text{(Eq A-98)}$$

$$\frac{\partial F_{11}}{\partial \beta} = 6\left[\frac{(1-DC)\times(1-\beta)\times\lambda^D\times TI}{2} + \frac{DC\times(1-\beta)\times\lambda^D\times DI}{2} + (1-\beta)\times\lambda^D\times MTTR\right]\times$$

$$\left[\left[\frac{(1-DC)\times(-1)\times\lambda^D\times TI}{2} + \frac{DC\times(-1)\times\lambda^D\times DI}{2} + (-1)\times\lambda^D\times MTTR\right]\right] +$$

$$\left[\frac{(1-DC)\times\lambda^D\times TI}{2} + \frac{DC\times\lambda^D\times DI}{2} + \lambda^D\times MTTR\right] \qquad \text{(Eq A-99)}$$

$$\frac{\partial F_{11}}{\partial DI} = 6\left[\frac{(1-DC)\times(1-\beta)\times\lambda^D\times TI}{2} + \frac{DC\times(1-\beta)\times\lambda^D\times DI}{2} + (1-\beta)\times\lambda^D\times MTTR\right]\times\left[\frac{DC\times(1-\beta)\times\lambda^D}{2}\right] +$$

$$\frac{DC\times\beta\times\lambda^D}{2} \qquad \text{(Eq A-100)}$$

$$\left[\frac{\partial F_{11}}{\partial \lambda^D}\right] = 6\left[\frac{(1-DC)\times(1-\beta)\times\lambda^D\times TI}{2} + \frac{DC\times(1-\beta)\times\lambda^D\times DI}{2} + (1-\beta)\times\lambda^D\times MTTR\right]\times$$

$$\left[\frac{(1-DC)\times(1-\beta)\times TI}{2} + \frac{DC\times(1-\beta)\times DI}{2} + (1-\beta)\times MTTR\right] +$$

$$\left[\frac{(1-DC)\times\beta\times TI}{2} + \frac{DC\times\beta\times DI}{2} + \beta\times MTTR\right] \qquad \text{(Eq A-101)}$$

Using the sensitivities for F_{11} calculated above, the variance in F_{11} due to uncertain or random variables is found as:

$$V(F_{11}) = \left[\frac{\partial F_{11}}{\partial DC}\right]^2 V(DC) + \left[\frac{\partial F_{11}}{\partial TI}\right]^2 V(TI) + \left[\frac{\partial F_{11}}{\partial \lambda^D}\right]^2 V(\lambda^D) +$$

$$\left[\frac{\partial F_{11}}{\partial MTTR}\right]^2 V(MTTR) + \left[\frac{\partial F_{11}}{\partial \beta}\right]^2 V(\beta) + \left[\frac{\partial F_{11}}{\partial DI}\right]^2 V(DI) \qquad \text{(Eq A-102)}$$

PROCEDURE FOR DETERMINATION OF VARIANCE FOR REPAIRABLE 1oo2, 1oo3 AND 2oo3 SYSTEMS

Based on the above derivations we may now write the formulas for the variance of various configurations of repairable equipment.

Step 1. Determine the configuration of equipment to be studied.

Step 2. Evaluate Function H using the expected value of all random or uncertain variables, equation A-51

Step 3. Evaluate Function Q using the expected value of all random or uncertain variables, equation A-52

Step 4. Evaluate the sensitivity of function H with respect to the random variables using equations A-65 thru A-70.

Step 5. Evaluate the sensitivity of function Q with respect to the random variables using equations A-72 thru A-77.

Step 6. Evaluate the variance as:

a. For 1oo2 systems use equation A-86 for $V(F_8)$
b. For 1oo3 systems use equation A-94 for $V(F_9)$
c. For 2oo3 systems use equation A-102 for $V(F_{11})$

The end result of the above calculations is the determination of the mean and variance of the configuration used in the sensor, logic solver or final control element systems. Table 4 of the main body of this paper presents a cross reference roadmap for the determination of the mean and variance of various of the probability of failure on demand (PFD) of various hardware configurations. The mean and variance for the particular configuration is then returned to the overall calculations of the PFD of the interlock.

APPENDIX B -- UNCERTAINTY EQUATIONS FOR CONFIGURATIONS NOT COVERED IN ISA TR 84.00.02

The ISA Technical report [Ref. 2] does not define the calculation equations for the PFDavg for non-repairable systems where duplicated devices create the potential for common cause failures. This appendix derives the equations needed to evaluate the uncertainty for non-repairable systems with the potential for common cause failures.

To derive the needed functions for the non-repairable systems with common cause failure (CCF) potential, we start with the equations for the repairable system with common cause failure allowed (Table A.4). First we note that a system composed of only one set of components (1oo1) without duplication has no potential for a common cause failure. In addition, a review of the equations of Table 2 finds that systems 1oo1, 2oo2 and 3oo3 do not contain the common cause failure (CCF) parameter β. Therefore, systems 1oo1, 2oo2, and 3oo3 can be eliminated from consideration as common cause failures cannot lead to failure of the specified interlocks.

For the other situations [1oo2, 1oo3 and 2oo3] we note that no repair will be allowed during the mission of the non-repairable system. This observation implies that the diagnostic coverage (DC), the diagnostic interval (DI) and the mean time for restoration (MTTR) are all zero (0). We will set DC = DI = MTTR = 0

Substitution of the zero value for DC, DI and MTTR into the equations of Table A.4 yields the non-repairable system PFD equations where CCF is allowed. Table B.1 presents the results of these algebraic substitutions.

Table B.1 Simplified PFD_{avg} formulas for Non-Repairable System Allowing for CCF

Configuration	Function	PFD based on "Average before" failure rate	Equation Number
1oo2	F_{13} =	$\left[\dfrac{(1-\beta)\times\lambda^D\times TI}{2}\right]^2+\left[\dfrac{\beta\times\lambda^D\times TI}{2}\right]$	B-1
1oo3	F_{14} =	$\left[\dfrac{(1-\beta)\times\lambda^D\times TI}{2}\right]^3+\left[\dfrac{\beta\times\lambda^D\times TI}{2}\right]$	B-2
2oo3	F_{15} =	$3\times\left[\dfrac{(1-\beta)\times\lambda^D\times TI}{2}\right]^2+\left[\dfrac{\beta\times\lambda^D\times TI}{2}\right]$	B-3

Where TI is the proof test interval, λ^D is the dangerous failure rate.

The expected value of the probability of failure on demand is found by calculating the PFD using the expected value (mean) of the random variables.

Let y a function of some random variables, x_i, as:

$$y = F(x_i) \quad \text{(Eq B-4)}$$

The mean of y can be estimated using the following approximation:

$$E(y) = F[E(x_i)] \quad \text{(Eq B-5)}$$

Where:

$E(y)$ = expected value of random variable y = mean of y

$E(x_i)$ = expected value of random variable x_i = mean of x_i

For the configurations defined in Table A.1, the corresponding expected PFD are presented in Table B.2.

Table B.2 Expected Value of PFD_{avg} formulas for Non-Repairable System Allowing CCF			
Configuration	**Function**	**Expected PFD based on "Average before" failure rate**	**Equation Number**
1oo2	$E(F_{13})$ =	$\left[\frac{(1-E(\beta))\times E(\lambda^D)\times E(TI)}{2}+\right]^2 + \left[\frac{E(\beta)\times E(\lambda^D)\times E(TI)}{2}\right]$	B-6
1oo3	$E(F_{14})$ =	$\left[\frac{(1-E(\beta))\times E(\lambda^D)\times E(TI)}{2}+\right]^3 + \left[\frac{E(\beta)\times E(\lambda^D)\times E(TI)}{2}\right]$	B-7
2oo3	$E(F_{15})$ =	$3\times\left[\frac{(1-E(\beta))\times E(\lambda^D)\times E(TI)}{2}+\right]^2 + \left[\frac{E(\beta)\times E(\lambda^D)\times E(TI)}{2}\right]$	B-8

Where E(x) is the expected value of random variable x.

The variance of the PFD is found by taking the weighted sum of the variance of the random variables. The weighting function is the sensitivity of the PFD with respect to random variable. In mathematical terms the variance of a function may be found in terms of the variance of the random variables as:

Once again let

$y = F(x_i)$ (Eq B-9)

The variance of Y can likewise be estimated as:

$$V(y) = \sum_{i=1}^{n} \left[\frac{\partial y}{\partial x_i}\right]^2 V(x_i) \qquad \text{(Eq B-10)}$$

Where:

V(y) = variance of random variable y as defined above

$V(x_i)$ = variance of random variable x_i as defined above

The sensitivity of y with respect to a random variable x_i is:

$$\text{Sensitivity of y with respect to } x_i = \left[\frac{\partial y}{\partial x_i}\right] \qquad \text{(Eq B-11)}$$

The variance of the PFD for the repairable cases 1oo2, 1oo3 and 2oo3 are presented in equations of Appendix A as:

- 1oo2, the variance is specified by equation A-86
- 1oo3, the variance is specified by equation A-94
- 2oo3, the variance is specified by equation A-102

These three equations are the basis upon which the equations for the non-repairable system with common cause failure (CCF) are developed.

<u>Variance of PFD for Non-Repairable 1oo2 with CCF</u>

To derive the variance for F_{13} we start with the general equation for the variance as:

$$V(F_{13}) = \left[\frac{\partial F_{13}}{\partial DC}\right]^2 V(DC) + \left[\frac{\partial F_{13}}{\partial TI}\right]^2 V(TI) + \left[\frac{\partial F_{13}}{\partial \lambda^D}\right]^2 V(\lambda^D) +$$

$$\left[\frac{\partial F_{13}}{\partial MTTR}\right]^2 V(MTTR) + \left[\frac{\partial F_{13}}{\partial \beta}\right]^2 V(\beta) + \left[\frac{\partial F_{13}}{\partial DI}\right]^2 V(DI) \qquad \text{(Eq B-12)}$$

The variance of a constant is zero. Thus, setting DC=DI=MTTR =0 results in the variance of these variables as becoming zero.

$$V(DC) = V(DI) = V(MTTR) = 0 \qquad \text{(Eq B-13)}$$

Thus, the variance for the 1oo2 case becomes:

$$V(F_{13}) = \left[\frac{\partial F_{13}}{\partial TI}\right]^2 V(TI) + \left[\frac{\partial F_{13}}{\partial \lambda^D}\right]^2 V(\lambda^D) + \left[\frac{\partial F_{13}}{\partial \beta}\right]^2 V(\beta) \quad \text{(Eq B-14)}$$

The evaluation of the sensitive's of the PFD may now be taken from appendix A equations as:

$\left[\frac{\partial F_{13}}{\partial TI}\right]$ from equation A-81

$\left[\frac{\partial F_{13}}{\partial \lambda^D}\right]$ from equation A-85

$\left[\frac{\partial F_{13}}{\partial \beta}\right]$ from equation A-83

Substituting DC=DI=MTTR=0 into these sensitivity equations yields:

$$\frac{\partial F_{13}}{\partial TI} = 2\left[\frac{(1-\beta)\times\lambda^D\times TI}{2} + \right]\times\left[\frac{(1-\beta)\times\lambda^D}{2}\right] + \frac{\beta\times\lambda^D}{2} \quad \text{(Eq B-15)}$$

$$\frac{\partial F_{13}}{\partial \lambda^D} = 2\left[\frac{(1-\beta)\times\lambda^D\times TI}{2} + \right]\times\left[\frac{(1-\beta)\times TI}{2}\right] + \left[\frac{\beta\times TI}{2}\right] \quad \text{(Eq B-16)}$$

$$\frac{\partial F_{13}}{\partial \beta} = 2\left[\frac{(1-\beta)\times\lambda^D\times TI}{2}\right]\times\left[\frac{(-1)\times\lambda^D\times TI}{2}\right] + \left[\frac{\lambda^D\times TI}{2}\right] \quad \text{(Eq B-17)}$$

Variance of PFD for Non-Repairable 1oo3 with CCF

To derive the variance for F_{14} we start with the general equation for the variance as:

$$V(F_{14}) = \left[\frac{\partial F_{14}}{\partial DC}\right]^2 V(DC) + \left[\frac{\partial F_{14}}{\partial TI}\right]^2 V(TI) + \left[\frac{\partial F_{14}}{\partial \lambda^D}\right]^2 V(\lambda^D) +$$

$$\left[\frac{\partial F_{14}}{\partial MTTR}\right]^2 V(MTTR) + \left[\frac{\partial F_{14}}{\partial \beta}\right]^2 V(\beta) + \left[\frac{\partial F_{14}}{\partial DI}\right]^2 V(DI) \qquad \text{(Eq B-18)}$$

The variance of a constant is zero. Thus, setting DC=DI=MTTR =0 results in the variance of these variables as becoming zero.

$$V(DC) = V(DI) = V(MTTR) = 0 \qquad \text{(Eq B-19)}$$

Thus, the variance for the 1oo3 case becomes:

$$V(F_{14}) = \left[\frac{\partial F_{14}}{\partial TI}\right]^2 V(TI) + \left[\frac{\partial F_{14}}{\partial \lambda^D}\right]^2 V(\lambda^D) + \left[\frac{\partial F_{14}}{\partial \beta}\right]^2 V(\beta) \qquad \text{(Eq B-20)}$$

The evaluation of the sensitive's of the PFD may now be taken from appendix A equations as:

$\left[\frac{\partial F_{14}}{\partial TI}\right]$ from equation A-89

$\left[\frac{\partial F_{14}}{\partial \lambda^D}\right]$ from equation A-93

$\left[\frac{\partial F_{14}}{\partial \beta}\right]$ from equation A-91

Substituting DC=DI=MTTR=0 into these sensitivity equations yields:

$$\frac{\partial F_{14}}{\partial TI} = 3\left[\frac{(1-\beta)\times\lambda^D\times TI}{2} + \right]^2 \times \left[\frac{(1-\beta)\times\lambda^D}{2}\right] + \frac{\beta\times\lambda^D}{2} \qquad \text{(Eq B-21)}$$

$$\frac{\partial F_{14}}{\partial \lambda^D} = 3\left[\frac{(1-\beta)\times\lambda^D\times TI}{2}\right]^2 \times \left[\frac{(1-\beta)\times TI}{2}\right] + \left[\frac{\beta\times TI}{2}\right] \qquad \text{(Eq B-22)}$$

$$\frac{\partial F_{14}}{\partial \beta} = 3\left[\frac{(1-\beta)\times\lambda^{D}\times TI}{2}\right]^{2}\times\left[\frac{(-1)\times\lambda^{D}\times TI}{2}\right]+\left[\frac{\lambda^{D}\times TI}{2}\right] \quad \text{(Eq B-23)}$$

Variance of PFD for Non-Repairable 2oo3 with CCF

To derive the variance for F_{15} we start with the general equation for the variance as:

$$V(F_{15}) = \left[\frac{\partial F_{15}}{\partial DC}\right]^{2} V(DC) + \left[\frac{\partial F_{15}}{\partial TI}\right]^{2} V(TI) + \left[\frac{\partial F_{15}}{\partial \lambda^{D}}\right]^{2} V(\lambda^{D}) +$$

$$\left[\frac{\partial F_{15}}{\partial MTTR}\right]^{2} V(MTTR) + \left[\frac{\partial F_{15}}{\partial \beta}\right]^{2} V(\beta) + \left[\frac{\partial F_{15}}{\partial DI}\right]^{2} V(DI) \quad \text{(Eq B-24)}$$

The variance of a constant is zero. Thus, setting DC=DI=MTTR =0 results in the variance of these variables as becoming zero.

$$V(DC) = V(DI) = V(MTTR) = 0 \quad \text{(Eq B-25)}$$

Thus, the variance for the 2oo3 case becomes:

$$V(F_{15}) = \left[\frac{\partial F_{15}}{\partial TI}\right]^{2} V(TI) + \left[\frac{\partial F_{15}}{\partial \lambda^{D}}\right]^{2} V(\lambda^{D}) + \left[\frac{\partial F_{15}}{\partial \beta}\right]^{2} V(\beta) \quad \text{(Eq B-26)}$$

The evaluation of the sensitive's of the PFD may now be taken from appendix A equations as:

$\left[\frac{\partial F_{15}}{\partial TI}\right]$ from equation A-97

$\left[\frac{\partial F_{15}}{\partial \lambda^{D}}\right]$ from equation A-101

$\left[\frac{\partial F_{15}}{\partial \beta}\right]$ from equation A-99

Substituting DC=DI=MTTR=0 into these sensitivity equations yields:

$$\frac{\partial F_{15}}{\partial TI} = 6\left[\frac{(1-\beta)\times\lambda^{D}\times TI}{2} + \right]\times\left[\frac{(1-\beta)\times\lambda^{D}}{2}\right] + \frac{\beta\times\lambda^{D}}{2} \qquad \text{(Eq B-27)}$$

$$\frac{\partial F_{15}}{\partial \lambda^{D}} = 6\left[\frac{(1-\beta)\times\lambda^{D}\times TI}{2}\right]\times\left[\frac{(1-\beta)\times TI}{2}\right] + \left[\frac{\beta\times TI}{2}\right] \qquad \text{(Eq B-28)}$$

$$\frac{\partial F_{15}}{\partial \beta} = 6\left[\frac{(1-\beta)\times\lambda^{D}\times TI}{2}\right]\times\left[\frac{(-1)\times\lambda^{D}\times TI}{2}\right] + \left[\frac{\lambda^{D}\times TI}{2}\right] \qquad \text{(Eq B-29)}$$

The evaluation of the functions for the variance are always done using the expected value of the variable. For example when the variable TI appears in an equation, the numerical evaluation of the function would be using E(TI) where E(TI) is the expected value of the random variable TI.

Comparison of Batch versus Continuous Process in Pharmaceutical Industry Based on Safety Consideration

Shiqi Chen
Mary Kay O'Connor Process Safety Center
Artie McFerrin Department of Chemical Engineering
Texas A&M University
College Station, Texas 77843-3122, USA Email address
shiqi_chen@tamu.edu

Yogesh Koirala, M. Sam Mannan
Mary Kay O'Connor Process Safety Center
Artie McFerrin Department of Chemical Engineering
Texas A&M University
College Station, Texas 77843-3122, USA Email address
ykoirala@tamu.edu, mannan@tamu.edu

Prepared for Presentation at
American Institute of Chemical Engineers
2017 Spring Meeting and 13th Global Congress on Process Safety
San Antonio, TX
March 26 – 29, 2017

Comparison of Batch versus Continuous Process in Pharmaceutical Industry Based on Safety Consideration

Shiqi Chen
Mary Kay O'Connor Process Safety Center
Artie McFerrin Department of Chemical Engineering
Texas A&M University
College Station, Texas 77843-3122, USA Email address
shiqi_chen@tamu.edu

Yogesh Koirala, M. Sam Mannan
Mary Kay O'Connor Process Safety Center
Artie McFerrin Department of Chemical Engineering

Keywords: Continuous Process, Hazard Identification, Pharmaceutical Manufacturing

Abstract

As opposed to the petro-chemical and bulk chemical industry, where continuous processes are widely applied, the pharmaceutical industry still primarily relies on traditional batch process due to the complexity of product, multi-step operation and low-volume production. Considering these conditions, the versatile batch process is more appropriate. Nowadays, driven by the contradiction between increasing demand for drugs and inefficient batch production mode, there is a trend in pharmaceutical industry, that is the transformation from traditional batch process to novel continuous process. Related projects and research that is aimed at analyzing this transition are conducted in worldwide, and the scale of these studies ranges from lab-scale reactions to overall arrangement of the factories.

Although continuous pharmaceutical process is thriving, the safety issue in this field is not promoted at the same time. Since the continuous pharmaceutical process is a novel technology, little information can be provided to evaluate its safety level. Moreover, process conditions are usually intensified for continuous process comparing to batch process. It also may bring potential risks and make continuous manufacturing inappropriate for some of pharmaceutical productions.

This research provides a comprehensive comparison for batch versus continuous pharmaceutical process by application of Dow's Fire and Explosion Index. In addition to this conventional safety evaluation, influences from production efficiency and specialties in pharmaceutical production are integrated into the comparison. Production of 2-methyl benzimidazole and peracetic acid via batch and continuous processes are conducted in this research. Via these integrative and systematic studies, safety levels of batch and continuous processes is analyzed and compared. The ways in which process conditions, production efficiency, and other requirements influence safety level for different production modes are illustrated.

1 Introduction

Continuous process is widely applied in petro-chemical and bulk chemical industry for its high production rate, automated operation and saving in cost[1]. In contrast, the pharmaceutical industry still primarily relies on traditional batch process. This aspect stems from intrinsic limitations in active pharmaceutical ingredients (APIs) production. Complexity of medical substances means multistep synthesis. An average of 8.1 steps is needed for one API synthesis[2]. Another limitation is production volume. Demand for drugs are not comparable with most of the bulk chemicals. These limitations make batch process dominate pharmaceutical industry for a long time.

However, due to the gradual expansion of the pharmaceutical market, batch process is behind the time because of its low production rate caused by low yield and inefficient operation mode. Driven by this contradiction between increasing demand for drugs and inefficient production mode, many pharmaceutical manufacturers are trying to shift from a traditional batch process to a novel continuous process. Experimental and industrial trials that are aimed at this transitioning have been done and exhibit the practicability of this new production mode. Several companies such as Lonza, Phoenix Chemical, and Corning make significant progress in this area[3-5]. Novel continuous process exhibits great potential for the pharmaceutical industry.

"Which process should be applied?" and "Is there enough benefit to move forward?" are the questions that are raised when an innovative process is introduced. To answer these questions, comprehensive comparison between batch process and continuous process is required.

This comparison should be specific in the early stages. For example, for a given substance, based on its intrinsic properties and demand in market, reasonable comparison is required to make a selection between batch and continuous process. After composing a conclusive summary that is based on number of different specific cases, the result of these comparisons can be modified and applied to generate a practical process screening methodology.

For the questions mentioned above, economic benefits, safety and environmental impacts should all be considered to get a reliable answer. In the batch versus continuous process comparisons, related economic analysis and environmental analysis have been done by many researchers[6-8]. However, there remains gaps in research and understanding of safety considerations in comparison of batch versus continuous process. Therefore, the need arises for related comparison within safety consideration and systematic methodology for safety evaluation.

In the following sessions, the development in continuous pharmaceutical process, the previous studies in process screening, selecting and comparing, and concept of inherent safety assessment are discussed. The statement of research objective, the description of methodology and results of case studies are illuminated as well.

2 Literature Survey

To decide whether to head to novel continuous process or not, evaluations of batch and continuous process are in demand for pharmaceutical companies.

In preliminary stage of the evaluation, fundamental information about reaction kinetics, heat transmission and mass transfer is collected and analyzed. The objective of this investigation is looking into the technical feasibility of chosen reaction. After these basic assessments for intrinsic properties of reaction, far-reaching analyze in economic, environmental and safety aspect can be achieved. In this more comprehensive comparison based on overall picture, more factors such as expenditure, wastewater treatment and loss control measure should be covered. Among these factors, which factor should be selected as criterion, how these factors interact, and what's the correlation between these factors and comparison results are questions require to be answered.

To answer these questions, some researchers put efforts in individual comparisons of manufacturing for specific active pharmaceutical ingredient (API). Most of these studies focus on economic advantage of continuous process. Spencer D. Schaber et al.[12] estimated the capital, operating and the overall cost for production of a given API from synthesis to tablet formulation by integrated continuous and batch pharmaceutical manufacturing respectively. Based upon the equal annual yield, the result shows that the capital expenditures for continuous production could be markedly lower than those of batch mode. Differences in operating and overall cost are not such significant but potential saving by material handling and labor in continuous process is explicitly illustrated in the study. Hessel, V. et al. compared continuous vs. batch pharmaceutical manufacturing for production of ibuprofen and artemisinin on the standpoint of cost and environmental impact. Capital expenditure and operating expenditure savings are up to 57.0% and 51.6% respectively for ibuprofen, while capital expenditure and operating expenditure savings are up to 19.6% and 29.3% respectively for artemisinin[6, 8]. These study shows the potential economic benefits in application of continuous pharmaceutical processes.

In addition to process comparison, some researchers developed general criterions or created systematic methodologies for process selection. Dominique M. Roberge et al.[13] classified reactions used in the fine and pharmaceutical industry into 4 categories according to kinetic properties. Based on this classification, the very fast reactions (reactions time less than 1s) and the fast reactions (reactions occurring from 1s to10min) would benefit the most from continuous process. These two types of reactions occupy 44% of all reactions investigated. Ryan L. Harman and Jonathan P. McMullen[14] developed a decision roadmap of flow process for use in lab-scale. Considering objectives of the study, the intrinsic properties of the chemistry and requirements for mixing, heat transferring, related questions are asked at the various nodes in this flow map to get the final choice. Gary S. Calabrese and Sergio Pissavini[15] established an conceptual methodology to evaluate applicability of continuous reactor in the form of reaction properties test. They measured applicability of flow reactor via series of questions. These questions covered issues in reaction properties and some requirements on process. Three "zones" that tagged with different level of applicability are defined. Different questions are included for each "zone". For any given reaction, the more answers with "yes" in the "zone", the higher possibility of this reaction to be at the level of applicability that the "zone" stands for.

3 Problem Statement

Although advantages of continuous process in pharmaceutical manufacturing are illustrated by many studies, it is still too early to claim that continuous process is the right direction to go forward. Several gaps in the literature survey are required to be focused on.

Firstly, most of process selection methodologies are confined to analyzing separate properties of reaction taken in process, rather than evaluation in a systematic way. For example, small inventory is usually identified as one of significant advantage of continuous process, however, it is possible that to achieve similar yield in the corresponding batch process, the reaction in continuous process requires more intensified conditions like higher temperature, higher pressure, and higher concentration of hazardous chemicals. Selection and comparison by independent and individual parameters without integration is not enough.

Secondly, existing comparisons between batch and continuous processes mainly focus on economic aspect, while lacking combination of safety issue and other important aspects from real production. Although the driving force for the application of this novel process is the economic interests, safety issues directly determine its feasibility. A choice between batch and continuous process needs to be made based on business performance within acceptable safety levels. Those comparisons that merely consider the profits are not enough.

Thirdly, the existing studies are limited to fixed production system. Potential problems in a different scale are not considered. In the pharmaceutical industry, gradually increasing production scales is required which is different from the bulk chemical industry. Hence, safety of process conditions needs to be guaranteed for all scale production. Comparison only based on fixed scale is not comprehensive.

Therefore, the primary objective of this study is to provide a comprehensive and systematic comparison between batch and continuous process used in pharmaceutical industry. Then on the basis of this analysis, this study will evaluate how main factors considered in Dow F&EI impact the safety level for these two different processes. Finally, this study will provide support to further improve existing methodology for process screening and selection for pharmaceutical industry.

It is worth mentioning that there are other risks such as chemical exposure for pharmaceutical manufacturing, however, this study mainly focuses on potential fire, explosion and reactivity incident. In addition, comparison in this study focuses on intrinsic properties of the reaction and process which are the major hazards of the reaction system. Therefore, other factors like loss control measures and emergency equipment are supposed to be effective for both batch and continuous systems.

4 Methodology

To measure the realistic fire, explosion and reactivity potential of batch and continuous process, Dow's Fire and Explosion Index (F&EI) methodology is applied as hazard index in this study, since it is the most widely used hazard index that recognized by the chemical industry[18].

From the first edition of Dow's Fire & Explosion Index Hazard Classification Guide in 1964, this index has been continuously improving though the past 50 years. The last version is the seventh edition that published in 1994. The purposes of this index include forecasting quantitative damage of potential explosion, reactivity incidents and fire; determining hazardous equipment which is possible to give rise to incidents in the system; conveying this F&EI risk potential to related managements and organizations. Since the aim of this study is to identify major hazard that affect selection, application of Dow F&EI will be mainly focus on its function of hazard identification.

In the Dow's Fire & Explosion Index Hazard Classification Guide, economic analysis is confined to financial loss after an incident. Correlation between safety and economic gain from production is not built. Actually, this kind of connection could be established via yield of reactions and geometry and inventory of units, since these factors take important role in both safety assessment and process gain. In the current study, this correlation between safety and profit is constructed on the foundation of Dow's Fire & Explosion Index.

5 Case Study I: Benzimidazole Derivative Synthesis

5.1 overview

Benzimidazole derivative is an important type of intermediates in pharmaceutical industry. It is widely exhibited in a variety of pharmaceutical substance, such as proton pump inhibitors, antimicrobials, antivirals, anticancer, anti-inflammatory, hormone modulators as well as depressants, antidiabetics and so on[22]. Famous drug with benzimidazole nucleus includes Omeprazole and Albendazole. Omeprazole is used to treat gastroesophageal reflux disease and peptic ulcer disease[23], and Albendazole is an effective drug in treatment of various parasitic worm infestations[24].

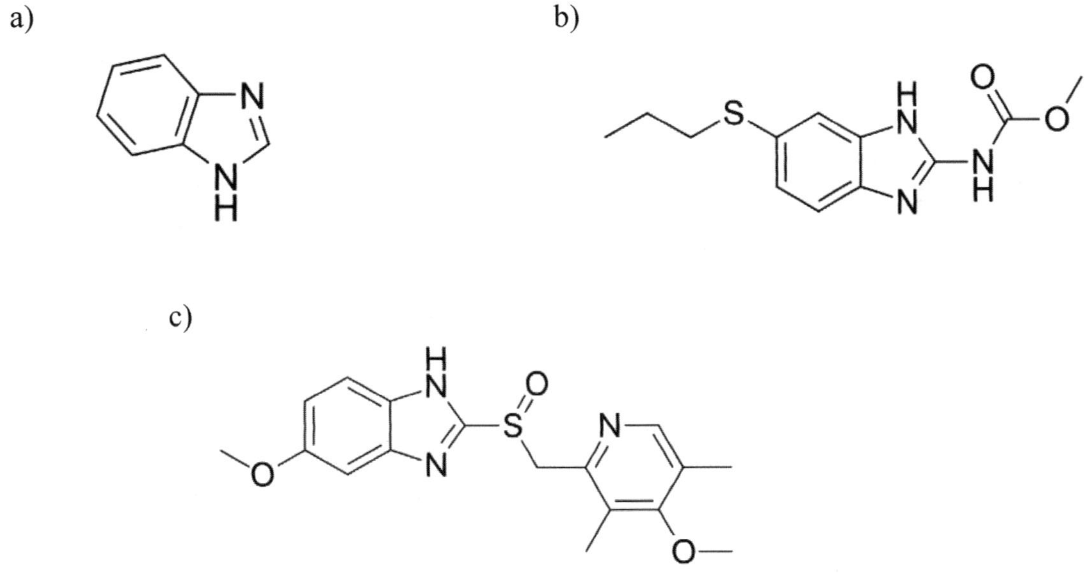

Figure 1. a) Molecular Structure of Benzimidazole, b) Molecular Structure of Albendazole, c) Molecular Structure of Omeprazole

2-methyl benzimidazole is a pharmaceutical-potential substance that belongs to this benzomidazole ring system. The condensation of o-phenylenediamines with carboxylic acids is one of the most common synthesis route for 2-methyl benzimidazole[25]. Conditions and production comparisons of o-phenylenediamines condensation by batch vs. continuous processes are showed in Table 1. These fundamental parameters are used for safety evaluation by Dow's F&EI in the following study.

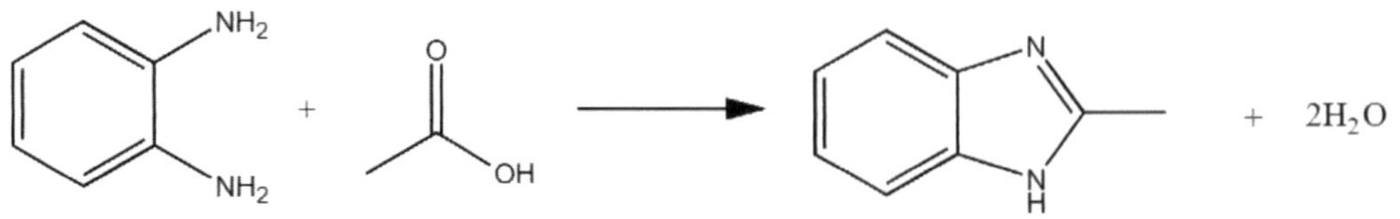

Figure 2. Synthesis of 2-methyl Benzimidazole by Condensation of o-phenylenediamine

Table 1. Production Conditions of Batch and Continuous Processes for Synthesis of 2-methyl benzimidazole

	Batch Process	Continuous Process
Temperature (℃)	200	270
Pressure (psi)	145	1885
Reaction Time (min)	5	0.5
Yield (%)	98	94
Substrate Concentration (M)	5	1
Reaction Volume (ml)	1000	4

In order to concentrate on study in intrinsic differences between two processes, and simplify the hazard identification in F&EI, several assumptions are adopted:

1. The reaction mass is perfect mixed in reactors;
2. Adjunct cooling systems in both processes are in perfect performances;
3. Emergency equipment is same and in perfect performances in both processes;
4. Scalabilities of both processes are good enough that reaction conditions remain same in all scales.

5.2. Safety Comparison Result

The comparison is conducted based on material properties and process features. Related explanation is stated below.

5.2.1. comparison on material

In this study, impacts from intrinsic properties of reactants are same for both processes. Material factors of all substances are listed in Table 2. According to Dow's Fire and Explosion Index Hazard Classification Guide, impact from the most hazardous material dominates the material factor of the whole process, therefore only impact from acetic acid is considered for both processes. In addition, although temperatures and composition are different in two processes and may lead to different chemical properties, these features are neglected in this part and contained in process comparison.

Table 2. Material Factor for Batch and Continuous Reactor

	Batch Reactor	Continuous Reactor
o-phenylenediamine	10	10
2-methylbenzimidazole	1	1
Acetic Acid	14	14

5.2.2. comparison on process

First of all, it's worth mentioning that albeit lots of distinctions exist between batch and continuous processes, comparisons in this study are confined to synthesis in reactor on account of its core role in the whole process: operation mode of reactor – continuously or intermittently – determines whether the process is continuous or not from the start. Therefore, the reactor is selected as the pertinent process unit without conventional unit screening procedure.

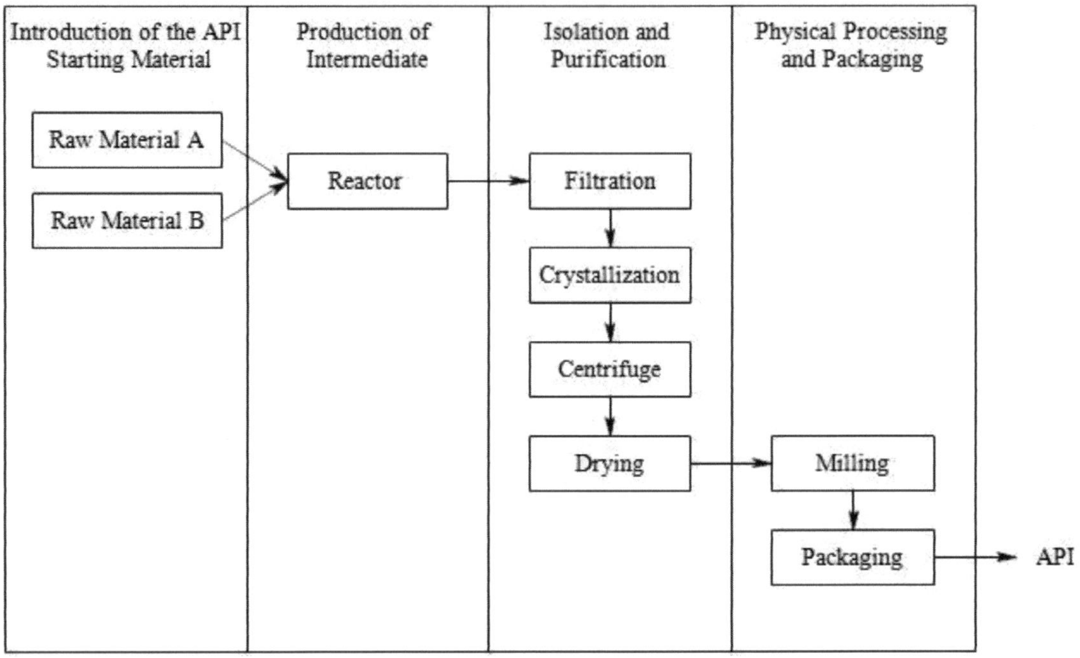

Figure 3. Procedure for Pharmaceutical Production

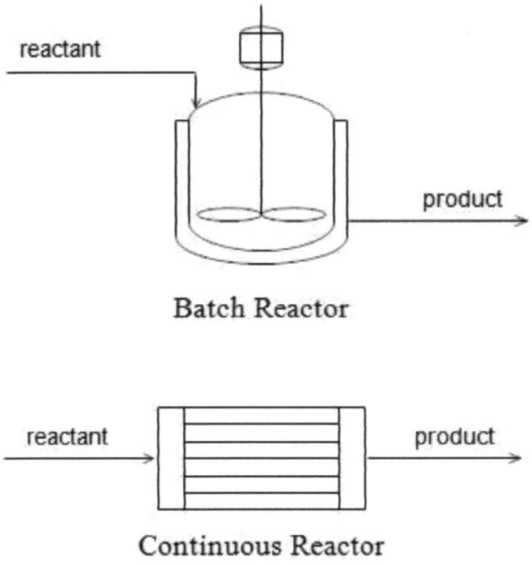

Figure 4. Differences in Structure between Batch and Continuous Reactors

On the foundation of safety evaluation by Dow's Fire & Explosion Index, quantitative judges between batch and continuous process are illustrated through comparison in sensitivity and productivity within safety consideration.

This sensitivity analysis focus on correlation between safety level and inventory of reactor. Inventory of unit plays an important role in the safety. In addition, this sensitivity analysis can also provide supportive information for scaling up procedure.

Inventory of unit plays an important part in the safety via quantity of flammable or unstable material. This potential is considered in both the General Process Hazards (F1) and the Special Process Hazards (F2) according to Dow's F&EI Hazard Classification Guide. In general, more the flammable or unstable material, higher the risk of fire and explosion within the process. For different unit (process unit or storage unit) and different phase (liquid, gas or solid), different correlation formula between inventory and penalty is given in the Special Process Hazards (F2) part of the Dow's F&EI Guide. Moreover, other issues such as whether these materials are processed indoor or outdoor, what's the temperature to handle these materials influences the safety level as well. These questions are included in the General Process Hazards (F1) part.

In this sensitivity analysis, F&EI that combines both the general and the special process hazards is expressed as a dependent variable that changes with the variation of total inventory of reactor. Other operation conditions remain the same for different inventory. Correlations between the F&EI and material inventory for batch and continuous process are shown in the Figure 5.

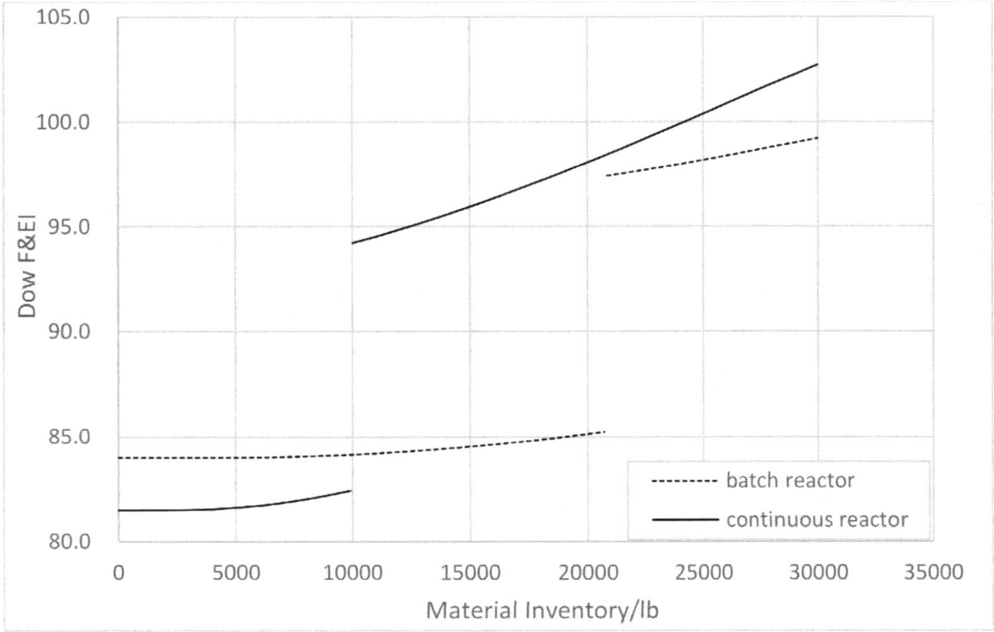

Figure 5. Analysis of Sensitivity to Inventory for Batch and Continuous Reactors Based on 2-methyl Benzimidazole Synthesis

for 0 lb < Inventory < 10000 lb:

F&EI of batch reactor is slightly greater than continuous reactor. This difference mainly comes from difference in usage of stirrer and reaction pressure. Due to requirement for homogeneous mixture, stirrer is necessary in batch reactor, while such a stirrer is not needed (or not allowed) in a continuous tubular reactor. Thus, stirrer, as a hazardous rotating equipment, adds extra penalty to batch process. As for difference in pressure, higher penalty is given to continuous process for its intensified condition. From the result, influence from rotating equipment is greater than that from pressure in this case.

Another observation is that, the magnitude of the difference in the F&EI between two reactors decreases with the growing inventory. This trend stems from the increasing impact from quantity of hazardous material. When inventory approaches zero, difference in F&EI is mainly caused by stirrer and distinct pressure aforementioned. As inventory increases, influence from unstable material extends. This influence depends on composition of mixture, or more specifically, concentration of acetic acid that defines the Material Factor (MF). Based on information of reactions, molar concentrations of acetic acid in batch and continuous reactor are 8.31M and 15.66M respectively. Thus, for similar total inventory, penalty from hazardous material in continuous reactor is greater than batch reactor.

when 10000 lb < Inventory < 20833 lb:

F&EI of continuous reactor is signally higher than batch reactor, and this difference grows along with the increasing inventory. In this range, part of influences come from rotating equipment, pressure and increasing inventory that mentioned above. More importantly, quantity of flammable and unstable liquids handled above boiling point exceeds threshold of 10,000lb for continuous reactor, therefore give a vertical rise to F&EI for continuous reactor, as shown in the Figure 5. This threshold is illustrated clearly in Dow's F&EI Hazard Classification Guide: when quantity of flammable and unstable liquids above its boiling point is less than 10,000 lb, a penalty of 0.60 is determined. A penalty of 0.90 is given when this liquids quantity above its boiling point is greater than 10,000 lb. In this case study, operation temperature of continuous reactor (270℃) is higher than boiling point of both o-phenylenediamine (257℃) and acetic acid (118℃), while operation temperature of batch reactor (200℃) merely higher than boiling point of acetic acid (118℃). Therefore, for inventory in the range of 10,000lb to 20833lb, total mass in continuous reactor is defined as unsafe, while merely acetic acid mass is defined as unsafe in the batch reactor. This divergence leads to different penalty in this range of inventory.

when 20833 lb < Inventory < 30000 lb:

F&EI of batch reactor drastically increases. However, F&EI of continuous reactor is still slightly greater than batch reactor, and the magnitude of this difference increases with the growing inventory.

This vertical rise in the F&EI of batch reactor is also caused by critical quantity of material above the boiling point. When total quantity in batch reactor goes beyond 20,833lb, quantity of flammable and unstable substance - acetic acid - also goes beyond threshold of 10,000 lb. Thus, penalty from material above boiling point is similar for both of processes.

The higher F&EI of continuous reactor originated from quantity of flammable and unstable material in this range of inventory. Since higher concentration of acetic acid is applied in continuous reactor, more proportion of substance is identified as unsafe for continuous reactor, thus higher penalty is given. This explanation is similar to that in range of low inventory (< 10,000 lb). However, unlike appearance in the low inventory range, influence from these flammable and unstable materials exceeds impact from pressure and stirrer. Therefore, continuous reactor becomes more dangerous for this range of total inventory.

It's worth mentioning that higher inventory is not considered here. Upper limit of inventory (30,000 lb) is set based on the general highest inventory of reactor used in pharmaceutical manufacturing.

Degree of hazard for batch versus continuous reactor used in this case study is shown in Table 3. In the range of inventory investigated, batch reactor is identified as moderate hazardous for a broader range of inventory, while continuous reactor is identified as intermediate hazardous for a broader range of inventory. Even so, It's worth mentioning that the F&EI values for both reactors are in a relative safe range (F&EI > 128 indicates heavy even severe hazard).

Table 3. Degree of Hazard for Batch and Continuous Reactors

Material Inventory X	Batch Reactor	Continuous Reactor
X < 15,152 lb	Moderate	Moderate
15,152 lb < X < 20,833 lb	Moderate	Intermediate
20,833 lb < X < 30,000 lb	Intermediate	Intermediate

In summary, F&EI for both batch and continuous reactors increase as material inventory increases. Moreover, F&EI of continuous reactor is more sensitive to change of inventory comparing to batch reactor. This higher sensitivity caused by comprehensive influence from higher concentration of unstable and flammable material, higher operation temperature and higher pressure it applied.

Evaluation in this part mainly focuses on safety of batch versus continuous process base on productivity considerations. Emphasis on productivity scale of this study primary comes from flexibility required for pharmaceutical production. Pharmaceutical industry is highly supply driven. Requirement on production rate is changed according to demand of market all the time. Therefore, better understanding on correlation between safety and productivity provides practicable support for research in process selection.

To approach this target, safety level is evaluated for the process at the same productivity capacity. In this part, F&EI is regarded as indicator for safety evaluation as above. Annual production is set as indicator for productivity. Related formula is shown below.

General calculations for productivity capacity are:

$$\text{Annual Production} = \text{Product Yield} \times \text{Annual Operation Time}$$

$$\text{Product Yield} = \text{Mass Flow of Reagent} \times \text{Conversion}$$

Mass flow is supposed to be proportional to reactor volume as set in the laboratory trial.

According to definition of fine chemical, annual production of fine chemical is limited to 1,000 tons per year[26]. Therefore, 1,000 tons per year is set as the upper limit in this evaluation.

As for product yield and annual operation time, the former is provided in Table 1, and expression for the latter is formed referring to the normal operation time[27].

Annual operation time for batch reactor is:

$$\text{Annual Operaton Time} = 300\ \text{day/yr} \times 24\ \text{hr/day}$$

Annual operation time for continuous reactor is:

$$\text{Annual Operation Time} = 8000\ \text{hr/yr}$$

Normally, in pharmaceutical industry, multi products are produced in one production line. Therefore, the same product cannot occupy the producing line for the whole year. In this study, 2-methyl benzimidazole is assumed to be produced in 2 months of one year.

In addition, circulating time in batch reactor need to be considered. According to the literature[11], the ratio of reaction time to total operation time for this reaction is shown below. Excepted reaction, other operations include filling, cleaning and cooling and other additional work.

$$\frac{\text{Reaction Time}}{\text{Total Operation Time}} = 0.185$$

Based on these formulas and assumptions, correlation between annual productions and F&EI is shown in the Figure 6.

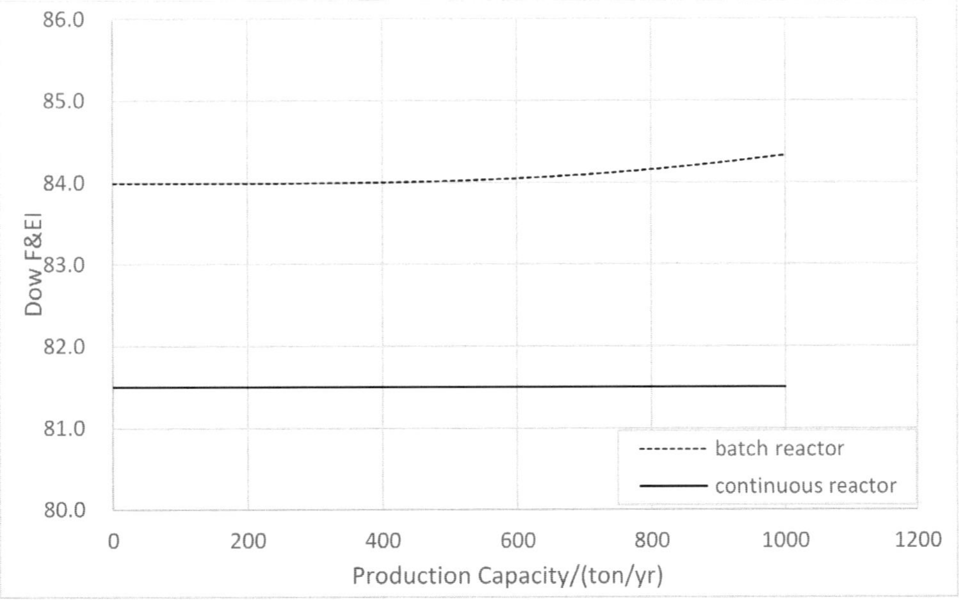

Figure 6. Analysis of Safety Based on Production for Batch and Continuous Reactors Based on 2-methyl Benzimidazole Synthesis

According to the Figure 6, F&EI value is constant in the range of practicable annual production (0 ~ 1000 tons per year). Plus, the F&EI for both reactors corresponds to a moderate level of hazard (61 < F&EI < 96).

This low F&EI resulted from relatively low demand for production. According to the sensitivity analysis, F&EI for both batch and continuous reactor increase as material inventory increases. Both processes could be intermediate hazardous once its inventory is higher than a specific value. However, these values are unapproachable in practical pharmaceutical manufacturing. Since demand for fine chemical, always less than 1,000 tons per year, confines scale of production.

Another observation is that, the F&EI of continuous reactor is always lower than that of batch reactor for same production. Plus, slight rise appears for batch reactor when production capacity exceeds 500 tons per year. These appearances predicate consistent superiority in safety for continuous reactor than batch reactor. Explanation is stated below:

Firstly, as stated in the sensitivity analysis, settled elements, such as operation pressure, temperature and use of rotating equipment, take major account for the difference at small-scale production. Advantages of continuous process in these aspects determine its superiority in small-scale pharmaceutical production.

Secondly, continuous reactor possesses higher space-time yield compared to batch reactor. This advantage stems from the intrinsic property of continuous operation mode - the low residence time and high flow throughput. That is to say, it costs fewer time and lower inventory for continuous reactor to produce same amount of product than batch reactor. These benefits in production efficiency influence safety level indirectly but significantly: Smaller inventory means fewer unstable materials in process, and shorter operation time means less chance to fail. This point is not considered in separate safety evaluation, but can be clearly illustrated by correlation between production capacity and F&EI. In this case, when demand on production growing, demand on inventory grows faster for batch reactor than continuous reactor. This faster rise in inventory lead to faster rise in amount of unstable material in process so as to the faster rise in F&EI value. It explains why slight rise appears for batch reactor when production capacity exceeds 500 tons per year.

In fact, the comparable batch and continuous reactor are rarely at the same level of volume. Referring to literature, for commercial production of fine chemical, the volumes of continuous reactor are usually in a range of 0.1 to 3 liter, while the volume of batch reactor is usually around the range of 2 to 10 cubic meters[6]. For this case, based on assumption of same annual production, ratio of batch reactor inventory and continuous reactor inventory is as much as one hundred.

6 Case Study II: Oxidation with Hydrogen Peroxide

6.1. overview

Hydrogen peroxide is one kind of widely used oxidation agent in pharmaceutical production. Typical reaction involving hydrogen peroxide includes Baeyer–Villiger oxidations for synthesis of lactone[9], N-oxidation of alkyl pyridine for synthesis of alkyl pyridine N-oxidant[28] and benzylic hydro-peroxide rearrangement for synthesis of Brivanib alaninate[29].

This common oxidation agent is also used in production of peracetic acid. Peracetic acid is produced via oxidation of acetic acid by hydrogen peroxide with sulfuric acid as the catalyst. Conditions for peracetic acid production by batch vs. continuous process are showed in Table 4[30].

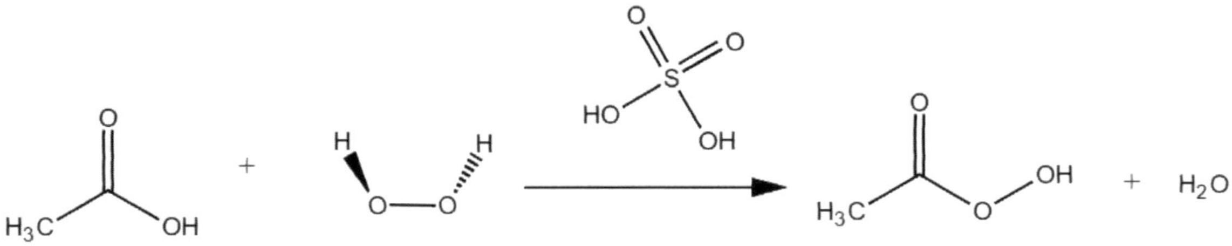

Figure 7. Synthesis of Peracetic Acid by Oxidation of Acetic Acid

Table 4. Production Conditions of Batch and Continuous Process for Synthesis of Peracetic Acid

	Batch Process	Continuous Process
Temperature (℃)	60	80
Pressure (psi)	14.5	14.5
Reaction Time (min)	30	15
Yield (kg/h)	170	20
Acetic Acid Concentration (M)	7.3	7.3
Reaction Volume (L)	4000	10

Assumption for this case study is listed below:

1. The reaction mass is perfect mixed in reactors.
2. Since the temperature for decomposition of hydrogen peroxide in peracetic acid synthesis reaction is higher than 120 ℃ [31], which is larger than operation temperature for both processes in the case study, decomposition of hydrogen peroxide is assumed to be inexistent.
3. Sulfuric acid is dissociated completely to H^+ and HSO_4^-, in addition, concentration of H^+ is assumed to be constant of 0.67 mol/L during the reactions for both processes.
4. Influence from sulfuric acid is not considered in safety evaluation.
5. Emergency equipment and cooling system are perfectly performed for both of processes.
6. Scalabilities of both processes are good enough that reaction conditions remain same in all scales.

6.2. safety comparison result

The comparison is conducted based on material property and process feature. Related explanations are stated in the following paragraph.

6.2.1. comparison on material

In this case, impact from material within the reactor is same for both of processes. Material factors for reactants involved are listed in Table 5. According to Dow's Fire and Explosion Index Hazard Classification Guide, influence from the most hazardous material dominates the material factor. Although material factor values are same for hydrogen peroxide and peracetic acid, since concentration of hydrogen peroxide is higher than peracetic acid, hydrogen peroxide is chosen to be the dominant material. Therefore, only impact from hydrogen peroxide is considered for both processes. Change on substances' property from temperatures are neglected in this part and reconsidered in process comparison.

Table 5. Material Factor for Batch and Continuous Reactor

	Batch Reactor	Continuous Reactor
Acetic Acid	14	14
Hydrogen Peroxide	24	24
Peracetic Acid	24	24

6.2.2. comparison on process

The major difference between these two production processes come from distinctions in reaction conditions and reactor constructions. These differences are shown in Table 4. Based on safety evaluation by Dow's Fire & Explosion Index, quantitative judges between batch and continuous process are illustrated through comparison in sensitivity and productivity within safety consideration.

Procedure for sensitivity comparison is the same as the first case study. Correlations between the F&EI and material inventory for batch and continuous process are shown in the Figure 8.

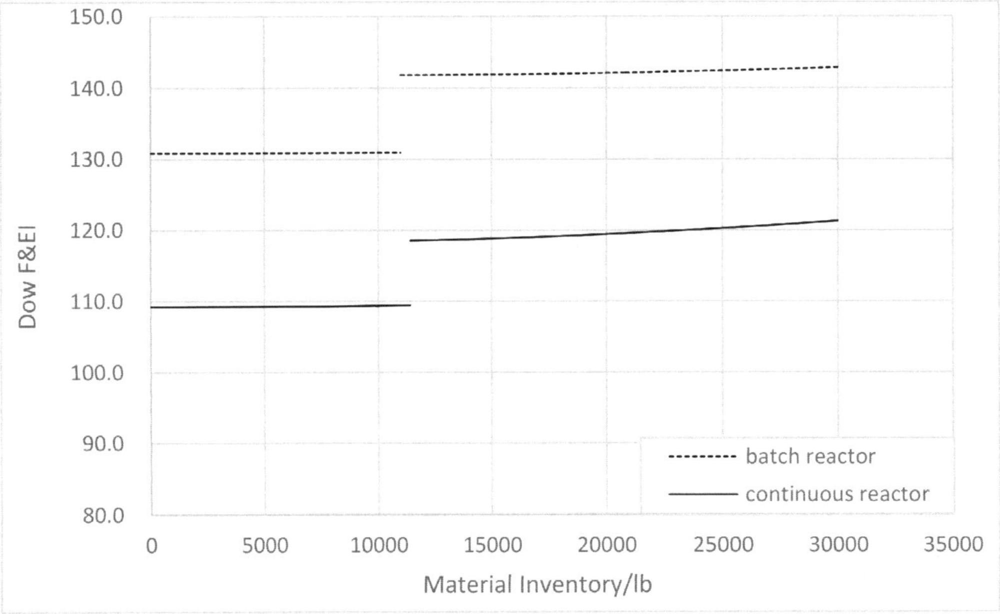

Figure 8. Analysis of Sensitivity to Inventory for Batch and Continuous Reactors Based on Peracetic Acid Synthesis

for 0 lb < Inventory < 11000 lb:

F&EI of batch reactor is larger than continuous reactor. This difference is almost caused by usage of stirrer in batch reactor. Application of such a rotating equipment increases penalty for batch process.

when 11000 lb < Inventory < 11430 lb:

F&EI of batch reactor is still higher than continuous reactor, but this difference grows drastically due to the sharp transition in F&EI of batch reactor. This transitions are caused by quantity of flammable liquids that handled above flashing point exceeds threshold of 10,000lb. This threshold is illustrated clearly in Dow's F&EI Hazard Classification Guide: when quantity of flammable liquids above their flashing point is less than 10,000 lb, a penalty of 0.30 is determined. A penalty of 0.45 will be given when this quantity of flammable liquids above their flashing point is larger than 10,000 lb. For peracetic acid production, operation temperature for batch reactor (60℃) and continuous reactor (80℃) is higher than flashing point of both acetic acid (39℃) and peracetic acid(40.5℃). Therefore, the value of inventory for higher penalty is determined by compositions of peracetic acid and acetic acid in reactors. Since concentrations of these two materials is slightly higher for batch comparing to continuous reactor, batch reactor approaches this transition in F&EI at a smaller total inventory.

when 11430 lb < Inventory < 30000 lb:

F&EI of continuous reactor exceed the transition caused by quantity of unsafe material above flashing point. However, it is still lower than F&EI of batch reactor.

Compared to the first case study, value of slope for both processes is similar low. It results in the consistent superiority of continuous reactor in safety level compared to batch reactor. This similarity in slopes comes from similarity in operation conditions. Differences only appear in operation temperature and hydrogen peroxide concentration. These differences are not such significant that lead to huge divergence between slopes. Thus, difference in F&EI value between two reactors does not change along with increasing inventory.

Degree of hazard of this case study is exhibited in Table 6. Although marked transitions appears for both reactor, safety levels for two reactors do not change based on scaling up of reactor size. The safety levels are majorly decided by settled parameters, such as operation pressure and usage of stirrer. These parameters do not change with inventory as variable. Hence, inventory does not play a significant role in judgement on degree of hazard.

Table 6. Degree of Hazard for Batch and Continuous Reactors

Reactor Type	Batch Reactor	Continuous Reactor
Degree of Hazard	Heavy	Intermediate

In sum, F&EI for both batch and continuous reactor increase as material inventory increases due to the increases of quantity above flashing point. However, since operation conditions are relatively mild and similar for these two reactors, both reactor do not exhibit high sensitivity when inventory is growing gradually.

For evaluation based upon productivity, assumption for practical production are set as following:

1. Confined by the conversion of this reaction route and volume of reactor, upper limit for annual production is settled to be 580 tons per year.
2. Same as the first case study, referring to normal operation time for this two operation mode[27], annual operation time for batch reactor is 8760 hour per day, and annual operation time for continuous reactor is 8000 hour per day.
3. Peracetic acid is assumed to be produced for 2 months in one year.
4. According to the literature[30, 32], the ratio of reaction time to total operation time for this reaction is 0.143

Based upon these formulas and assumptions, correlation between annual productions and F&EI is shown in the Figure 9.

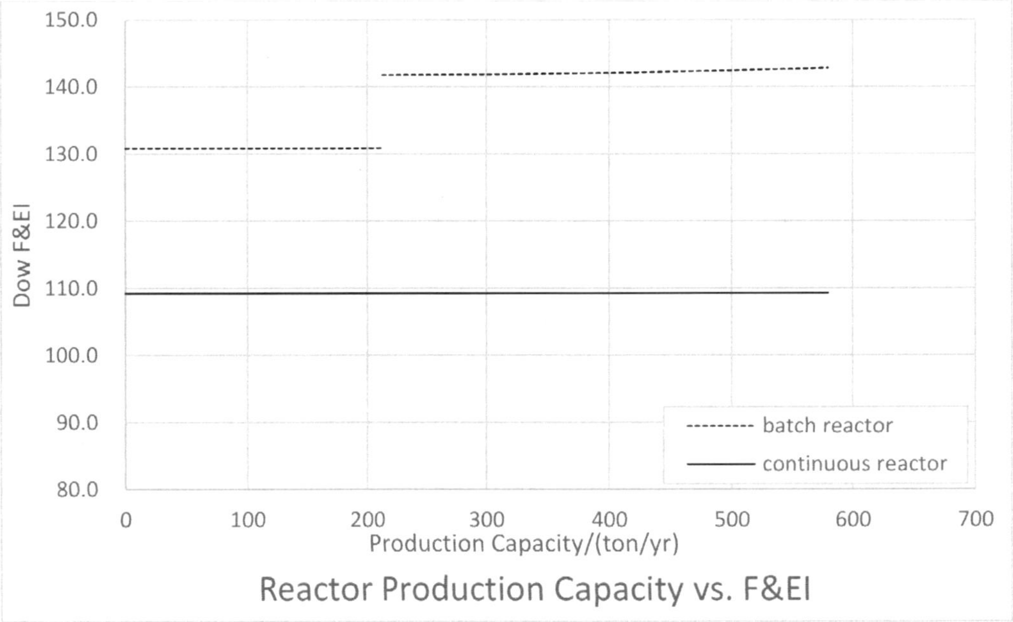

Figure 9. Analysis of Safety Based on Production for Batch and Continuous Reactors Based on Peracetic Acid Synthesis

According to the Figure 9, F&EI is stabilized at 109 in the range of practicable annual production for continuous reactor. For batch reactor, the F&EI value rise drastically when the annual production is higher than 220 tons per year.

For the batch production mode, relatively low production efficiency leads to substantial increase in reactor volume when demand on production grows. This increase in reactor volume is so significant that exceed the threshold for large quantity of flammable and unstable material above flashing point. Therefore, the F&EI value jump from 131 to 143 when the annual production is higher than 220 tons per year.

For continuous production mode, high production efficiency causes only a few change in volume hence only few change in F&EI in as well. This difference in F&EI between batch and continuous reactor means that, to produce the same amount of peracetic acid in the given period, the volume of continuous reactor is much smaller than batch reactor.

7 Result Summary

Figure 10 represents the comparisons of two case studies. The X-axis represents two different case studies, and the Y-axis represents the average Dow's F&EI value which can be used to define the degree of hazard.

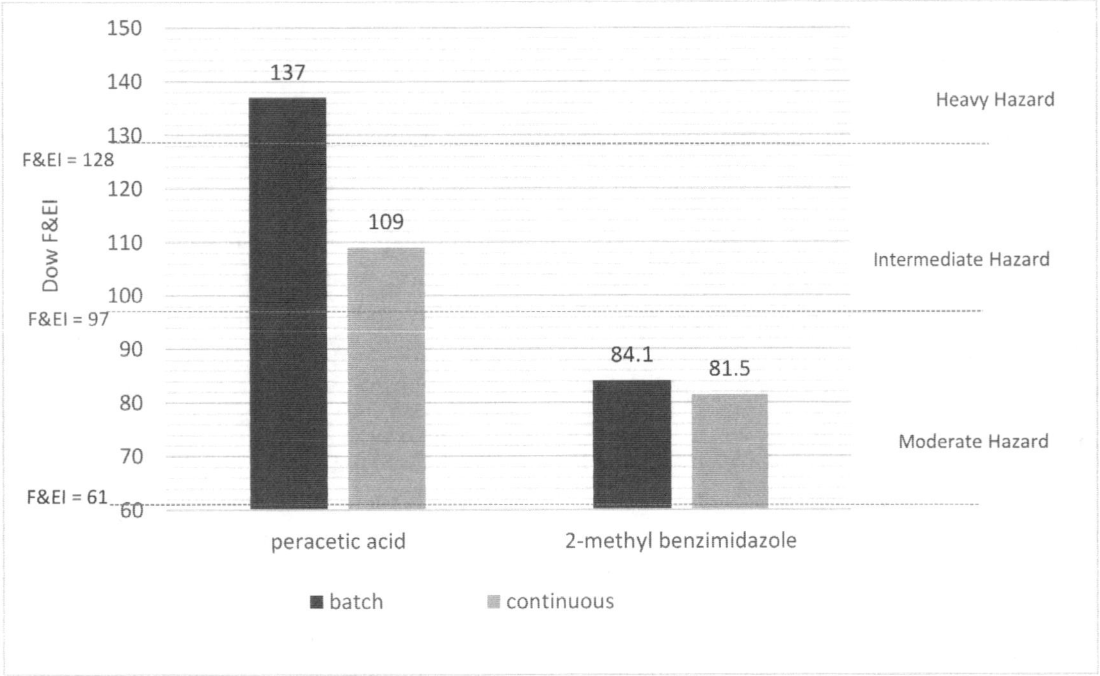

Figure 10. Comparison of Safety Level by Batch and Continuous Productions for Two Case Studies

Regardless of the reactor type, the safety level is predetermined by properties of the synthesis reaction. Although distinctions in F&EI appear when using different reactors for both cases, peracetic acid production is generally more hazardous than 2-methyl benzimidazole production, no matter which reactor it uses. This higher level of danger comes from the intrinsic properties, which consist of the flammability and reactivity of the chemicals involved in the synthesis of peracetic acid.

Taking batch and continuous reactors into consideration, distinction in F&EI comes from the differences in operation conditions and production efficiency. These varying factors such as temperature, pressure, and yield do not influence F&EI as individual factors. Instead, they are correlated and impact the F&EI as an ensemble. Difference in these ensemble decides the variance in F&EI value between reactors, then determines to which extent the selection of reactor can make a difference. For this study, the variance in F&EI value between batch and continuous reactor is greater for peracetic acid production (variance in F&EI = 28) than 2-methyl benzimidazole production (variance in F&EI = 2.6). Therefore, selecting a reactor is more important for peracetic acid comparing to 2-methyl benzimidazole production.

The importance of choosing a reactor is different for the two cases, the F&EI value for a continuous reactor is lower than a batch reactor for both of these cases. This consistency exhibits the superiority of a continuous reactor in safety.

8 Conclusion

A comprehensive comparison between the batch and continuous process used in the pharmaceutical industry based upon Dow's F&EI is made in this study. Two typical pharmaceutical productions, the production of 2-methyl benzimidazole and the production of peracetic acid are analyzed. Safety levels for batch and continuous reactors are compared based on both inventory and annual production being used as variables. The conclusions on the foundation of these two case studies are stated below:

1. The comparison of batch versus continuous processes in the pharmaceutical industry is complex. It's determined by reaction properties, operation conditions, structures of equipment, production efficiency, scale of production, and many other factors. To get a conclusive comparison between these two production modes, more systematic studies based on the integrated batch and continuous process need to be conducted for safety consideration.

2. Just focusing on the reactor in which part of API synthesis occurs, the continuous reactor is more sensitive to inventory in general. The magnitude of this sensitivity is determined by operation conditions such as operation temperature, operation pressure and concentration of hazardous material. This sensitivity to inventory might lead to problems for continuous production in a relatively large scale.

3. However, when considering practical pharmaceutical production, which is always in a small scale (no larger than 1000 tons per year), hazardous large-volume continuous reactor is not necessary. In addition, the space time yield of continuous production mode is usually higher than the batch production mode, application of a continuous reactor is relatively safer when compared to batch reactors.

9 References

1. Plumb, K., Continuous processing in the pharmaceutical industry: changing the mind set. *Chemical Engineering Research and Design* **2005,** *83* (6), 730-738.

2. Laird, T., What Type of Reactions Do Process Chemists Use on Scale? *Organic Process Research & Development* **2006,** *10* (5), 851-852.

3. Braune, S.; Poechlauer, P.; Reintjens, R.; Steinhofer, S.; Winter, M.; Lobet, O.; Guidat, R.; Woehl, P.; Guermeur, C., Selective nitration in a microreactor for pharmaceutical production under cGMP conditions. *Chimica oggi* **2009,** *27* (1), 26-29.

4. Proctor, L. D.; Warr, A. J., Development of a Continuous Process for the Industrial Generation of Diazomethane1. *Organic process research & development* **2002,** *6* (6), 884-892.

5. Roberge, D. M.; Gottsponer, M.; Eyholzer, M.; Kockmann, N., Industrial design, scale-up, and use of microreactors. *Chemistry Today* **2009,** *27* (4), 8-11.

6. Hessel, V.; Vural Gürsel, I.; Wang, Q.; Noel, T.; Lang, J., Potential analysis of smart flow processing and micro process technology for fastening process development: use of chemistry and process design as intensification fields. *Chemical Engineering & Technology* **2012,** *35* (7), 1184-1204.

7. Schaber, S. D.; Gerogiorgis, D. I.; Ramachandran, R.; Evans, J. M.; Barton, P. I.; Trout, B. L., Economic analysis of integrated continuous and batch pharmaceutical manufacturing: a case study. *Industrial & Engineering Chemistry Research* **2011,** *50* (17), 10083-10092.

8. Jolliffe, H. G.; Gerogiorgis, D. I., Plantwide design and economic evaluation of two continuous pharmaceutical manufacturing (CPM) cases: ibuprofen and artemisinin. *Computers & Chemical Engineering* **2016,** *91*, 269-288.

9. Gutmann, B.; Cantillo, D.; Kappe, C. O., Continuous - Flow Technology—A Tool for the Safe Manufacturing of Active Pharmaceutical Ingredients. *Angewandte Chemie International Edition* **2015,** *54* (23), 6688-6728.

10. Wakami, H.; Yoshida, J.-i., Grignard exchange reaction using a microflow system: from bench to pilot plant. *Organic process research & development* **2005,** *9* (6), 787-791.

11. Damm, M.; Glasnov, T. N.; Kappe, C. O., Translating high-temperature microwave chemistry to scalable continuous flow processes. *Organic Process Research & Development* **2009,** *14* (1), 215-224.

12. Schaber, S. D.; Gerogiorgis, D. I.; Ramachandran, R.; Evans, J. M. B.; Barton, P. I.; Trout, B. L., Economic Analysis of Integrated Continuous and Batch Pharmaceutical Manufacturing: A Case Study. *Ind Eng Chem Res* **2011,** *50* (17), 10083-10092.

13. Roberge, D. M.; Ducry, L.; Bieler, N.; Cretton, P.; Zimmermann, B., Microreactor technology: A revolution for the fine chemical and pharmaceutical industries? *Chem Eng Technol* **2005,** *28* (3), 318-323.

14. Hartman, R. L.; McMullen, J. P.; Jensen, K. F., Deciding Whether To Go with the Flow: Evaluating the Merits of Flow Reactors for Synthesis. *Angew Chem Int Edit* **2011,** *50* (33), 7502-7519.

15. Calabrese, G. S.; Pissavini, S., From Batch to Continuous Flow Processing in Chemicals Manufacturing. *Aiche J* **2011,** *57* (4), 828-834.

16. Kletz, T. A., What you don't have, can't leak. *Chemistry and Industry* **1978,** *6*, 287-292.

17. Kletz, T. A., *Plant design for safety: a user-friendly approach*. Taylor & Francis: 1991.

18. Fire, D., Explosion Index Hazard Classification Guide. *AICHE New York* **1994**.

19. Lawrence, D. Quantifying inherent safety of chemical process routes. © Duncan Lawrence, 1996.

20. Heikkilä, A.-M., *Inherent safety in process plant design: an index-based approach*. VTT Technical Research Centre of Finland: 1999.

21. Gentile, M.; Rogers, W.; Mannan, M., Development of a fuzzy logic-based inherent safety index. *Process Safety and Environmental Protection* **2003,** *81* (6), 444-456.

22. Bansal, Y.; Silakari, O., The therapeutic journey of benzimidazoles: a review. *Bioorganic & medicinal chemistry* **2012,** *20* (21), 6208-6236.

23. omeprazole. https://www.drugs.com/monograph/omeprazole.html.

24. albendazole. https://www.drugs.com/monograph/albendazole.html.

25. Patil, A.; Ganguly, S.; Surana, S., A systematic review of benzimidazole derivatives as an antiulcer agent. *Rasayan J Chem* **2008,** *1* (3), 447-460.

26. Patt, R.; Kordsachia, O.; Süttinger, R.; Ohtani, Y.; Hoesch, J.; Ehrler, P.; Eichinger, R.; Holik, H.; Hamm, U.; Rohmann, M., Ullmann's encyclopedia of industrial chemistry. Wiley: 2002.

27. Annual Production. http://www.eng-tips.com/viewthread.cfm?qid=389883.

28. Pineda-Solano, A.; Saenz-Noval, L.; Nayak, S.; Waldram, S.; Papadaki, M.; Mannan, M. S., Inherently safer reactors: Improved efficiency of 3-picoline N-oxidation in the temperature range 110–125° C. *Process Safety and Environmental Protection* **2012,** *90* (5), 404-410.

---

29. LaPorte, T. L.; Spangler, L.; Hamedi, M.; Lobben, P.; Chan, S. H.; Muslehiddinoglu, J.; Wang, S. S., Development of a Continuous Plug Flow Process for Preparation of a Key Intermediate for Brivanib Alaninate. *Organic Process Research & Development* **2014,** *18* (11), 1492-1502.

---

30. Ebrahimi, F.; Kolehmainen, E.; Turunen, I., Safety advantages of on-site microprocesses. *Organic Process Research & Development* **2009,** *13* (5), 965-969.

---

31. Kadla, J. F.; Chang, H., The reactions of peroxides with lignin and lignin model compounds. ACS Publications: 2001.

---

32. Ebrahimi, F.; Kolehmainen, E.; Oinas, P.; Hietapelto, V.; Turunen, I., Production of unstable percarboxylic acids in a microstructured reactor. *Chemical engineering journal* **2011,** *167* (2), 713-717.

478238 Risk Based Decision Making Methods for Evaluating Complex Technologies, a Comparative Analysis on Pipelines and Railroad for Crude Oil Transport

Abstract:

Decision making for long-term capital intensive projects involve large number of uncertainties. Traditional risk assessments are based on probabilities which are estimated from limited data sets and provide limited flexibility for complex technologies. Further, the environment and specific conditions under which the data is obtained may be unknown and yield to erroneous extrapolations. The method proposed in this work utilizes a novel method for estimation of the risk based on the net present value of technologies and provides the basis for a decision support tool based on the expected value theorem. This is demonstrated using the case of the safety and economic crisis in onshore crude oil transport in the continental United States. The exponential growth of crude oil production in the continental United States has led to crisis in the mid-stream industry with existing pipeline infrastructure under pressure to service the increased production. This fact is reflected in the 4000% increase in the carloads in a span of five years from 2009 to 2013. This has been compounded by the volatile fluctuation in crude oil price from $100 to $30 over the past few years. This has put enormous constraints on capital decision making. Since railroad infrastructure albeit aged, already exists in many regions, loading and unloading crude oil is relatively easy and their connection to the refining centers along the Gulf of Mexico apart from sites in the east and west coast make it an option to many producers. In this paper a novel decision making model is proposed using the concept of 'Net present value' of a technology to evaluate the option between railroad and as suitable mid-stream alternatives for the transportation of crude oil. The model is built on using a prescriptive decision making technique for the quantification of factors affecting the decision, incorporating also the utility of wealth of the decision maker. Monte-Carlo simulation is further used to obtain results and trends based on variation of initial parameters. This analysis can provide a basis for making decisions given the complex environment of crude oil transport operations and can be extended to a variety of geographies given that they satisfy the requirements of the model. The method can also be further extended to other technology evaluation given appropriate factors can be identified to build the model.

479492 Key Modeling Parameters and Process Safety Information for Effective Atmospheric Dispersion Modeling

Abstract:

Dispersion modeling is an important part of Quantitative risk assessment (QRA) documentation. Dispersion modeling can be performed using commercially available software. Variety of modeling parameters and Process Safety Information (PSI) data are needed for atmospheric dispersion modeling. Some of the parameters include weather, facility geographical location, flammability and toxicity data of the service fluid, radiation intensity, software tool, and model. Some PSI data needs include process and instrumentation diagrams (PIDs), Process Flow Diagrams (PFDs), relief system evaluation documentation, Heat and Material Balance (H&MB), and equipment data sheets. This paper describes the key parameters and their selection for effective dispersion modeling.

479373 Implementation of Hazard Mitigation Solutions: Consequence Vs. Risk-Based Building Specifications

Abstract:

Implementation of Hazard Mitigation Solutions: Consequence vs. Risk-Based Building Specifications

Travis Holland, Joshua Bruce-Black

Baker Engineering and Risk Consultants

3330 Oakwell Court, Suite 100

San Antonio, TX 78218

tholland@bakerrisk.com, jbblack@bakerrisk.com

Abstract

There are various strategies to mitigate occupant vulnerability and risk after completion of a consequence based facility siting study (FSS) or quantitative risk assessment (QRA). One potentially cost-effective mitigation strategy is to focus on reducing the occupant vulnerability and risk to personnel by upgrading existing buildings or designing new buildings. A building specification is commonly used to relay a small set of building design parameters based on the results of a consequence based FSS to engineers or contractors. However, with the industry shift towards risk based studies (QRAs), there has been a growing perspective that the information required to develop an appropriate building specification is more complicated given the increased number of variables involved in a QRA. Using the development of a new control room specification as an example, this paper will compare and contrast the steps and benefits/drawbacks involved in developing an appropriate building design basis employing a consequence and risk based approach. This paper will highlight how establishing tolerance criteria, emergency response strategy, personal protective equipment (PPE), and building occupancy early in the project definition stage promotes effective streamlined communication in the specification and reduces the differences between consequence and risk-based building specifications.

13TH GLOBAL CONGRESS ON PROCESS SAFETY

Generalization of Modeling Along-Wind Dispersion

Jessica M. Morris
Tom Spicer, Ph.D., P.E.
Ralph E. Martin Department of Chemical Engineering
University of Arkansas, Fayetteville, AR
Jmm065@uark.edu

Prepared for Presentation at
American Institute of Chemical Engineers
2017 Spring Meeting and 13th Global Congress on Process Safety
San Antonio, TX
March 26 – 29, 2017

Generalization of Modeling Along-Wind Dispersion

Jessica M. Morris
Tom Spicer, Ph.D., P.E.
Ralph E. Martin Department of Chemical Engineering
University of Arkansas, Fayetteville, AR
Jmm065@uark.edu

Keywords: along-wind, dispersion, finite duration, time-varying, atmosphere, releases, modeling

Abstract

The method to model along-wind dispersion is an important aspect of currently used commercial atmospheric dispersion models when considering time-limited or time-varying releases of toxic or flammable materials. Past wind tunnel experimental data has shown the validity of this approach. Inclusion of the effects of along-wind dispersion has proven to be critically important to modeling available field-scale data with current models for such time-varying releases. Many investigators have considered modeling along-wind dispersion from a theoretical point of view as well as basic parameterization of available field test data. This paper reviews past work and proposes a non-dimensionalized parameterization which shows improved agreement with available field-scale data.

1 Introduction

Continued world-wide industrial development has brought the wide spread use of industrial chemicals, many of which are hazardous. Whether released accidentally (or intentionally), air borne hazardous chemicals effect large populations quickly as evidenced by several industrial chemical accidents (e.g., Bhopal[1], Flixborough[2], Seveso[3]). Understanding and modeling hazardous chemical releases is vital to protecting the public. Many researchers have worked to assess these hazards using physical and computational models. Predicting the potential impact of hazardous gas releases in the atmosphere is imperative to properly determine where transport, storage, and use of such chemicals should be located. In the case of an actual loss of containment, understanding where air borne chemicals can travel is important to determine the proper response to an emergency.

While models exist to address these questions, there are opportunities to improve modeling methods particularly for (realistic) finite duration and time varying hazardous gas releases, which depend strongly on parameterizing the effects of along-wind dispersion (the elongation of the cloud in the along-wind direction). Modeling methods for along-wind dispersion are frequently based on established theoretical approaches which provide guidance for analyzing currently available field scale data. This paper considers experimental data and established theoretical methods to consider how the effects of along-wind dispersion can be improved.

2 Theoretical Description

2.1 Theoretical Models

Along-wind dispersion can be calculated under many different conditions and assumptions. There have been many established theories accounting for different conditions and assumptions. Scientists choose the along-wind dispersion coefficient as either a function of time or distance. Majority of scientists choose the along-wind dispersion coefficient as a function of the friction velocity. Some of the most prominent theories about along-wind dispersion are discussed below.

2.2 Batchelor

Batchelor (1964)[4] determined the Lagrangian Similarity Hypothesis. This hypothesis relates Eulerian and Lagrangian quantities when deriving theoretical solutions to problems. Applying known properties of fluid flow to the atmosphere, Batchelor was able to establish a generalized equations under neutral stability conditions, with no restraints on maximum height, for center line concentration at ground level and wind speed dependence with height. Batchelor derived many equations that are the foundation for modern theoretical equations commonly utilized. Batchelor's relationship between turbulence in a constant stress region on a boundary layer is a function of the friction velocity and time. This relationship and the Lagrangian Similarity Hypothesis serve as a template for along-wind dispersion equations to come.

Release height can be interpreted differently on its effect on dispersion. Restating Batchelor's similarity theory, "the statistical properties of the velocity of a marked fluid particle at time, t, after release at the ground depend only on friction velocity, u_* and t." This theory is then extended to the case of an elevated release: "The statistical properties of the velocity of a marked particle at time, t, after release at height, h, are the same as those of a particle released at the ground (z=0) at the instance $-t_1$, provided $t \gg t_1$ where t_1 is expected to be of the order of magnitude of the time-scale of the turbulence at height h, that is, of order h/u_*". Under Batchelor's similarity theory, after a sufficient amount of time, the height of release can be ignored.

2.3 Chatwin

Chatwin (1968)[5] was the first mathematician to theoretically derive a full equation with numerical values for the along-wind dispersion coefficient. He also derived the shear stress component of the along-wind dispersion equation. Chatwin used Batchelor's similarity theory along with the assumptions of the ability of turbulence to be described as an eddy diffusivity to solve for along-wind dispersion, σ_x. He makes further assumptions of neutral stability, ground release, and an unbounded boundary layer. The generalized equations for both overall along-wind dispersion, σ_x , and wind shear, σ_{xs} , by Chatwin are as followed

$$\sigma_x = 0.803\frac{u_* t}{\kappa} \quad (1)$$

$$\sigma_{xs} = 0.596\frac{u_* t}{\kappa} \quad (2)$$

where u_* is the friction velocity and κ is the van Karman's constant, approximately equal to 0.41.

2.4 Van Ulden

Van Ulden (1992)[6] derived an equation under slightly different dispersion definitions. Van Ulden made the distinction between absolute and relative dispersion. Wheatley and Chatwin were able to authenticate their equations with the assumption of a hypothetical atmosphere: where no large horizontal eddies exist. Taking into consideration this distinction of absolute and relative dispersion, van Ulden derived a new equation for σ_x, the total variance of the concentration distribution in the wind direction

$$\sigma_x{}^2 = \sigma_c{}^2 + \sigma_d{}^2 \quad (3)$$

where σ_c is the deterministic part (or centroid variance), and σ_d is the random part (or diffusive variance). The total longitudinal variance, σ_x, is dependent on three separate quantities: centroid variance, diffusive variance due to shear effects, and direct dispersion. These three variables were able to be solved for shear, centroid, and diffusive effects respectively as a function of friction velocity and time

$$\sigma_s = 0.5\frac{u_* t}{\kappa} \quad (4)$$

$$\sigma_c \simeq 0.54\frac{u_* t}{\kappa} \quad (5)$$

$$\sigma_d = 0.67\frac{u_* t}{\kappa}. \quad (6)$$

Under the assumption that $\sigma_c \simeq \sigma_s = 0.5\frac{u_* t}{\kappa}$, these variables can be substituted into equation (3) to get the total longitudinal variance. An exact equation for σ_x is obtained:

$$\sigma_x = 0.84\frac{u_* t}{\kappa} \quad (7)$$

Van Ulden not only derived an along-wind dispersion equation under neutral conditions from a ground release in an unbounded boundary layer, but also for an along-wind dispersion equation for very stable conditions. Under very stable conditions, the wind speed profile and eddy diffusivities change. When applying the new constraints for wind speed and eddy diffusivities in a different stability class, the total variance of concentration distribution in the wind direction is shown to be:

$$\sigma_x = \{(\frac{3}{4} - \frac{16}{9\pi})\frac{5u_*{}^3 t^3}{kL}\}^{1/2} \simeq 0.43(\frac{5u_*{}^3 t^3}{kL})^{1/2}. \quad (8)$$

where L is the Monin-Obukov length.

2.5 Summary

The relationship between advection cloud speed, time, and distance to determine the along-wind diffusion coefficient is complex. Changing conditions such as stability class, surface roughness, wind speed, function of distance or time, release height, and other parameters all have an effect on determining the along-wind diffusion coefficient. With the varying conditions, not all the equations are able to be integrated together into an overarching equation meeting all conditions. Different assumptions or criteria result in diverse equations for the same diffusion coefficient. An equation that takes into account all of these parameters is desired.

3 Correlation of Field Data by Hanna and Franzese

The 12 field and wind tunnel tests mentioned prior were analyzed by Hanna and Franzese to determine a suitable along-wind dispersion coefficient that resulted in an equation comparable to Chatwin's theoretical equation[5Error! Bookmark not defined.]. Results of the data and the predicted equation are shown below in Figure 1.

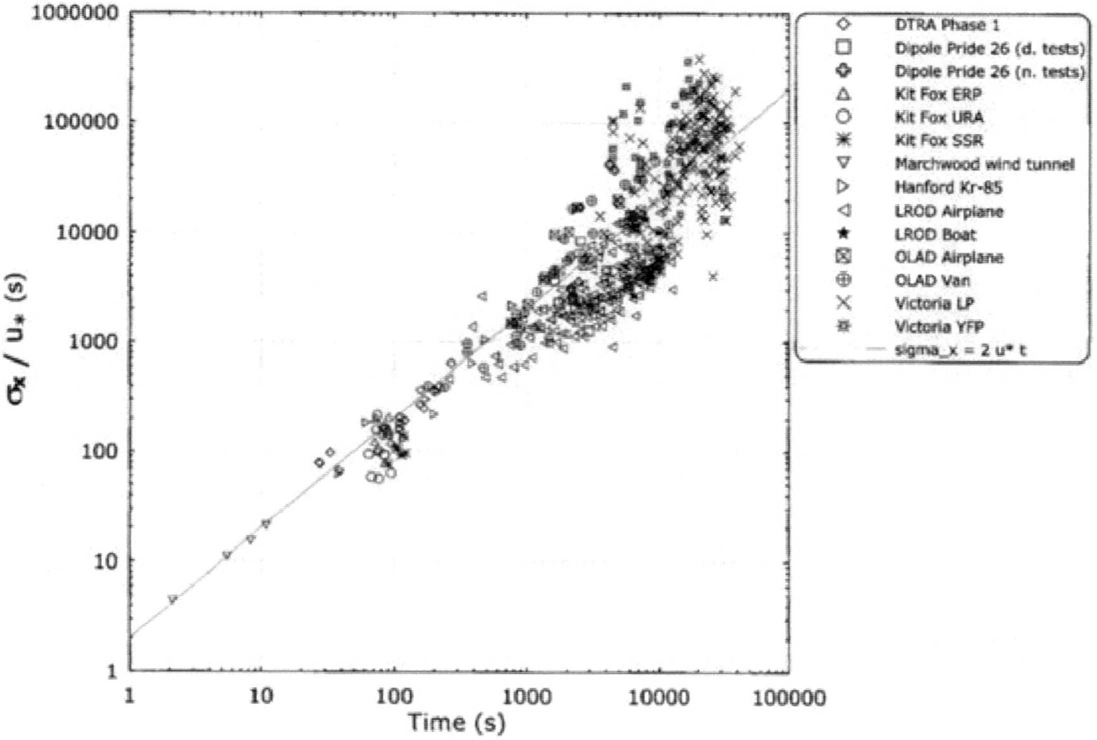

Figure 1. Hanna and Franzese results of along-wind dispersion coefficient divided by friction velocity and plotted vs time. Line represents $\sigma_x = 2u_*t$.[8]

4 Reconsideration of Field Data Using Dimensionless Groups

Figure 1 shows a correlation of the along-wind dispersion coefficient with the friction velocity as a function of time. While this is linear, their data can be up to a factor of 10 away from the predicted equation. There is also a noticeable trend of data beneath the predicted line when time is less than 10,000 seconds and above the predicted line when time is greater than 10,000 seconds. This is believed to be an effect of the near field vs. the far field. This equation for the along-wind dispersion coefficient does not take into effect the near field and far field which results in a very large portion data over predicted from 1,000-10,000 seconds and a large portion of data under predicted from 10,000 seconds and above. An improved equation to improve the correlation accounting for the error and trends is desired.

A non-dimensionalized equation was created to better predict the along-wind dispersion coefficient. The parameters of the non-dimensionalized equation were based off theory and assumptions regarding trends seen with atmospheric dispersion and along-wind dispersion. The non-dimensionalized equation allows for the along-wind dispersion coefficient to be predicted based on wind speed, distance, and stability class for each test in addition to the previous parameters of time and friction velocity.

The non-dimensionalized equation takes into account multiple theories and trends. Batchelor's Lagrangian Similarity Theory predicts σ_x to be a function of friction velocity, u_* , and time, t, while ignoring the release height[4]. Van Ulden predicts separate equations for both neutral and stable conditions[6]. Surface roughness has been shown to have an effect on wind speed[7]. The overall non-dimensionalized equation for along-wind dispersion is shown below

$$\frac{\sigma_x}{x} = \left(\frac{ut}{x}\right)^p \left(\frac{u_*}{u}\right)^q C_1 \qquad (9)$$

where x is the distance in meters and p, q, and C_1 are all dimensionless constants. The $\left(\frac{ut}{x}\right)$ term accounts for the effect of wind speed, u, time, t, and distance, x. The $\left(\frac{u_*}{u}\right)$ term is to take into effect the wind speed, u, and friction velocity, u_*, which ultimately represents the stability class.

Data was accumulated and analyzed by Stephen Hanna and Pasqual Franzese in 1999 from 11 field tests and 1 wind tunnel test[8]. The accumulation of these 12 tests produced 446 data sets over varying conditions. Other tests were accumulated and stored in the Modeler's Data Archive[9] for future reference. The following tests were analyzed in hopes of a better fit for along-wind dispersion in atmospheric modeling. Of the 446 data sets, only six data points were not used in the analysis due to lack of recorded friction velocity and/or wind speed. Reference heights for recording the wind speed from each of the data sets vary. A constant reference height of 10m was implemented across all data sets for a more standardized data analysis. Data sets used in this analysis are described below.

4.1 DTRA Phase 1

The DTRA Phase 1 tests were carried out from the 9th -26th of September in 1996 at the Dugway Proving Ground on Pad 11 producing 23 ensemble averages[10]. Propylene was instantaneous released from air cannons at ground level. Each air cannon produced 0.36kg of propylene in a period of 0.2-0.4 seconds. DigiPIDs and Ultra-Violet Ion Collectors were both used at distances of 200 m, 300 m, 400 m, 800 m, and 1200 m along a fixed sampling line 1.5m above the ground. Increasing distances resulted in an increased number of air cannons to result in an increased mass of propylene for the increased distance. 1 air cannon was used for distances of 200 m and 300 m, 2 air cannons for 400 m, 4 air cannons for 800 m, and 8 air cannons for 1200 m. Wind speed is referenced at a height of 2m which ranged from 2-7 m/s. The Great Salt Lake Desert produced a surface roughness of less than 1mm.

4.2 Dipole Pride

Dipole Pride tests were grouped into day and night tests from the 4th -20th of November in 1996 at the Department of Energy Nevada Test Site – Yucca flat. SF_6 was released from 2 vertically mounted, pressurized cylinders on the back of a 5-ton flatbed truck. These point sources released 8-21.6 kg of mass in a duration of 1-3 seconds[11]. 90 whole air samplers, with 15 minute time averaging, and 6 continuous TGA-4000 SF_6 analyzers were placed in 3 lines at distances of 1 km, 10 km and 20 km[12]. Wind speeds were referenced at a height of 10 m resulting in wind speeds of 3-5 m/s during the day and 2-5 m/s during the night. Surface roughness at the Yucca flat was around 32 mm. The 23 total tests produced 10 adequate day releases and 5 adequate night releases.

4.3 Kit Fox (ERP, URA, SSR)

All Kit Fox releases took place at the National Spill Test Facility at the Nevada Test Site. The Equivalent Roughness Pattern (ERP) and Uniform Roughness Array (URA) tests took place from the 24th -31st of August in 1995 whereas the Smooth Surface Rough (SSR) tests took place the 11th -15th of September in 1995.[13] The tests were similar in all conditions other than surface roughness. The Kit Fox ERP tests had a surface roughness of 0.2 m, the Kit Fox URA tests had a surface roughness of 0.02 m, and the Kit Fox SSR tests had a surface roughness of 2 mm[14]. Carbon Dioxide was released from a ground level area source with dimensions of 1.5 m x 1.5 m x 1 m at a rate of 1 m/s resulting in 1.5 kg/s – 4 kg/s being released[15]. Concentrations were measured at distances of 25 m, 50 m, 100 m, and 225 m. 84 high and low infrared concentration sensors, solid state infrared CO_2 sensors, and sample bag Nova sensors, were all used to detect concentrations of CO_2 with an averaging time of around 20 seconds[16]. Wind speeds were referenced at a height of 2 m which resulted in wind speeds of 2-3 m/s for the ERP and SSR test and 3-4 m/s for the URA tests. All tests took place right before sunset which resulted in lower wind speeds with neutral to stable stability classes. The ERP and SSR tests resulted in 6 sufficient trials each and the URA tests resulted in 9 sufficient trials to be used in this analysis.

4.4 Hanford Kr-85

The Hanford Fr-85[17] tests took place from 18 August to 8 November 1967 at the Battelle Pacific Northwest Laboratory. A vial containing Krypton 85 at ambient pressure was crushed by a guillotine-like device at ground level, instantaneously releasing a 1 m puff of the radioactive gas. Concentration was detected by 64 halogen-quenched Geiger-Muller tubes at a distance of 200 m and an elevation of 1.5 m above the ground. 24 halogen-quenched Geiger-Muller tubes were positioned on 6 towers at a distance of 800 m at elevations of 0.8 m, 4.6 m, 10.7 m and 21.3 m. The time averaging associated with the Geiger-Muller tubes was 4.8 seconds. 8 trials were performed, but only 6 releases remained within the boundary of the array of sensors. The wind speed was taken at a reference height of 1.5 m which resulted in wind speeds of 1-8 m/s. The roughness was around 30 mm due to the sagebrush and steppe grass.

4.5 LROD

The Long Range Overwater Dispersion (LROD)[18] tests took place from 13th -26th of July in 1993 at the Pacific Missile Range Facility in Kauai, Hawaii. A crosswind line source of pressurized, liquid SF_6 was released from a high pressure hose from the back of a plane at an elevation of 90 m at a rate of 9.6 g/meter and a speed of 129 m/s over water. A total duration of the release of SF_6 was around 13 minutes for the 100 km distance. Concentration was measured at ground level from boats along with planes at various elevations using electron capture detectors (ECD) and TGA-4000 whole air samplers. The averaging time was around 30-120 seconds for the sequential bag samples. Noise associated with the continuous analyzer resulted in limited use. Concentration was detected from the plane by flying parallel to the dissemination line and 45 km south of the sampling line (5 km, 15 km, 30 km, 60 km, and 100 km). The plane, equipped with a continuous SF_6 analyzer, would fly to the dissemination line then back to 25 m. It would slowly ascend from 25 m to 2500 m for various heights. Testing over water resulted in a surface roughness of 0.02-0.86 mm. A reference height of 10 m was used for wind speed which resulted in speeds of 3-11 m/s. The LROD tests produced 11 ensemble averages.

4.6 OLAD

The Over Land Along-wind Dispersion (OLAD)[19] tests were similar to the LROD tests, except over land. These tests took place from the 8th -25th of September in 1997 at the Dugway Proving Ground, West Desert Test Center. SF_6 was released from both plane and truck line sources. The plane released 100 kg of SF_6 at an elevation of 100 m from a N_2 pressurized tank at a rate of 33.3 kg/min for a distance of 20 km. The truck released 15 kg of SF_6 at an elevation of 3 m from a mounted tank at a rate of 1.5 kg/min for a distance of 10 km. Concentrations were recorded with TGA-4000 whole air samplers, with time averaging of 15 minutes, and 2 continuous electron capture detector (ECD) analyzers at the end of each line at distances of 2 km, 5 km, and 10 km for the truck releases and 10 km, 15 km, and 20 km for the plane releases. Wind speed was measured at a height of 2 m which resulted in speeds of 0.5 – 7 m/s. The surface roughness was measured around 30 mm. These tests resulted in 3 ensemble averages for the aircraft release and 8 surface ensemble averages for the truck release.

4.7 Marchwood wind-tunnel

The Marchwood[20] wind tunnel tests are the only tests that took place in a wind tunnel. A mixture of propane and helium, resulting in neutral buoyancy, was released from a 15 mm ground level point source. The mixture was released for a duration of 0.5 – 12 seconds. 150-300 releases were summarized into 4 ensemble averages. A single FID measured the concentration with a 5 ms sampling interval at locations of 1 m, 3 m, 5 m, and 7 m. Wind speeds ranged from 1-4 m/s at neutral stability class. Surface roughness was measured at 0.3 mm for the wind tunnel.

4.8 Oceanside

The Oceanside[21] tests took place at the shoreline of Oceanside and Del Mar in Southern California. 20 μm particles of green and yellow fluorescent pigment (FP) were released by plane, boat, and land with standardized Dugway Proving Ground equipment. For the plane releases, the DPG model D-1 released 1500 g/min of FP for a duration of 10 minutes at an elevations of 61 m, 122 m, and 152 m over a 40 km distance. For the boat releases, a DPG Mark IV was used to release 110 g/min for a duration of 2 hours at sea level over 40 km. For instantaneous land releases, the DPG Mark IX was used at three different inland locations to release 10-50 g of FP instantaneously at ground level. Concentration of FP were recorded from rotorod samplers, sequential samplers, vertical samplers, 100 m square grids, dense sampling lines, and metronics drum impact samplers. Time averaging ranged from 5 minutes at coastal regions to 15 minutes at inland conditions. Surface roughness ranged from 30-50 m due to the coastal shoreline. The wind speed reference height varied from 8-120 m which resulted in wind speeds ranging from 1-5 m/s.

4.9 Fort Wayne

Fort Wayne[22] releases took place in the Summer and Fall of 1963 in Fort Wayne, Indiana. An aerosol of fluorescent pigment (FP) was released via 2 aircrafts in the form of an elevated, 91-244 m, line source. Dry green and yellow FP were released from the planes via FP dispensers supplied by the Dugway Proving Ground. The amount of release was measured by the difference in weight, before and after. The release was upwind of the 180,000 population urban setting with a river running through the city and agriculture surrounding. Rotorods, Gelman throw-away plastic filter holders, and Gelman paper tape samplers were used for concentration measurements of the FP. The tape provided an averaging time of 5 minutes where the filter holders had an averaging time of 30 minutes. Wind speed was measured at the release height and resulted in speeds ranging from 5-16 m/s.

4.10 Short Range Diffusion Experiment Series

The Short Range Diffusion Experiment Series[23] took place in 1979 in Japan. A 200 m x 200 m vacant lot covered in turf and surrounded by trees was used for the releases. NO_x was released instantaneously from a balloon at an elevation of 1.5 m. Chemiluminescence NO_x meters were used to detect concentration with a sampling time of <2 minutes. Sample lines were chosen at distances of 60 m and 100 m. Wind speeds were referenced at heights of 0.5 m, 3 m, and 9 m which resulted in speeds of 1-2 m/s.

4.11 Results and Discussion

A multivariable regression analysis, also known as multiple regression, is an extension of linear regression[24]. A regression analysis determines the relationships among variables.

With a multivariable regression analysis, the relationships between multiple independent variables can be determined. A regression analysis is a useful tool in understanding the changes of one variable with respect to others. The form of a typical multi-variable regression analysis is shown below

$$Y = a + b_1 x_1 + b_2 x_2 + \cdots + b_n x_n \qquad (10)$$

A multi-variable regression analysis can be applied to the non-dimensionalized equation. The non-dimensionalized equation is rearranged with the natural log of each variable to perform the multi-variable regression analysis. This equation is shown below

$$ln\left(\frac{\sigma_x}{x}\right) = p\, ln\left(\frac{ut}{x}\right) + q\, ln\left(\frac{u_*}{u}\right) + C_2 \qquad (11)$$

where $C_2 = \ln(C_1)$. A multivariable regression analysis with known data is used with this equation to determine the constants for p, q, and C_1. An analysis is done for the overall data and categorized into stability classes using macros in Excel[25].

Multiple parameters were analyzed for optimal results using the multivariable regression analysis. Data was quantified and analyzed based on individual tests, surface roughness, friction velocity, Monin-Obukhov length, distance, time, near field, and far field.

Data from all 12 experimental data sets referenced above were analyzed with a multi-variable regression analysis to determine appropriate exponents and a constant for an optimum non-dimensionalized equation. While many parameters were originally analyzed for the optimum non-dimensionalized equation, the quantification of near field and far field as a function of time was chosen as the most prominent parameter. The optimum non-dimensionalized equation was parameterized by near field with time less than 10,000 seconds and far field with time greater than 10,000 seconds. These equations are shown below in Table 1. The ratio, α, of the along-wind dispersion coefficient, σ_x, for the data vs σ_x for the predicted non-dimensionalized equation is used for comparison of the difference in equations. A non-parametric confidence interval was determined for the non-Gaussian distribution of α.[26] Hanna and Franzese's along-wind dispersion coefficient was analyzed in the same manner in Table 1.

Table 1. Comparison of along-wind dispersion equations with 90% confidence intervals

Model	Equation	90% Confidence Interval of $\alpha = \frac{\sigma_x\ (predicted)}{\sigma_x\ (data)}$	Range of Confidence Interval
Hanna and Franzese	$\sigma_x = 2u_* t$	1.63 < α < 1.88	26%
Near Field Non-Dimensionalized	$\frac{\sigma_x}{x} = 1.72\left(\frac{ut}{x}\right)^{1.18}\left(\frac{u_*}{u}\right)^{1.10}$	1.16 < α < 1.31	16%
Far Field Non-Dimensionalized	$\frac{\sigma_x}{x} = 0.05\left(\frac{ut}{x}\right)^{0.67}\left(\frac{u_*}{u}\right)^{-0.27}$	0.85 < α < 1.05	20%

From Table 1, the range of the 90% confidence intervals for each of the equations is shown. The non-dimensionalized equations have a tighter correlation of the predicted values. The 90% confidence intervals shows that Hanna's predicted along-wind dispersion equation has a trend over predict the σ_x values. An interesting result from the multi-variable regression analysis is the determination of the coefficient for the $\left(\frac{u_*}{u}\right)$ term. The -0.27 exponential value essentially confirms that the stability class has no effect on the along-wind dispersion coefficient after a specific amount of time. A discontinuity is understood between the near field and far field non-dimensionalized equation and is currently being evaluated.

The results of the non-dimensionalized predicted σ_x values were divided by the friction velocity and plotted against the original data to determine their relationship. This relationship is shown below in Figure 2.

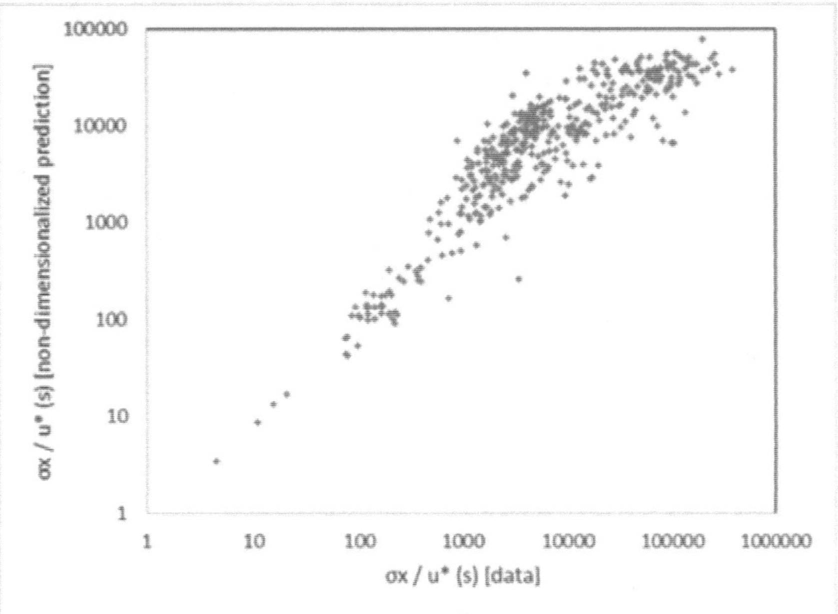

Figure 2. Along-wind dispersion coefficient divided by the friction velocity of the non-dimensionalized modeled values vs the original data.

As shown in Figure 2, the correlation between the data and non-dimensionalized predicted values shows a 45° line. This correlates a good fit for the predicted σ_x values when compared to the original σ_x values.

A visual representation of the non-dimensionalized predicted σ_x values, Hanna and Franzese's σ_x values, and the original σ_x values as a function of time are all shown below in Figure 3.

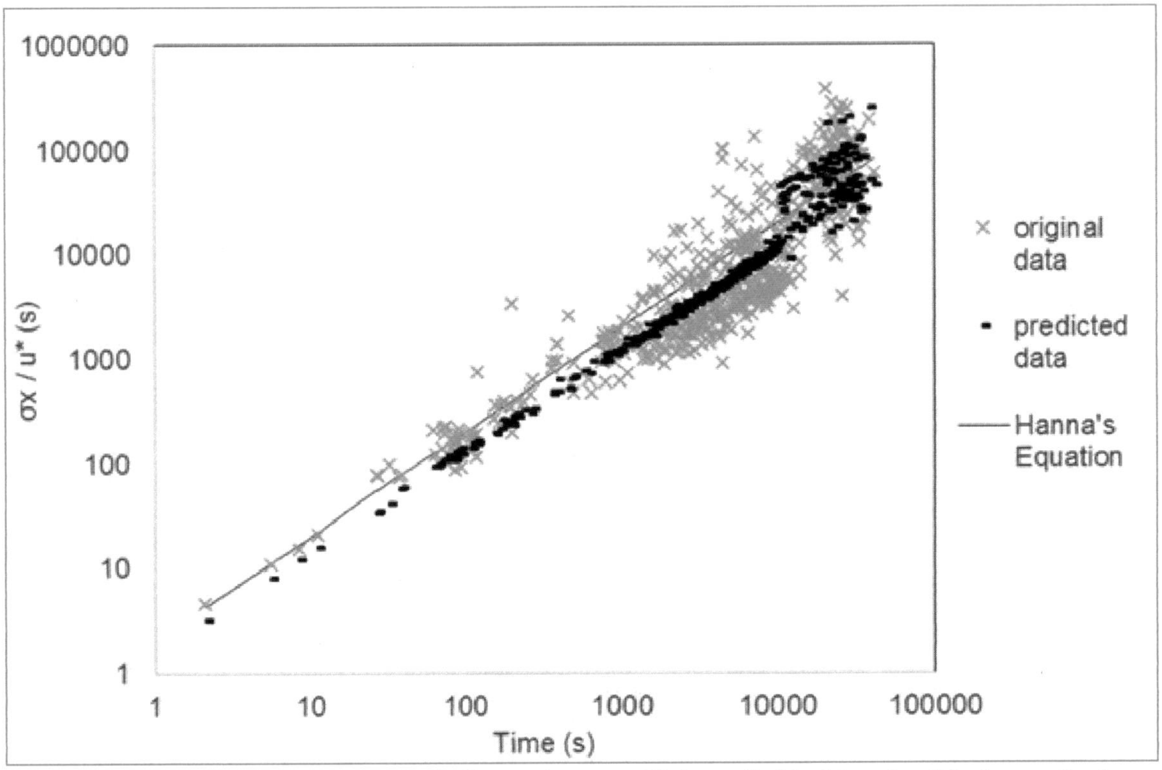

Figure 3. Time vs. the along-wind dispersion coefficient divided by the friction velocity as a function of the original data, non-dimensionalized predicted data, and Hanna's predicted data.

From this figure, the predicted values from both Hanna and Franzese's predicted data and the non-dimensionalized predicted data are shown in comparison with the original data. The non-dimensionalized equation takes into account additional parameters, other than time, which results in the spread of σ_x values as a function of time and additional parameters in comparison to the single line by the $\sigma_x = 2u_*t$ along-wind dispersion equation. The non-dimensionalized equation also takes into effect the near field and far field. This is clearly seen at times greater than 10,000 seconds where the σ_x values are spread over a wider range. Overall, the non-dimensionalized equation for along-wind dispersion proves better than the previous $\sigma_x = 2u_*t$ correlation.

5 Summary

Along-wind dispersion is critical when modelling toxic, flammable, or hazardous time-limited or finite-duration releases into the atmosphere. While many models exist from a theoretical standpoint or parameterization of experimental data, there is no overarching correlation to predict the along-wind dispersion coefficient. This paper took into effect established theoretical equations validated with experimental data to determine a better equation to model the effects of along-wind dispersion. A successful non-dimensionalized equation parameterized by the near field and far field was determined and validated against available field-scale data which showed improvement in comparison to previous equations. The non-dimensionalized equation for along-wind dispersion solved for in this paper will improve the existing atmospheric modeling with along-wind dispersion.

6 Acknowledgments

We gratefully thank our colleagues from DNV GL, Henk Witlox and Maria Fernandez, who provided insight and expertise that greatly assisted the research. We also thank Steven R. Hanna and Pasquale Franzese for sharing their accumulated data for further analysis.

7 References

[1] Boybeyi, Zafer, Sethu Raman, and Paolo Zannetti. "Numerical investigation of possible role of local meteorology in Bhopal gas accident." *Atmospheric Environment* 29.4 (1995): 479-496.

[2] Høiset, S., et al. "Flixborough revisited—an explosion simulation approach." *Journal of hazardous materials* 77.1 (2000): 1-9.

[3] De Marchi, Bruna, Silvio Funtowicz, and Jerome Ravetz. "Seveso: A paradoxical classic disaster." *The long road to recovery: Community responses to industrial disaster* (1996): 86-120.

[4] Batchelor, G. K. "Diffusion from sources in a turbulent boundary layer." *Arch. Mech. Stosow* 3.6 (1964): 661-670.

[5] Chatwin, P. C. "The dispersion of a puff of passive contaminant in the constant stress region." Quarterly Journal of the Royal Meteorological Society 94.401 (1968): 350-360.

[6] Van Ulden, A. P. "A surface-layer similarity model for the dispersion of a skewed passive puff near the ground." *Atmospheric Environment. Part A. General Topics* 26.4 (1992): 681-692.

[7] DNV GL. DNV GL Theory Manual for Unified Dispersion Model. London UK 2015.

[8] Hanna, Steven R., and Pasquale Franzese. "Alongwind dispersion—a simple similarity formula compared with observations at 11 field sites and in one wind tunnel." *Journal of Applied Meteorology* 39.10 (2000): 1700-1714.

[9] Chang, Joseph C., and Steven R. Hanna. "Modelers' Data Archive – A Collection of Atmospheric Transport and Dispersion Data Sets." (2010).

[10] Biltoft, C., Phase I of DSWA transport and dispersion model validation study. DPG-FR-97-058, (1997): 156.

[11] Biltoft, C., Dipole Pride 26: Phase II of DSWA transport and dispersion model validation study. DPG-FR-98-001, (1998): 130.
[12] Watson, T. B, R. E. Keislar, B. Reese, D. H. George, and C. A. Biltoft, The DSWA Dipole Pride 26 field experiment. NOAA Tech. Memo. ERL ARL-225, (1998): 99.
[13] King, S. Bruce. The Kit Fox Field Demonstration Project and Data Set. *International Conference and Workshop on Modeling the Consequences of Accidental Releases of Hazardous Materials.* (1999).
[14] Spicer, T. and Havens, J. Description and Analysis of Atmospheric Dispersion Tests Conducted by EPA. *International Conference and Workshop on Modeling the Consequences of Accidental Releases of Hazardous Materials.* (1999).
[15] Sheesley, D. et al.. Chemical Dispersion Modeling Project, 1995 Kit Fox Field Demonstration DOE Nevada Hazardous Spill Center. Volume I – Experiment Description and Data Processing. (1998).
[16] Hanna, S. R., and K. W. Steinberg, Studies of dense gas dispersion from short-duration transient releases over rough surfaces during stable conditions. *Air Pollution Modeling and Its Applications XI,* Plenum Press, (1996): 481–490.
[17] Nickola, P. W., J. D. Ludwick, and J. V. Ramsdell Jr., An inert gas tracer system for monitoring the real-time history of a diffusing plume or puff. *J. Appl. Meteor.,*9, (1970): 621–626.
[18] Bowers, J. F., G. E. Start, R. G. Carter, T. B. Watson, K. L. Clawson, and T. L. Crawford, Experimental design and results for the Long-Range Overwater Diffusion (LROD) Experiment. DPG/JCP-94/012, (1994): 531.
[19] Biltoft, C., S. D. Turley, T. B. Watson, G. H. Crescenti, and R. G. Carter, Over-Land Atmospheric Dispersion (OLAD) test summary and analysis. WDTC Doc. WDTC/JCP-99/048, (1999): 38.
[20] Robins, A. G., and J. E. Fackrell, An experimental study of the dispersion of short-duration emissions in a turbulent boundary layer. *Air Pollution VI,* C. A. Brebbia, C. F. Ratto, and H. Powers, Eds., WIT Press, (1998): 697–707.
[21] Smith, and B. L. Niemann, Shoreline diffusion program, Oceanside, CA. Vols. I and II. MRI-64-FR-860, (1969): 230 and 456.
[22] Hilst, G. R., and N. E. Bowne, A study of the diffusion of aerosols released from aerial line sources upwind of an urban complex. Vols. I and II. Rep. DA-42-007-AMC-38®, (1966): 231 and 583.
[23] Sato, J., F. Kimura, and T. Yoshikawa, The longitudinal spread of puff in the short range diffusion experiment. *Pap. Meteor. Geophys.,*32, (1981): 155–162.
[24] Draper, Norman Richard, Harry Smith, and Elizabeth Pownell. *Applied regression analysis*. Vol. 3. New York: Wiley, (1966).
[25] Davis, Richard A. "Practical Numerical Methods for Chemical Engineers: Using Excel with VBA". Lexington, KY, (2015).
[26] Conver, W. J. "Practical Nonparametric Statistics". Wiley, New York, (1971).

478500 Review and Analyze the Effects of Parameters for H_2s Gas Mixture Dispersion Models: Degadis and CFD Models

Abstract:

Due to the increasing demands of energy, natural gas explorers and developers are interested in exploring sour gas fields, which have a high concentration of hydrogen sulfide (H_2S). Because H_2S is highly corrosive and could reduce the heating values of natural gas, H_2S needs to be removed from natural gas before transfer. Then the removed H_2S gas will be converted to sulfide as the products in the Claus process. The H_2S mixture (*e.g.*, H_2S and CO_2) may cause corrosion to pipelines and thus, could lead to H_2S gas mixture release.

H_2S mixture behaves as a heavy gas, which travels down to the ground causing harm to the local population and the environment. In this paper, the DEnse GAs DISpersion (DEGADIS) model and the Computational Fluid Dynamics (CFD) model (*e.g.*, ANSYS Fluent) are implemented for the H_2S mixture simulation, and assumptions in the models are also studied. The effects of topography and meteorological parameters such as atmospheric temperature, wind speed, and humidity are investigated. Finally, the simulation results from the DEGADIS model and the CFD model for flat terrain with constant wind velocity are compared.

The simulation results show that the DEGADIS model, which is suitable for flat terrain and constant wind conditions, can be applied to simple terrain geography and stable atmospheric conditions. Whereas the CFD model, which considers unsteady wind speed and turbulence near the obstacles, can be utilized for complex topography. The series of simulation results can provide guidance for the risk assessment of H_2S mixture releases under different topographies and meteorological conditions.

478998 Considerations When Justifying the Purchase of an Operator Training Simulator

Abstract:

Highly skilled plant operators ensure higher levels of plant safety, reliability and profitability. An Operator Training Simulator accelerates both the time to competence and raises the competency level of an operator. The ARC Advisory Group recently surveyed companies who have invested in Operator Training Simulators (OTS). Two critical results from the survey are that over sixty five percent of those companies report fewer abnormal situations, fewer trips, and incidence avoidance attributable to the use of OTS. Over sixty percent of these companies report improved safety and reduced risk and lower liability attributable to the use of the OTS.

Many leading companies have embraced the use of Operator Training Simulators across their Enterprise however many facilities and companies have yet to embrace the idea. The most common reason given for not investing in such a solution is that the cost of an OTS cannot be justified. The nuclear industry mandates that a nuclear power plant must have an OTS whereas the American Petroleum Institute only recommends the use of Training Simulators. The Process industries overall will benefit by the greater adoption of training simulators and this paper discusses the different ways that can be used to justify the investment in an OTS.

479500 Consequence Analysis of Gas Explosion in Underground Common Utility Tunnels

Abstract:

The purpose of this study is to figure out the effects of blast overpressure in gas piping leakage in underground common utility tunnels through the explosion simulation tool FLACS (Flame Acceleration Simulator) and find and apply methods that can reduce the risk.

Common utility tunnels are facilities to jointly bury pipes in underground spaces. Placing gas piping in common utility tunnels has the advantages of reducing the risk of corrosion and preventing the physical damage due to civil engineering works and excavating work on the ground. However, when an explosion has occurred in a common utility tunnel, the consequences could be larger in that it is a closed space. Nevertheless, studies on underground common utility tunnels are currently limited to fire accidents. Therefore, risks due to explosions should be quantitatively studied.

Therefore, in this study, the effects of explosions were analyzed assuming cases where gases leaked from gas piping in underground common utility tunnels exploded. In addition, blast overpressures were evaluated for diverse scenarios regarding ignition sources, leakage locations, and discharge rates and the results were statistically derived. Blast overpressures high enough to cause dangerous situations of the underground common utility tunnels were observed. Methods that can reduce the risk, that is, methods that can reduce the frequency of explosions, and methods for designs of internal structures of common utility tunnels that can mitigate the consequences will be presented.

479158 Air Cooler Explosion Analysis Using CFD

Abstract:

Air coolers or fin-fans are located in most process plants. They cool process fluids using forced air flow over radiators containing the fluid. Air coolers initially appear to offer many features that would result in a significant explosion source should a flammable cloud reach the cooler. The radiator coils are closely spaced pipes which will accelerate flames. The metal box represents confinement and the turbulent air flow produced by the fans will accelerate flames similar to how congestion produces turbulence ahead of a flame front. Also, the flow of air can draw rich flammables near the ground up into the air cooler.

This study examines the severity of an air cooler explosion using computational fluid dynamics (CFD). The commercial software FLACS is used to examine the parameters mentioned above (congestion, confinement, and turbulence) and their effect on explosion severity. Comparisons between FLACS and the Multi-Energy Method are made and recommendations provided for assessing air coolers in non-CFD facility siting studies.

13TH GLOBAL CONGRESS ON PROCESS SAFETY

Revisions to the QMEFS Vapor Cloud Explosion Model

Jeffrey D. Marx and Benjamin R. Ishii
Quest Consultants Inc.®
908 26th Avenue N.W.
Norman, OK 73069
jdm@questconsult.com / bri@questconsult.com

Prepared for Presentation at
American Institute of Chemical Engineers
2017 Spring Meeting and 13th Global Congress on Process Safety
San Antonio, TX
March 26 – 29, 2017

Revisions to the QMEFS Vapor Cloud Explosion Model

Jeffrey D. Marx and Benjamin R. Ishii
Quest Consultants Inc.®
908 26th Avenue N.W.
Norman, OK 73069
jdm@questconsult.com / bri@questconsult.com

Keywords: vapor cloud explosion (VCE), Baker-Strehlow-Tang (BST) model, estimating flame speeds

Abstract

The Quest model for estimating flame speeds (QMEFS) was introduced in 2007, has since been published, and has been used by Quest and other companies to provide a detailed method for vapor cloud explosion (VCE) predictions. QMEFS provides a set of parameters that predict a flame speed, which then is used with the Baker-Strehlow-Tang (BST) blast curves to predict overpressure and impulse. The original intent of the QMEFS model was to provide better characterization of confined or congested areas, as well as fuel reactivities, which fit between the low, medium, and high categories offered by the BST model, so that the large differences in predictions between categories could be eliminated. In this way it provided a more refined system for predicting the consequences of a VCE, while remaining anchored within the flame speeds provided by the BST matrix.

In the years since QMEFS' publication, Baker has added to the BST flame speed matrix. In addition, use of the model has highlighted some of the areas where the model could be improved. This paper shows several ways in which the model has been enhanced to encompass a wider range of conditions than are covered by the BST model.

1 Introduction

Any release of a flammable fluid in a petrochemical facility has the potential to generate a flammable vapor cloud that, if ignited, could produce a vapor cloud explosion (VCE). Most VCEs that generate damaging levels of overpressure originate in what is called a potential explosion site (PES), where confinement and congestion contribute to flame acceleration and the generation of overpressure. When flammable vapors enter the PES and are ignited, the possibility of human injury/death, asset damage, or event escalation becomes a concern. The concern for human injury or death is most often addressed in the form of a building siting study. Because people are somewhat less likely to be injured or killed when outside, as compared to when inside a building, the siting study focuses on the potential impacts to buildings within and around petrochemical facilities. It then becomes the task of process safety professionals to estimate the potential for VCE events, and their resulting overpressure impacts on buildings.

The Baker-Strehlow-Tang (BST) model and the TNO Multi-energy (ME) model use blast curves to estimate the overpressure impacts generated by an exploding vapor cloud. Both approaches require the user to estimate the strength of the explosion as a function of the reactivity of the flammable material and the degree of confinement or congestion present in the cloud. This information is then used to determine which strength curve (in the case of the TNO model) or flame speed curve (in the case of the BST model) is used to calculate the overpressure of the explosion as a function of distance from the center of the explosion. It is not clear, however, how best to choose the correct curve when modeling an explosion.

There have been several papers published regarding blast curve selection for the ME model. These methods are generally presented as a qualitative ruleset which acts as an index into which of the 10 curves should be used. The BST model provides a simple set of guidelines that result in a selection of a flame speed, which corresponds to an overpressure vs. distance relationship. Three parameters are used to determine the flame speed: reactivity of the flammable gas, the degree of confinement of the flammable cloud, and the degree of obstruction due to obstacles within the flammable cloud. These parameters are represented by a set of what are effectively low, medium, and high choices for each parameter.

While the BST model provides reasonable guidance on choosing the explosion parameters, there are still areas that could be improved. One problem is that the BST flame speed matrix includes large jumps in flame speed between categories of confinement, reactivity, or congestion. This was first addressed by Baker in a 1998 paper [1] where a new confinement category, 2½-D, was added for those situations that were more confined than the 3-D case, but less so than the 2-D case. The flame speeds used for the 2½-D confinement were simply the arithmetic average of the flame speeds for the 2-D and 3-D cases for a given reactivity and congestion class. Even with this extension, large discontinuities remained among the categories for flame speed according to the prescribed methodology. These discontinuities drive the need for a new method to determine flame speed based upon quantifiable properties of the flammable gas and its surroundings that varies smoothly across the range of conditions that are found in actual process plants.

The creation of the QMEFS model [2] improved upon the basic BST scheme by combining it with the GAMES correlations [3] and providing a numerical resolution to the BST matrix. Integration of the GAMES model allows the strength of the explosion to be defined by numeric input parameters and avoids the low/medium/high categorizations.

2 Parameters Affecting Flame Speed

With the incorporation of the GAMES correlations, the resulting explosion flame speed is characterized by five parameters:

- Laminar burning velocity (LBV) is used as the characteristic property for fuel reactivity

- Volume blockage ratio (VBR) and average obstacle diameter define the level of congestion
- Confinement is characterized as the number of confining planes, or walls, that are available to confine the expanding gases.
- Flame run-up distance is used to characterize the flame acceleration path

The last part of explosion characterization is the explosive energy of the fuel/air mixture, which is defined by the gas concentration, the combustion energy of the fuel, and the volume occupied by the fuel/air mixture.

2.1 Reactivity

Fuel reactivity is a measure of the propensity of the flame front to accelerate and create overpressures or potentially undergo a deflagration-to-detonation transition (DDT). The BST model originally classified reactivity as high, medium, or low based on the fuel in the fuel-air mixture. The original rule set [1] defined propylene oxide, ethylene oxide, and hydrogen (or materials with a 0.8 m/s LBV or larger) as high reactivity; methane and carbon monoxide were classified as low reactivity, and everything else as medium reactivity. According to the current BST model, materials having a laminar flame speed greater than 0.75 m/s are considered high reactivity (to accommodate ethylene being high reactivity) while those having a laminar flame speed below 0.4 m/s are considered low reactivity [1], although some BST publications list the low boundary as 0.45 m/s. In 2016, the BST categories were expanded to include a very-low category for fuels like ammonia, which has an LBV in the range of 0.07 m/s to 0.15 m/s [4].

In the experimental studies that form the basis of the BST model, ammonia, methane, propane, and ethylene were used to define the very low, low, medium, and high reactivity categories. These categories produce confusion due to how they create boundaries between the BST categories. For example, the LBV of ethylene, depending on the cited reference, ranges between 0.64 and 0.83 m/s. With the current definition of 0.75 m/s as the high reactivity boundary, this makes ethylene a "borderline" fuel – one that may either be defined as medium or high reactivity. However, ethylene is explicitly categorized as a high reactivity material, as their experimental work used ethylene as the high reactivity material [5]. This then begs the question: what about ethylene oxide (LBV 1.0 m/s) or acetylene (LBV 1.6 m/s)? These two materials are clearly more reactive than ethylene, but are still categorized as high reactivity. The BST model effectively says that an ethylene explosion will be as severe as an acetylene explosion, when all other parameters are held constant.

2.2 Congestion

In the original BST model, low obstacle density was defined as having an area blockage ratio (ABR) of less than 10%, while high obstacle density was defined as an ABR of 40% or greater, and everything in between is considered medium [6]. Later BST guidelines have used the volume blockage ratio (VBR) as the parameter for classifying obstruction [5]. In practice both blockage ratios will affect how fast a flame accelerates, but, for a uniform obstacle field, the two are related simply by the pitch-to-diameter ratio of the obstacles. If the obstacles included in an ABR are not repeated quasi-uniformly throughout the obstacle field, then their effect would be more accurately portrayed by VBR.

The classifications of low, medium, and high are often too coarse for proper assignment of congestion to a particular PES. In the experimental programs that helped to define the BST model, VBRs of 1.5%, 4.3%, and 5.7% were used to represent the low, medium, and high categories. As with other VCE parameters, the differences between these categories leaves significant room for variation between the categories. In addition, no definition of 0% VBR is provided, nor is any relationship to VBRs higher than 5.7%.

2.3 Confinement

The effect of the confinement of the flammable cloud is taken into account by determining the number of dimensions in which the burning gas may expand. 3-D expansion allows the burning cloud to expand freely in all directions and results in the slowest flame acceleration and lowest overpressures. 2-D expansion, such as a flame between two flat plates, generates higher overpressures because the combustion gases have fewer directions in which to expand resulting is a higher flame acceleration.

Confinement is included in the BST model by identifying the number of dimensions that are available to the products of combustion for expansion. Expansion into free space (except for the plane of the ground) is considered 3-D expansion, expansion between two parallel planes is considered 2-D expansion, and expansion in a pipe is considered 1-D expansion. The 1-D case was removed in the more recent publications discussing the BST model [5]. To handle the case of a frangible or partially-confining plane, such as a very closely spaced pipe rack, the BST model added a 2.5-D classification which simply averaged the 2-D and 3-D flame speed results [5].

While the classification is fairly straightforward, especially after the 2.5-D category was added, some situations are still difficult to define. These include partial walls, pieces of equipment that may form more confinement than congestion, or surfaces that are porous such as grated decks.

2.4 Other Factors

Several researchers [5, 7, 8] have acknowledged that the overall dimensions of the flammable vapor cloud before ignition directly affect the final flame speed and consequent overpressures of the vapor cloud explosion. Since a flame will accelerate until it reaches a maximum sustainable value (or undergoes a deflagration to detonation transition, a DDT), the maximum dimension of a flammable cloud is expected to be an important variable in the creation of overpressure. As a point of reference, the length available for flame acceleration in the more recent BST tests was 15 meters [5].

In addition, research [7, 8] suggests that the scale of obstacles within the congested area also affects the maximum flame speed that may be achieved. As a flame front moves around and past obstacles, the turbulence that is generated can increase the flame surface area and accelerate combustion. The characteristic obstacle diameter for the BST tests was 2 inches (0.0508 meters); other test series (including the MERGE/EMERGE tests, varied the average obstacle diameter).

2.5 *Parameters Comparison*

The BST parameters discussed above are listed in Table 1, along with the corresponding numerical values that are equivalent to those applied in the QMEFS model. When the numerical values in Table 1 are applied in the QMEFS model, the results will match the predictions of the BST model.

Table 1. BST Flame Speed Parameters

BST Parameter	BST Category	Model Parameter	Corresponding Parameter Value
Reactivity	Very-Low (e.g., Ammonia)	S_l	0.08 m/s
	Low (e.g., Methane)		0.37 m/s
	Medium (e.g., Propane)		0.43 m/s
	High (e.g., Ethylene)		0.75 m/s
Obstacle Density	Low	*VBR*	0.015
	Medium		0.043
	High		0.057
Expansion	3-D	*nPlanes*	1
	2.5-D		1.5
	2-D		2

3 Model for Estimating Flame Speed

The QMEFS model was created to overcome some of the limitations described above. By incorporation of the GAMES correlations and the overpressure/impulse curves from the BST method, a model that provides the capability for more detailed descriptions of explosion scenarios has been created. Thus, QMEFS incorporates the modeling methodologies and experimental data sets that have been used in both the BST model and in the GAMES project.

3.1 *Fundamental Correlations*

The fundamental correlation is based on correlations developed from the GAMES project [3]. These correlations related overpressure to parameters based on the MERGE and EMERGE experiments [7,8]. These correlations can be combined with acoustic theory to relate flame speed to explosion parameters [2] to produce the equations below.

For no confining surfaces other than the ground (3D expansion):

$$2.4 * \frac{M_f^2}{1+M_f} * p_0 = 0.84 * \left(\frac{VBR*L_f}{D}\right)^{2.75} * S_l^{2.7} * D^{0.7} \tag{1}$$

For confinement between two flat planes (2D expansion):

$$2.4 * \frac{M_f^2}{1+M_f} * p_0 = 3.38 * \left(\frac{VBR*L_f}{D}\right)^{2.25} * S_l^{2.7} * D^{0.7} \tag{2}$$

where:

VBR = volume blockage ratio
L_f = maximum distance a flame may propagate in the obstructed region (i.e., the run-up distance), m
D = average obstacle diameter, m
S_l = laminar burning velocity (LBV) of the flammable gas, m/s
p_0 = the ambient pressure, bara
M_f = the flame speed relative to a fixed observer, expressed as a Mach number

To correct the flame speed to account for confinement in spaces other than one or two planes, a curve fit of the ratio of burned cloud to unburned cloud radii versus the number of confining planes was made. A curve fit was generated with an expansion ratio (α) equal to 7, a typical expansion ratio for common hydrocarbon fuels in air. This is also the value assumed in the derivation of portions of the BST model [9]. The equation resulting from this curve fit is:

$$NPF = \frac{1}{0.5035 - 0.0757 \cdot nPlanes} - 1.338 \tag{3}$$

where:

NPF = flame speed correction factor
$nPlanes$ = number of confining planes

For true 3-D spherical expansion, *nPlanes* = 0. For a typical explosion at grade level, *nPlanes* = 1. For 2-D expansion, such as a flame propagating between a floor and a ceiling, *nPlanes* = 2. For 1-D expansion, such as a flame front in a pipe, *nPlanes* = 4. Planes that are not completely solid or rigid may be accounted for using a fraction of a plane.

The flame speed correction factor, *NPF*, does not fully account for increases in flame speed as the number of confining planes increases. This is accomplished by applying the following factor to the flame speed when *nPlanes* is equal to 2:

$$\beta = \frac{M_{f,2D}}{M_{f,3D} \cdot NPF(nPlanes = 2)} \tag{4}$$

This can be implemented for nPlanes =1 to nPlanes = 4:

nPlanes = 1

$$M_f = M_{f\,3D} * NPF \tag{5}$$

1 > *nPlanes* < 2

$$M_f = M_{f\,3D} * NPF * \beta \tag{6}$$

nPlanes >= 2

$$M_f = M_{f\,3D} * NPF * [1 + (\beta - 1) * (nPlanes - 1)] \tag{7}$$

3.2 Flame Speed Limits

The equations above are implemented in the QMEFS model to define the flame speed (M_f) and then coupled with the latest BST flame speed matrix [4]. One limitation to the GAMES correlations is that they can predict unrealistic flame speeds. This is a direct result of the correlations fitting the data within the scope of the MERGE and EMERGE tests, and not being designed to extend beyond that scope. By coupling with the BST model, this problem is partially resolved because the BST matrix provides a set of upper limits to flame speeds (and by extension, the explosion strength).

Upon examination, it can be seen the GAMES correlations will predict higher and higher flame speed values as parameters such as L_f or S_l, increase. In reality, for a given obstacle configuration, geometry, and flammable gas mixture, there is a limit to the flame speed that can be achieved. This flame speed may be either subsonic or the cloud may undergo a deflagration-to-detonation transition (DDT) in which the flame front becomes a detonation propagating at roughly the Chapman-Jouguet (C-J) detonation velocity for the flammable mixture. For most common flammable hydrocarbons in air, this speed is roughly 1800 m/s or M_f= 5.2. For purposes of this model, any flame that accelerates to a calculated velocity greater than M_f = 3.0 is considered to have undergone DDT [10, 11] and the flame speed is set to the C-J velocity, M_f= 5.2.

The flame speeds suggested by the latest BST model are applied as the upper limit flame speeds to be used in this model, as they have been "scaled up" to account for industrial scale flammable clouds [5] and account for DDTs.

The result of these correlations is a model that estimates the flame speed based on *VBR*, *D*, S_l, L_f and the number of confining planes. The resulting flame speed is compared to the published BST model, which provides a matrix of flame speeds as a function of reactivity, obstacle density, and flame expansion. The BST matrix values are used as maximum flame speed values for the set of parameters given in Table 1. Values between the BST matrix elements are calculated by linear interpolation (as was done for Baker's 2½-D flame expansion category). This methodology provides an extended, systematic approach for estimating flame speeds resulting from the combustion of a flammable cloud in an obstructed and/or confined region. Predicted flame speeds are used with the existing BST blast curves to produce estimates of overpressure at a distance from the explosion source.

3.3 Expansion of the BST Matrix

The range of the QMEFS model is limited by the upper and lower bounds of the BST matrix elements for congestion, confinement, and fuel reactivity. Through application of this model to various petrochemical projects, the need to expand the matrix became obvious. As described above, the model is built to provide predictions for the *nPlanes* (confinement) parameter in the range of 1.0 to 4.0 (the model is not validated nor trusted above this value). Values of average obstacle diameter and flame run-up distance are built into the GAMES correlations, so naturally are accounted for within reasonable limits. This leaves the congestion (VBR) and fuel reactivity parameters as places to potentially expand the model.

The intent of model expansion is to widen the range of congestion and reactivity inputs so that it removes questions about its use in the "below low" and "above high" regions. With proper expansion of the model described above, the predicted flame speed following ignition of a flammable cloud may then be used with the BST blast curves to estimate the overpressure impacts in the area surrounding a vapor cloud explosion.

To accomplish this expansion, additional categories were added to the BST matrix for congestion and reactivity. This follows Baker's own expansion into the **very-low reactivity** category when ammonia explosion experimental results were published [4].

For a **very-high reactivity** category, hydrogen was assumed as the characteristic fuel, and represents a true upper bound for LBV. Expansion to the very-high category, for parts of the matrix with high flame speeds is simple because DDTs are expected in these conditions already. For other areas, published data was required for the basis of the expansion.

In the case of the **very-low congestion** category, open air (3-D flame expansion) explosion tests were performed by Harris and Wickens [12] as well as Sato and Iwabuchi [13] for a variety of materials. The **very-high congestion** category, several of the BFETS series of tests were used [14] due to their use of a VBR in the 10% range.

The expanded matrix applied in the new QMEFS model is presented in Table 2. The latest BST flame speed matrix values are outlined by bold lines in Table 2. Values outside these areas have been added as part of the QMEFS expansion.

While a 1-D confinement is not expected to be applied in large-scale industrial scenarios, the *nPlanes* = 5 (1-D) confinement category is included in this model to provide an interpolation point beyond *nPlanes* = 2 (2-D flame expansion).

Table 2. Expanded Flame Speed Matrix

Confinement	Reactivity	Congestion				
		Very-Low	Low	Medium	High	Very-High
1D	Very-High	DDT	DDT	DDT	DDT	DDT
	High	DDT	DDT	DDT	DDT	DDT
	Medium	1.029	1.029	1.765	2.265	DDT
	Low	0.5	0.5 *	1.029	2 *	DDT
	Very-Low	0.023	0.023	0.124	0.168	0.303
2D	Very-High	DDT	DDT	DDT	DDT	DDT
	High	0.588	0.588	DDT	DDT	DDT
	Medium	0.47	0.47	0.66	1.6	DDT
	Low	0.079	0.079	0.47	0.66	1.244
	Very-Low	0.008	0.008	0.049	0.069	0.130
3D	Very-High	0.5	DDT	DDT	DDT	DDT
	High	0.06	0.36	DDT	DDT	DDT
	Medium	0.03	0.11	0.44	0.5	1.084
	Low	0.025	0.026	0.23	0.34	0.678
	Very-Low	0.003	0.003	0.024	0.036	0.073

* Values adjusted from BST publication for model consistency

4 Application

It is helpful to test the QMEFS model outside of the published BST matrix, in the areas where the QMEFS matrix has been expanded. Large scale hydrogen experiments were performed in a hemispherical test rig with no congestion. The test volume was 300 m^3 with a base of 10.4 meters diameter [13] and center ignition. The tests were modeled with the latest BST model and the expanded BST model. The model parameters listed in Table 3 were applied to these calculations. Figure 1 presents the overpressure in relation to distance from center of the fuel-air mixture, for the test results and both BST and QMEFS predictions. The BST model does not have a category for a hydrogen vapor cloud with no congestion, thus the scenario falls into the high fuel reactivity and low congestion portion of the matrix. The result is an underprediction of the overpressure. The BST high fuel reactivity works well for fuel such as ethylene (LBV = 0.76 m/s), but will often underpredict the impacts of a hydrogen (LBV = 3 m/s) explosion. The expanded matrix allows a higher fuel reactivity category and a lower congestion category, and consequently a better prediction of the explosion impacts. As seen in Figure 1, the QMEFS result is a slight overprediction of the overpressure.

Table 3. Modeling Parameters for an Unconfined Hydrogen Explosion

Parameter	BST	Expanded QMEFS
Volume [m^3]	300	300
nPlanes	1	1
Reactivity	High	Very-High
VBR	N/A	0
Congestion	low	N/A
Run-up distance [m]	N/A	5.2

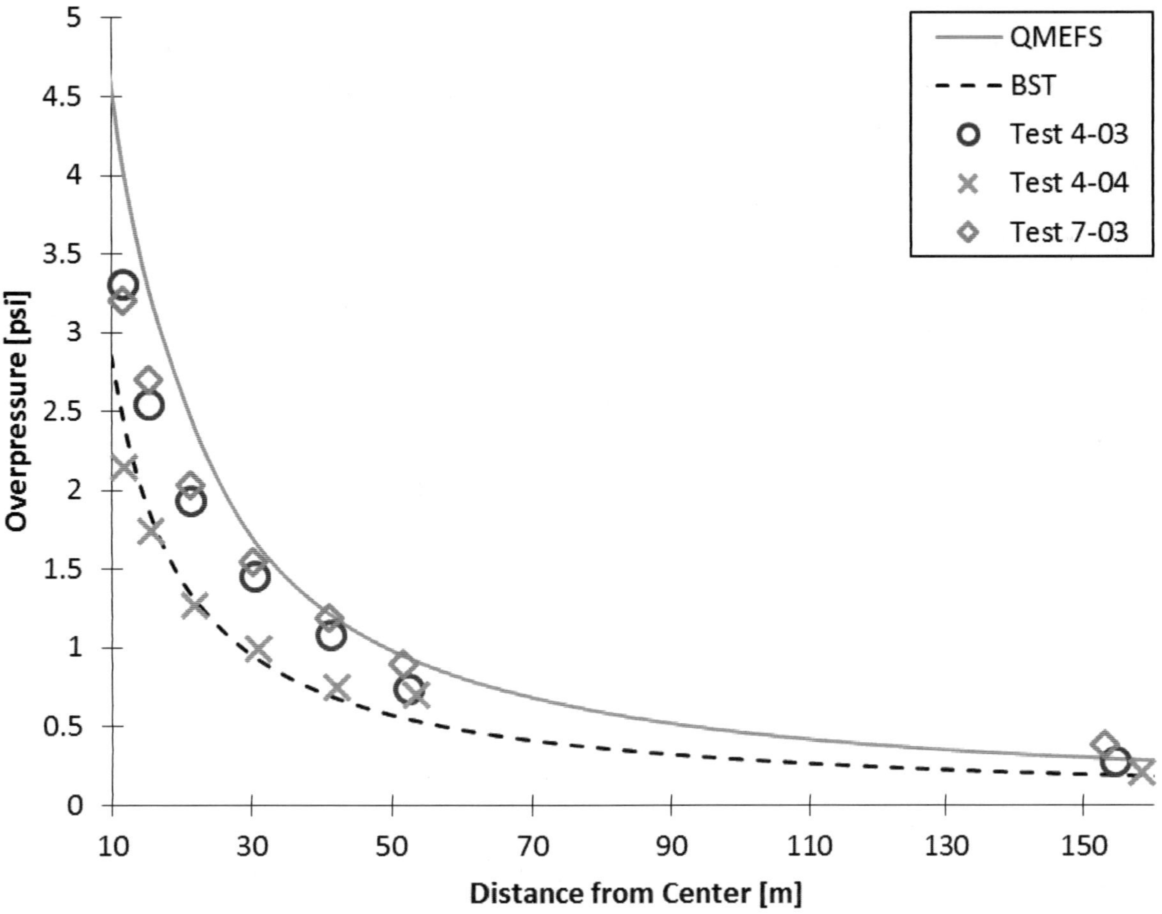

Figure 1. Overpressure vs. Distance for Open-Air Hydrogen Tests

As discussed in the paper, the distance that is available for the flame to accelerate, i.e. the run-up distance, can have a significant effect on the flame speed attained in a flammable gas cloud. The flame speeds presented in the original Baker-Strehlow (BS) model [6] were generated in a test rig whose largest dimension was less than 6 feet (1.8 meters) [5]. The flame speeds presented in the 2005 BST paper are based on tests where the largest dimension was 48 feet (14.6 meters) with the published flame speed results "scaled up" to account for the maximum size of a typical industrial plant [5]. One test of the QMEFS model is to determine the flame speed versus run-up distance for the test configurations used in the newest BST model when specific parameters are varied.

Figure 2 shows the results for the 3-D flame expansion (*nPlanes* = 1), medium reactivity (S_l = 0.43 m/s) case. The curves plotted in Figure 2 show six VBR values – the high (5.7%), medium (4.3%), and low (1.5%) congestion values used in the BST experiments, one intermediate value to show how the model behaves as the VBR is adjusted, a fifth value (1%) below the BST low category to demonstrate the behavior in the "less-than-low" congestion range, and the sixth value (10%) above the BST high category to demonstrate the behavior in the "higher-than-high" congestion range.

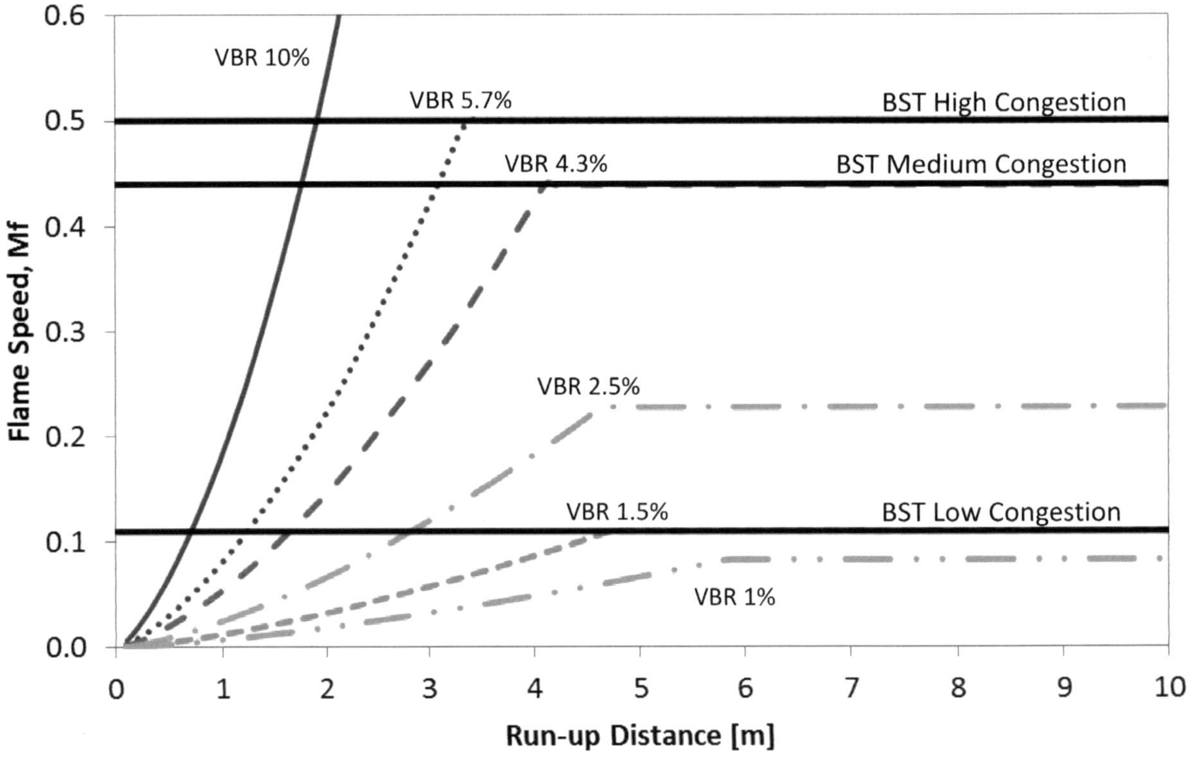

Figure 2. Flame Speed vs. Run-Up Distance; 3-D, Propane, D = 2"

Figure 3 demonstrates how flame speed varies with *nPlanes* and LBV. The basis of the figure is an explosion involving a medium VBR (0.043) and a 4-inch average obstacle diameter. The *nPlanes* parameter is varied between 1.0 (3-D flame expansion) and 4.0 (the maximum value recommend for use in the QMEFS model). The curves plotted in Figure 3 show five materials with various LBV values – ammonia (0.08 m/s), methane (0.37 m/s), propane (0.43 m/s), ethylene (0.76

m/s) and ethylene oxide (0.91 m/s). In the BST model ethylene oxide and ethylene are grouped together into the high category, but these two fuels have different burning properties as can be seen with their LBV values. Thus, by accounting for the LBV of the fuel, the new model provides a finer resolution for flame speed. In addition, as seen in Figure 3, the QMEFS model provides more resolution for the flame expansion parameter through the use of the *nPlanes* variable.

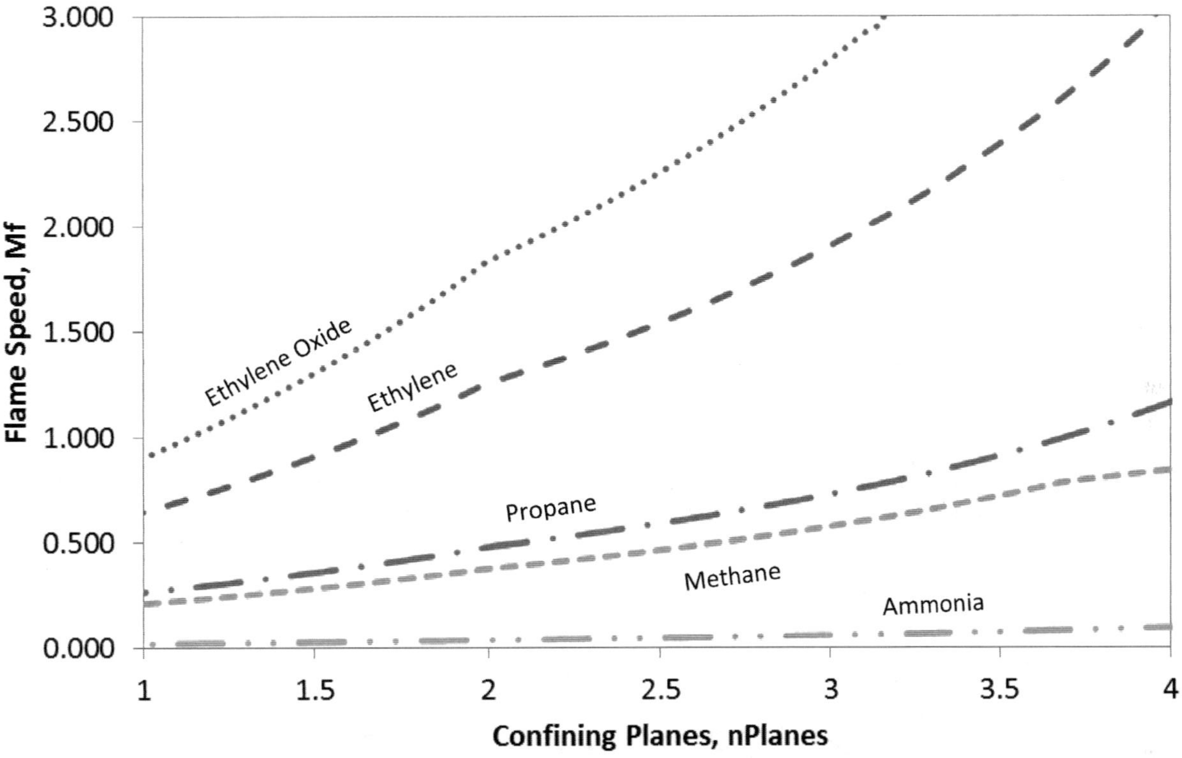

Figure 3. Flame Speed vs. Confining Planes; Medium VBR, D = 4"

5 Conclusions

The model described in this paper provides an improved method to describe the characteristics of a vapor cloud explosion over a wide range of conditions. It combines elements of the work conducted by Baker to develop the BST model, as well as the European MERGE/EMERGE projects; thus incorporating data from the two largest, modern vapor cloud explosion test projects.

Because the BST categories are resolved into numeric ranges, a higher resolution solution space and a clearer set of measureable inputs are achieved. With the ability to model the reactivity of a flammable gas cloud based on laminar flame speed instead of a low, medium, or high classification, the model is able to more accurately describe a wide range of flammable gases and mixtures. Expansion of the constraining flame speed categories to very-low and very-high then provides an even wider range of possibilities.

With the QMEFS matrix expanded to congestion (VBR) categories of very-low and very-high, the model provides a wider range of predictive capability for explosion scenarios. These expansions and enhancements are still confined by the flame speeds in the BST system, and are applied within the set of curves for overpressure and impulse. Thus, with the QMEFS model presented here, process safety experts have a more versatile tool to predict VCE impacts in petrochemical plants.

6 References

[1] Baker, Q.A., C.M. Doolittle, G.A., Fitzgerald, and M.J.Tang, "Recent Developments in the Baker-Strehlow VCE Analysis Methodology," *Process Safety Progress*, v.17, No.4 (1998).

[2] Melton. T.A. and Marx J.D., "Estimating Flame Speeds for Use with BST Blast Curves," Process Safety Progress, v. 28, March, (2009).

[3] Mercx, W.P.M., A.C. van den Berg, and D. van Leeuwen, A*pplication of correlations to quantify the source strength of vapour cloud explosions in realistic situations*. Final report for the project: 'GAMES', TNO report PML 1998-C53, Rijswijk, The Netherlands (1998)

[4] Thomas, K.J., Malik. D.R., Downes, A.M., Rodges, S., "Ammonia Vapor Cloud Explosion Testing and Implications for Facility Siting Studies, Consequence Analyses and Risk Assessments," *50th Loss Prevention Symposium*, AIChE, April, (2016).

[5] Pierorazio, A.J., J.K.Thomas, Q.A.Baker, and D.E.Ketchum, "An Update to the Baker-Strehlow-Tang Vapor Cloud Explosion Prediction Methodology Flame Speed Table," *Process Safety Progress*, v.24, no.1 (2005).

[6] Baker, Q.A., M.J. Tang, E. Scheier, and G.J. Silva, "Vapor Cloud Explosion Analysis," *28th Loss Prevention Symposium*, AIChE, April, (1994).

[7] Mercx, W.P.M., editor, *Extended Modelling and Experimental Research into Gas Explosions*, Final summary report of the project EMERGE, CEC contract EV5VCT930274, European Commission, Directorate General XII, Brussels, Belgium (1997).

[8] Mercx, W.P.M., A.C. van den Berg, C.J. Hayhurst, N.J. Robertson, and K.C. Moran (2000), *Developments in Vapour Cloud Explosion Blast Modeling*, Journal of Hazardous Materials 71 (2000) 301-319.

[9] Tang, M.J. and Q.A.Baker, "A New Set of Blast Curves from Vapor Cloud Explosion," *Process Safety Progress*, v.18, no.4 (1999).

[10] Oran, E.S. and V.M. Gamezo, "Origins of the deflagration-to-detonation transition in gas-phase combustion," *Combustion and Flame*, 148 (2007).

[11] Vasil'ev, A.A., "Estimation of Critical Conditions for the Detonation-to-Deflagration Transition," *Combustion, Explosion, and Shock Waves*, v.42, no. 2 (2006).

[12] Harris, R.J. Wickens, M.J., "Understanding Vapor Cloud Explosions – An Experimental Study," The Institute of Gas Engineers 55th Autumn Meeting, (1989).

[13] Sato, Y., Iwabuchi, H., Groethe, M., Merilo, E., Chiba, S., "Experiments on Hydrogen Deflagration," Journal of Power Sources v.159 (2006).

[14] Health & Safety Executive, "Explosions in Full Scale Offshore Module Geometries Main Report," Offshore Technology Report – OTO 1999 043. Prepared by BG Technology, April, (2000).

479444 Do We Need a Quantitative Risk Assessment (QRA) If My Consequence Analysis Facility Siting Presents a Problem?

Abstract:

There are three steps that a regular facility siting study (FSS) can go through. The first step is the basic separation table approach descrived in API 752. The second step is the consequence analysis facility siting and the third and final one is the quantitative risk assessment. It turns out, however, that the vast majority of the consequence analysis FSS studies quickly demonstrate that the impact to buildings can be more significant that what have been defined as acceptable. For such cases, the analysis normally turns into the QRA, exceedance curves, building structural assessment, etc. The idea of this paper is to cover a intermediate step in early stages to use computational fluid dynamics (CFD) technique in the early stage of the analysis without compromising the time schedule and the budget, this would include dispersion, radiation and explosion modeling. The more accurate results obtained with CFD modelling lead to cloud sizes which are mainly less conservative than traditional methods. The objective of this paper is to go a setp beyond a previous analysis for the dispersion using CFD, but rather make a full evaluation of a FSS using CFD.

Proposed Method for Investigating the Reactivity of Chlorine with Environmental Materials under Relevant and Controlled Conditions

Audrey Feuvrier, PhD.
Ralph E. Martin Department of Chemical Engineering
University of Arkansas, Fayetteville, AR
audrey.feuvrier@uark.edu

Tom Spicer, Ph.D., P.E.
Ralph E. Martin Department of Chemical Engineering
University of Arkansas, Fayetteville, AR
tos@uark.edu

Prepared for Poster Presentation at
American Institute of Chemical Engineers
2017 Spring Meeting and 13th Global Congress on Process Safety
San Antonio, TX
March 26 – 29, 2017

Proposed Method for Investigating the Reactivity of Chlorine with Environmental Materials under Relevant and Controlled Conditions

Audrey Feuvrier, PhD.
Ralph E. Martin Department of Chemical Engineering
University of Arkansas, Fayetteville, AR
audrey.feuvrier@uark.edu

Tom Spicer, Ph.D., P.E.
Ralph E. Martin Department of Chemical Engineering
University of Arkansas, Fayetteville, AR

Keywords: Chlorine reactivity, chlorine release, dry deposition, Jack Rabbit II

Abstract

Many common toxic industrial chemicals (TICs) such as chlorine and ammonia are very reactive with commonly encountered materials in the environment. The amount of TIC material reacted in the environment is an important factor in the impact of a release. Typically characterized with a deposition rate in atmospheric dispersion models, the reaction rate and total deposition quantities have predominately been studied in field scale experiments. Laboratory experiments have also been undertaken, but past investigations have inadequately quantified the available turbulence which can importantly impact surface reaction rates. The state of the art description of surface deposition represents the process as a series of resistances including the effect of atmospheric turbulence and surface reaction(s) depending on the surface material (substrate). Sponsored by the Chemical Security Analysis Center (CSAC) of the U.S. Department of Homeland Security and Transport Canada, an experimental program is underway to test potential substrates under known inlet gas concentrations with levels of turbulence that are comparable to those of the atmosphere. The objective is to develop a test methodology for subjecting a substrate to a reactive gas flow of known TIH concentration with turbulence levels consistent with those in the atmosphere. With data from repeat experiments, deposition of a reactive TIH chemical can be quantified with the proposed experimental protocol.

1 Introduction

Estimating the quantity of a Toxic Industrial Chemical (TIC) removed from the environment through chemical processes is critical to proper consequence assessment of accidental (or intentional) releases, particularly in urban areas [1]. TIC removal mechanisms play an important role in the rational assessment of risk, as these phenomena may significantly and quickly reduce the TIC concentration near a release [2,3], particularly for a reactive chemical such as chlorine. Typically characterized as dry deposition, reactions of chlorine with the environment was not included in the computational models used to predict the consequences of large releases of chlorine in the past [2], but new efforts are underway to characterize chlorine reactivity and incorporate such effects in consequence and risk assessments.

While species like sulfur dioxide and iodine have been the subject of extensive experimental studies [3], studies about deposition involving chlorine are scarce and show a great deal of variability. The high reactivity of chlorine as well as its toxicity make it difficult to study in a laboratory environment, and consequently, many of the studies are conducted as field studies [4]–[6], which restricts test conditions to a limited set of atmospheric conditions. In one set of laboratory experiments, Hill [7] studied the effect of deposition of several TICs on an alfalfa canopy in a chamber where humidity, light, and wind speed were controlled. He showed that chlorine deposition is rapid on alfalfa and that chamber air velocity has a significant effect on the chlorine uptake, but the level of turbulence in the chamber is unknown. Other experimental studies gave promising results such as [8]–[10], but these studies involved no other movement of the flow than the one produced by the mixing device or the circuit used to monitor the concentration level. The state of the art description of surface deposition, [11], represents the process as a series of resistances including the effect of atmospheric turbulence and surface reaction(s) depending on the surface material (substrate).

In this context, an experimental program, sponsored by the Chemical Security Analysis Center (CSAC) of the U.S. Department of Homeland Security and Transport Canada is underway to test potential substrates in a controlled environment. The project includes the design and characterization of a mixing chamber as well as development of a protocol for the measurement of chlorine removal rate for different samples of soil and vegetation. Concentration, velocity, and turbulence measurements will be made in the working section to confirm that the flow is uniform and consistent. It is also desired that the flow in the working section of the apparatus is consistent with the atmospheric flow with target turbulence intensities on the order of 10%. Working section flow speed will be variable, with a maximum in the section of 8 m/s, and expected velocities between 0.5 to 3 m/s. Passive flow conditioning will be used to eliminate non-uniformities in the flow. Recognizing that chlorine gas concentrations as high as 10% or 100,000 ppm may be used, apparatus materials of construction will be carefully chosen to resist damage and also to not remove significant amounts of chlorine during a test. Finally, to limit the hazards involved with the use of chlorine at high concentration level, the apparatus will need to fit under a 6 ft long (about 1.8 m) fume hood.

In the following, the proposed designs for the apparatus and its main elements are introduced. The choice of the mixer, fan, and materials will be discussed. In the second part of the article, the future protocol of measurement will be presented. Focus will be put on the means of measurement used and the characterization of the chamber. The steps to determine chlorine removal rates for several samples types under different environmental conditions will be considered.

2 Proposed design

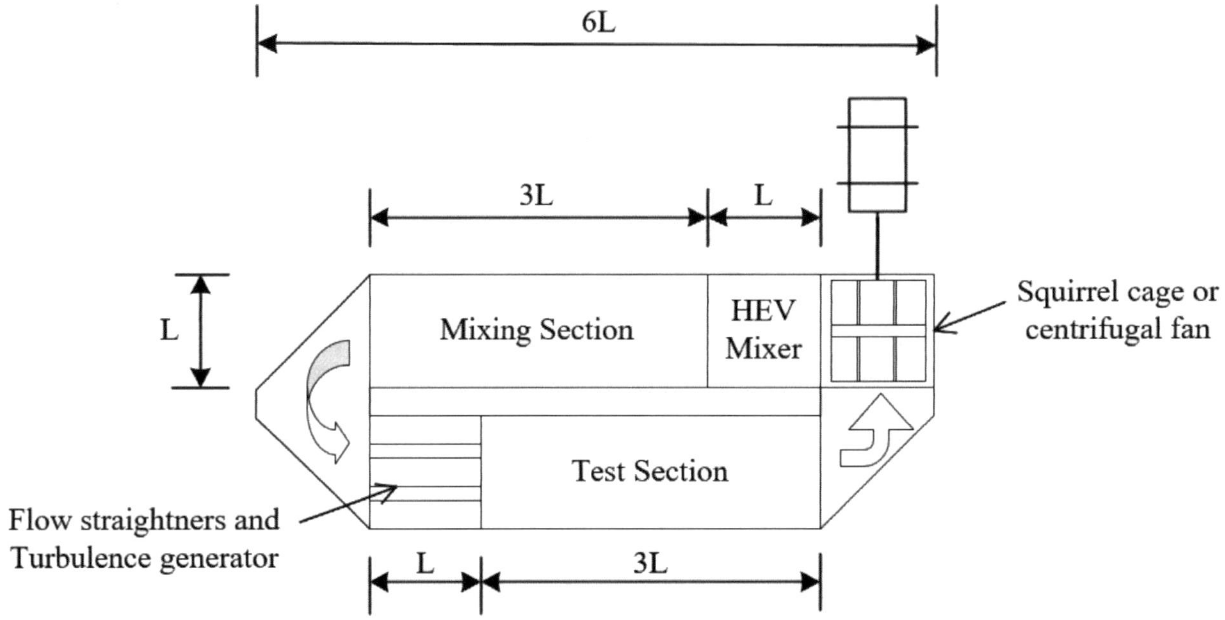

Figure 1. Preliminary apparatus design

Figure 2 shows the preliminary design of the apparatus following the criteria laid down previously using a simple closed return design. The test or working section is placed under the mixing section to facilitate access to the test section through a lab hood sash. The characteristic length, L, chosen to fit in a standard 6 ft hood, will be between 0.25 and 0.3 m, giving a total volume for the apparatus of between 0.16 and 0.28 m^3. A standard lecture bottle contains 454 g of chlorine, which would be sufficient to provide chlorine for 8 to 14 tests depending on the length L based on a maximum concentration of 100,000 ppm. The use of lecture bottles limits the quantity of chlorine required in the lab during the experiments and effectively eliminates the hazard associated with the storage of large quantities of chlorine on site. These volume values are approximate because it may be convenient to use a circular conduit for mixing and a square cross-section for the working section. A circular conduit reduces the required volume of the mixing section (with some minor reduction in power consumption), and a working test section with flat surfaces facilitates optical measurements during characterization and sample placement in the testing phase.

2.1 Mixer and flow conditioner

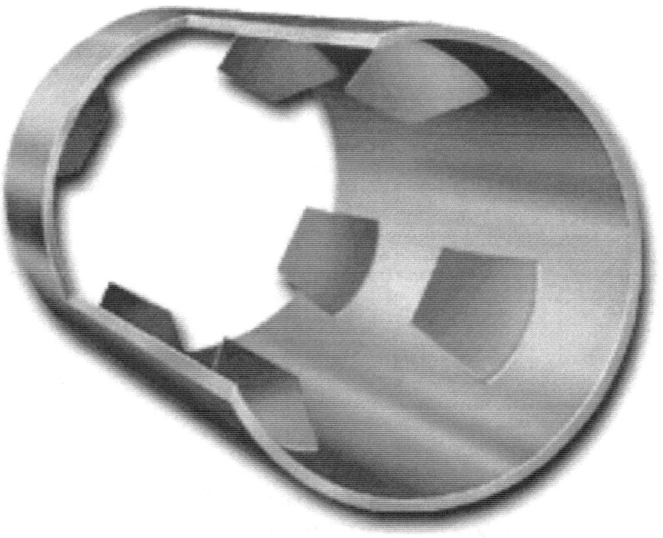

Figure 2. Typical High Efficiency Vortex (HEV) mixer [12].

Figure 3 shows a typical High Efficiency Vortex (HEV) mixer which is placed downstream of the fan. This type of mixer was selected because it provides efficient mixing with turbulent flow with minimal pressure drop. For a Kenics^{TM} HEV mixer, Figure 4 shows the relative coefficient of variation, COV_R , defined as the quotient between the outlet and inlet coefficient of variation, as a function of the Reynolds number (defined as $U_\infty \, L/\upsilon$ where U_∞ represent the characteristic velocity in the tunnel, L is the characteristic length and υ is the kinematic viscosity of the gas). The coefficient of variation for the mixing of two fluid represents the quotient of the standard deviation of the concentration and the mean concentration. Therefore, the lower the COV_R, the better the mixture quality. The gray zone on Figure 4 represents the range of Reynolds number of this study. Based on Myers et al [13], the pressure drop for our (custom made) apparatus was estimated to be between 1.7 and 17.3 Pa, making it negligible. The HEV mixing process rely on the production of hairpin vortices downstream of the blades of the mixer combined with a pair of counter rotating longitudinal vortices [14]. A mixing section downstream of the device and equal to three times the length of the mixer is recommended to provide for sufficient time for the vortices to complete the mixing process and consequently obtain the predicted COV_R. Additional turbulence generated between the mixing section and the test section will only improve the COV_R before the flow enters the test section..

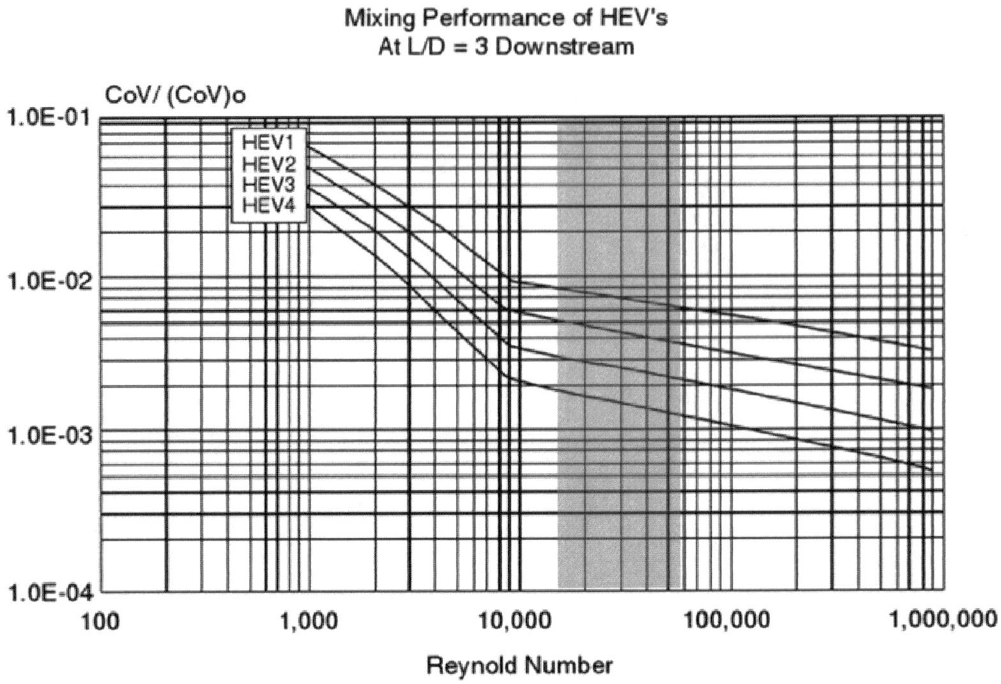

Figure 3. Relative coefficient of variation (COV_R) vs. Reynolds Number and number of tabs row for the Kenics HEV, [12].

The corners currently consist of two 45° bends with a short duct between them, and the 45° bends are preferred to 90° bends to prevent the accumulation of gases in the dead space. To help maintain good flow distribution in the channel, sets of vanes will be placed in each bend. Operating at low wind speed allows us to consider simple circular cross sectional shapes for the vanes [15].

After the two first corners, the flow conditioner section is planned to eliminate large eddies that might be present in the flow ensuring uniformity of the flow and simultaneously raising the turbulence intensity up to 10%. To achieve this, we plan to use a combination of screens or perforated plates. Samples will be placed in the test section through a removable side with provision for a tray in the floor of the test section to accommodate soil and/or plant roots. After experiments, ventilation of the apparatus will be achieved through two hatches on each bend between the mixing and flow conditioner sections.

2.2 The fan

On Figure 2, the motor of the fan is represented outside of the apparatus because chlorine would be expected to react and corrode any of its exposed surfaces. The motor shaft will be supported by Nylon or Teflon bearings located in the side of the apparatus to isolate the motor from chlorine exposure. A centrifugal (squirrel cage) fan is being considered to allow placement of the fan above the apparatus. The fan blades and shaft exposed to chlorine will be treated to limit damage due to chlorine exposure as discussed below.

2.3 Building materials

During the first phase of the experimentation, the apparatus will only use a mixture of air and an inert gases will be used for optic velocity and turbulence measurements and concentration measurements (including a propylene tracer). During this preliminary phase, the structure of the test section will be made with plywood, but two of its walls will be made with polycarbonate to allow visualization and optical measurements of the velocity in the section.

In the second phase of the experimentation involving chlorine, all elements in contact with the test mixture must be made with a material or coated with a material capable of resisting high chlorine concentrations. In addition to extending the workable lifetime of the apparatus, reducing the potential for chlorine to react with materials in the test chamber is also desired so that the reduction in chlorine gas concentration can be attributed to reaction with the sample tested as opposed to the chamber itself. In the case of the fan blades and shaft, a commercial coating is planned (such as Cerakote or titanium). Titanium is known to be highly reactive with dry chlorine, however 1% of water may be enough for passivation after mechanical damage to titanium under normal conditions [16]. For the mixing section, a metal or plastic (PVC) duct will be used since no optical access is required. The structure of the corners and the test section will be made with plywood. In the second phase, polycarbonate may be replaced by plywood. The flow straighteners will likely be made of a suitable metal. All surfaces except the fan blades and shaft are planned to be coated with Viton™, a coating material widely used in applications with chlorine in the chemical industry. We are currently evaluating whether the Viton™ coating under consideration will adhere sufficiently to plywood, and if it does not, then the entire apparatus may need to be fabricated from sheet metal with all internal surfaces painted with Viton™.

3 Protocol for future experimentations

3.1 Experimental resources and characterization of the mixing chamber

Several instruments will be used during the characterization of the chamber and concentration measurement. During both phases of the experimental measurements, the temperature, pressure and humidity of the ambient air will be monitored.

For the characterization of the chamber, the first step will be verifying that the flow entering the test section is uniform and fully mixed through visualization. The working section will be equipped with LED lights and a ROSCO 1500 fogger machine will provide the fog. The fog will be introduce downstream of the fan, at the same point planned for injection of the gases. Video recordings of the fog in the working section will be made with several velocities. After visualization experiments, velocity measurements will be made with a 2D-LDV (Laser Doppler Velocimetry) system using an Argon Ion 300 mW air-cooled laser. This single point technique is non-intrusive and allows simultaneous measurement of two components of the velocity with a good spatial and temporal resolution. Finally, the concentration measurements will be made using an Aurora Scientific mini-PID (photo ionization detectors) The PID measurement method relies on photoionization of a volatile organic compound (in this case a propylene tracer). Small access points will have to be made in the test section for the PID. In the characterization phase, a simulant gas mixture (using sulfur hexafluoride) with be used to reproduce the molecular weight of chlorine. Sulfur hexafluoride is non-toxic, non-flammable, non-reactive and can be used in the PID system.

In conclusion, the characterization of the mixing chamber will assess the quality of the mixing and the flow in the working section, the steps are listed below :

1. Visualization of the flow going in the test section.
2. Velocity profiles taken in the test section will ensure the uniformity of the flow and that the turbulence intensity level is respected.
3. Concentration profiles will provide information on the uniformity of the concentration of the simulant gases in the chamber.

3.2 Concentration measurements and evaluation of the reactivity of chlorine with different samples

After the characterization of the working section, the apparatus will be modified to undergo tests with high concentration of chlorine. The changes have been described in section 2.3. Furthermore, the tunnel will have to allow the access of the JAZ sensors in order to measure the concentration of chlorine. As with characterization of the chamber, the evaluation of the reactivity of chlorine with different samples will request several steps listed hereunder.

1. Characterization of the response of the apparatus to chlorine : Tests will be done without samples, so that the chlorine concentration decay due to the apparatus itself can be measured. A correction factor will be calculated and implemented in the following concentration measurements. Multiple tests will be done, in order to ensure of the repeatability of the measurements.
2. Removal rate estimation : Samples will be placed in the tray, allowing the removal rate of these samples to be estimated through concentration measurements.
3. Analysis of the results : The measurement of the concentration decay in the chamber and the study of the samples put in contact with chlorine will provide valid information in order to calculate the removal rate of chlorine for these different samples. Analytical techniques for determining the amount of chlorine uptake by the test material will depend on the material.

4 Conclusion

In this article, the design of an innovative apparatus or mixing chamber and its main elements have been presented in details. One of the main difficulties encountered was to scale the device to fit under a six feet long fume hood to limit the hazards involved with the use of chlorine at high concentration level in the laboratory. One other obstacle was to find materials and coating suitable and able to resist chlorine in the long run. The future means and procedures of characterization and measurement have been introduced as well. Once build and characterized, this apparatus will provide new data sets for chlorine deposition on vegetal samples and small objects in conditions similar to the atmospheric environment. In the end, this study aim at the improvement of the understanding and computational modeling of toxic inhalation hazards and the consequences to affected population and infrastructure.

5 Aknowledgment

This project is supported by the Chemical Security Analysis Center (CSAC) of the U.S. Department of Homeland Security and Transport Canada. The authors wishes to thank Katherine Elmore for her assistance with this project.

6 References

[1] M. B. Dillon, "The role of deposition in limiting the hazard extent of dense-gas plumes," *J. Hazard. Mater.*, vol. 164, no. 2–3, pp. 1293–1303, 2009.

[2] S. Hanna, S. Dharmavaram, I. Zhang, J., Sykes, H. Witlox, S. Khajehnajafi, and K. Koslan, "Comparison of Six Widely-Used Dense Gas Dispersion Models for Three Actual Railcar Accident," in *Process Safety Progress*, 2008, vol. 27, no. 3, p. 248–259.

[3] S. P. Arya, *Air pollution meteorology and dispersion*, 6th ed. 1999.

[4] J. P. Dimbour, A. Dandrieux, and G. Dusserre, "Reduction of chlorine concentrations by using a greenbelt," *J. Loss Prev. Process Ind.*, vol. 15, no. 5, pp. 329–334, 2002.

[5] G. Dusserre and J. Ollivier, "Small scale field experiments of chlorine dispersion," *J. Loss Prev. Process Ind.*, vol. 15, pp. 5–10, 2002.

[6] J. D. Hearn, R. Weber, R. Nichols, M. V. Henley, and S. Fox, "Deposition of Cl_2 on soils during outdoor releases," *J. Hazard. Mater.*, vol. 252, pp. 107–114, 2013.

[7] A. C. Hill, "Vegetation: A Sink for Atmospheric Pollutants," *J. Air Pollut. Control Assoc.*, vol. 21, no. 6, pp. 341–346, Jun. 1971.

[8] J. Hearn, J. Eichler, C. Hare, and M. Henley, "Cl_2 deposition on soil matrices," *J. Hazard. Mater.*, vol. 237–238, pp. 307–314, 2012.

[9] J. Hearn, J. Eichler, C. Hare, and M. Henley, "Effect of soil moisture on chlorine deposition," *J. Hazard. Mater.*, vol. 267, pp. 81–87, 2014.

[10] D. Brown, "The removal of toxic gases from the air by water, plants and soils.," 2006.

[11] M. L. Wesely and B. B. Hicks, "A review of the current status of knowledge on dry deposition," *Atmos. Environ.*, vol. 34, pp. 2261–2282, 2000.

[12] J. R. Couper, W. R. Penney, and J. R. Fair, *Chemical process equipment revised 2E: selection and design.* Gulf Professional Publishing, 2009.

[13] K. J. Myers, A. Bakker, and D. Ryan, "Avoid agitation by selecting static mixers," *Chem. Eng. Prog.*, vol. 93, no. 6, pp. 28–38, 1997.

[14] W. J. Gretta and C. R. Smith, "The Flow Structure and Statistics of a Passive Mixing Tab," *J. Fluids Eng.*, vol. 115, no. 2, p. 255, 1993.

[15] R. H. Perry, D. W. Green, and J. O. Maloney, *Perry's chemical engineers' handbook.* McGraw-Hill, 2008.

[16] Timet corporation, "Corrosion resistance of titanium," 1997.

Fire Proofing Cost Reduction Measures for LNG Plant

Raminaidu Girada
Senior Process Safety Engineer, Bechtel Oil, Gas & Chemicals, Inc.
3000 Post Oak Boulevard, Houston, TX 77056
rgirada@bechtel.com

Jegan Babu Arumugakan Thuraiswamy
Senior Process Safety Engineer, Bechtel Oil, Gas & Chemicals, Inc.
3000 Post Oak Boulevard, Houston, TX 77056
jbarumug@bechtel.com

Nino Nicotra
Process Safety Engineering Chief, Bechtel Oil, Gas & Chemicals, Inc.
3000 Post Oak Boulevard, Houston, TX 77056
anicotra@bechtel.com

Prepared for Presentation at
American Institute of Chemical Engineers
2017 Spring Meeting and 13th Global Congress on Process Safety
San Antonio, TX
March 26 – 29, 2017

Fire Proofing Cost Reduction Measures for LNG Plant

Raminaidu Girada
Senior Process Safety Engineer, Bechtel Oil, Gas & Chemicals, Inc.
3000 Post Oak Boulevard, Houston, TX 77056
rgirada@bechtel.com

Jegan Babu Arumugakan Thuraiswamy
Bechtel Oil, Gas & Chemicals, Inc.
3000 Post Oak Boulevard, Houston, TX 77056
jbarumug@bechtel.com

Nino Nicotra
Process Safety Engineering Chief, Bechtel Oil, Gas & Chemicals, Inc.
3000 Post Oak Boulevard, Houston, TX 77056
anicotra@bechtel.com

Keywords: Consequence Modeling, Liquefied Natural Gas (LNG) & Liquefied Petroleum Gas (LPG) and Safe Design

Abstract

The current industry practice in designing fireproofing requirements for LNG facilities follows the traditional approach of "Liquid Pool Fire". The common industry practice for fire proofing design criteria defines a fire potential equipment (FPE) as the equipment that handles flammable liquid above a specific threshold volume. The FPE generates fire scenario envelope whose dimension ranges are recommended by API 2218. Equipment and structural supports that falls inside the fire envelope are considered for fire proofing.

LNG plants are handling products with very low boiling point, and any leak to atmosphere results in an instant flashing into vapor phase or aerosols. The results from our recent works on pool fire consequence modelling has proved that most of the currently considered fire potential equipment seldom create pool fire. This consequence modeling results clearly demonstrates that the current fire proofing design approach is overdesign, that is not effective. The apparent saving in man-hours for not modeling the liquid release will result in a considerable amount of engineering efforts and material cost being spent on the design and installation of fire proofing for equipment and structure with potentially no real benefits to the inherent safety of the process plant. Considering the current economic trend, a change in approach using consequence modelling will effectively result in considerable cost savings in major LNG projects.

1 Introduction

Currently in LNG projects, fireproofing is considered for equipment supports, pipe rack support, structural supports and other auxiliary critical hydrocarbon carrying pipe supports within a fire proofing zone to improve its structural integrity during a fire. The principal value of fire proofing is realized during the early stages of a fire by providing additional time to fire fighters to control or extinguish a fire thus by delaying additional loss of containment which may escalate the incident. The fireproofing requirement is determined by identifying the fire potential equipment and the associated fireproofing zones. Fire potential equipment is defined as an equipment holding flammable liquid above a specific threshold volume, which will have the potential to create a liquid pool and subsequent pool fire if ignited. Per API 2218, the fireproofing zone is defined for every fire potential equipment, that extends 20 ft. to 40 ft. (6 m to 12 m) horizontally and 20 ft. to 40 ft. (6 m to 12 m) vertically from the source of liquid fuel. As a typical approach most projects consider fireproofing zones that extend 30 ft. (9m) horizontally and 30 ft. (9m) vertically from the source.

The above mentioned qualitative approach does not reflect the realistic possibility of pool fire scenario. Therefore, a quantitative analysis may be performed to analyze whether a pool fire is realistic for equipment handling multiple refrigerants and LNG. This analysis has been performed on some process equipment handling multiple refrigerants and LNG using Canary software by Quest Consultants.

2 General Technical Guidance

2.1 Scope of Fire Proofing

Fire scenario envelopes for an LNG plant are based on pool fire scenarios. Jet fires are typically not considered as credible for defining fire scenario envelops for fireproofing as described below:

- Jet fires are handled by Prevention, Intervention and Protection. Prevention is by designing the facility in compliance with general fire prevention activities such as those included in NFPA 59A and in the broad application of Process Safety Management(PSM). Once a pressurized jet fire occurs, the intervention and protection can be achieved by a high flow rate firewater stream directed on the point of impingement. This has been proved as an effective mitigation measure in preventing escalation with Jet fires. This will be achieved by strategically locating the fire water monitors around high risk process area, where jet fire is foreseen as a credible scenario.

- Emergency isolation and depressurization system will be provided to lower the pressure of specific isolated sections of the plant that may be involved in a localized jet fire scenario and to de-inventory of gas to flare.

- Efforts on similar LNG projects to extend the fire proofing systems for jet fire scenarios led to enormous impractical level of fireproofing requirement. This could also introduce other hazards such as corrosion.

- Jet fire represents a significant risk or concern for personnel life, working on offshore installations whereas property protection is the basis of pool fire for on-shore LNG facility or petroleum refinery or petrochemical facilities.

2.2 Pool Fire Formations

A pool fire may result via a number of scenarios and it begins typically with the release of flammable material from process equipment. If the material is liquid, stored at a temperature below its normal boiling point, the liquid will collect in a pool. The geometry of the pool is dictated by the surroundings (i.e., diking), but an unconstrained pool in an open, flat area is possible, particularly if the liquid quantity spilled is inadequate to completely fill the diked area. If the liquid is stored under pressure above its normal boiling point, then a fraction of the liquid will flash into vapor, with un-flashed liquid remaining to form a pool in the vicinity of the release.

The pool of flammable liquid may be ignited creating a pool fire. If the release of flammable material from a pressurized process equipment continues, then a jet fire is also likely. If the ignition source occurs at the very beginning of the release, then inadequate time is available for the liquid to form a pool and only a jet fire will result.

Pressurized liquefied gases such as LNG, methane, ethylene and propane will in general disperse as mists or vapor, but may possibly form a small pool if the source operates at low pressure conditions. An exception has to be made for butanes, pentanes and heavier hydrocarbons, which always create pool.

Refer to the following **Figure 1** for pictorial view of the pool fire formation.

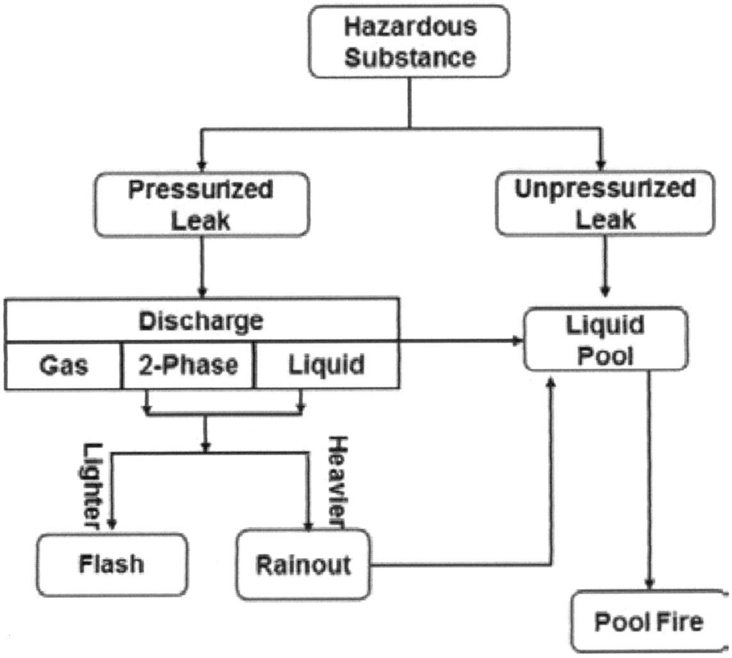

Figure 1: Pool Fire Formation

2.3 Determination of Pool Fire scenario

This section provides an overview of the methodology for determining pool fire existence.

In case of accidental release of flammable materials such as Pentanes and heavier hydrocarbons create liquid pool irrespective of their operating conditions. However, the lighter hydrocarbons such as butane and propane may create liquid pool only under specific operating conditions. The potential for creating liquid pool is based on certain parameters as described below:

- **Butanes**

  Butane releases potentially create liquid pool except for high pressure releases and slight sub cooling of release. Further consequence analysis is required to check if a liquid pool is created from the release.

- **Propane and Ethylene**

  Ethylene and propane liquids may or may not pool, depending on conditions. It has been found that a pool formation can be correlated on several parameters: (1) vessel operating pressure, (2) size of rupture hole for self-leakage vessel, (3) the amount of subcooling, which is also important but the assumption is the liquid will be slightly subcooled, and (4) confinement of *fire hazard area*. Together, these variables establish the size of the pool.

  Higher pressure release are more likely to vaporize and mist, and low pressure releases are more likely to pool. Based on a slightly subcooled liquid at pressure, the analysis is divided into three regimes (1) no liquid pools, (2) highly likely to form a significant pool fire, and (3) possible pool fire. Depending on the size of the release, slope, temperature and composition, a consequence model may be necessary to determine if a pool fire is warranted.

- **LNG**

  Similar to propane and ethylene liquids, LNG may or may not pool depending on several parameters: (1) vessel operating pressure, (2) size of rupture hole, (3) the amount of subcooling, and (4) confinement.

Therefore, a quantitative approach was performed in the following section.

3 Pool Fire Case Study

In this case study, the following light flammable fluids are modelled: methane, propane, ethylene and LNG. Based on the current fire proofing design approach a total of 28 equipment of the LNG plant that handles the above fluids are qualified as Fire Potential Equipment. All those leakage scenarios are modelled using Canary modelling tool, based on its operating condition.

3.1 Leak Scenarios

To provide a spill rate using CANARY, one has to postulate a leak scenario.

Typical hole size choices range from small (3-10mm) to full bore ruptures (diameter of pipe). Hole size selection can define the general level of risk (larger, high consequence releases are significantly less frequent than small, low consequence release).

CCPS book of "Guideline for Fire Protection in Chemical, Petrochemical, and Hydrocarbon Processing Facilities" defined hole sizes as small (3-10 mm), medium (10-50 mm), large (50-150 mm), and full-bore ruptures. Discussion around these hole sizes also points out that the large and rupture releases are not always the worst-case events due to release duration and dosage effects that should be discounted for.

The Canadian Standard Association (CSA) standards Z276 specifies design spills which define the potential extent of hazard zones that are not allowed to impact the public. Z276 defines, among other (smaller) release scenarios, that the design spill should be "The maximum flow of LNG that could be spilled, from the largest branch connection, 60 mm (2-inch) or smaller..."

Based on the above references, 2-inch is considered as an appropriate hole size for consequence modeling. Therefore, the pool fire modeling has been performed for 2-inch hole size. However, sensitivity analysis has been done considering 6-inch hole to check if larger hole size can result in pool formation. The modeling results for both 2-inch and 6-inch are presented below.

The release elevation was considered to be constant at 1 m height (the lowest height is deemed to result in higher consequences). The release directions in all scenarios were assumed to be horizontal. The consequence modeling input data shown below were used for the calculations. All scenarios were modeled using CANARY software developed by Quest. The software is widely used in industrial consequence modelling studies.

3.2 Atmospheric Conditions

The following more stringent atmospheric conditions that favors/sustains the pool scenario are considered:

Wind Speed: 2 m/s

Stability Class: F

Ambient temperature: 80 °F

3.3 Liquid Methane Release at High Operating Pressure

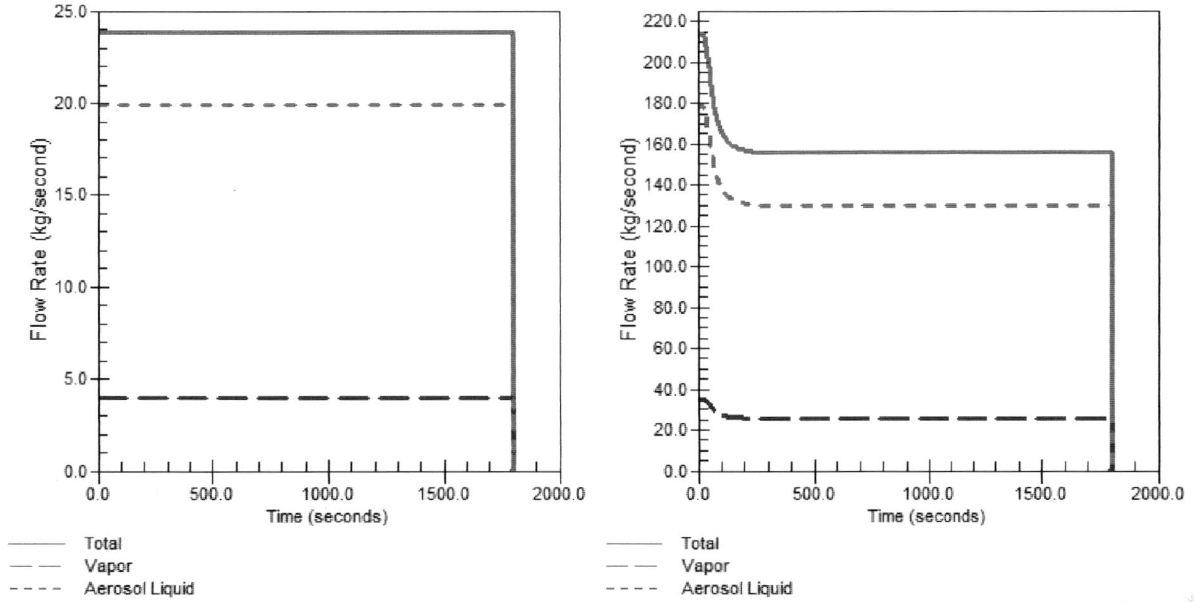

Figure 2: Liquid Methane Release – 2-inch (left figure) & 6-inch (right figure)

Consequence modeling for a liquid methane release was performed at specific process conditions for 2-inch and 6-inch hole size. As shown in the above figures, most of the release is flashing into aerosol liquid and remaining release is in vapor phase. No liquid pool is generated from the liquid methane release.

3.4 Liquid Propane Release

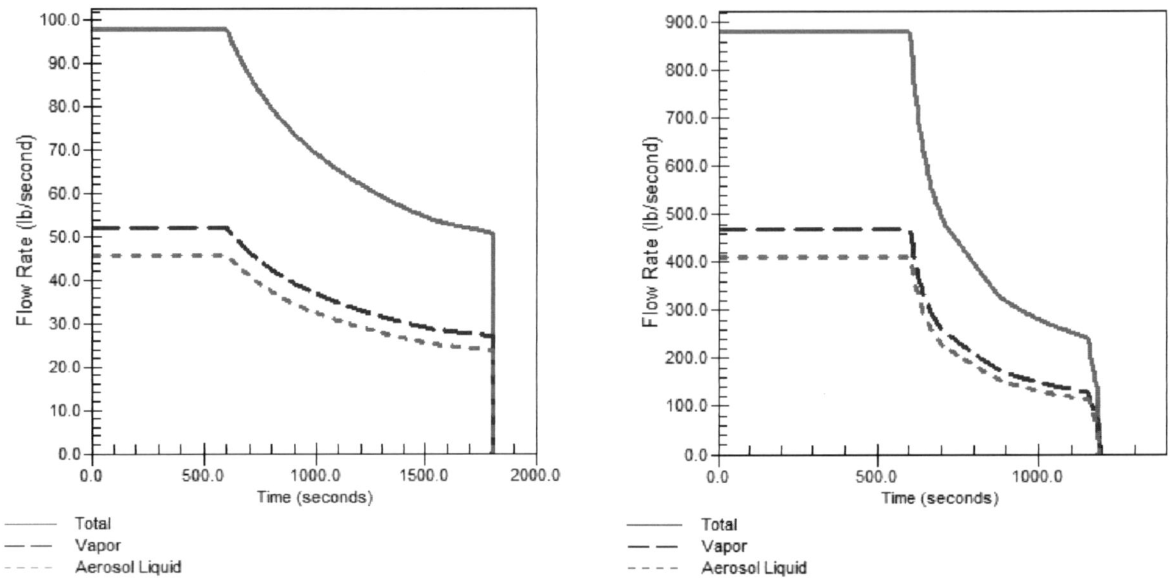

Figure 3: Propane Release – 2-inch (left figure) & 6-inch (right figure)

Consequence modeling for a propane release was performed at specific process conditions for 2-inch and 6-inch hole size. As shown in the above figures, the release is flashing into aerosol liquid and vapor phase. No liquid pool is generated from the propane release.

3.5 Liquid Ethylene Release

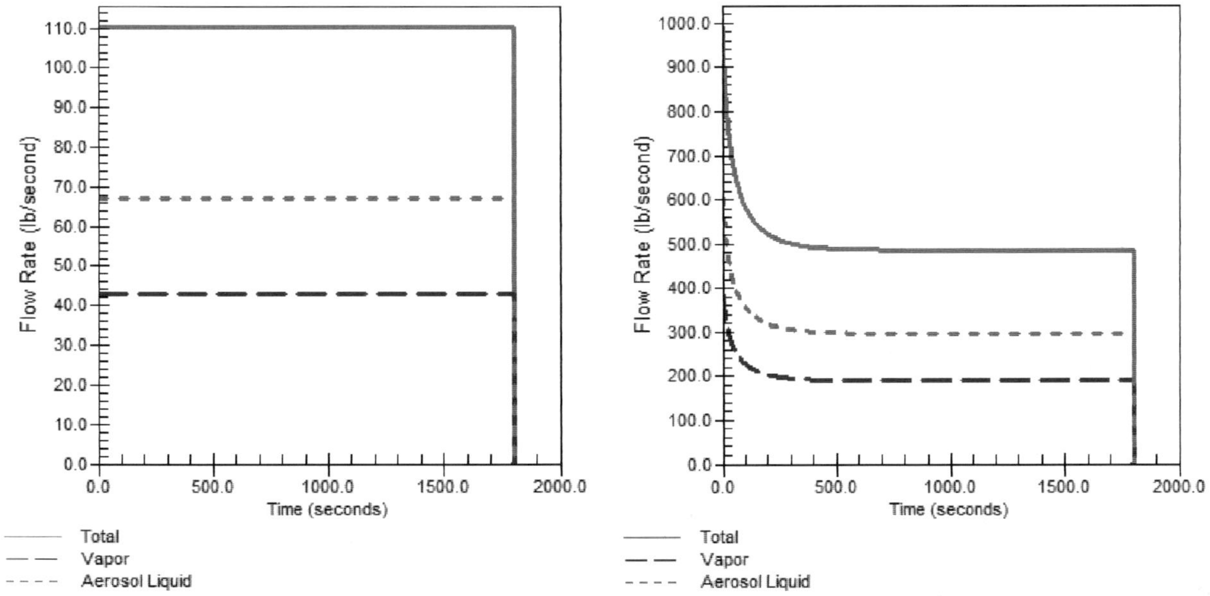

Figure 4: Ethylene Release – 2-inch (left figure) & 6-inch (right figure)

Consequence modeling for ethylene release was performed at specific operating process conditions for 2-inch and 6-inch hole size. As shown in the above figures, the release is flashing into aerosol liquid and vapor phase. No liquid pool is generated from the refrigerant release i.e. Ethylene release.

3.6 LNG Release at Low Operating Pressure condition

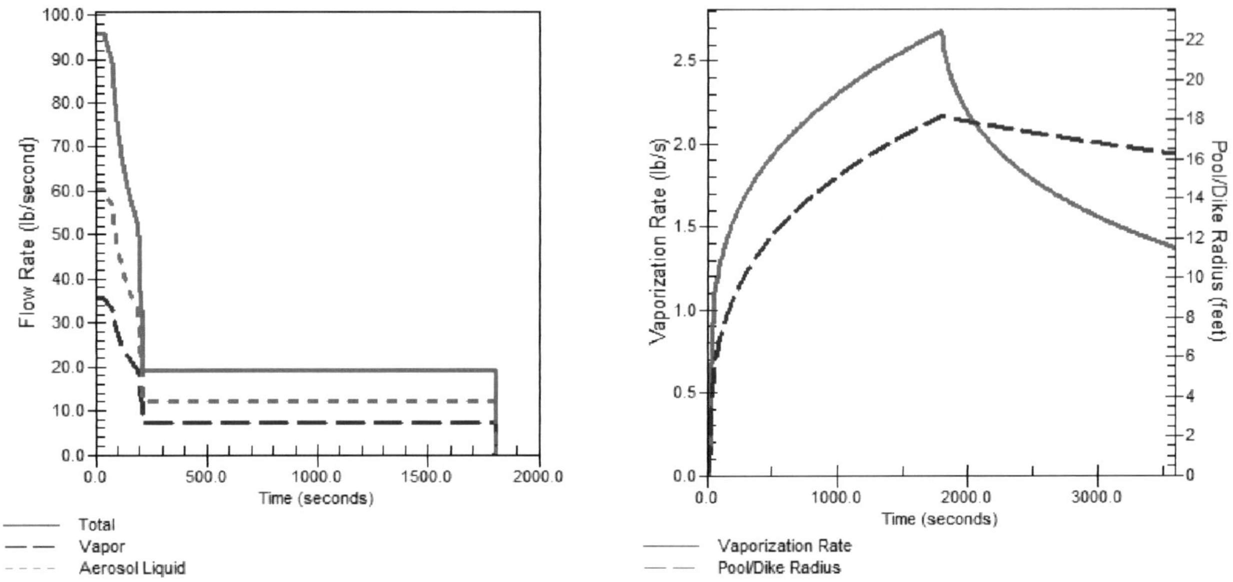

Figure 5: Liquefied Natural Gas Release – 2-inch (left figure) & 6-inch (right figure)

Consequence modeling for a liquefied methane was performed at the process conditions of an equipment for 2-inch. and 6-inch hole size. As shown in the above left side figure, the release for 2-inch hole size is flashing into aerosol liquid and vapor phase, without creating any pool. However, as shown in the right side figure, a release from 6in. hole size from the same equipment is creating a liquid pool of 16 ft. radius.

4 Conclusions

Consequence modeling results showed that out of the total twenty-eight (28) equipment of LNG liquefaction plant that were normally considered as Fire Potential Equipment (FPE) based on specific threshold volume, thirteen (13) equipment that handles methane, propane, ethylene at some specific operating conditions (e.g. low or medium or high pressure) did not create any liquid pool. Instead they flash along forming a large fraction of aerosol liquids, which never creates liquid pool.

The pool fire modeling was performed for a 2-inch release and a sensitivity analysis was carried out for 6-inch release to check the possibility of liquid pool formation for larger size releases for all the 28 equipment. The 6-inch release did not make a major difference in pool fire modeling results except for one equipment which contain LNG at low operating pressure condition. As stated earlier, thirteen (13) out of twenty-eight (28) equipment had not produced a liquid pool irrespective of the release size.

One (1) out of the remaining fifteen (15) equipment, that handles LNG at lower pressure condition did not generate liquid pool at 2-inch release, however it has generated liquid pool for a larger size (6-inch) release. Therefore, it is recommended to perform the consequence modeling for larger hole size to check the liquid pool formation.

The consequence modeling proved that 46% of the total equipment containing liquid such as methane or propane or ethylene or LNG had not produced a liquid pool. Therefore, considering the current economic trend, consequence modeling based approach for determining fireproofing requirements in conjunction with API 2218 requirement can be used in major LNG projects to justify considerable cost savings.

5 References

[1] J.D. Marx and A. Nicotra, "Is a Two-Inch Hole Adequate for Siting Study?", 12th Global Congress on Process Safety, American Institute of Chemical Engineers, Houston, Texas, 2016.

[2] API. "Fireproofing Practices in Petroleum and Petrochemical Processing Plants", American Petroleum Institute, 2013.

[3] CCPS (2003), "Guideline for Fire Protection in Chemical, Petrochemical, and Hydrocarbon Processing Facilities", Center for Chemical Process Safety of the American Institute of Chemical Engineers, 3 Park Avenue, New York, 10016-5991. (ISBN 0-8169-0898-2).

[4] CAN/CSA Z276, "Liquefied natural gas (LNG) - Production, Storage, and Handling", National Standard of Canada, 2015.

479398 Depressurization of Process Equipment during Emergency or Planned Shutdown on a High Potential CO2-Solid-Forming Stream Utilizing the Latest Dynamic Simulation and Blowdown Models

Abstract:

Natural gas produced from many major reservoirs can contain significant amounts of carbon dioxide (CO2) and must be treated to meet typical specifications for pipelines or liquefaction plant feed. These treatment processes often require high-pressure operations forming highly concentrated CO2-rich streams. Pressure protection for these systems has been challenging to date because of the potential for CO2 solids generation upon pressure let down and the consequent potential for plugging.

Blowdown, the emergency or planned depressuring of process equipment, is a critical process safety operation. It may be necessary, in the event of a fire, leak, pipe rupture or other hazardous situation, as well as for a planned shutdown. Devices such as control valves, relief valves, restriction orifices, rupture disks, and safety valves transfer the potentially dangerous contents of process equipment to a safe lower-pressure location, or to the flare system for controlled combustion.

To ensure blowdown can be executed safety and effectively, a number of design concerns must be addressed, such as solid CO2 identification, low temperature (for both process and equipment material). Rapid depressuring and gas expansion can potentially putting equipment at risk of brittle fracture and if the construction material goes below its ductile-brittle transition temperature as well as potential plugging, due to CO2 solid formation. In addition, the entire pressure relief system, including safety valves, relief orifices, flare piping and knockout drums, must be sufficiently sized to handle the flowrates that occur during blowdown, in addition to the piping and capacity of the flare system.

For new installations, accurately predicting the minimum vessel wall temperature during blowdown is important for selecting the appropriate construction material, for helping reduce overdesign and consequently for lowering project cost. Similarly, having an accurate prediction of the maximum flow rate during blowdown reduces overdesign associated with the relief valve/network, without compromising on safety.

The paper will address the potential of solid CO2 formation based on commercially available software for blowdown and propose some mitigation plan with respect to solid CO2 formation.

Inherently Safer Piping – Best Practices

Steve Streblow
ACS Engineering
10914 Milano Ct.
Richmond, TX 77406
streblowsr@yahoo.com

Prepared for Presentation at
American Institute of Chemical Engineers
2017 Spring Meeting and 13th Global Congress on Process Safety
San Antonio, TX
March 26 – 29, 2017

Inherently Safer Piping – Best Practices

Steve Streblow
ACS Engineering
10914 Milano Ct.
Richmond, TX 77406
streblowsr@yahoo.com

Keywords: 49 CFR 192, 49 CFR 195, API 1102, ASME B31, buried, midstream, pipe, piping, pressure, relief, safety, upstream

Abstract

This paper presents lesser known items in the upstream and midstream industries important to onshore metal piping stress and overpressure protection.

Well known RAGAGEP presented are 49 CFR 195, 49 CFR 192 and the ASME piping codes B31.1/3/4/8. Specific items include corrosion allowance, equations, manufacturing tolerance, maximum allowable pressures, overpressure protection devices, stress values and thread depth. Inspection, testing and maintenance of over pressure protection devices is also discussed. The implications of B31.8 references to B31.1 for water and steam and B31.3 for air, lube oil and hydraulic power are discussed. Many times, design pressures are set below the code minimum design pressures, so those are presented for steam, feedwater, blow piping and drains.

Items detailed outside the CFRs and ASME piping code that should be taken into account, such as surface dead loads, surface live loads, and cyclic loads on buried pipe fatigue are also presented. Failure mechanisms, equations, and acceptance criteria for deflection, ovality, through wall bending, crushing, ring buckling, fatigue, longitudinal, thermal expansion and fluid transients are described. Sources of this information are API Recommended Practice 1102, Steel Pipelines Crossing Railroads; and Highways and American Lifelines Alliance, Guidelines for the Design of Buried Steel Pipe.

Using these presented methods results in inherently safer piping and safer pressure relief system design.

1 Introduction

In having worked with and for many upstream and midstream engineering and operating companies, I have seen that the onshore piping design (for both gases and liquids) is often not done in enough detail. I have always seen that the pipe wall thickness is checked for maximum internal operating pressure, but even then, there are some factors that are often times overlooked.

For gas piping, the category (production, gathering, transmission or distribution) is first determined, and then a determination if the piping is regulated or not is made. The main regulations for gas piping, 49 CFR 192, the minimum requirement, is discussed.

The main regulation for liquid piping, 49 CFR 195, the minimum requirement, is discussed.

Then, some best practices are presented, starting with ASME B31.8 for Gas Piping and Stations such as Compression. The best practice for liquid piping, ASME B31.4 is presented. The sections of ASME B31.1 and B31.3 which are best practices for Compressor station piping are presented.

Some best practices for piping, regardless of fluid and location are presented, starting with API 1102 and then finishing with Guidelines for the Design of Buried Steel Pipe.

Using these documents allows for many more factors such as corrosion allowance, manufacturing tolerance, stress values and thread depth to be taken into account.

Surface dead loads, surface live loads, and cyclic loads on buried pipe fatigue are also presented. Failure mechanisms, equations, and acceptance criteria for deflection, ovality, through wall bending, crushing, ring buckling, fatigue, longitudinal, thermal expansion and fluid transients are described.

Using these presented methods results in inherently safer piping and safer pressure relief system design.

2 What is the Gas Piping Category?

For natural gas piping, one of the first steps is to determine if the pipeline is production, gathering, transmission or distribution. Reference 1, section 2.2, is used to determine the split between production and gathering.

The gathering classification can't extend beyond the first natural gas processing plant, when it switches to transmission. Transmission then eventually switches to distribution.

3 Regulated Gas Gathering or Not?

In reference 2, there is a parameter called the class location, which accounts for risk due to buildings and occupancy near the pipeline. The class location ranges from 1, for lowest risk to 4 for highest risk. Details in reference 2, paragraph 192.5.

If the gas piping is gathering, one must then determine if it is regulated, per reference 2, paragraph 192.8. Regulated gathering depends on the hoop stress, class location and safety buffer. Regulated gathering lines have to follow certain aspects of reference 2.

4 CFR 192 for Gas Piping – Part 1

So, one needs to follow reference 2 for regulated gathering and transmission piping.

Per reference 2, paragraph "192.103 General. Pipe must be designed with sufficient wall thickness, or must be installed with adequate protection, to withstand anticipated external pressures and loads that will be imposed on the pipe after installation." However, reference 2 does not provide any formulas or methods to satisfy this requirement. This is one area that I have commonly seen overlooked.

Per reference 2, paragraph "192.105 Design formula for steel pipe. (a) The design pressure for steel pipe is determined in accordance with the following formula:"

P = (2 S x t/D) x F x E x T, where

P =Design pressure in pounds per square inch gage.
S =Yield strength in pounds per square inch determined in accordance with §192.107.
t = Nominal wall thickness of the pipe in inches. Additional wall thickness required for concurrent external loads in accordance with §192.103 may not be included in computing design pressure.
D =Nominal outside diameter of the pipe in inches.
F =Design factor determined in accordance with §192.111.
E =Longitudinal joint factor determined in accordance with §192.113.
T =Temperature derating factor determined in accordance with §192.115.

Per reference 2, paragraph "192.107 Yield strength (S) for steel pipe."…"The yield strength to be used in the design formula in §192.105 is the SMYS stated in the listed specification."

Per reference 2, paragraph "192.111 Design factor (F) for steel pipe.

Table 1. Design factor (F) for steel pipe.

Class location	Design factor (F)
1	0.72
2	0.60
3	0.50
4	0.40

A design factor of 0.60 or less must be used in the design formula in paragraph 192.105 for steel pipe in Class 1 locations that: (1) Crosses the right-of-way of an unimproved public road, without a casing: (2) Crosses without a casing, or makes a parallel encroachment on, the right-of-way of either a hard surfaced road, a highway, a public street, or a railroad: (3) Is supported by a vehicular, pedestrian, railroad, or pipeline bridge: or (4) Is used in a fabricated assembly, (including separators, mainline valve assemblies, cross-connections, and river crossing headers) or is used within five pipe diameters in any direction from the last fitting of a fabricated assembly, other than a transition piece or an elbow used in place of a pipe bend which is not associated with a fabricated assembly. (c) For Class 2 locations, a design factor of 0.50, or less, must be used in the design formula in §192.105 for uncased steel pipe that crosses the right-of-way of a hard surfaced road, a highway, a public street, or a railroad. (d) For Class 1 and Class 2 locations, a de-sign factor of 0.50, or less, must be used in the design formula in §192.105 for– (1) Steel pipe in a compressor station, regulating station, or measuring station." These more specific design factors for specific locations and types of facilities is one area that I have commonly seen overlooked.

Per reference 2, "§192.113 Longitudinal joint factor (E) for steel pipe. The longitudinal joint factor to be used in the design formula in §192.105 is determined in accordance with the following table:"

5 E factor comparison

The E factor table from 49 CFR 192 was combined with the E factor table from 49 CFR 195 and the E factor from ASME B31.8 and is presented immediately below. Comparison of the three tables shows that there are a few differences.

Table 2. Factor (E) for steel pipe.

Specification	Pipe Class	49 CFR 192 Longitudinal Joint Factor (E)	49 CFR 195 Seam Joint Factor (E)	ASME B31.8
ASTM A53/A53M	Seamless	1.00		
	Electric resistance welded	1.00		
	Furnace lap welded	NA	0.80	NA
	Furnace butt welded	0.60		
ASTM A106	Seamless	1.00		
ASTM A134	Electric-fusion arc-welded	NA		0.80
ASTM A135	Electric resistance-welded	NA		1.00
ASTMA139	Electric-fusion arc-welded	NA		0.80
ASTM A333/A333M	Seamless	NA	1.00	1.00
	Electric resistance welded	1.00	NA	1.00
	Welded	NA	1.00	NA
ASTM A381	Double submerged arc welded	1.00		
ASTM A671	Electric-fusion welded Classes 13, 23, 33, 43, 53	1.00		0.80
	Electric-fusion welded Classes 12, 22, 32, 42, 52	1.00		
ASTM A672	Electric-fusion welded Classes 13, 23, 33, 43, 53	1.00		0.80
	Electric-fusion welded Classes 12, 22, 32, 42, 52	1.00		

Table 2. Factor (E) for steel pipe. (continued)

Specification	Pipe Class	49 CFR 192 Longitudinal Joint Factor (E)	49 CFR 195 Seam Joint Factor (E)	ASME B31.8
ASTM A691	Electric-fusion welded Classes 13, 23, 33, 43, 53	1.00		0.80
	Electric-fusion welded Classes 12, 22, 32, 42, 52	1.00		
ASTM A984	Electric-resistance-welded	NA		1.00
ASTM A1005	Double submerged-arc-welded	NA		1.00
ASTM A1006	Laser beam welded	NA		1.00
API 5L	Seamless	1.00		
	Electric resistance welded	1.00		
	Electric flash welded	1.00		
	Submerged arc welded	1.00		
	Furnace lap welded	NA	0.80	NA
	Furnace butt welded	0.60		
Other	Pipe over 4 inches (102 millimeters)	0.80	NA	
Other	Pipe 4 inches (102 millimeters) or less	0.60	NA	

6 CFR 192 for Gas Piping – Part 2

Per reference 2, "192.115 Temperature derating factor (T) for steel pipe. The temperature derating factor to be used in the design formula in §192.105 is determined as follows:

Table 3. Temperature derating factor (T) for steel pipe.

Gas Temperature in degrees Fahrenheit	Temperature derating factor (T)
250 or less	1.000
300	0.967
350	0.933
400	0.900
450	0.867

For intermediate gas temperatures, the derating factor is determined by interpolation."

Per reference 2, "§192.149 Standard fittings. (a) The minimum metal thickness of threaded fittings may not be less than specified for the pressures and temperatures in the applicable standards referenced in this part, or their equivalent." However, reference 2 does not provide any formulas or methods to satisfy this requirement. This is one area that I have commonly seen overlooked.

Per reference 2, "§192.159 Flexibility. Each pipeline must be designed with enough flexibility to prevent thermal expansion or contraction from causing excessive stresses in the pipe or components, excessive bending or unusual loads at joints, or undesirable forces or moments at points of connection to equipment, or at anchorage or guide points." However, reference 2 does not provide any formulas or methods to satisfy this requirement. This is one area that I have commonly seen overlooked.

Per reference 2, "§192.161 Supports and anchors.

(a) Each pipeline and its associated equipment must have enough anchors or supports to: (1) Prevent undue strain on connected equipment; (2) Resist longitudinal forces caused by a bend or offset in the pipe; and, (3) Prevent or damp out excessive vibration.

(b) Each exposed pipeline must have enough supports or anchors to protect the exposed pipe joints from the maximum end force caused by internal pressure and any additional forces caused by temperature expansion or contraction or by the weight of the pipe and its contents.

(c) Each support or anchor on an exposed pipeline must be made of durable, noncombustible material and must be designed and installed as follows: (1) Free expansion and contraction of the pipeline between supports or anchors may not be restricted. (2) Provision must be made for the service conditions involved." However, reference 2 does not provide any formulas or methods to satisfy this requirement. This is one area that I have commonly seen overlooked.

It is important to note that reference 2 describes minimum standards, not best practices. Read on to learn about best practices.

7 CFR 195 for Liquid Piping – Part 1

Hazardous liquids transportation pipelines are often required to follow reference 3 for several aspects, including design.

From reference 3, "§195.1 Which pipelines are covered by this part?" "(2) Transportation through any pipeline, other than a gathering line, that has a maximum operating pressure (MOP) greater than 20-percent of the specified minimum yield strength;" "(4) Transportation of petroleum in any of the following onshore gathering lines: (i) A pipeline located in a non-rural area; (ii) To the extent provided in §195.11, a regulated rural gathering line defined in §195.11."

From reference 3, "(b) Excepted. This part does not apply to any of the following:" "(8) Transportation of a hazardous liquid or carbon dioxide through onshore production (including flow lines), refining, or manufacturing facilities or storage or inplant piping systems associated with such facilities;"

From reference 3, "§195.11 What is a regulated rural gathering line and what requirements apply?...(a) Definition. As used in this section, a regulated rural gathering line means an onshore gathering line in a rural area that meets all of the following criteria—(1) Has a nominal diameter from 6⅝ inches (168 mm) to 8⅝ inches (219.1 mm);...(3) Operates at a maximum pressure established under §195.406 corresponding to—(i) A stress level greater than 20-percent of the specified minimum yield strength of the line pipe"

From reference 3, "§195.106 Internal design pressure. (a) Internal design pressure for the pipe in a pipeline is determined in accordance with the following formula:
P = (2 S x t/D) x E x F", where

P = Internal design pressure in p.s.i. gage.
S = Yield strength in pounds per square inch determined in accordance with paragraph (b) of this section.
t = Nominal wall thickness of the pipe in inches.
D = Nominal outside diameter of the pipe in inches.
E = Seam joint factor determined in accordance with paragraph (e) of this section.
F = A design factor of 0.72

8 Comparing Wall Thickness for Gas and Liquid Piping

Comparing the internal design pressure formulas of 49 CFR 192 and 195 shows that the formula are slightly different. 49 CFR 192 design factors range from 0.72 to 0.4, while 49 CFR 195 design factor is set at 0.72. Therefore it can be seen that for a given internal design pressure, with the same piping, the gas piping oftentimes needs to have a thicker wall. 49 CFR 192 temperature derating factor ranges from 1.000 to 0.867. while 49 CFR 195 does not have a temperature derating factor, therefore, for a given operating temperature with the same piping, the gas piping needs to have a thicker wall.

9 CFR 195 for Liquid Piping – Part 2

From reference 3, "(b) The yield strength to be used in determining the internal design pressure under paragraph (a) of this section is the specified minimum yield strength."

From reference 3, "(d) The minimum wall thickness of the pipe may not be less than 87.5 percent of the value used for nominal wall thickness in determining the internal design pressure under paragraph (a) of this section. In addition, the anticipated external loads and external pressures that are concurrent with internal pressure must be considered in accordance with §§195.108 and 195.110 and, after determining the internal design pressure, the nominal wall thickness must be increased as necessary to compensate for these concurrent loads and pressures." However, reference 3 does not provide any formulas or methods to satisfy this requirement. This is one area that I have commonly seen overlooked.

From reference 3, "(e) The seam joint factor used in paragraph (a) of this section is determined in accordance with the following table:" See Table 2 above.

Per reference 3, "§195.108 External pressure. Any external pressure that will be exerted on the pipe must be provided for in designing a pipeline system." However, reference 3 does not provide any formulas or methods to satisfy this requirement. This is one area that I have commonly seen overlooked.

Per reference 3, "§195.110 External loads. (a) Anticipated external loads (e.g.), earthquakes, vibration, thermal expansion, and contraction must be provided for in designing a pipeline system. In providing for expansion and flexibility, §419 of ASME/ANSI B31.4 must be followed. (b) The pipe and other components must be supported in such a way that the support does not cause excess localized stresses. In designing attachments to pipe, the added stress to the wall of the pipe must be computed and compensated for." However, reference 3 does not provide any formulas or methods to satisfy this requirement. I This is one area that I have commonly seen overlooked.

It is important to note that reference 3 describes minimum standards, not best practices. Read on to learn about best practices.

10 Best Practice For Gas Piping and Stations such as Compression – ASME B31.8

What if the gas pipeline or facility, such as production or non-regulated gathering, is not subject to reference 2? How does one account for factors and loads not detailed in CFR 192? One can then follow reference 4, which has the following applicability "piping transporting products that are predominately gas between sources and terminals, including compressor, regulating, and metering stations; gas gathering pipelines."

From reference 4, paragraph "802.1 Scope…(b) This Code does not apply to…(4) piping in oil refineries or natural gasoline extraction plants, gas treating plant piping other than the main gas stream piping in dehydration, and all other processing plants installed as part of a gas transmission system, gas manufacturing plants, industrial plants, or mines (See other applicable sections of the ASME Code for Pressure Piping, B31.) (5) vent piping to operate at substantially atmospheric pressures for waste gases of any kind."

From reference 4, paragraph "833.2 Calculation of Longitudinal Stress Components
(a) The longitudinal stress due to internal pressure in restrained pipelines is
$S_P = 0.3S_H$, where S_H is the hoop stress, psi
(b) The longitudinal stress due to internal pressure in unrestrained pipeline is
$S_P = 0.5S_H$, where S_H is the hoop stress, psi
(c) The longitudinal stress due to thermal expansion in restrained pipe is $S_T = E\ a\ (T_1 - T_2)$, where
E = the elastic modulus, psi, at the ambient temperature, which can be found in reference 4, Table 832.5-1.
T_1 = the pipe temperature at the time of installation, tie-in, or burial, °F
T_2 = the warmest or coldest pipe operating temperature, °F
a = the coefficient of thermal expansion, 1/°F, which can be found in reference 4, Table 832.1-1
If a section of pipe can operate either warmer or colder than the installed temperature, both conditions for T_2 may need to be examined." This is one area that I have commonly seen overlooked.

From reference 4, paragraph "833.3 Summation of Longitudinal Stress in Restrained Pipe
(a) The net longitudinal stresses in restrained pipe are $S_L = S_P + S_T + S_X + S_B$
Note that S_B, S_L, S_T, or S_X can have negative values.
(b) The maximum permitted value of $|S_L|$ is 0.9ST, where S is the specified minimum yield strength, psi, per para. 841.1.1(a), and T is the temperature derating factor per para. 841.1.8." This is one area that I have commonly seen overlooked.

From reference 4, paragraph "833.4 Combined Stress for Restrained Pipe
(a) The combined biaxial stress state of the pipeline in the operating mode is evaluated using the calculation in either (1) or (2) below. (1) max. $\{|S_H - S_L|, |S_H|, |S_L|\}$ or (2) $[S_L^2 - S_LS_H + S_H^2]^{1/2}$
The maximum permitted value for the combined biaxial stress is kST where S is the specified minimum yield strength, psi, per para. 841.1.1(a), T is the temperature derating factor per para. 841.1.8, and k is defined in (b) and (c) below. (b) For loads of long duration, the value of k shall not exceed 0.90. (c) For occasional nonperiodic loads of short duration, the value of k shall not exceed 1.0. (d) S_L in (a) above is calculated considering both the tensile and compressive values of S_B. (e) Stresses induced by loads that do not occur simultaneously need not be considered to be additive. (f) The biaxial stress evaluation described above applies only to straight sections of pipe." This is one area that I have commonly seen overlooked.

From reference 4, paragraph "833.6 Summation of Longitudinal Stresses in Unrestrained Pipe
(a) The net longitudinal stress in unrestrained pipe is $S_L = S_P + S_X + S_B$, psi (b) The maximum permitted longitudinal stress in unrestrained pipe is $S_L \leq 0.75ST$, where S is the specified minimum yield strength, psi, per para. 841.1.1(a), and T is the temperature derating factor per para. 841.1.8." This is one area that I have commonly seen overlooked.

From reference 4, paragraph "833.8 Flexibility Stresses and Stresses Due to Periodic or Cyclic Fatigue Loading (b) The cyclic stress range $S_E <= S_A$, where
$S_A = f\,[1.25\,(S_c + S_h) - S_L]$, where
$f = 6N^{-0.2} \leq 1.0$
N = equivalent number of cycles during the expected service life of the piping system
$S_c = 0.33\ S_uT$ at the minimum installed or operating temperature
$S_h = 0.33\ S_uT$ at the maximum installed or operating temperature
S_L = longitudinal stress calculated according to para. 833.6(a), psi
S_u = specified minimum ultimate tensile strength, psi
T = temperature derating factor per para. 841.1.8"
This is one area that I have commonly seen overlooked.

Per reference 4, paragraph "833.9 Local Stresses (b) The maximum allowable sum of circumferential stress due to internal pressure and circumferential through-wall bending stress caused by surface vehicle loads or other local loads is 0.9ST, where S is the specified minimum yield strength, psi, per para. 841.1.1(a), and T is the temperature derating factor per para. 841.1.8." This is one area that I have commonly seen overlooked.
Per reference 4, paragraph "833.10 Buckling and Lateral Instability. In order to prevent buckling in the form of wrinkling of the pipe wall or lateral instability, the maximum allowable net compressive stress is 2⁄3 of the critical buckling stress estimated using a suitable stability criterion." This is one area that I have commonly seen overlooked.

Per reference 4, see "Table 841.1.6-2 Design Factors for Steel Pipe Construction". Consider these in addition to those from reference 2.

From reference 4, paragraph "841.1.9 Additional Design Information or Instructions…
(g) Pressure/Flow Control Facilities…
(1) Overpressure protection shall be provided by the use of (-a) a monitor regulator in series with a controlling regulator (each regulator run). (-b) adequately sized relief valve(s) downstream of the controlling regulator(s). (-c) overpressure shutoff valve(s) upstream or downstream of the controlling regulator(s). Installation of alarm devices that indicate primary (controlling) regulator failure is useful and should be considered for monitoring regulator systems…
(k) Other (Nonmandatory) Considerations for Pressure Cycle Fatigue of Longitudinal Pipe Seams. Line pipe longitudinal seams are not generally regarded as susceptible to fatigue due to operational pressure cycles in most natural gas service. Evaluation of the potential for fatigue crack growth due to pressure cycles is recommended for pipe containing longitudinal seams where the expected lifetime accumulation of full MAOP cycles may exceed the following number of occurrences: $N = 6.0\text{x}10^{17} / ((F \text{ x } S)^3 \text{ x } t^{0.5})$ where
F = design factor from Table 841.1.6-1
N = equivalent number of cycles during the expected service life of the piping system
S = specified minimum yield strength, psi
t = nominal wall thickness, in."
This is one area that I have commonly seen overlooked.

From reference 4, paragraph "843.3.4 Pressure-Limiting Requirements in Compressor Stations

(a) Pressure Relief. Pressure relief or other suitable protective devices of sufficient capacity and sensitivity shall be installed and maintained to ensure that the maximum allowable operating pressure of the station piping and equipment is not exceeded by more than 10%.
(b) Pressure Relief Valves. A pressure relief valve or pressure-limiting device, such as a pressure switch or unloading device, shall be installed in the discharge line of each positive displacement transmission compressor between the gas compressor and the first discharge block valve. If a pressure relief valve is the primary overprotection device, then the relieving capacity shall be equal to or greater than the capacity of the compressor.
(c) Venting. Vent lines provided to exhaust the gas from the pressure relief valves to atmosphere shall be extended to a location where the gas may be discharged without undue hazard. Vent lines shall have sufficient capacity so that they will not inhibit the performance of the relief valve. For additional design considerations, see para. 841.1.9(d)." This is one area that I have commonly seen overlooked.

From reference 4, paragraph "843.4.3 Air Piping System. (a) All air piping within gas compressing stations shall be constructed in accordance with ASME B31.3". This is one area that I have commonly seen overlooked.

From reference 4, paragraph "843.4.4 Lubricating Oil Piping. All lubricating oil piping within gas compressing stations shall be constructed in accordance with ASME B31.3." This is one area that I have commonly seen overlooked.

From reference 4, paragraph "843.4.5 Water Piping. All water piping within gas compressing stations shall be constructed in accordance with ASME B31.1." This is one area that I have commonly seen overlooked.

From reference 4, paragraph "843.4.6 Steam Piping. All steam piping within gas compressing stations shall be constructed in accordance with ASME B31.1." This is one area that I have commonly seen overlooked.

From reference 4, paragraph "843.4.7 Hydraulic Piping. All hydraulic power piping within gas compressing stations shall be constructed in accordance with ASME B31.3." This is one area that I have commonly seen overlooked.

From reference 4, paragraph "844.5 General Provisions Applicable to Both Pipe-Type and Bottle-Type Holders...Relief valves shall be installed in accordance with provisions of this Code that will have relieving capacity adequate to limit the pressure imposed on the filling line and thereby on the storage holder to 100% of the design pressure of the holder or to the pressure that produces a hoop stress of 75% of the specified minimum yield strength of the steel, whichever is the lesser." This is one area that I have commonly seen overlooked.

Paragraph "845.3 Requirements for Design of Pressure Relief and Pressure-Limiting Installations. (c) The discharge stacks, vents, or outlet ports of all pressure relief devices shall be located where gas can be discharged into the atmosphere without undue hazard." This is one area that I have commonly seen overlooked.

From reference 4, paragraph "845.4.1 Required Capacity of Pressure-Relieving and Pressure-Limiting Stations…(a) Each pressure relief station, pressure-limiting station, or group of such stations installed to protect a piping system or pressure vessel shall have sufficient capacity and shall be set to operate to prevent the pressure from exceeding the following levels: (1) Systems With Pipe or Pipeline Components Operating at Hoop Stress Levels Over 72% of the SMYS. The required capacity is the maximum allowable operating pressure plus 4%. (2) Systems With Pipe or Pipeline Components Operating at Hoop Stress Levels at or Below 72% of the SMYS Other Than in Low-Pressure Distribution Systems. The required capacity is the lesser of the following two items: (-a) the maximum allowable operating pressure plus 10% (-b) the pressure that produces a hoop stress of 75% of the specified minimum yield strength." This is one area that I have commonly seen overlooked.

From reference 4, paragraph "845.4.2 Proof of Adequate Capacity and Satisfactory Performance of Pressure-Limiting and Pressure Relief Devices. Where the safety device consists of an additional regulator that is associated with or functions in combination with one or more regulators in a series arrangement to control or limit the pressure in a piping system, suitable checks shall be made. These checks shall be conducted to determine that the equipment will operate in a satisfactory manner to prevent any pressure in excess of the established maximum allowable operating pressure of the system, should any one of the associated regulators malfunction or remain in the wide open position." This is one area that I have commonly seen overlooked.

From reference 4, paragraph "849.2 Steel Service Lines. (b) Steel service pipe shall be designed in accordance with the requirements of paras. 841.1.1 and 841.1.9(a). Where the pressure is less than 100 psig, the steel service pipe shall be designed for at least 100 psig pressure." This is one area that I have commonly seen overlooked.

From reference 4, paragraph "853.1 Compressor Station Maintenance…853.1.2 Inspection and Testing of Relief Valves. All pressure-relieving devices in compressor stations shall be inspected and/or tested in accordance with para. 853.3, and all devices except rupture disks shall be operated periodically to determine that they open at the correct set pressure. Any defective or inadequate equipment found shall be promptly repaired or replaced. All remote control shutdown devices shall be inspected and tested at least annually to determine that they function properly." This is one area that I have commonly seen overlooked.

From reference 4, paragraph "853.3 Maintenance of Pressure-Limiting and Pressure-Regulating Stations…853.3.1 Condition and Adequacy. All pressure limiting stations, relief devices, and other pressure regulating stations and equipment shall be subject to systematic, periodic inspections and suitable tests, or reviewed to determine that they are…(2) At least once each calendar year, a review shall be made to ensure that the combined capacity of the relief devices on a piping system or facility is adequate to limit the pressure at all times to values prescribed by this Code. This review should be based on the operating conditions that create the maximum probable requirement for relief capacity in each case, even though such operating conditions actually occur infrequently and/or for only short periods of time. If it is determined that the relieving equipment is of insufficient capacity, steps shall be taken to install new or additional equipment to provide adequate capacity." This is one area that I have commonly seen overlooked.

From reference 4, paragraph "853.3.3 Stop Valves (a) An inspection and/or test of stop valves shall be made to determine that the valves will operate and are correctly positioned. (Caution shall be used to avoid any undesirable effect on pressure during operational checks.) The following shall be included in the inspection and/or test: (1) station inlet, outlet, and bypass valves (2) relief device isolating valves (3) control, sensing, and supply line valves (b) The final inspection procedure shall include the following: (1) a check for proper position of all valves. Special attention shall be given to regulator station bypass valves, relief device isolating valves, and valves in control, sensing, and supply lines. (2) restoration of all locking and security devices to proper position." This is one area that I have commonly seen overlooked.

From Reference 4, see "MANDATORY APPENDIX D, SPECIFIED MINIMUM YIELD STRENGTH FOR STEEL PIPE, COMMONLY USED IN PIPING SYSTEMS"

11 Best Practice for Liquid Piping – ASME B31.4

What if the liquid pipeline or facility, such as production or non-regulated gathering, is not subject to reference 3? How does one account for factors and loads not detailed in CFR 195? One can then follow reference 5, which has the following applicability "piping transporting products that are predominately liquid between plants and terminals and within terminals, pumping, regulating, and metering stations."

Reference 5 shows that various categories of loads need to be considered, such as paragraph "401.1.2 Sustained Loads…401.1.3 Occasional Loads…401.1.4 Construction Loads and 401.1.5 Transient Loads." However, reference 5 does not provide any formulas or methods to satisfy this requirement. This is one area that I have commonly seen overlooked.

Reference 5 shows some subcategories of loads needing consideration, such as paragraph "401.2.2.2 Internal Design Pressure…401.2.2.3 External Hydrostatic Pressure…401.2.2.4 Weight Effects…401.2.2.5 Residual Loads…401.2.2.6 Subsidence…401.2.3.1 Earthquakes, 401.2.3.2 Wind Loads…401.2.3.3 Ice Loads…401.2.3.4 Road and Rail Traffic…401.2.3.5 Vibration…401.2.3.6 Waves and Currents…401.2.3.7 Temperature Effects…401.2.4.1 Installation Loads and 401.2.4.2 Hydrostatic Testing." However, reference 5 does not provide any formulas or methods to satisfy this requirement. This is one area that I have commonly seen overlooked.

From reference 5, paragraph "401.3 Combining of Loads" needs to be considered. This is one area that I have commonly seen overlooked.

From reference 5, paragraph "402 CALCULATION OF STRESSES…402.1 General…Circumferential, longitudinal, shear, and equivalent stresses shall be considered, taking into account stresses from all relevant sustained, occasional, construction, and transient loads, including vibration, resonance, and subsidence." This is one area that I have commonly seen overlooked.

From reference 5, paragraph "403.2 Criteria for Pipe Wall Thickness and Allowances…403.2.1 Criteria. The nominal wall thickness of straight sections of steel pipe shall be equal to or greater than t_n determined in accordance with the following equation: $t_n \geq t + A$, where
A = sum of allowances for threading, grooving, corrosion, and erosion as required in paras. 403.2.2 through 403.2.4, and increase in wall thickness if used as protective measure in para. 403.1
t_n = nominal wall thickness satisfying requirements for pressure and allowances
t = pressure design wall thickness"

This is one area that I have commonly seen overlooked.

In reference 5, see "Table 403.3.1-1 Allowable Values for Pipeline System Stresses." This is one area that I have commonly seen overlooked.

From reference 5, Paragraph "452.2 Controls and Protective Equipment (a) Controls and protective equipment, including pressure-limiting devices, regulators, controllers, relief valves, and other safety devices, shall be subjected to systematic periodic inspections and tests, at least annually, except as provided in para. 452.2(b), to determine that they are (1) in good mechanical condition (2) adequate from the standpoint of capacity and reliability of operation for the service in which they are employed (3) set to function at the correct pressure (4) properly installed and protected from foreign materials or other conditions that might prevent proper operation (b) Relief valves on pressure storage vessels containing LPG or liquid anhydrous ammonia shall be subjected to tests at least every 5 yr." This is one area that I have commonly seen overlooked.

12 Best Practice for Compressor Station Piping- ASME B31.1

From reference 4, paragraph "843.4.5 Water Piping. All water piping within gas compressing stations shall be constructed in accordance with ASME B31.1." This is one area that I have commonly seen overlooked.

From reference 4, paragraph "843.4.6 Steam Piping. All steam piping within gas compressing stations shall be constructed in accordance with ASME B31.1." This is one area that I have commonly seen overlooked.

So, to follow the best practice for compressor station water liquid and vapor piping, refer to reference 7, which has the following applicability "piping typically found in electric power generating stations, in industrial and institutional plants, geothermal heating systems, and central and district heating and cooling systems."

From reference 7, paragraph, "102.4.2 Threading and Grooving. The calculated minimum thickness of piping (or tubing) which is to be threaded shall be increased by an allowance equal to thread depth; dimension h of ASME B1.20.1 or equivalent shall apply. For machined surfaces or grooves, where the tolerance is not specified, the tolerance shall be assumed to be 1⁄64 in. (0.40 mm) in addition to the specified depth of cut. The requirements of para. 104.1.2(C) shall also apply." This is one area that I have commonly seen overlooked.

From reference 7, see "Table 102.4.3 Weld Joint Efficiency Factors"

From reference 7, paragraph "104.1.2 Straight Pipe Under Internal Pressure, Design pressure shall not exceed $P = 2SE(t_m - A) / (D_o - 2y(t_m - A))$, where
P = internal design pressure, psig
SE = maximum allowable stress in material due to internal pressure and joint efficiency at the design temperature, psi. The value of SE shall not exceed that given in Appendix A, for the respective material and design temperature. These values include the weld joint efficiency E.
tm = minimum required wall thickness, in. If pipe is ordered by its nominal wall thickness, the manufacturing tolerance on wall thickness must be taken into account. After the minimum pipe wall thickness tm is determined, this minimum thickness shall be increased by an amount sufficient to provide the manufacturing tolerance allowed in the applicable pipe specification or required by the process. The next heavier commercial wall thickness shall then be selected from thickness schedules such as contained in ASME B36.10M or from manufacturers' schedules for other than standard thickness. (A.1.4) Where ends are subject to forming or machining for jointing, the wall thickness of the pipe, tube, or component after such forming or machining shall not be less than tm minus the amount provided for removal by para. 104.1.2 (A.6.1).
A = additional thickness, in. to compensate for material removed in threading, grooving, etc., required to make a mechanical joint.
Do = outside diameter of pipe, in. For design calculations, the outside diameter of pipe as given in tables of standards and specifications shall be used in obtaining the value of tm. When calculating the allowable working pressure of pipe on hand or in stock, the actual measured outside diameter and actual measured minimum wall thickness at the thinner end of the pipe may be used to calculate this pressure.
y = coefficient having values as given in Table 104.1.2(A)". This is a temperature derating factor.

From reference 7, "(C.1) Where steel pipe is threaded and used for steam service at pressure above 250 psi or for water service above 100 psi with water temperature above 220°F, the pipe shall be seamless having the minimum ultimate tensile strength of 48,000 psi and a weight at least equal to Schedule 80 of ASME B36.10M." This is one area that I have commonly seen overlooked.

From reference 7, paragraph "104.1.3 Straight Pipe Under External Pressure. For determining wall thickness and stiffening requirements for straight pipe under external pressure, the procedures outlined in UG-28, UG-29, and UG-30 of Section VIII, Division 1 of the ASME Boiler and Pressure Vessel Code shall be followed." This is one area that I have commonly seen overlooked.

From reference 7, see paragraph "104.8.1 Stress Due to Sustained Loads formula." This is one area that I have commonly seen overlooked.

From reference 7, see paragraph "104.8.2 Stress Due to Occasional Loads formula." This is one area that I have commonly seen overlooked.

From reference 7, see paragraph "104.8.3 Thermal Expansion Stress Range." This is one area that I have commonly seen overlooked.

From reference 7, paragraph "122.1.2 Steam Piping. The design pressure shall not be taken at less than 100 psig for any condition of service or material." This is one area that I have commonly seen overlooked.

From reference 7, paragraph "122.1.3 Feedwater Piping. For piping from the boiler to and including the required stop valve and the check valve, the minimum value of P except as permitted in para. 122.1.3(A.4) shall exceed the maximum allowable working pressure of the boiler by either 25% or 225 psi, whichever is the lesser. (A.3) The value of P in the formula shall not be taken at less than 100 psig for any condition of service or material, and shall never be less than the pressure required to feed the boiler." This is one area that I have commonly seen overlooked.

From reference 7, paragraph "122.1.4 Blowoff and Blowdown Piping. (A.1) The value of P to be used in the formulas in para. 104 shall exceed the maximum allowable working pressure of the boiler by either 25% or 225 psi whichever is less, but shall be not less than 100 psig." This is one area that I have commonly seen overlooked.

From reference 7, paragraph "122.1.5 Boiler Drains. (D) Drain piping from the drain connection, including the required valve(s) or the blanked flange connection, shall be designed for the temperature and pressure of the drain connection. The remaining piping shall be designed for the expected maximum temperature and pressure. Static head and possible choked flow conditions shall be considered. In no case shall the design pressure and temperature be less than 100 psig and 220°F, respectively." This is one area that I have commonly seen overlooked.

From reference 7, see "NONMANDATORY APPENDIX II - RULES FOR THE DESIGN OF SAFETY VALVE INSTALLATIONS….II-2.3 Reaction Forces From Valve Discharge" and "II-3.0 BENDING MOMENT COMPUTATIONS". These are two areas that I have commonly seen overlooked.

From reference 7, see "NONMANDATORY APPENDIX VII - PROCEDURES FOR THE DESIGN OF RESTRAINED UNDERGROUND PIPING." This is one area that I have commonly seen overlooked.

13 Best Practice for Compressor Station Piping- ASME B31.3

From reference 4, paragraph "843.4.3 Air Piping System. (a) All air piping within gas compressing stations shall be constructed in accordance with ASME B31.3". This is one area that I have commonly seen overlooked.

From reference 4, paragraph "843.4.4 Lubricating Oil Piping. All lubricating oil piping within gas compressing stations shall be constructed in accordance with ASME B31.3." This is one area that I have commonly seen overlooked.

From reference 4, paragraph "843.4.7 Hydraulic Piping. All hydraulic power piping within gas compressing stations shall be constructed in accordance with ASME B31.3." This is one area that I have commonly seen overlooked.

So, to follow the best practice for compressor station air, lube oil and hydraulic piping, refer to reference 6, which has the following applicability "piping typically found in petroleum refineries; chemical, pharmaceutical, textile, paper, semiconductor, and cryogenic plants; and related processing plants and terminals."

From reference 6, see "NONMANDATORY APPENDIX II" for piping material information.

From reference 6, see "Table 302.3.4 Weld Quality Factor."

From reference 6, paragraph "304.1.1 General (a) The required thickness of straight sections of pipe shall be determined in accordance with tm = t + c, where

tm = minimum required thickness, including mechanical, corrosion, and erosion allowances
t = pressure design thickness, as calculated in accordance with para. 304.1.2 for internal pressure or as determined in accordance with para. 304.1.3 for external pressure
c = sum of the mechanical allowances (thread or groove depth) plus corrosion and erosion allowances. For threaded components, the nominal thread depth (dimension h of ASME B1.20.1, or equivalent) shall apply. For machined surfaces or grooves where the tolerance is not specified, the tolerance shall be assumed to be 0.5 mm (0.02 in.) in addition to the specified depth of the cut."

From reference 6, paragraph "304.1.2 Straight Pipe Under Internal Pressure
(a) For $t < D/6$, the internal pressure design thickness for straight pipe shall be not less than that calculated in accordance with eq. $t = PD / (2(SEW + PY))$ where
P = internal design gage pressure
D = outside diameter of pipe
S = stress value for material from Table A-1
E = quality factor from Table A-1A or A-1B
W = weld joint strength reduction factor in accordance with para. 302.3.5(e)
Y = coefficient from Table 304.1.1, valid for $t < D/6$ and for materials shown. The value of Y may be interpolated for intermediate temperatures. For $t \geq D/6$, $Y = (d + 2c) / (D + d + 2c)$ where d = inside diameter of pipe. For pressure design calculation, the inside diameter of the pipe is

the maximum value allowable under the purchase specification.
(b) For $t \geq D/6$ or for $P/SE > 0.385$, calculation of pressure design thickness for straight pipe requires special consideration of factors such as theory of failure, effects of fatigue, and thermal stress."

From reference 6, See "Table 314.2.1 Minimum Thickness of External Threaded Components." This is one area that I have commonly seen overlooked.

14 Best Practice for Piping – API 1102

API 1102, reference 9, is RAGAGEP that I have commonly seen overlooked.

From reference 9, "Table 2" is a good flowchart of the design process considering loads, stresses and fatigue.

From reference 9, paragraph "4.7.2.1 Stresses Due to Earth Load" has a calculation formula (1). Use Figures 3, 4 and 5 to get values to use in formula (1).

From reference 9, paragraph "4.7.2.2.1 Surface Live Loads", has guidance on rail road and highway road loads. Use equation (2) to determine the applied surface design pressure.

From reference 9, paragraph "4.7.2.2.2 Impact Factor" has guidance on impact factor to be used with live loads.

From reference 9, paragraph "4.7.2.2.3.1" has a has a calculation formula (3) for cyclic circumferential stress due to rail load. Use Figures 7, 8 and 9 to get the values to use in formula (3).

Reference 9, paragraph "4.7.2.2.3.2" has a has a calculation formula (4) for cyclic longitudinal stress for rail load. Use Figures 10, 11, 12 and 13 to get the values to use in formula (4).

Reference 9, paragraph "4.7.2.2.4.1" has a has a calculation formula (5) for cyclic circumferential stress for highway load. Use Table 2, Figures 14 and 15 to get the values to use in formula (5).

Reference 9, paragraph "4.7.2.2.4.2" has a has a calculation formula (6) for longitudinal circumferential stress for highway load. Use Figures 16 and 17 to get the values to use in formula (6).

Reference 9, paragraph "4.7.3" has a has a calculation formula (7) for stress from internal load.

Reference 9, paragraph "4.8.1.1" has a has a calculation formula (8a) to compare the circumferential stress due to internal gas pressure to the factored specified minimum yield strength. It also has a has a calculation formula (8b) to compare the circumferential stress due to internal liquid pressure to the factored specified minimum yield strength.

Reference 9, paragraph "4.8.1.2" has a has a calculation formula (9) to determine the maximum circumferential stress. It also has a calculation formula (10) to determine the maximum longitudinal stress. It also has a calculation formula (11) to determine the maximum radial stress. It also has a calculation formula (12) to determine the total effective stress. It also has a has a calculation formula (13) to compare the total effective stress to the factored specified minimum yield strength.

Reference 9, paragraph "4.8.2.1" has a has a calculation formula (14) to compare design to the girth weld fatigue. Table 3 is used with formula 14.

Reference 9, paragraph "4.8.2.1.1.1" has a has a calculation formula (15) to compare design to the girth weld fatigue for a railroad crossing for distance less than 5 feet.

Reference 9, paragraph "4.8.2.1.1.2" has a has a calculation formula (16) to compare design to the girth weld fatigue for a railroad crossing for distance 5 feet and greater. Figures 18A and B are used with formula 16.

Reference 9, paragraph "4.8.2.1.2" has a has a calculation formula (17) to compare design to the girth weld fatigue for a highway crossing.

Reference 9, paragraph "4.8.2.2.1" has a has a calculation formula (18) to compare design to the longitudinal weld fatigue.

Reference 9, paragraph "4.8.2.2.2" has a has a calculation formula (19) to compare design to the longitudinal weld fatigue for a railroad crossing.

Reference 9, paragraph "4.8.2.2.3" has a has a calculation formula (20) to compare design to the longitudinal weld fatigue for a highway crossing.

Reference 9, "Appendix B" has example design problems.

15 Best Practice for Piping – Guidelines for the Design of Buried Steel Pipe

Guidelines for the Design of Buried Steel Pipe, reference 10, is RAGAGEP that I have commonly seen overlooked.

Reference 10 has some additional useful information on piping design. Section "4.1" covers loads due to airport. Section "4.2" covers ovality, through wall bending, crushing of side walls and ring buckling. Section "5.1" covers maximum impact load. Section "5.2" covers penetration. Section "6" covers buoyancy. Section "11" covers earthquake. Section "12" covers nearby blasting. Section "13" covers fluid transients. "Appendix A" covers suggested acceptance criteria for stress and strain for 13 different loading conditions.

16 Conclusions

This paper presented design portions of CFRs for onshore metal pipelines. It guided the reader through the selection process of production/gathering and regulated/non-regulated in conjunction with API 80. It them went through design aspects of ASME B31.1/3/4/8, some of the equations and factors and the interrelationships between the ASME pipe codes, in particular for compressor stations. There are various loads and stresses that the CFRs and ASME pipe codes require to be considered, but many of the equations and methods are not detailed in those documents. Therefore, API 1102 and Guidelines for the Design of Buried Steel were discussed to fill in some of the gaps in the CFRs and ASME pipe codes, such as surface dead loads, surface live loads, and cyclic loads, failure mechanisms, equations, and acceptance criteria for deflection, ovality, through wall bending, crushing, ring buckling, fatigue, longitudinal, thermal expansion and fluid transients.

Using these presented methods results in inherently safer piping and safer pressure relief system design.

17 References

[1] API, RP 80, Guidance for the Definition of Onshore Gas Gathering Lines, 1st Edition, April 1, 2000, Reaffirmed January 2013.

[2] 49 CFR Part 192, Transportation of Natural and Other Gas By Pipeline: Minimum Federal Safety Standards, Revision 2/11.

[3] 49 CFR Part 195, Transportation of Hazardous Liquids by Pipeline, Revision 2/10.

[4] ASME, B31.8-2014, Gas Transmission and Distribution Piping Systems, ASME Code for Pressure Piping, B31, American Society of Mechanical Engineers, New York, NY, 2014.

[5] ASME, B31.4-2012, Pipeline Transportation Systems for Liquids and Slurries, ASME Code for Pressure Piping, B31, American Society of Mechanical Engineers, New York, NY, 2012.

[6] ASME, B31.3-2014, Process Piping, ASME Code for Pressure Piping, B31, American Society of Mechanical Engineers, New York, NY, 2015.

[7] ASME, B31.1-2004, Power Piping, ASME Code for Pressure Piping, B31, American Society of Mechanical Engineers, New York, NY, 2004.

[8] ASME, B1.20.1-2013, Pipe Threads, General Purpose (Inch), American Society of Mechanical Engineers, New York, NY, 2013.

[9] API, Recommended Practice 1102, Steel Pipelines Crossing Railroads and Highways, 6th Edition, American Petroleum Institute, Washington, D.C., April 1993, Reaffirmed July 2002.

[10] American Lifelines Alliance, Guidelines for the Design of Buried Steel Pipe, American Lifelines Alliance, July 2001.

13TH GLOBAL CONGRESS ON PROCESS SAFETY

Understanding Tolerable Risk Criteria – Considering the Growth of LNG Transportation

Ryan J. Hart, Ph.D., P.E.
Exponent, Inc.
4580 Weaver Parkway, Suite 100
Warrenville, IL
rhart@exponent.com

Delmar "Trey" Morrison, Ph.D., P.E.
Exponent, Inc.
4580 Weaver Parkway, Suite 100
Exponent, Inc.
Warrenville, IL
tmorrison@exponent.com

Prepared for Presentation at
American Institute of Chemical Engineers
2017 Spring Meeting and 13th Global Congress on Process Safety
San Antonio, TX
March 26 – 29, 2017

Understanding Tolerable Risk Criteria – Considering the Growth of LNG Transportation

Ryan J. Hart, Ph.D., P.E.
Exponent, Inc.
4580 Weaver Parkway, Suite 100
rhart@exponent.com

Delmar "Trey" Morrison, Ph.D., P.E.
Exponent, Inc.
tmorrison@exponent.com

Keywords: risk assessment; tolerable risk criteria; transportation risk

Abstract

Cheap natural gas and the aim to lower emissions from combustion systems such as stationary power generation and heavy haul vehicle engines has led to a multi-industry effort to convert to natural gas equipment. Natural gas pipelines can reach most of the U.S., but there are many areas where shipping liquefied natural gas (LNG) via truck or train may be the most economical solution. This article focuses on shipping LNG via rail and benchmarking the risk this poses. Although similar to other hazardous materials, LNG shipping via rail is a new endeavor in the U.S., and the regulations do not currently allow this to happen. The Federal Rail Administration (FRA) is the government entity that regulates rail shipping, and has implemented a requirement for a risk-based analysis in order to get approval to ship high hazard flammable trains (HHFT) after a series of crude oil train accidents over the last decade. Further, the FRA has issued specific guidance related to risk assessment for LNG trains. Quantitative Risk Assessment (QRA) is one tool that can be applied to evaluate the risks posed by shipping LNG. This falls into a special class of QRA called Transportation Risk Analysis.

But how does one evaluate the quantified risk for shipping LNG? How can it be demonstrated that the risk of shipping LNG is acceptable to society? Currently, the U.S. federal government has not explicitly provided risk criteria for evaluating acceptability of shipping LNG via specific routes. In contrast, international regulators and industry organizations have implemented tolerable risk criteria for determining acceptability of new technologies, but their underlying bases may not be apparent. In this paper, we will demonstrate how existing stationary LNG plant tolerable risk criteria can be used to develop criteria for LNG transportation risk. The inappropriateness of directly applying these existing stationary plant risk criteria to transportation will be explained, and another existing HHFT commodity (propane (LPG) for example) will be used to benchmark against the stationary LNG plant criteria and to show how these criteria can be modified to apply to transportation risk. Ultimately, the risks will be quantified for LNG HHFTs and compared to existing HHFTs to show that the newer fuel, LNG, poses reasonable and equivalent risks to many of the other HHFTs currently shipped through the U.S. From this work, a tolerable risk criterion for LNG shipping via rail will be proposed.

1 Introduction

The growth in domestic natural gas production is expected to continue to grow over the next several decades (Figure 1) [1]. The surplus of natural gas is expected to be handled through exports as well as domestic consumption. More and more we will start seeing LNG transport containers being shipped over the road and along the rail. To what extent is yet to be determined, but an increase in transportation of LNG within the U.S. is an almost certainty.

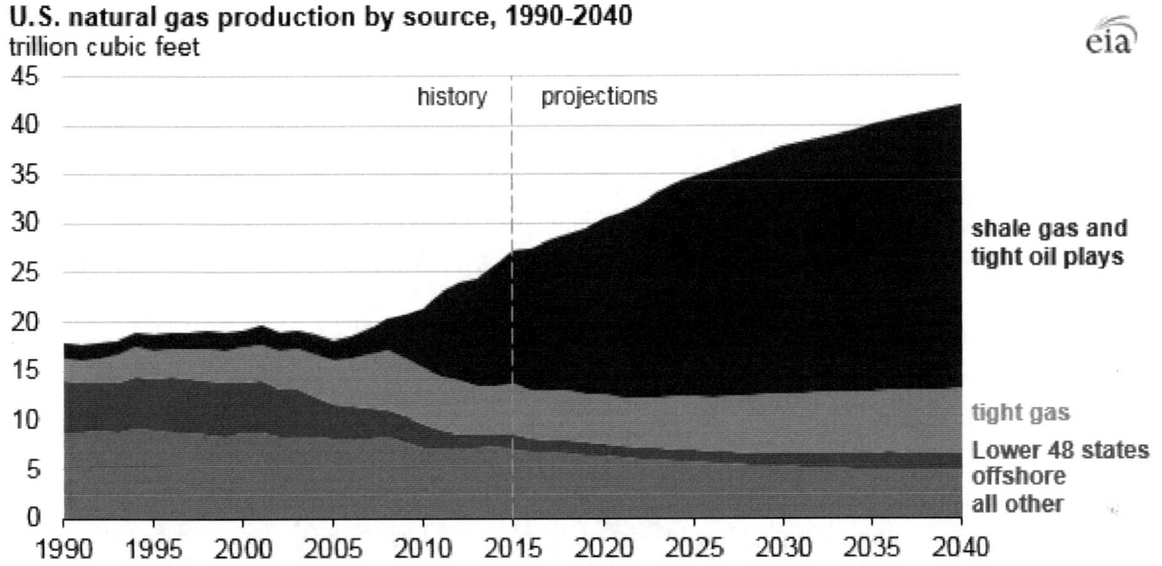

Figure 1. Project domestic natural gas production [1].

Although LNG transportation has existed for decades within the U.S., the anticipated surge in transportation is not without concern. Given the high visibility of transportation incidents involving crude oil, much attention is being paid to LNG transport by rail. Ensuring that the development and implementation of LNG transportation is performed safely is of critical importance to industry, communities, and regulators.

Much of the concern surrounding LNG transport may be attributed to a general lack of understanding regarding the unique and particular hazards of LNG. Quantitative risk assessment (QRA) is a tool that has been used successfully in the past to evaluate the potential risks of emerging technologies and proposed transportation routes. QRA is not a common public-policy tool in the U.S., but it is beginning to grow in popularity with private companies, regulatory agencies, and standards committees alike. Most QRA has been performed for stationary facilities, which are inherently different than transportation projects. In this paper, we discuss tolerable risk criteria for use in evaluating the QRA of transport projects. We anticipate that this approach may be useful for evaluating the blossoming LNG transport in the U.S.

2 Understanding Quantitative Risk Assessments

Risk, simply defined, is the potential to lose something of value. Risk is calculated by taking the product of event likelihood with the event outcome severity. A simplified flow diagram depicting the process for determining risk is provided in Figure 2.

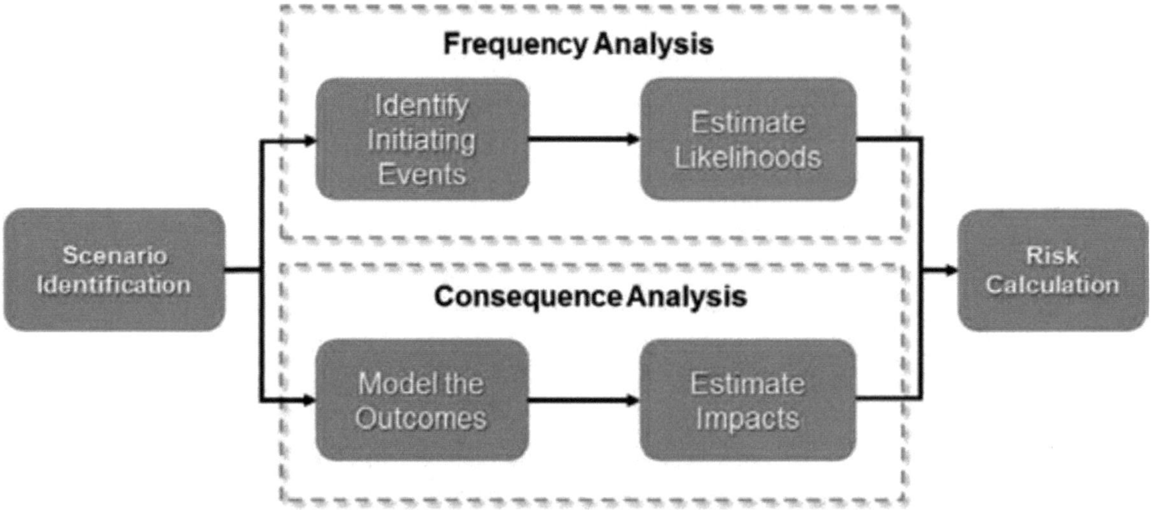

Figure 2. Flow diagram depicting the process for calculating risk.

The likelihood of an event can be estimated using experience relating to given equipment in similar service, industry data, or engineering approximations. A challenge of quantifying risk, or affixing a number to a particular risk level, is determining how to quantify the event outcome portion of the equation. For quantifying risk at industrial facilities and transportation operations, the outcome of an event is typically evaluated as the potential for a fatality or multiple fatalities.

In evaluating the potential for fatality, two metrics are utilized to yield the risk: (1) Individual Risk (IR) and (2) Societal Risk (SR). IR is the frequency (yr^{-1}) where an individual with continuous potential exposure may be expected to sustain a serious or fatal injury. When calculating IR, it is assumed that persons are constantly occupying all space where an incident may express an undesired impact. As such, IR is largely independent of population density except to the extent that the likelihood component of the risk is dependent on population (e.g. likelihood of ignition calculation). IR results are typically presented as frequency contours on aerial maps that illustrate the risk to individuals positioned within those contours. An example of an IR result is presented in Figure 3.

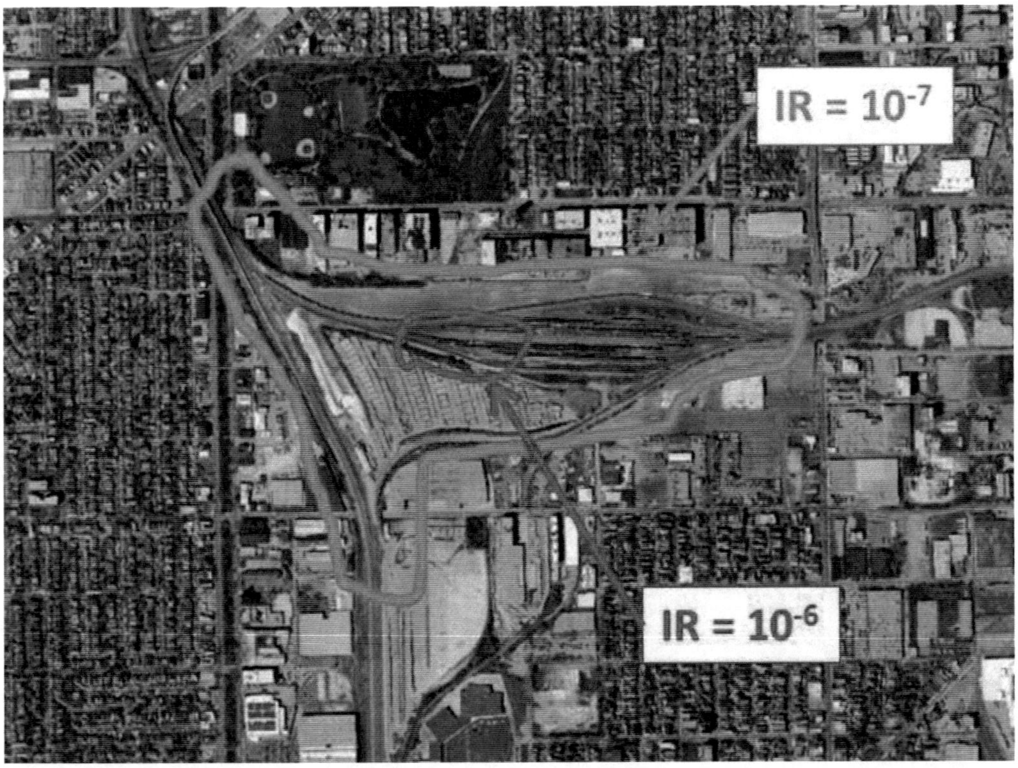

Figure 3. Example of an IR result, presented around a rail yard.

Societal Risk (SR) is another metric for evaluating the risk of a given facility or operation. Unlike IR, the SR calculation considers the relationship between the cumulative number of potential fatalities and the likelihood from a series of potential events. The outcome of a SR analysis is a graph depicting annual frequency on the y-axis and N fatalities on the x-axis, where N is the cumulative number of potential fatalities for all scenarios represented by the corresponding cumulative frequency of events. Unlike IR, the SR calculation takes into account the differences in population density in the vicinity surrounding the incident. Whereas the IR calculation gives insight into the probability of having a fatality at a specific location, the SR calculation gives the likelihood of a number of potential fatalities covering the entire operation. This is especially important for evaluating scenarios with a large potential impact for loss of life, such as train derailments of flammable materials. An example of an SR result for two separate scenarios is provided in Figure 4.

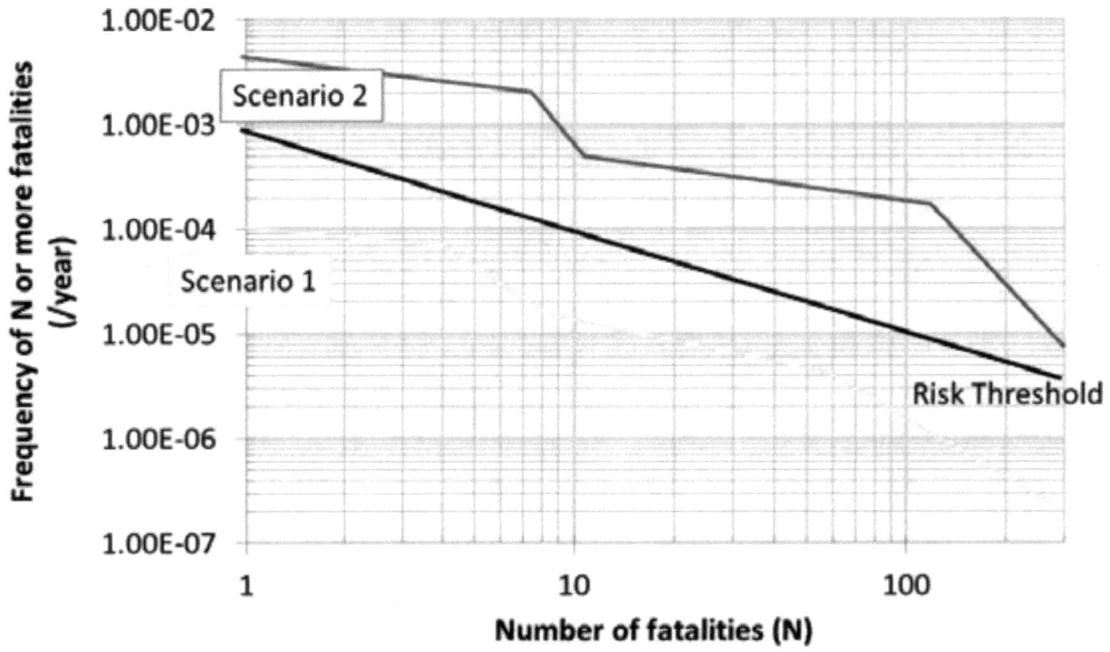

Figure 4. Example of an SR result for two different scenarios.

Quantitative risk assessments may be performed for several reasons:

1. Comparing the relative risk of two or more competing options (e.g. comparing route options for transporting hazardous materials)
2. Identifying high risk aspects of a specific operation
3. Demonstrating that the risks of a certain operation are tolerable or acceptable

When analyzing risks to demonstrate their acceptability or tolerability, the selection of appropriate risk tolerability criteria is paramount. Selection of inappropriate tolerability criteria could potentially result in a false-confidence in the perceived safety (in instances where the risk tolerability may be too high) or unnecessary elimination of a proposed project or operation (in cases where the risk tolerability may be too low).

3 Quantitative Risk Tolerability Criteria

After quantifying risk and presenting the calculations as IR and SR for a given operation or process, the results are evaluated for tolerability (or acceptability). Typically, stakeholders (e.g., government agencies, investors, communities) have a threshold risk level that is deemed acceptable—known as quantitative risk criteria. Currently, the U.S. federal regulatory agencies have not codified quantitative risk criteria for hazardous materials transportation scenarios [2] and QRA analyses are not common regulatory requirements in the U.S. Several foreign regulatory agencies, such as the Dutch government, have been international leaders in utilizing QRA techniques for determining acceptability of fixed facilities and transportation routes.

However, no broadly-accepted risk criteria are employed by domestic communities or industries. There are several foreign and several domestic examples of quantitative risk criteria (see [3]-[5]) and these will be summarized in the next sections for both IR and SR.

Recommendations for QRA of LNG plants were issued in the National Fire Protection Association (NFPA) standard, NFPA 59A *Standard for the Production, Storage, and Handling of Liquefied Natural Gas (LNG)* [6]. In addition to including QRA as a risk assessment tool in NFPA 59A, the standard also includes quantitative risk criteria for fixed LNG facilities. NFPA 59A explicitly applies to LNG plants and stationary facilities; it does not apply to LNG transportation or portable LNG containers. Thus, the tolerable quantitative risk criteria proposed in the standard are not directly applicable to the transportation of LNG, as will be discussed below.

There are significant differences between stationary facilities and transportation projects, for example:

- Size scale of the enterprise. The risk calculation is directly related to the number of vessels or components and the length of piping systems or a route. In a fixed facility, the length scale is fixed and on the order of hundreds to thousands of feet. In transport, the length scale may be hundreds to thousands of miles. Thus, transport is anticipated to create higher risks for similar hazmat.
- Site location. Stationary facilities that are yet to be built typically have the opportunity to survey multiple potential sites, whereas transportation routes options are more restricted (in some cases, only one transportation route may be an option).
- Siting. For stationary facilities, consideration can be given where on the site to place hazardous material operations in order to minimize potential off-site impacts or reduce the likelihood of suffering cascading events. Transportation of hazardous materials has little flexibility in placement along a given route and has no control over adjacent hazards (e.g. a hazmat transport carrier cannot prevent another hazmat transport carrier from overtaking them on the highway).
- Standoff distances. Proposed facilities may have the opportunity to acquire more land for a given project in order to increase the distance from hazmat operations to the property boundary. The separation distances between a transport carrier and the public are already established due to the zoning of the transport route and adjacent properties. Additionally, in some transport modes (like over the road trucking) the transport carrier has no choice but to be a few feet from a public receptor.
- Control of ignition sources. Facilities operating with hazmat have complete control over the abundance and location of ignition sources on their site. Transport routes have no control over ignition sources, and nearly every public transporter can be assumed to be a competent ignition source.
- Training/emergency response. Facilities have the opportunity to train personnel and prepare for an onsite release of a hazmat. Transport carriers may have a highly trained crew for their specific carrier, but the public in the immediate vicinity of the hazard is presumed to have no training with respect to the hazards presented by hazmat. Additionally, hazmat transport relies upon the emergency response of the jurisdiction through which it travels, which may change dozens or hundreds of times on a given

route. Facilities, by comparison, usually have only 1 or 2 emergency responder agencies to coordinate with.

For the reasons stated above, the quantitative risk criteria for stationary facilities should be different than the quantitative risk criteria for transportation routes. The risks to the onsite and offsite receptors associated with transporting hazmat are inherently different than the risks to those populations surrounding a stationary facility. To account for these differences, transportation risks should be evaluated differently than stationary facility risks.

3.1 Individual Risk Criteria

Among the available references for quantitative risk criteria used domestically and internationally, there is a wide disparity in risk criteria for public exposure, with acceptable IR fatality probabilities ranging from 10^{-4} yr^{-1} (or a fatality per 10,000 years) to 10^{-8} yr^{-1} (or a fatality per 100,000,000 years). NFPA 59A identifies three "Zones" representing ranges of quantitative risk criteria for evaluating IR at stationary LNG facilities. Each risk zone reflects general types of public occupancies recommended to be permitted. As the magnitude of the calculated risk increases, the type of occupancy becomes more restrictive. The quantitative risk criteria for IR of LNG plants, according to NFPA 59A, are reproduced in Table 1.

Based on Zone 3 being the most restrictive zone, any IR values less than 3×10^{-7} yr^{-1} are not of concern for the analysis. The IR ranges and associated criteria reported in NFPA 59A appear to be based on guidance provided by the Health and Safety Executive in the U.K. for QRA [7] and do not account for the factors typically considered in a transportation risk analysis. Conversely, a commonly acceptable level of IR for transportation risks on sensitive populations is 10^{-6}, which is the upper threshold for Zone 3 [5].

Table 1. Quantitative risk criteria for IR contours as provided by NFPA 59A (2016).

Criterion Annual Frequency (yr^{-1})	Remarks
Zone 1 $IR > 10^{-5}$	Not permitted: Residential, office, and retail Permitted: Occasionally occupied developments (e.g., pump houses, transformer stations)
Zone 2 $10^{-6} \leq IR < 10^{-5}$	Not permitted: Shopping centers, large-scale retail outlets, restaurants, etc. Permitted: Work places, retail and ancillary services, residences in areas of 7,250 to 23,300 persons/mile^2 density
Zone 3 $3 \times 10^{-7} \leq IR < 10^{-6}$	Not permitted: Churches, schools, hospitals, major public assembly areas, and other sensitive establishments Permitted: All other structures and activities

For flammable material release scenarios, the magnitude of the risk generally increases as the observation point is moved closer to the transport route. The distance to each risk level identified in the table is a result of the compilation of the outcomes calculated from an event tree of many potential fire and explosion events. An example of calculated distances to various IR thresholds for the transportation of LNG and LPG have been reported **Error! Reference source not found.**, reproduced in Table 2. LPG represents an inherently acceptable hazmat commodity, which has been shipped on railroads for a long time. Table 2 shows that the IR of LNG is similar or less than LPG for both road and rail transport.

Table 2. Example calculated IR distances for the road and rail transport of LNG and LPG.

Material	Transport Mode	Pop Density (/mile2)	IR Distance to... 10^{-5} IR (m)	10^{-6} IR (m)	$3x10^{-7}$ IR (m)
LPG	Road	10,000		118	177
		1,000		86	115
		100		84	108
LNG	Road	10,000		119	169
		1,000		62	95
		100		56	84
LPG	Rail	10,000	186	290	334
		1,000	178	290	334
		100	174	290	334
LNG	Rail	10,000	92	193	213
		1,000		145	184
		100		104	132

It should be noted that the analysis in **Error! Reference source not found.** was conservative, especially in the instances of rail risks where it was assumed that a train carrying a hazmat would involve the hazmat rail car in all instances. Despite the conservativeness, the data illustrates that the NFPA 59A Zone 3 criteria for IR (3×10^{-7} yr^{-1}) is much more restrictive than the Zone 2 criteria (1×10^{-6} yr^{-1}). Additionally, the use of the Zone 3 criteria would indicate that the inherently acceptable risks of transporting LPG may not satisfy this criterion in instances where there are sensitive targets within 334 m of railroads and 177 m of roads where LPG is transported.

3.2 Societal Risk Criteria

As with IR, the international and domestic quantitative risk criteria for SR can vary greatly. Most SR criteria present two tolerability lines with three distinct regions for which to evaluate the resulting FN curve. For discussion purposes, the SR criteria from NFPA 59A is provided in Figure 5. There are two lines presented in the figure:

(1) an upper risk criteria line (solid red line), commonly referred to as the maximum risk line, and

(2) a lower risk criteria line (dashed line), commonly referred to as the tolerable risk line.

For calculated FN curves that fall above the upper risk criteria line, the risk is considered "unacceptable." Calculated FN curves that fall below the lower risk criteria line are determined to be acceptable, and FN curves between the two criteria lines fall within the "As Low As Reasonably Practicable (ALARP)" region. Risks falling within the ALARP region are typically evaluated in more detail for opportunities to either mitigate or reduce the perceived risk of the operation, in the form of additional engineering or administrative controls.

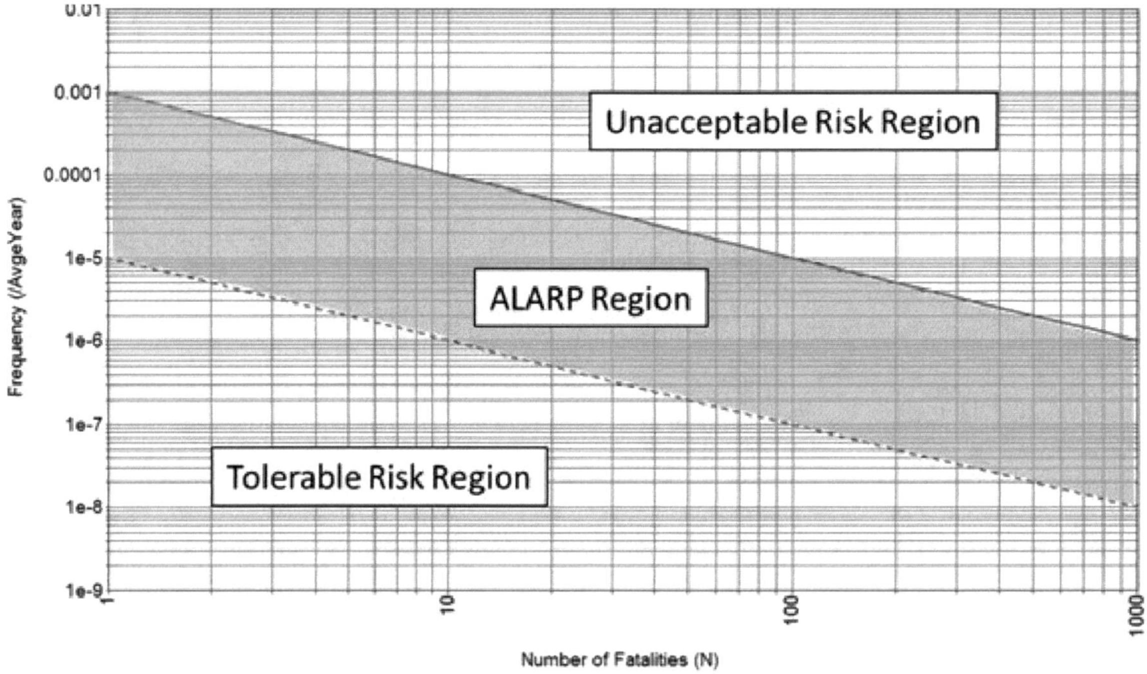

Figure 5. NFPA 59A (2016) SR tolerability criteria.

Most international and domestic examples of SR criteria follow the same general format with both an upper and lower risk bound. Although different regulatory agencies, standards committees, and private companies may have different definitions of upper and lower risk lines, nearly all examples of SR criteria apply to only stationary facilities.

Using a previous reported QRA study for the transport of LNG and LPG [8], we have adopted the data here to demonstrate calculated FN curves. Figure 6 depicts the calculated FN curves for the rail and road transport of LNG and LPG along a 50-mile long route surrounded by a population with a density of 1,000 people/mile^2. The figure also includes the maximum and minimum risk lines from NFPA 59A, and the ALARP region highlighted in grey. The figure illustrates the potential issues when directly applying SR criteria intended for stationary facilities to transportation risks. The risks for all four transport routes fall within the "Unacceptable Risk" region as defined in NFPA 59A. LPG is certainly transported by truck and rail through similar distances and population densities, but the risk to society is acceptable.

Although most SR criteria available both domestically and internationally are applied to stationary facilities, methods have been developed to apply the stationary SR criteria to transportation routes. These methods involve normalizing the scale of the transport project risks and modifying the risk criteria lines. To correct the fixed facility SR criteria for application to a transportation project that may be considerably larger in scale, the SR for transportation routes are calculated for a fixed unit of length. This allows for similar hazard scales to be compared between transport routes and fixed facilities. The common unit for normalizing the FN curves of transportation projects is either 1 mile or 1 km, depending on the common engineering units in the project's location. This can be accomplished by either (1) calculating the FN curve for the entire transportation route and dividing the y-axis (frequency) by the route length or (2) conservatively assuming that the highest-risk section of the transport route bounds the risk along the entire line. Such an approach is sensible because any potential incident can only affect a limited portion of the society along the route for these flammable materials; thus, using a per-mile or per-kilometer basis for SR is reasonable. The same approach may not be acceptable for toxic materials, e.g., chlorine gas, which could project significant hazards over a greater distance. An additional approach is to compare the aggregate SR curve for the route for the new hazmat (LNG) to an existing hazmat (LPG). If the new curve is similar or less, then it poses less risk along the route than the reference material.

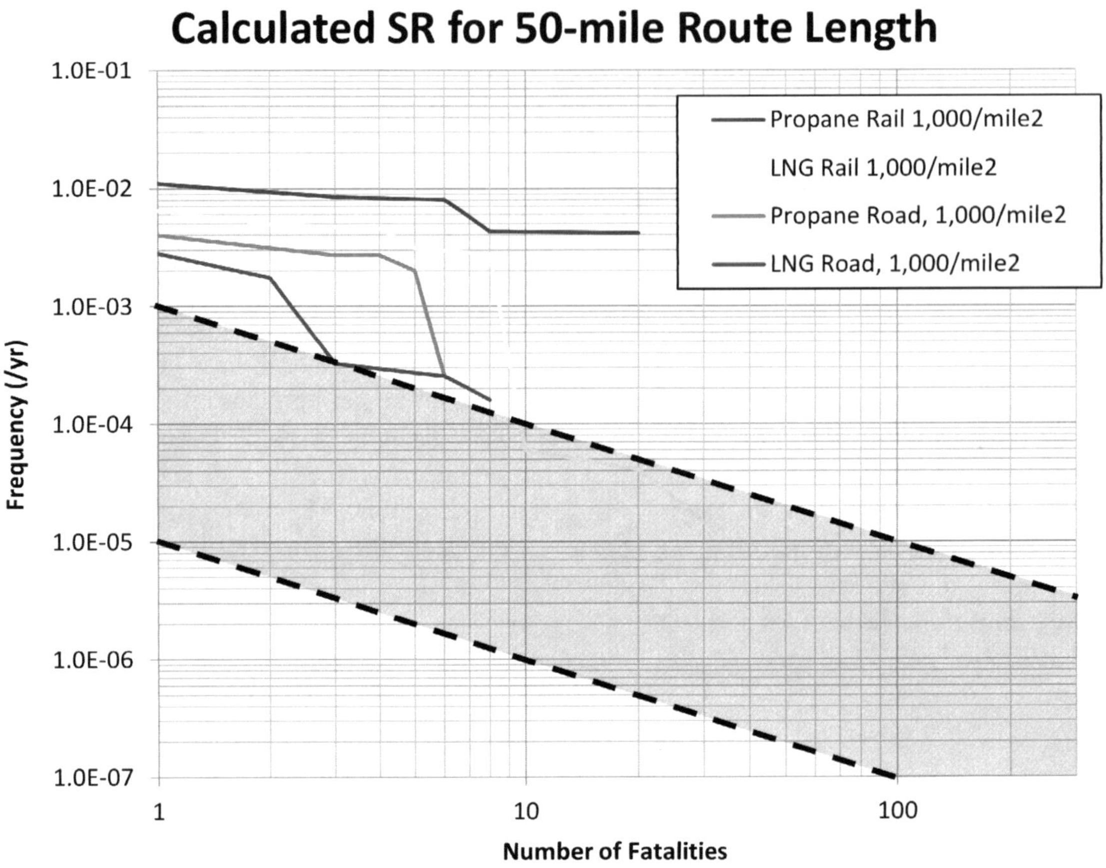

Figure 6. Calculated FN curves for the road and rail transport of LNG and LPG along a 1-mile long route surrounded by a population with a density of 1,000 people/mile^2.

Figure 7 provides the FN curves for the same transport routes above, but normalized on a 1-mile length. All other inputs are identical between the two graphs. When scaling the transportation FN curves to be comparable to the length scale of stationary facilities, the application of the stationary facility SR criteria becomes more reasonable. However, it can be seen that one of the analyzed transportation routes, propane rail transport, still exhibits a portion of the FN curve with the NFPA 59A defined "unacceptable risk" region. As previously mentioned, there are other considerations between project scales that distinguish the risks of stationary facilities from transportation projects. In order to consider for these other differences (such as ignition source control, training, siting), some international organizations increase the maximum and minimum risk lines by an order of magnitude – the implication here is that the tolerability of risks associated with transportation are inherently higher that for exposure to a stationary facility. If this approach were to be taken for the FN curves presented in Figure 7, none of the FN curves would lie within the "unacceptable risk" region (in fact, the LNG Road scenario would lie in the "acceptable risk" region).

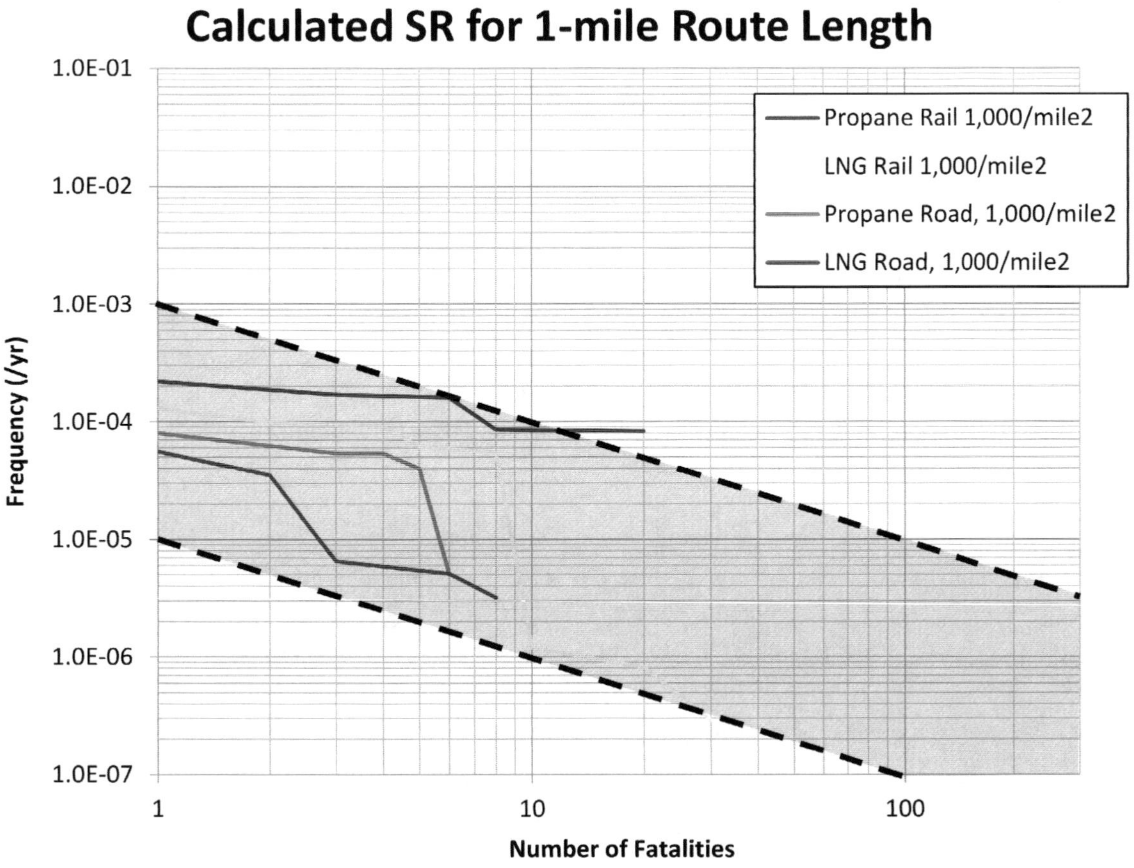

Figure 7. Calculated FN curves for the road and rail transport of LNG and LPG along a 1-mile long route surrounded by a population with a density of 1,000 people/mile^2.

4 Conclusions

Quantitative risk assessments are powerful tools for assessing the risks of emerging and existing hazardous material operations within stationary facilities and along transportation routes. The risks can be assessed through either comparative analyses or by application of tolerable risk criteria. Many international and domestic organizations, both public and private, have identified tolerable risk criteria for evaluating both Individual Risk (IR) and Societal Risk (SR), but considerable variability exists. Additionally, most tolerable risk criteria have been developed for stationary facilities.

As shown in this paper, application of risk criteria developed for stationary facilities to transportation routes may not be appropriate. The risks of transporting hazardous materials is different from fixed facilities, as are the thresholds for risk tolerability that society is inherently willing to accept. The application of improper risk criteria may prevent or delay the proposed transportation of emerging materials, such as LNG.

Based on previously reported IR and SR values for LNG and LPG transport, we have demonstrated that although the risks of LNG transport are less than the risks of transporting LPG, certain risk tolerability criteria may indicate that the risks are unacceptable. This conflicts with the inherently acceptable risks of LPG transport, which were demonstrated to be higher than those for comparable LNG transport.

A survey of domestic and international risk tolerability was performed. Most risk tolerability criteria have been developed for and applied to stationary facilities. The risks of stationary facilities are inherently different that transportation risks for two primary reasons: (1) transportation routes can be significantly larger than stationary facilities and (2) the inherent tolerability of transportation risks is different than for stationary facilities. Based on the analysis of representative LNG and LPG transport risks, compounded with the typical treatment of risk tolerability criteria by international organizations, methods for using stationary facility risk criteria to evaluate transport risks are proposed:

- IR Tolerability Criteria – for transport projects, the use of 10^{-6} yr^{-1} as the IR frequency threshold for impacts to sensitive targets is consistent with an abundance of international sources. This value is approximately one order-of-magnitude higher than the minimum sensitive target tolerability criteria presented in NFPA 59A for stationary LNG facilities (3×10^{-7} yr^{-1}).
- SR Tolerability Criteria – for transport projects, normalizing the FN curves for a representative length (1 mile or 1 km) can allow for re-scaling of transportation risks to a stationary facility. Further, to account for the higher risk tolerability for transportation relative to stationary facilities, the maximum and minimum risk criteria lines may be increased by an order of magnitude.

5 References

[1] U.S. Energy Information Administration, "Annual Energy Outlook 2016," July 7, 2016.

[2] Strang J, "Federal Railroad Administration Risk Reduction Programs," United States Army Corps of Engineers Workshop on Tolerable Risk, March 18-19, 2008, Alexandria, Virginia.

[3] Appendix B: Survey of Worldwide Risk Criteria Applications, *Guidelines for Developing Quantitative Safety Risk Criteria*. Center for Chemical Process Safety, AIChE (2009).

[4] Cornwell JB and MM Meyer, "Risk Acceptance Criteria or 'How Safe is Safe Enough?'" presented at II Risk Control Seminar in Puerto La Cruz, Venezuela, October 13, 1997.

[5] Ham JM, M Struckl, AM Heikkila, E Krausmann, C DiMauro, M Christou, JP Nordvik, "Comparison of Risk Analysis Methods and Development of a Template for Risk Characterisation," Institute for the Protection and Security of the Citizen, European Commission, Directorate-General Joint Research Center (2006).

[6] NFPA 59A, Standard for the Production, Storage, and Handling of Liquefied Natural Gas (LNG), 2016 edition, National Fire Protection Association.

[7] "B.1 Evolution of Land Use Planning Criteria in the UK," in Guidelines for Developing Quantitative Safety Risk Criteria, American Institute of Chemical Engineers, Center for Chemical Process Safety (2009).

[8] Hart RJ, Morrison DR. The hazard we know: Comparing transportation risk of LPG and LNG. American Institute of Chemical Engineers, 2015 Spring National Meeting and 11th Global Congress on Process Safety, Austin, TX, April 26-30, 2015.

The Capability Engine, Analyzing the Past to Protect the Future

Lawrence Short
DuPont Sustainable Solutions
500 Studio Drive, Virginia Beach, Virginia
Lawrence.S.Short@dupont.com

Alfonsius Ariawan
DuPont Sustainable Solutions
1919 Minnesota Court, Mississauga, Ontario
Alfonsius.B.Ariawan@dupont.com

Prepared for Presentation at
American Institute of Chemical Engineers
2017 Spring Meeting and 13th Global Congress on Process Safety
San Antonio, TX
March 26 – 29, 2017

The Capability Engine, Analyzing the Past to Protect the Future

Lawrence Short
DuPont Sustainable Solutions
500 Studio Drive, Virginia Beach, Virginia
Lawrence.S.Short@dupont.com

Alfonsius Ariawan
DuPont Sustainable Solutions
1919 Minnesota Court, Mississauga, Ontario
Alfonsius.B.Ariawan@dupont.com

Keywords: Learning, Capability Building

Abstract

Hazardous operational environments require the ability to identify and minimize process and personal risk variables. In Northern Alberta, this is challenging, with many upstream oil companies sharing a large common population of contract workers. The Personal Safety Collaboration brought owners together to maximize the effects of a comprehensive learning model. Through this initiative, the Capability Engine was conceived and designed. Using theoretically sound best practices, descriptive assessment data is examined with our analytics model combined with field based performance and environmental data. This combination renders an inferential and predictive tool enabling prescriptive action while alerting of the historical leading indicators of incidents.

1. Introduction

Through our work with a client in Northern Alberta, we have developed a Learning Effectiveness model to enhance operators' learning journey. We did this by incorporating more structures and learning assurances throughout an operator's learning cycle. The model is depicted in Figure 1 and is explained elsewhere in more details[1].

An important element of our model is the analysis of learning measures, specifically the Kirkpatrick K3 measures, that link field performance to training variables. The K3 measures are made up of field performance indicators, such as those related to operational discipline, audit performance, incidents and near-misses, as related to specific tasks (It is differentiated from K1 and K2 measures which focus on the perceived effectiveness of training tools and delivery, and the understanding of knowledge through testing and field demonstrations, respectively. The reader is referred to another paper by the same authors for further explanation[1]). Our instructional designers combine both the practical experience that we have obtained through client work and instructional theories to design the various assessment protocols in order to determine the right Kirkpatrick measures. This allows us to link the resulting data to the various critical points of the operator's learning journey. With some basic analysis, we can then determine where failure modes have happened in the journey. For example, we are able to determine if the way the course content is delivered to operators lacks impact, or if there is a lack of supervisory reinforcement in the field. This then allows the organization to address issues to further enhance learning vs. investing in more training session.

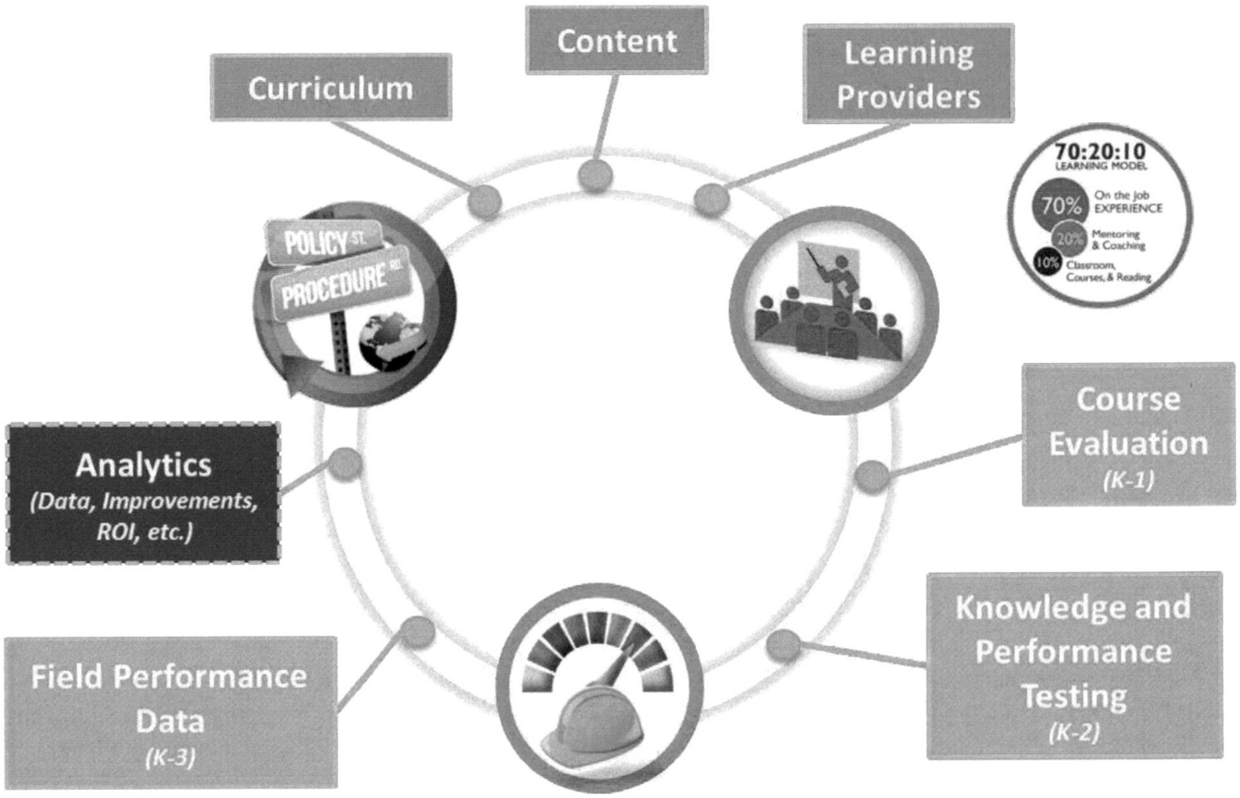

Figure 1. Learning Effectiveness model

In more recent applications of this model, we have also incorporated environmental data to the analysis. Our premise is that environmental factors may influence how operators learn, and create a bias in their field performance.

2. The Capability Engine

The capability engine of the learning effectiveness model refers to the curriculum, content and delivery mechanisms that facilitate knowledge transfer. These are continuously improved based on K1-K3 assessment results and the subsequent data analysis. For example, recall that K1 data tells us about the perceived effectiveness of training and its delivery. It is based on participants' rating of the facilities, instructor, quality of content, applicability of work, etc., complemented with our own measures of the same. K2 data, on the other hand, measures the transfer of knowledge through testing and field demonstrations. Lackluster K1 and K2 data, but an acceptable K3 outcome or performance suggest that learners do not think much of the training session and do not necessarily obtain the knowledge that is intended to be transferred. However, learners are able to practice in the field what are expected of them, and do so in an acceptable consistency. This is certainly a possibility. For example, this could be due to effective field supervision and strong peer support/coaching. However, this begs the question on the return of investment of the training sessions, i.e. what is the value of the training sessions? Or, more positively, one may want to ask: if we were to improve the quality of the training sessions, would we be able to further improve performance? If so, how?

On the flip side, if K1 and K2 measures are deemed to be acceptable, but the organization does not see the expected outcome or performance in terms of improvement in incident rate, near-misses, audit infractions, etc., then different questions may need to be asked. The issue could be in the conversion of training materials to practical application. Have learning objectives been defined correctly? Can we assure the fidelity of the course contents to these learning objectives? There may also be field variables or external factors at play that need to be addressed.

Before drawing any conclusion, the design and quality of the assessment questions/protocols (through which K1 – K3 data are measured) must also be analyzed. Poorly designed questions can result in data bias and/or in results that are not sensitive enough to input variables. These questions need to be continually checked and upgraded as new learnings are obtained, possibly complemented with greater number of input variables, such as environmental variables, to strengthen correlation analysis (big data analysis). In designing these assessment questions/protocols, we also consider the three domains of learning to ensure comprehensiveness:

- **Psychomotor:** The ability to perform tasks by manipulating objects and / or moving one's body in space to complete a task
- **Affective:** The internalization of attitudes, beliefs that are best measured through observation of behavior
- **Cognitive:** knowledge, comprehension, application, analysis, synthesis, evaluation / judgment

3. A New Way of Learning

One of the more significant environmental variables that influences learning performance is the demographic shift in workforce. The rapid replacement of operators with younger personnel has challenged our traditional method of knowledge transfer, primarily due to the way the younger generations are learning. Analysis of K1 – K3 data forces us to rethink how we should design and deliver learning contents. The reality is that technology advancement has changed the way information is transmitted and therefore, consumed. The ease with which information is transmitted often put us in a constant state of distraction and make us the victim of information overload. In response to this new reality, we adjust. We now tend to consume information in small 'chunks'. We do not need to look far for evidence: YouTube, Tweeter, Ted Talks and Instagram are just a few examples of platform that transfer information in short bursts to their intended audience – effectively. So, why do we still expose our operators to days-long training sessions, where information typically flows one way – from the instructor to the participants? Analysis of K1 – K3 data suggests that micro-learnings are also more effective in knowledge transfer in operational setting. The use of multi-medias and interactive exercises have also been found to enhance learning, especially if they are designed and delivered to affect the learners' emotions.

5. Summary

In this paper, we describe a learning model that is designed to be continuously improving through the analysis of K1 – K3 measures. These measures need to be thoughtfully designed in order to ensure comprehensiveness, sensitivity to input variables and soundness in their cause-effect relationships to the various elements of the capability engine. The results of the analysis allow us to pin point and address critical failures in our operators' learning journey. A finding from our work with clients in this space suggests the need to focus on micro-courses as a possibly preferred mechanism to transfer knowledge, with the use of multi-media and interactive exercises. Such approach anecdotally leads to better knowledge retention, especially if designed and delivered to reach the learners' emotional level.

6. References

[1] American Institute of Chemical Engineers (2017). *Ensuring Learning (Not Just Training) – Measuring the Impact of Learning on Business Performance*. New York, NY: Lawrence Short, Alfonsius Ariawan

479452 *"Getting More Out of Your PHAs with Data Analytics"*

Abstract:

Process Hazard Analysis (PHA) studies are required to identify process and operations risks for process plants. PHAs enable management and process safety staff to make decisions that improve safety performance, minimize downtime, and optimize utilization.

Operating companies also must make these key decisions under financial capital constraints and ensure maximum return on investment for safety. These decisions require a clear understanding of the interconnectivity of the risks associated with a facility's operations which are captured in a PHA. Most PHA studies generate recommendations based on residual risk and consequence levels, but they do not directly address how safeguards and recommendations are linked to the hazard scenarios. By improving the way PHA data are analyzed, using data analytics, additional insight is generated that is typically hidden, or not considered, when traditional PHA approaches are followed. These insights include more intuitive identification of critical safeguards (barriers), better prioritization of recommendations from a cumulative risk perspective, and visual aids to improve awareness and communication of risk drivers to operational staff and decision makers.

Traditional PHAs require a substantial investment in time and resources, which results in a potential gold mine of useful data that can be further extracted and exploited. With an ever shifting focus on new technology and innovation, data analytics is the next key step for improving risk management practices within organizations and the industry as a whole. Taking PHA data analytics further, "Big Data" data mining techniques can be utilized to extract information from the collected data and expose trends and patterns in industry PHA data. These trends can be benchmarked across various industries providing an advanced level of decision making insight not available from a single PHA. Statistical analysis of the trends and deviations, along with analysis of incident data can also predict increasing levels of risk, thus creating strong leading indicators to aid in decision making before incidents occur. This paper showcases a case study for a unit within a refinery which demonstrates how application of PHA with data analytics can greatly improve decision making.

479294 Prevent the Next Catastrophe

Abstract:

With the increasing complex manufacturing processes and environmental regulations, and the challenges to retaining knowledge from retiring employees and transferring it to the newer generation, operations managers face multiple challenges such as staying competitive, insuring safety, maximizing production, and reducing operational risks and breakdowns. In the world of legacy, traditional manufacturing practices rely on repetitive procedures, overwhelming human interventions leading to costly errors. The fourth industrial revolution presents an unprecedented mind shift in Manufacturing Operations Management with the emergence of Cyber-Physical Production Systems, Augmented Reality Technologies, Big Data Analytics and Artificial Intelligence. New practices consist in creating systems that learn from data and deploy the extracted knowledge in real time so the end-user can have a single version of the truth, rather than abstract mathematical certainties. Thus, sudden equipment failures and process disturbance are being predicted, and costly damages are prevented. Integration Objects provides an Industrial Internet of Things platform that allows end-users to remotely and securely assess and monitor the reliability of their assets while maximizing profitability. Our innovative approach uses data driven hybrid models combined with expert rules and fault propagation models that can be deployed online for real time execution. Our IIoT platform is an integrated, flexible, scalable and user friendly graphical environment that allows the deployment of industrial scale applications.

In order to address the market shortage of resources, Integration Objects has made the technology easy-to-use, affordable, and accessible to the average user for a quick deployment. It allows increasing safety by avoiding false warnings, and boost gains by seizing any optimization opportunity.

13TH GLOBAL CONGRESS ON PROCESS SAFETY

Effectively Addressing the Results of Damage Mechanism Reviews in a Process Hazard Analysis: A Checklist Approach

Tazim S. Rehmat, ABSG Consulting Inc., 300 Commerce Drive, Suite 200, Irvine, CA, 92602 (714) 734-2526 TRehmat@absconsulting.com

Gary Carrithers, ABSG Consulting Inc., 10301 Technology Drive, Knoxville, TN, 37932 (502) 592-9386 GCarrithers@absconsulting.com

Randal Montgomery, ABSG Consulting Inc., 16855 Northchase Drive, Houston, TX 77060 (865) 671-5832 RMontgomery@absconsulting.com

David Whittle, ABSG Consulting Inc., 10301 Technology Drive, Knoxville, TN, 37932 (865) 671-5819 DWhittle@absconsulting.com

Sayed Termah, ABSG Consulting Inc., 16855 Northchase Drive, Houston, TX 77060 (281) 994-0859 STermah@abs-group.com

Prepared for Presentation at
American Institute of Chemical Engineers
2017 Spring Meeting and 13th Global Congress on Process Safety
San Antonio, TX
March 26 – 29, 2017

Effectively Addressing the Results of Damage Mechanism Reviews in a Process Hazard Analysis: A Checklist Approach

Tazim S. Rehmat, ABSG Consulting Inc., 300 Commerce Drive, Suite 200, Irvine, CA, 92602 (714) 734-2526 TRehmat@absconsulting.com

Gary Carrithers, ABSG Consulting Inc., 10301 Technology Drive, Knoxville, TN, 37932 (502) 592-9386 GCarrithers@absconsulting.com

Randal Montgomery, ABSG Consulting Inc., 16855 Northchase Drive, Houston, TX 77060 (865) 671-5832 RMontgomery@absconsulting.com

David Whittle, ABSG Consulting Inc., 10301 Technology Drive, Knoxville, TN, 37932 (865) 671-5819 DWhittle@absconsulting.com

Sayed Termah, ABSG Consulting Inc., 16855 Northchase Drive, Houston, TX 77060 (281) 994-0859 STermah@abs-group.com

Keywords: Damage mechanism review, process hazard analysis, mechanical integrity, checklist, Cal/OSHA

Abstract

Performing damage mechanism reviews (DMRs) to determine credible degradation modes and susceptibilities of processing equipment has become an important step in developing effective mechanical integrity plans in the refining and chemical processing industries. DMRs aid in formulating inspection plans to mitigate risk from loss of containment and/or unplanned outages and are the subject of the corresponding API Recommended Practice (RP) 571. In addition, California OSHA's (Cal/OSHA's) recently revised *Process Safety Management for Petroleum Refineries* regulation now requires that DMRs be performed and process hazard analyses (PHAs) address DMR reports that are applicable to the process unit being analyzed.

A refinery PHA team subject to that regulation will now need to consult with individuals with expertise in damage mechanisms when performing the review of applicable DMR reports. To help effectively address this requirement, the primary PHA methodology can be supplemented with a checklist to help the PHA team identify DMR issues that may be applicable to the process unit. This paper describes the application and contents of a checklist the PHA team can use to help identify DMR issues such as (1) identifying and demonstrating that known corrosion and damage mechanisms (high temperature hydrogen attack, chloride stress corrosion cracking, etc.) are understood and controlled, (2) confirming a DMR report appropriately documents applicable damage mechanisms to the process unit, and (3) confirming a DMR is performed for incident investigations where a damage mechanism is identified as a contributing factor.

1. Introduction and Regulatory Background

In 2012, the governor of California formed an Interagency Working Group on refinery safety to examine ways to improve public safety and worker safety at petroleum refineries. The objective of the group was to identify a means to improve refinery and agency performances. The Interagency Working Group published a report in February 2014 that identified (1) regulatory gaps, (2) areas of overlap between regulatory agencies, and (3) areas in need of improved coordination.

Subsequent to the findings of the Interagency Working Group, California announced it would implement changes to the Cal/OSHA process safety management (PSM) regulation and strengthen the California Accidental Release Prevention program (CalARP) regulations as they apply to petroleum refineries.

2. Regulatory Revisions

As a result of the above initiative, a new section, 5189.1 *Process Safety Management for Petroleum Refineries*) [1], was added to the Cal/OSHA regulation that specifically targets petroleum refineries. The latest version of the change to the Cal/OSHA regulation was published in July 2016 and is expected to be finalized in 2017.

The regulation includes revision of existing PSM elements and addition of new PSM elements and topics (see Figure 1). Damage mechanism review (DMR) is a new process safety element that has specific requirements and needs to be considered and incorporated into other PSM elements (see Figure 2). This paper focuses on implementing the DMR requirement into a process hazards analysis (PHA).

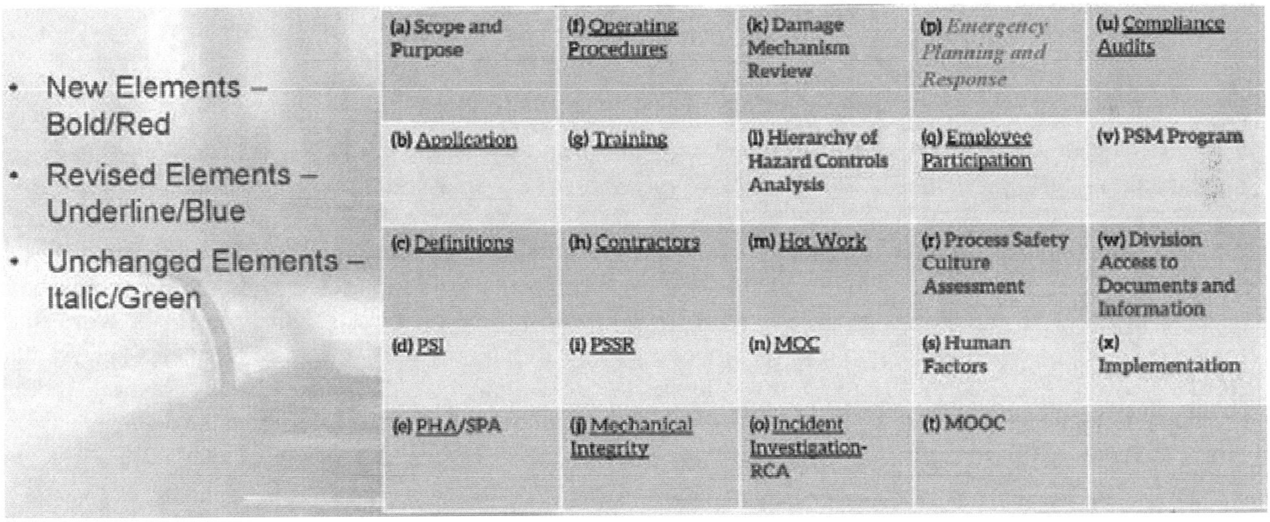

Figure 1. Regulatory Revisions

Figure 2. Incorporating DMR into Other PSM Elements

3. What Is a DMR?

Cal/OSHA's new regulation, *Process Safety Management for Petroleum Refineries,* defines a damage mechanism as *a mechanical, chemical, physical, or other process that results in equipment degradation.*

A DMR is an analysis process designed to determine credible damage mechanism susceptibilities of fixed equipment. Such equipment includes process vessels, heat exchangers, process piping, storage tanks, and process heaters. The analysis process is performed circuit-by-circuit by a multi-disciplinary team that typically includes operations, maintenance, mechanical, materials, and inspection personnel. The results from a DMR can be used to formulate inspection plans to mitigate the risk of a loss of containment and/or unplanned outages.

4. Importance of Performing a DMR

Many refining and chemical processes are not always steady state operations. This may be due to process unit feedstock variations and/or upsets in operation. When processes and operating parameters change, new damage mechanisms may be introduced as well as acceleration of existing damage mechanisms. Thus, an understanding of the process/operating conditions and the potential resulting damage mechanisms is often needed in order to establish and maintain a robust mechanical integrity program (see NACE MR0103, API 510, API 570, API RP 580, and API RP 581) [4, 6, 7, 8, 9].

Understanding damage mechanisms is important for several reasons, including:

- Selecting appropriate equipment inspection intervals/due dates, locations, and techniques,
- Making decisions (e.g., modifications to a process, materials selection, monitoring frequencies) that can help eliminate or reduce the probability of a specific damage mechanism, and
- Determining the type of failures that could occur.

5. How to Perform a DMR

DMRs are typically performed using a systematic approach (see Figure 3). The six steps are (1) obtain data, (2) develop information, (3) conduct DMR, (4) identify DMs, (5) document DMR, and (6) use DMR results.

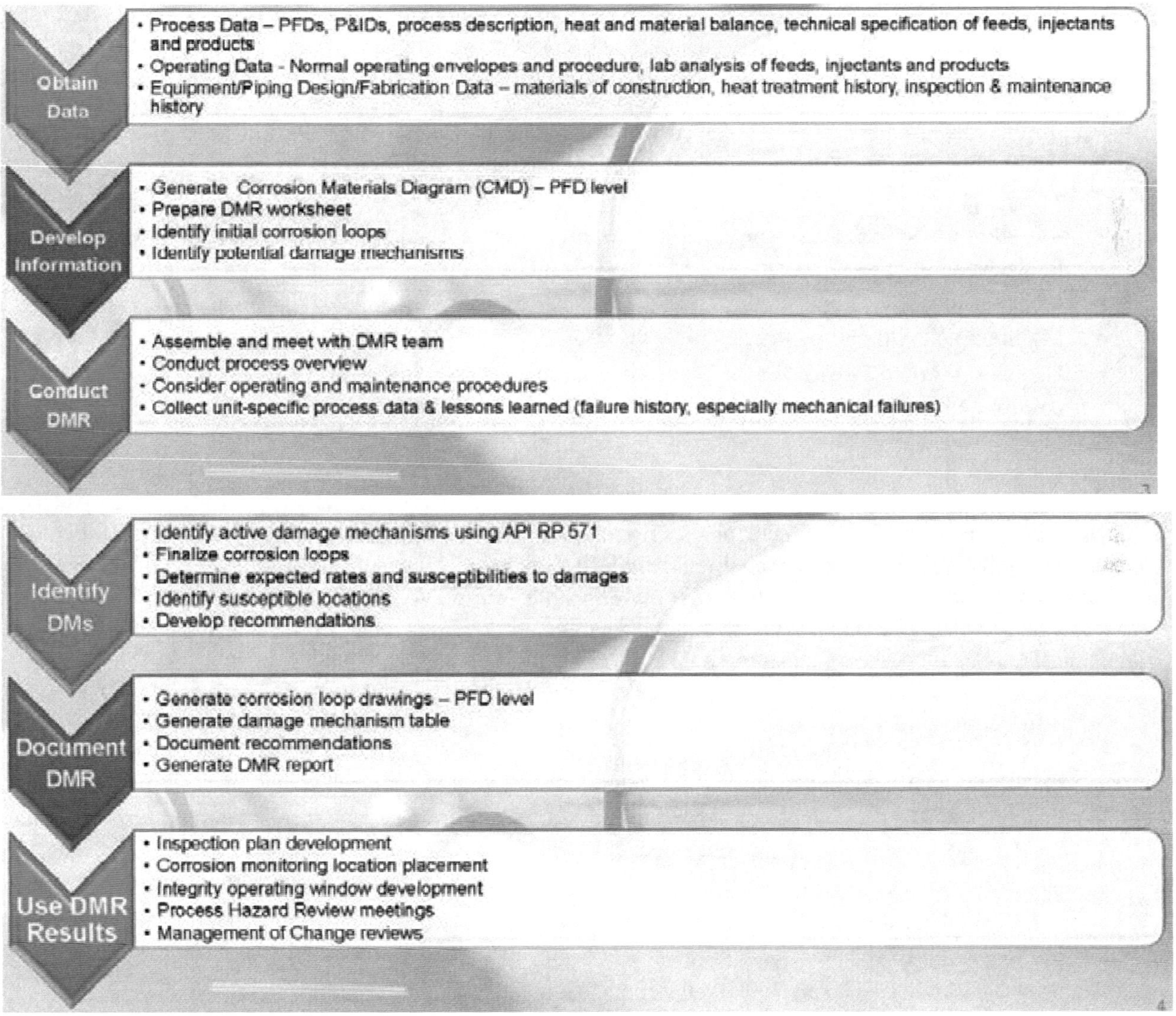

Figure 3. How to Perform a DMR

1. Obtain Data

Data required in performing a DMR include process data, operating data, and equipment/piping design/fabrication data. Site-specific information associated with risk-based inspection studies, preventive maintenance inspections, results from inspections, and failure history should be available electronically for reference during the study.

2. Develop Information

A DMR includes (1) generating a corrosion materials diagram, (2) preparing a DMR worksheet, (3) identifying initial corrosion loops, and (4) identifying potential damage mechanisms.

Developing corrosion loops is an important step in performing a DMR. A corrosion loop is defined as a section of a process that shares common damage mechanism susceptibilities with similar anticipated rates of damage. A corrosion loop is typically comprised of similar materials exposed to similar process/environmental conditions and asset characteristics (materials of construction, insulation type, coatings, heat tracing present, buried, etc.).

Corrosion loops are developed using the following guidelines:

- Maintain a single process chemical (for example, a major indication of a loop boundary is when the service that is flowing through the piping undergoes a change that greatly affects the composition),
- Maintain a single phase, and
- Use the inlet/outlet of a major piece of equipment (pressure vessel, separator, heat exchanger, column, etc.) as a loop boundary.

3. Conduct DMR

A DMR team typically includes a corrosion engineer/specialist, unit inspector, unit process engineer, unit operations representatives, unit mechanical/reliability engineer, unit subject matter experts for licensed technologies (if needed), and a facilitator knowledgeable in the DMR process. The team (1) conducts a process overview, considering operating and maintenance procedures, and (2) collects unit-specific process data and discusses mechanical failure history.

The steps in conducting a DMR include:

1. Developing color-coded process flow diagrams (PFDs) to illustrate various corrosion loops in each process. PFDs, P&IDs, piping isometric drawings, and process stream heat and material balances are normally used to generate color-coded PFDs.
2. Collecting materials of construction and fabrication records to be added to the PFDs so corrosion materials diagrams (CMDs) can be developed. The CMD is a modified process flow diagram or database containing relevant equipment and piping corrosion mechanisms, operating conditions, materials of construction, and corrosion circuits.

4. Identify Damage Mechanisms

Damage mechanisms are identified from a review of established corrosion/materials literature, company-specific studies, and industry guidance documents such as API RP 571 [3].

API RP 571 classifies damage mechanisms into the following major groups:

- Mechanical and Metallurgical Failure Mechanisms,
- Uniform or Localized Loss of Thickness,
- High Temperature Corrosion (>400°F [204°C]), and
- Environment – Assisted Cracking.

Expected damage rates, susceptible locations, and susceptibilities to damages are documented, and recommendations are developed as needed.

5. Document DMR

A DMR report contains the information needed to understand materials damage issues in a specific type of operating process unit at a plant site. The detailed DMR report includes corrosion loop drawings, corrosion damage mechanism tables, and documented findings and recommendations from the DMR. It is recommended that the report include the team member composition, the DMR approach, and data used in performing the review.

6. Use DMR Results

Once completed, the DMR report can be used in a variety of ways, including (1) inspection plan development, (2) corrosion monitoring location (CML) placement, (3) integrity operating window development, (4) PHA meetings, (5) management of change reviews, and (6) incident investigations.

6. How to Address the Results of a DMR in the PHA

Cal/OSHA's new regulation states that …"*the PHA shall address applicable DMR reports*." Thus, a PHA conducted to meet that regulation now requires consultation with personnel (can be a part-time team member) with expertise in damage mechanisms during PHA meeting sessions. The DMR report serves as additional process safety information for reference during the PHA. To help effectively address this requirement, the primary PHA methodology (e.g., HAZOP and/or what-if analysis) may be supplemented with a DMR-specific checklist to help the PHA team identify hazardous scenarios, while considering the results of any DMR report(s) applicable to the process unit. The checklist discussion during a PHA can serve as an effective transfer of knowledge of unit-specific damage mechanisms to unit personnel.

It should be noted that some damage mechanisms may be identified using the primary hazard evaluation technique (HAZOP and/or what-if analysis) during the PHA. However, supplementing the primary hazard evaluation with a specific checklist allows the PHA team to further address damage mechanisms in a systematic fashion to ensure more thorough coverage than that afforded by the primary analysis technique(s) alone.

The checklist approach also helps the PHA team (1) discuss the contents of the DMR report(s) prepared for the process unit and (2) communicate how the damage mechanisms are being managed by the mechanical integrity group for the process unit. The checklist can also guide the discussion of hazards and potential major incidents associated with any identified damage mechanisms.

Table 1 illustrates a suggested corrosion and materials DMR checklist that can be used by a PHA team that has expertise in damage mechanisms. The application and contents of the checklist can help the PHA team confirm items such as whether (1) known corrosion and damage mechanisms (e.g., high temperature hydrogen attack, chloride stress corrosion cracking) are understood and controlled, (2) the existing DMR report appropriately documents applicable damage mechanisms in the process unit, (3) a DMR is performed for incident investigations where a damage mechanism is identified as a contributing factor, and (4) a process unit change (i.e., management of change) has an effect on damage mechanisms or integrity operating windows.

The contents of the checklist include a set of questions that allows the PHA team to confirm that the DMR requirements of Cal/OSHA's regulation as it applies to the PHA element are appropriately met along with questions that deal with the specific damage mechanisms known to occur in petroleum refineries. The questions are related to specific types of damage mechanisms (e.g., high temperature, low temperature corrosion and stress corrosion cracking, external damage, mechanical damage, and other) and include questions developed from industry standards as well as incidents related to mechanical integrity issues [2,4,5].

The checklist is constructed to include topic-specific items and questions to help a PHA team leader elicit responses from the team. The intent of the questions is to allow a PHA team to brainstorm and discuss the damage mechanism report prepared for the process unit being analyzed, along with ensuring that the report was prepared to include damage mechanisms applicable for all potential operating modes (e.g., normal startup and shutdown). This is similar to a PHA team using a facility siting checklist in a PHA to help identify any facility siting issues while considering the results of any separate facility siting study report conducted for the process unit (per API RP 752 and/or API RP 753) [10, 11].

Potential responses to the questions could include an acknowledgement that the PHA team understands which specific damage mechanisms may be present in the process unit and a discussion of how the damage mechanisms are managed (i.e., inspection programs, defined integrity operating windows, etc.). If a corrosion mechanism listed in the checklist is not present in the process unit being analyzed, the team can respond by indicating it is not applicable.

The checklist discussion may also help to (1) identify other damage mechanisms that may not have been identified during preparation of the DMR report that could occur from a process parameter deviation (flow, temperature, pressure) and (2) ensure mechanical integrity plans are robust and have taken into consideration any process changes identified by the PHA team.

7. PHA Documentation Tips

To confirm that a DMR was adequately addressed and was consistent with Cal/OSHA's expectations, consider using these guidelines when preparing the PHA report:

- Explicitly describe "how" the PHA team addressed/considered damage mechanisms in the process unit being analyzed,
- Reference the damage mechanism report prepared for the process unit for which the PHA is being performed,
- List the team members, including any individuals with expertise in damage mechanism, in the PHA report,
- Include the DMR checklist used for the evaluation,

- Use language in the PHA documentation that shows damage mechanisms were considered during the meeting(s), and
- When applicable, write recommendations to help mitigate any hazards related to damage mechanisms identified by the PHA team.

8. Conclusion

While conducting DMRs and incorporating their results in PHAs and other elements of a process safety management framework are not currently required by most regulatory agencies, state regulators in California have begun requiring them for refineries. As industry continues to recognize the value of conducting and documenting DMRs (per guidance in API RP 571) to enhance mechanical integrity programs and mitigate the risk of loss of containment events, the results of these DMRs can also provide additional relevant information for assisting PHA (and management of change) review teams in identifying potential hazardous scenarios for processes. A checklist, such as the one presented in Table 1, can be an effective method for PHA teams to efficiently identify and discuss DMR issues identified in separate, detailed DMR reports when conducting their hazard evaluations.

9. References

1. Title 8, Division 1, Chapter 4 Subchapter 7. General Industry Safety Orders Group 16. Control of Hazardous Substances Article 109. Hazardous Substances and Processes, §5189.1. *Process Safety Management for Petroleum Refineries* (Draft Regulation), July 2016.
2. *Corrosion Control Documents*, Downstream Segment, API Recommended Practice 970, Proposed First Edition, March 2016.
3. *Damage Mechanisms Affecting Fixed Equipment in the Refining Industry*, Downstream Segment, API Recommended Practice 571, Second Edition, April 2011.
4. NACE Standard MR0103-2003, *Materials Resistant to Sulfide Stress Cracking in Corrosive Petroleum Refining Environments*, Houston, Texas.
5. ABSG Consulting Inc. Course 110 *Process Hazard Analysis Leadership Training*, Resource Material Tab Mechanical Integrity Issues Checklist.
6. *Pressure Vessel Inspection Code: In-service Inspection, Rating, Repair, and Alteration,* API 510, 10th Edition, May 2014.
7. *Piping Inspection Code: In-service Inspection, Repair, and Alteration of Piping Systems,* API 570, 4th Edition, February 2016.
8. *Risk-Based Inspection,* API RP 580, 3rd Edition, February 2016.
9. *Risk-Based Inspection Methodology*, API RP 581, 3rd Edition, April 2016.
10. *Management of Hazards Associated with Location of Process Plant Permanent Buildings*, API RP 752, 3rd Edition, December 2009.
11. *Management of Hazards Associated with Location of Process Plant Portable Buildings*, API RP 753, 1st Edition, June 2007.

Table 1. Process Hazards Analysis Corrosion and Materials Damage Mechanism Review (DMR) Checklist

Item No.	Question	Response	Recommendations
I. General			
1	Does a DMR report exist that addresses the applicable damage mechanisms for the process unit? If not, was there a determination that no damage mechanisms exist and was the rationale documented?		
2	Was the DMR report for the process unit available to the PHA team?		
3	Are the known corrosion and materials damage mechanisms understood and controlled? (Note: Refer to sections II through V of this checklist)		
4	Was a DMR conducted where a damage mechanism exists on (1) new processes and/or (2) major changes that have been implemented in the process unit prior to initial startup and/or implementation of the change?		
5	Have any incident investigations for the unit identified a damage mechanism as a contributing factor of the incident?		
II. High Temperature Damage Mechanisms			
1	Is there a potential for carbon steel and/or 0.5Mo steel to exceed the Nelson curve and be exposed to high temperature hydrogen attack (HTHA)? Include consideration of piping in a normally closed bypass being overheated.		
2	Is there a potential for accelerated creep due to operating outside of the operating window (e.g., higher temperature)?		

Table 1. Process Hazards Analysis Corrosion and Materials Damage Mechanism Review (DMR) Checklist (cont'd)

Item No.	Question	Response	Recommendations
3	Is there a potential for liquid metal embrittlement?		
4	Is there a potential for accelerated sulfidic corrosion from a gradual increase in sulfur content?		
5	Is there a potential for accelerated corrosion from increased concentration of naphthenic acids in crude oil or various distillation cuts?		
6	Is there a potential for accelerated corrosion as a result of organic chloride contamination in crude oil or other feeds?		
7	Is there a potential for spontaneous ignition of materials (including titanium, zirconium, structured packing, iron sulfide, etc.) from HCl, Cl_2, O_2, and/or air?		
8	For alloys operating above 700 °F, are there any high temperature aging embrittlement phenomena that might lead to brittle behavior when equipment is pressurized at lower temperatures?		
III. Low Temperature (Aqueous) Corrosion and Stress Corrosion Cracking Damage Mechanisms			
1	Is there a potential for severe corrosion downstream of alloyed equipment due to process upsets that cause unneutralized corrosive materials or contaminants to flow downstream into systems constructed of materials not designed for such corrosive conditions?		

Table 1. Process Hazards Analysis Corrosion and Materials Damage Mechanism Review (DMR) Checklist (cont'd)

Item No.	Question	Response	Recommendations
2	Is there a potential for rapid corrosion at or downstream of injection or process mixing points due to heating/cooling, condensation/evaporation, reaction, between the injecting and mixing streams, etc.?		
3	Is there a potential for rapid corrosion at or near an injection point due to a change in flow rate, a change in flow pattern, or failure of the injection system to perform as intended?		
4	Is there a potential for rapid corrosion due to a change in flow rate or flow regime (e.g., vaporization, flashing, or other multiphase flow conditions)?		
5	Is there a potential for localized corrosion or cracking of austenitic stainless steels during shutdowns where polythionic acids may be present?		
6	Is there a potential for rapid localized corrosion at hot spots as a result of direct contact between heat tracing and process piping (e.g., lack of standoff or improper use of heat transfer material)?		
7	Is there a potential for accelerated corrosion from fluids and/or solids trapped in deadlegs?		
8	Is there a potential for process upsets to introduce moisture into moisture-free environments (e.g., anhydrous HCl or Cl_2) or to result in insufficient moisture in environments reliant on moisture for the protection of the system (e.g., anhydrous NH_3 and alcohols)?		

Table 1. Process Hazards Analysis Corrosion and Materials Damage Mechanism Review (DMR) Checklist (cont'd)

Item No.	Question	Response	Recommendations
9	Is there a potential for accelerated corrosion by the activation or deactivation of a piping segment (e.g., a bypass line or other alternate flow path) or the installation of a temporary flow path?		
10	Can changes in pump or compressor capacity (including the use of a spare pump/compressor) lead to increased corrosion rates?		
11	Can increase or decrease in pH (inadvertent or undetected) lead to increased corrosion or stress corrosion cracking?		
12	Can minor changes in feed compositions (e.g., loss of trace amounts of corrosion inhibitors, presence of organic chlorides, etc.) adversely affect corrosion rates?		
13	Can non-post weld heat treated equipment containing caustics or amines be steam cleaned/purged leading to a potential for cracking?		
14	Is there a potential for air and/or moisture ingress into otherwise oxygen-free and/or dry environments (e.g., breather valves on tanks) that could cause accelerated corrosion or cracking?		
15	Can solids be present causing increased erosion and/or corrosion (e.g., catalyst carryover, accumulation of corrosion products, etc.)?		

Table 1. Process Hazards Analysis Corrosion and Materials Damage Mechanism Review (DMR) Checklist (cont'd)

Item No.	Question	Response	Recommendations
16	Is there a potential for increased corrosion or cracking downstream of piping specification breaks due to process changes or upsets?		
17	Is there potential for increased corrosion in a pipe or tube inlet zone or at a vapor/liquid interface?		
18	Is there a potential for caustic cracking of bolted connections from small leaks of boiler feed water or other caustic solutions? Is there potential for caustic cracking of non-stress relieved equipment from heat exchanger tube leaks or caustic carryover?		
19	Is there a potential for bolting to be exposed to process environments that could lead to catastrophic fracture due to stress corrosion cracking such as caustic, amine and wet H_2S, etc.?		
20	Is there a potential for process contamination that could cause stress corrosion cracking (e.g., wet H_2S, caustic, amines, chlorides, polythionic acids) or corrosion fatigue of deaerators?		
21	Is there a potential for accelerated corrosion from the installation of incorrect materials of construction in highly hazardous systems?		
22	Is there a potential for localized corrosion or cracking of heat-affected zones in sensitized stainless steels?		
23	Is there a potential for hydrogen embrittlement or hydrogen blistering from increased concentrations of hydrogen sulfide or cyanides?		

Table 1. Process Hazards Analysis Corrosion and Materials Damage Mechanism Review (DMR) Checklist (cont'd)

Item No.	Question	Response	Recommendations
24	Is there a potential for sulfuric acid concentrations falling below critical limits due to process upsets or changes and causing accelerated corrosion?		
25	Is there a potential for temperature, pressure, or other process changes that cause shifts in dew point, and therefore, create or shift areas where corrosive compounds condense leading to accelerated corrosion?		
IV. External Damage Mechanisms			
1	Is there a potential for unmitigated and undetected external corrosion that can lead to rupture (e.g., soil-to-air interface, downstream of cooling water sprays, etc.)?		
2	Have there been changes in the monitoring or maintenance of cathodic protection systems that could lead to increased corrosion of buried piping or tank bottoms?		
3	Can changes in process conditions lead to increased corrosion under insulation, (e.g., idling of normally hot equipment, equipment in cyclic service above and below 250 °F, or exposure of stainless steel equipment to external chloride cracking)?		

Table 1. Process Hazards Analysis Corrosion and Materials Damage Mechanism Review (DMR) Checklist (cont'd)

Item No.	Question	Response	Recommendations
V. Mechanical and Other Damage Mechanisms			
1	Is there any potential for low temperatures causing freezing of water in piping deadlegs, especially in light hydrocarbon systems, resulting in piping rupture?		
2	Is there a potential for liquid slugging of piping (including flare lines) that could cause piping failure due to hydraulic shock, water hammer, and other transient overstress conditions?		
3	Is there a potential for accelerated corrosion, fouling, or plugging of inlet or outlet piping of relief devices or flare systems?		
4	Has the impact on relief capacity been reviewed for any increases in unit throughput?		
5	Can temperature changes on piping systems cause mechanical overload because of lack of expansion capability in piping design?		
6	Is there a potential for overstressing or shocking brittle materials of construction (e.g., cast iron, aged or embrittled steels)?		
7	Is there a potential for overpressuring equipment or piping when using high pressure fluids and/or positive displacement pumps to unplug a line?		
8	Is there a potential for brittle fracture of heavy wall equipment from being fully pressurized before the metal temperature has reached approximately 200 °F?		

Table 1. Process Hazards Analysis Corrosion and Materials Damage Mechanism Review (DMR) Checklist (cont'd)

Item No.	Question	Response	Recommendations
9	Is there a potential for pressurization or impact during low temperature conditions that may cause brittle fracture of equipment that is not designed for such conditions (i.e., stress applied below minimum design metal temperature)?		
10	Is there a potential for vibration (either applied or induced) that could lead to fatigue fracture of vibrating piping, over-stressed threaded connections, unsupported or over-hung weight, or exchanger tubes?		
11	Is there a potential for thermal fatigue cracking due to sudden or severe swings in temperature?		
12	Is there a potential for cavitation from rapid pressure or temperature changes?		
13	Could heat tracing be inadvertently shut off on critical equipment or relief systems and result in impairment (e.g., plugging leading to failure of a relief system to function on demand)?		
14	Are there check valves in the process that can lead to a process safety incident if they fail to function .properly?		
15	Are there any vents and drains downstream of block valves that may require additional protection from pipe plugs/caps to prevent releases?		
16	Are temporary repairs and clamps identified for scheduled removal at the next maintenance opportunity?		

Table 1. Process Hazards Analysis Corrosion and Materials Damage Mechanism Review (DMR) Checklist (cont'd)

Item No.	Question	Response	Recommendations
17	Can the failure or malfunction of internal elements such as exchanger bundles, trays, distributors, steam rings, etc., lead to overpressure, accelerated corrosion, and/or metallurgical damage to equipment and piping?		
18	Is there a potential for temperature changes that might cause brittle fracture of equipment constructed of materials with low toughness due to (1) thermal shock during steady-state operation or (2) transient conditions experienced in normal startup or shutdown of heavy walled vessels?		
19	Is there a potential for excessive temperatures or hot spots that could cause rupture from short-term overheating of furnace tubes, transfer lines, or catalyst-containing vessels?		
20	Is there a potential for presence of molten salts or liquid metals that could cause very high rates of corrosion or cracking?		
21	Is there a potential for liquid carryover into gas streams or velocity changes of mixed phase streams causing accelerated corrosion-erosion of elbows, tees, and other areas subject to downstream turbulence (e.g., downstream of control valves)?		
22	Is there a potential for changes in feed compositions where trace amounts of certain chemical compounds previously provided corrosion inhibition?		

Table 1. Process Hazards Analysis Corrosion and Materials Damage Mechanism Review (DMR) Checklist (cont'd)

Item No.	Question	Response	Recommendations
23	Is there potential for accelerating corrosion rates in sour hydrocarbon service from gradual increase of sulfur content or inadvertent increase in content of other sulfur compounds such as H_2S or mercaptan?		
24	Is there a potential for changes in water treating chemistry or procedures that could accelerate water-side corrosion?		
25	Is there any potential for ammonium hydrosulfide, ammonium chloride, or amine salt deposition and resultant corrosion rate acceleration with increased concentrations of ammonia, sulfides, or chlorides or insufficient water washing?		
26	Are there any other known corrosion mechanisms that may be of concern during normal, abnormal, or non-routine operating modes that are unique to the process unit? If so, are they currently managed by the mechanical integrity group?		
27	Is there a potential for brittle fracture (e.g., hydrogen embrittlement, temper embrittlement, 0.5Mo steel) of heavy wall hydro processing equipment from rapid heating/cooling or pressurizing when the equipment temperature is below the minimum pressurization temperature?		
28	Is there a potential for catastrophic brittle fracture (hydrogen embrittlement) of hydro processing equipment from rapid cooling?		

The Integration of Inherently Safer Design and Control

Denis Su-Feher
Mary Kay O'Connor Process Safety Center, Texas A&M University
College Station, 77843-3122, USA
denis@tamu.edu

Yogesh Koirala[1,2], Efstratios Pistikopoulos[2], & M. Sam Mannan[1,2]
[1] Mary Kay O'Connor Process Safety Center, Texas A&M University
[2] Artie McFerrin Department of Chemical Engineering, Texas A&M University
College Station, 77843-3122, USA
ykoirala@tamu.edu
stratos@tamu.edu
mannan@tamu.edu

Prepared for Presentation at
American Institute of Chemical Engineers
2017 Spring Meeting and 13th Global Congress on Process Safety
San Antonio, TX
March 26 – 29, 2017

The Integration of Inherently Safer Design and Control

Denis Su-Feher
Mary Kay O'Connor Process Safety Center, Texas A&M University
College Station, 77843-3122, USA
denis@tamu.edu

Denis Su-Feher[1,2], Yogesh Koirala[1,2], Efstratios Pistikopoulos[2], & M. Sam Mannan[1,2]
[1] Mary Kay O'Connor Process Safety Center, Texas A&M University
[2] Artie McFerrin Department of Chemical Engineering, Texas A&M University

Keywords: Process Design & Development, Safe Design, Process Automation & Control

Abstract

Current inherently safer design strategies in the conceptual design stage focus on reducing the overall hazard of a new refinery without considering the operability of the process. The process is first designed to be inherently safer with respect to a nominal, steady-state case and then, after the process is designed, layers of protection are added and operability issues are addressed. However, this sequential design approach does not account for the impact of the design itself on the operability of the refinery. A particular design may be safer with respect to its steady-state operation, but may suffer from operability issues. For example, an intensified process may contain less of a hazardous substance and thus be inherently less hazardous, but the design may restrict the controllability of the process, thus making the design have a higher risk and be less safe overall. A considerable amount of research has been done to simultaneously optimize the design and control system of refineries, but no such approach has integrated inherent safety, only seeking to produce an economically optimal design rather than a safer one.

The objective of this research is to implement a strategy to simultaneously design and control an inherently safer refining plant. Parametric Optimization and Control (PAROC), a simultaneous design and control framework will be used as a basis for the design and control of a refining process. Different inherently safer design indices will be integrated into PAROC, and the operability, safety, and economic profitability of the results will be compared.

An extensive literature review has identified indices by which controllability of a refining process can be quantitatively characterized and optimized along with the design, as well as inherently safer design indices that can be implemented into the PAROC framework. Simultaneous use of these inherently safer design and controllability indices will be compared to create a new index for the integration of inherently safer design and control.

With refinery systems and their dynamics becoming increasingly complex, consideration of operability issues in the design stage becomes even more necessary to prevent incidents. The integration of inherently safer design and control will substantially reduce operability issues that result from an uncontrollable process design and allow for greater tolerance and ease of control.

1 Introduction

The objective of an industrial refinery is to be both profitable and safe. Traditionally, the oil and gas industry has optimized the design of refineries primarily for profitability, accepting the existence of hazards in a design and managing the risk involved by implementing prevention, mitigation, and response methods after the design has been implemented. However, in 1943, Ziegler and Nichols determined that the design of a process can influence how well process control can reduce the effects of deviations and uncertainty in the process [1]. As a result of this realization, it became necessary to optimize both profitability and operability, as failing to consider operability could lead to a shorter refinery life and potentially unsafe conditions.

A number of operability characteristics were identified since then, such as controllability and flexibility [2, 3]. Controllability is the process's ability to reach and maintain a steady-state. Flexibility is the process's ability to stay at steady-state operation despite disturbances and uncertainty. Each of these characteristics deal with how well the process control system can maintain control on the design, and as such they are affected by the design itself and must be accounted for even in the initial stages of the design.

If a design has high operability, then it is likely to be a safer design. A process with high operability is capable of operating within its boundaries and uncertainties and deviations are unlikely to have an effect on a high operability design. If deviations to the process can be accounted for and an appropriate process control system is put in place, then process safety incidents are less likely to occur.

Unfortunately, high operability of the process control system does not guarantee safety. The plant safety depends on both consequence and probability of incidents that may occur. Although increasing operability will decrease the likelihood of incident occurrence, operability fails to incorporate the consequences involved with the failure. As a result, although the total number of failures are minimized, high consequence failures are not treated with the severity they require. Therefore, it is necessary to incorporate consequence measurement into the initial refinery design.

The aim of the present work is to define a process to integrate inherently safer design and control in order to create simultaneously profitable, operable, and safe processes. Several inherent safety indices are introduced to quantify consequence, and several controllability indices are introduced to quantify operability. These safety and controllability indices are used simultaneously with the Parametric Optimization and Control (PAROC) Framework as a case study to demonstrate how safety can be integrated into optimization for inherently safer design and control.

1.1 Relevant Works

1.1.1 Inherently Safer Design

Inherently Safer Design (ISD) is the idea that the level of safety in a refinery can be improved by modifying the base design of a process by reducing or removing hazards. The main principles that it defines and quantifies are shown in Table 1 [4].

Table 1. Inherent Safety Principles and Definitions [4]

Inherent Safety Principle	Definition
Intensification	Using smaller quantities of hazardous substances
Substitution	Using less hazardous materials in the process
Attenuation	Using less hazardous conditions; using the same material in a less hazardous form
Simplification	Reducing complexity in operational design to make operational errors less likely
Limitation of Effects	Reducing the severity of consequences associated with process design
Error Tolerance	Making the process resilient to disturbances
Ease of Control	Reducing the amount and complexity of control necessary to operate the process

Although there are methods such as Hazard and Operability Studies (HAZOP), Fault Tree Analysis (FTA), and Failure Modes and Effects Analysis (FMEA) that can be used to determine the consequence of incidents, all of them require detailed information that may not be available in early stages of plant design. Quantitative inherently safer design indices, on the other hand, require relatively little information and can be used to assist risk-based decision-making in the early stages of refinery design.

Furthermore, ISD is ideal for the initial design stage because the number of available inherent safety options is most abundant then, and will only decrease as the design proceeds, as shown in Figure 1 [5]. Although opportunities for prevention and mitigation appear in the middle of the design process, preventative and mitigative barriers later are often costly and not as effective at reducing the hazard inherent to the design. Much like traditional operability, inherent safety must be considered as early as possible, or costlier and/or less effective prevention and mitigation methods will have to be used instead.

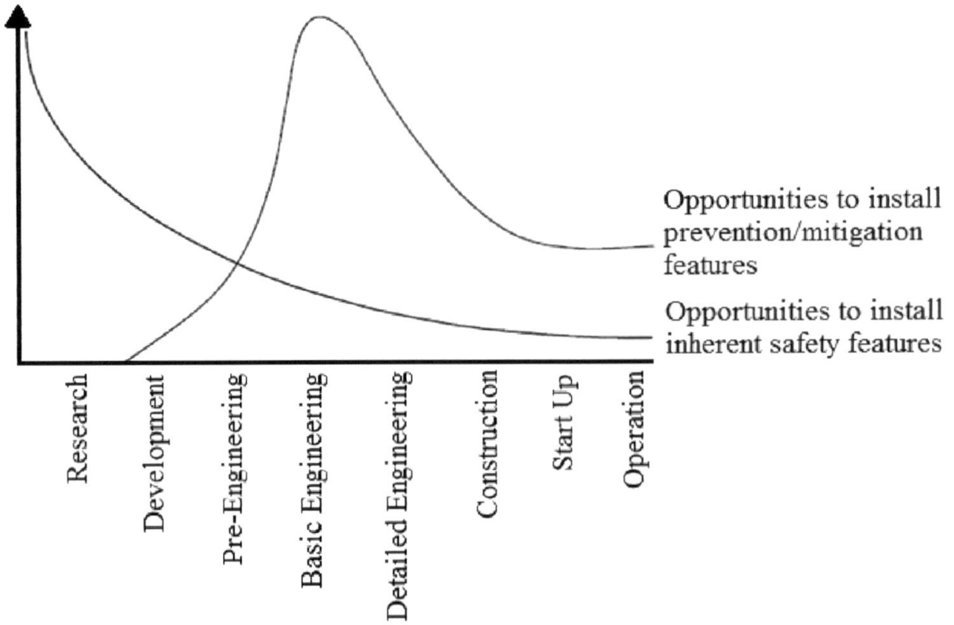

Figure 1. Inherent vs preventative safety in various stages of plant design [5].

The need for better risk assessment methods for the safer design of chemical processes has driven a substantial amount of research for comprehensive ISD indices. Ever since Trevor Kletz popularized ISD in his 1978 article entitled, "What you don't have can't leak," there have been numerous indices published that attempt to quantify different principles of inherent safety [6]. Table 2 lists these indices and summarizes some of these indices and their contributions.

Most of the current literature on inherent safety has focused on identifying the effects of intensification, substitution, attenuation, simplification, and, to an extent, limitation of effects. These principles quantify the hazard, or the potential consequences that might occur in a safety incident.

Ease of control and error tolerance, on the other hand, represent the effect of the design on operability, and the complexity of the controller needed to maintain control over the process. These principles quantify the likelihood of the hazard causing a process safety incident, and can be quantified with controllability indices.

Table 2. Inherently Safer Design Indices and their Contributions

Inherently Safer Design Indices	Contributions
Dow F&EI [7]	First safety index reported in literature, and most commonly used in industry. Evaluates material properties and process conditions.
Dow CEI [8]	Evaluates toxic effects and distance to neighbouring plants and communities to assess acute health hazard potential to them.
Mond Index [9]	Extends Dow F&EI by evaluating toxicity of chemicals.
INSET Toolkit [10]	Provides a guide to creating an inherent health and safety index.
HIRA [11]	Integrated material factor with operating conditions based on thermodynamic properties and expert opinions to find a damage radius under given overall conditions.
PIIS [12]	First inherent safety index reported to analyse the predesign of a process. Quantifies substitution by measuring the reaction route hazards.
ISI [13]	Expanded PIIS. Quantifies minimization, moderation, substitution and simplification in terms of a ranking.
I2SI [14]	Extends ISI by weighing thermodynamic data and expert opinions to determine a risk ranking in terms of damage radius.
PSI [15]	Quantifies inherent safety of process streams during simulation work in the preliminary design stage that influences explosions.
RISI [16]	Extends I2SI by evaluating frequency as well as consequence to obtain a risk ranking.

1.1.2 Controllability

Controllability is measure of a design's operability, specifically its ability to achieve and maintain a desired equilibrium value despite disturbances and uncertainties. Certain process designs can have regions of instability that an inherently more controllable design may not have [1]. For example, if there is a region of instability near a steady-state, then attempting to control the system at that steady-state might be impossible, and another, less economically feasible steady-state would have to be used instead [17]. These unstable regions would have to be addressed through either more control action, or less optimal set points, making the process system more complex, costlier, and potentially less reliable as a whole. A number of methods have been developed to measure different aspects of the controllability of both linear and nonlinear processes. They are displayed in Table 3. All of these methods take into account dynamic aspects of the process.

Table 3. Controllability Indices

Controllability Index	Summary	Linearity
RGA & BRGA [18] [19]	RGA defines a matrix that measures the effect of manipulated variables on controlled variables. BRGA extends it to block decentralized control.	Linear
NBRGA [20]	Extends BRGA to nonlinear systems.	Nonlinear
IMC [21]	Creates an upper bound of controllability by assuming that the ideal controller is the inverse of the chemical process transfer function.	Linear
RHTP zeros [22]	RHTP zeros define areas where instabilities in control response may occur. The only way to move a RHTP zero is to change the process design. A controllable process must remove as many RHTP zeros as possible.	Linear
Input-Output Controllability [23]	Defines eight rules for a controllable process.	Linear
Nonlinear Zero Dynamics [24]	Quantifies the effect of the inverse responses in nonlinear systems for limited types of systems.	Nonlinear
Relative Order [25]	Defines the pole excess of the transfer function. Quantifies how inputs directly affect outputs.	Nonlinear

It has been shown that for nonlinear systems, controllability analysis based on a linearized form of the system in many cases provides a close estimate, even if the system is strongly nonlinear [26]. However, this does not hold for all cases. At start-up, shutdown, and in batch or semi-batch processes, the controllability of highly nonlinear processes may not be accurately represented, and the accuracy of different control methods have to be validated. In this case, simulation may be needed to validate the controllability index.

The Parametric Optimization and Control (PAROC) framework, as shown in Figure 2, is a tool that provides the means for the simultaneous design and control of process systems [27]. ODE's and PDE's are implemented into optimization software with a first principle based, high-fidelity model analysis of the refinery design. However, because the design of many refineries is too complex to optimize with an economically feasible computation time, model reduction techniques are used to simplify the high-fidelity model into an approximate model. Then, the control system is implemented using multiparametric receding horizon policies. Finally, the difference between these policies and the original high-fidelity model are used to validate the process. With exception of the model reduction of the design and multi-parametric programming to determine the control structure, both of which only need to be performed once, the process is iterated until the error is minimized.

PAROC allows for rapid, repeatable simulation of realistic control structures and the creation of optimal process designs under various conditions. PAROC has been implemented in the gPROMS software, and can be used to solve the simultaneous design and control problem to obtain the optimal process design and controller.

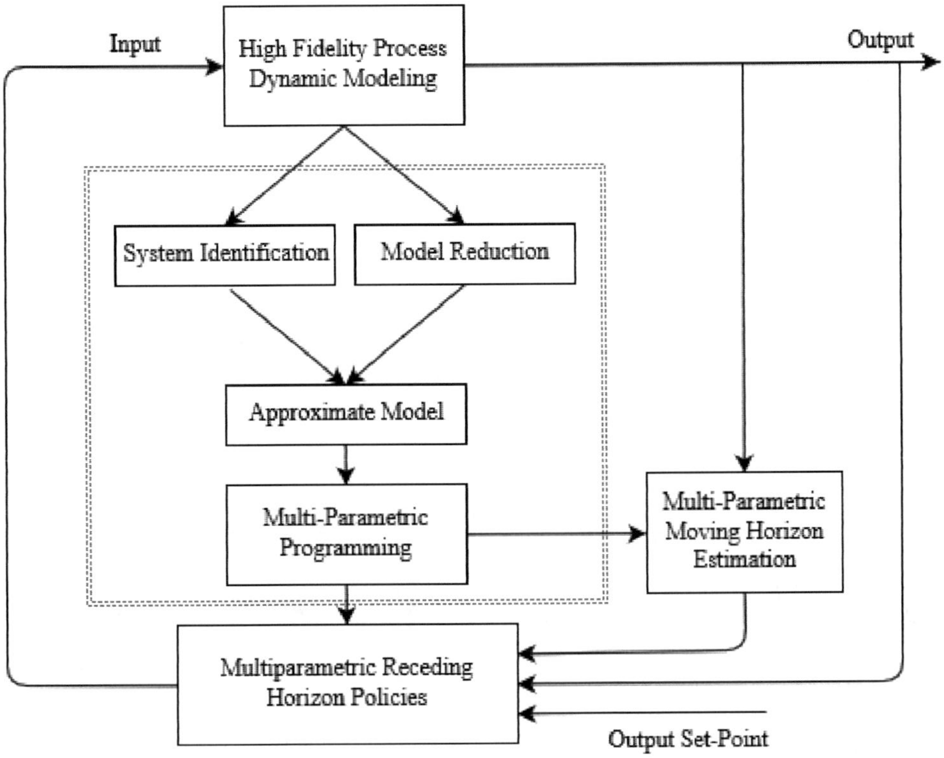

Figure 2. The PAROC Framework [27]. Actions in the dotted box happen once, offline.

2 The Integration of Inherently Safer Design and Control

The integration of ISD and control can be seen as a multi-objective optimization problem, where the objective function is given by (1), and is subject to design constraints.

$$\min\ TAC + w_C I_C + w_S I_S \tag{1}$$

TAC is the design's total annual cost and is represented by a net present value. I_C is the measured controllability based on controllability indices, and is scaled with its weighting factor, w_C. I_S is the measured safety index value, and is scaled with its weighting factor, w_S. Each of these factors respectively represent the utility cost of having a more expensive, less controllable, or higher risk design. Since the range of values for each index is different, and each designer places a different value on each different index, it is difficult to decide on the weighting factors for such a formulation.

To solve this multi-objective optimization problem, ϵ-constraint optimization can be used to reduce the multi-objective problem to a single objective function with two new constraints in addition to the original design constraints [28]. The objective becomes (2) controllability constraints become (3), and the safety constraints become (4).

$$\min TAC \tag{2}$$

subject to new constraints:

$$I_C \leq \epsilon_C \tag{3}$$

$$I_S \leq \epsilon_S \tag{4}$$

By varying the controllability and safety parameters, ϵ_C and ϵ_S, a pareto-optimal solution set can be generated. However, if the system poses high degrees of nonlinearity, then the controllability index will need to be validated. Furthermore, the ease of control factor must be quantified based on the complexity of the required controller. Figure 3 shows how this factor may be quantified.

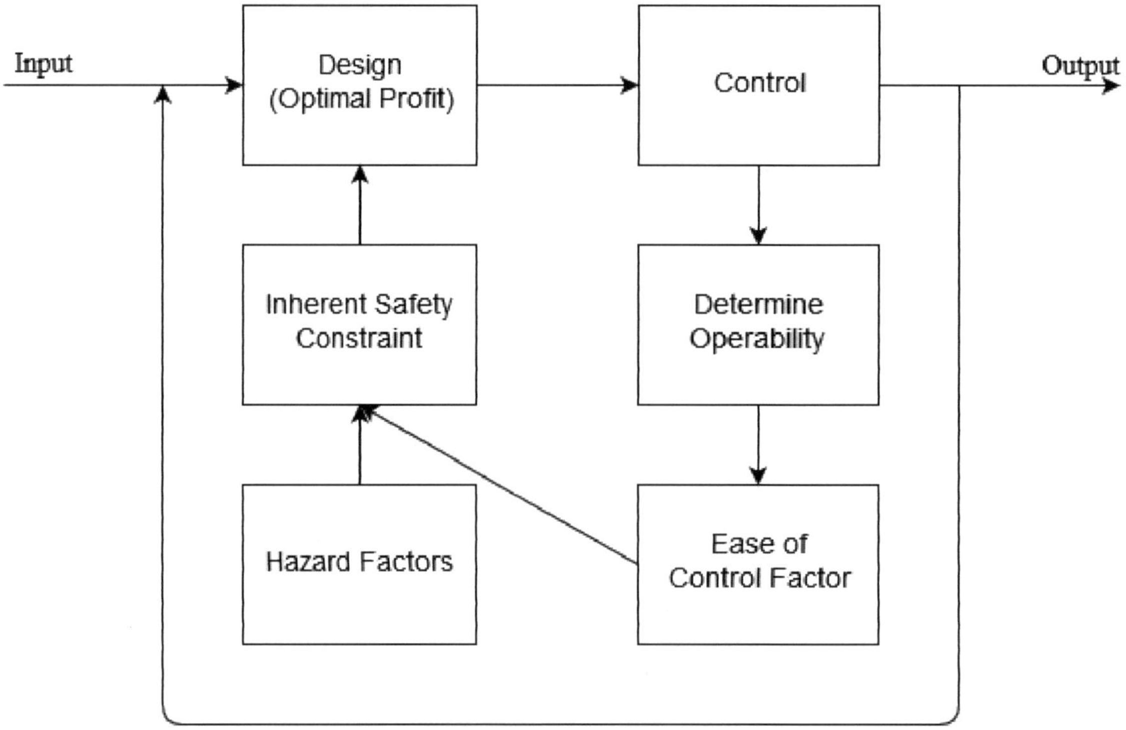

Figure 3. Process for Integrating Design, Control, and Safety

Since a nonlinear system may be too complex to be analyzed accurately with controllability indices, an estimation of the range of control must be performed with variations of the design. The effectiveness of the resulting controllers should be compared with the expected value given by the controllability index, and the controllability index should be evaluated and weighed based on its accuracy. The optimal control scheme of each design must then be analyzed for its complexity, which should be factored into the inherent safety index to quantify the ease of control of the design, a factor of the inherent safety of the design.

3 Case Study in Simultaneous Design and Control: PAROC

Simultaneous design and control guided with controllability and safety indices can be used to produce more optimal designs. The process for integrating safety into simultaneous design and control with PAROC can be seen in Figure 4. The objective function of the design remains the same (2), and controllability (3) and inherent safety (4) index variables and constraints are added to the design of the high-fidelity model of the process.

Model reduction is performed, and an optimal controller is designed from the reduced model. The controller's effectiveness is used to validate the controllability index, and its complexity is used to define the ease of control for the inherent safety index. The weighing factors of the design constraints are updated, and the process repeats until the optimal design is found.

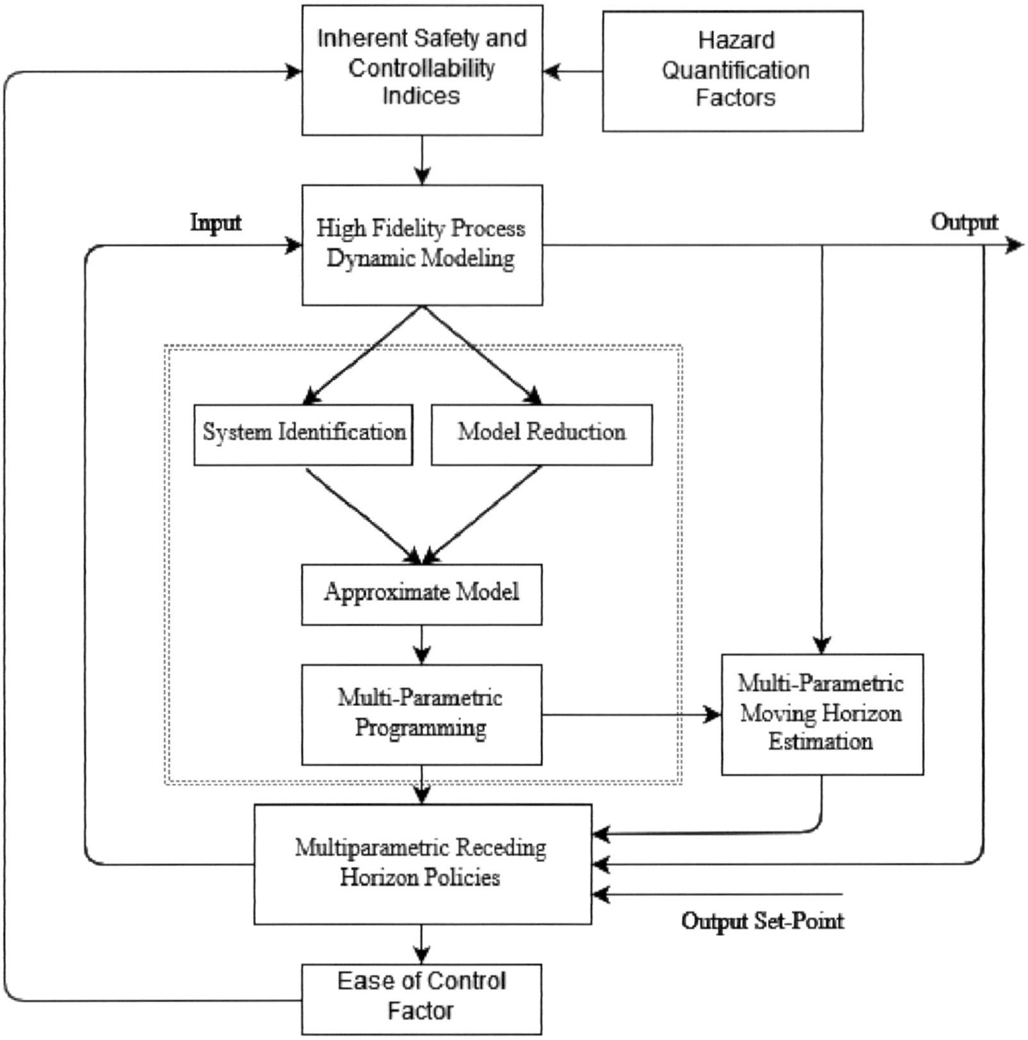

Figure 4. Integrating Safety into the PAROC Framework [27].

4 Conclusions

The background and process for integrating inherently safer design and control are discussed in this paper. A list of inherent safety indices and controllability indices are presented. A multi-objective formulation for integrating inherently safer design with controllability is presented. A case study is performed where inherent safety is integrated into the PAROC framework for simultaneous design and control.

5 References

[1] J. G. Ziegler and N. B. Nichols, "Process Lags in Automatic Control Circuits," *Trans. ASME,* vol. 65, pp. 433-444, 1943.

[2] A. M. Blanco and J. Bandoni, "Interaction between process design and process operability of chemical processes: An eigenvalue optimization approach," *Comp Chem Eng,* vol. 27, pp. 1291-1301, 2003.

[3] V. Sakizlis, J. D. Perkins and E. N. Pistikopoulos, "Recent advances in optimization-based simultaneous process and control design," *Comp Chem Eng,* vol. 28, pp. 2069-2086, 2004.

[4] T. Kletz, Process Plants: A handbook for inherently safer design, Bristol, PA: Taylor & Francis, 1998.

[5] M. Hurme and M. Rahman, "Implementing inherent safety throughout the process lifecycle," *Journal of Loss Prevention,* vol. 18, pp. 238-244, 2005.

[6] T. Kletz, "What you don't have, can't leak," *Chemistry & Industry,* vol. 9124, pp. 289-292, 1978.

[7] AIChE, Dow's Fire and Explosion Index Hazard Classification Guide, 7th ed., New York: American Institute of Chemical Engineers (AIChE), 2016.

[8] AIChE, Dow's Chemical Exposure Index Hazard Classification Guide, 1st ed., New York: American Institute of Chemical Engineers (AIChE), 1994.

[9] B. J. Tyler, A. R. Thomas, P. Doran and T. R. Greig, "The Mond fire, explosion and toxicity index - A development of the Dow Index.," Manchester, 1979.

[10] D. Mansfield, "The INSET Toolkit, INherent SHE Evaluation Tool," London, 1997.

[11] F. I. Khan and S. A. Abbasi, "Multivariate Hazard Identification and Ranking System," *Process Safety Progress,* vol. 17, no. 3, pp. 157-170, 1998.

[12] D. W. Edwards and D. Lawrence, "Assessing the inherent safety of chemical processes," *Trans IChemE,* vol. 71B, pp. 252-258, 1993.

[13] A.-M. Heikkila, "Inherent Safety in Process Plant Design, An Index-Based Approach," Helsinki University of Technology, Espoo, 1999.

[14] F. I. Khan and P. R. Amyotte, "Integrated Inherent Safety Index (I2SI): A Tool for Inherent Safety Evaluation," *Process Safety Process,* vol. 23, no. 2, pp. 136-148, 2004.

[15] A. M. Shariff, C. T. Leong and D. Zaini, "Using process stream index (PSI) to assess inherent safety level during preliminary design stage," *Safety Science,* vol. 50, pp. 1098-1103, 2012.

[16] F. K. P. A. Samith Rathnayaka, "Risk-based process plant design considering inherent safety," *Safety Science,* vol. 70, pp. 438-464, 2014.

[17] R. Aris and N. R. Undson, "An analysis of chemical reactor stability and control-I The possibility of local control, with perfect or imperfect control mechanisms," *Chemical Engineering Science,* vol. 7, no. 8, pp. 121-131, 1958.

[18] E. H. Bristol, "On a new measure of interactions for multivariable process control," *IEEE Trans. Automat. Control,* Vols. AC-11, no. 1st, pp. 133-134, 1966.

[19] V. Kariwala, J. F. Forbes and E. S. Meadows, "Block Relative Gain: Properties and Pairing Rules," *Ind. Eng. Chem. Res. ,* vol. 42, no. 20th, pp. 4564-4574, 2003.

[20] C. Manousiouthakis and M. Nikolaou, "Analysis of Decentralized Control Structures for Nonlinear Systems," *AIChE Journal,* vol. 35, no. 4th, pp. 549-558, 1989.

[21] C. E. Garcia and M. Morari, "Internal Model Control 1. A Unifying Review and Some New Results," *Ind. Eng. Chem. Process Des. Dev.,* vol. 21, no. 2nd, pp. 308-323, 1982.

[22] B. R. Holt and M. Morari, "Design of resilient processing plants—VI, The effect of deadtime on dynamic resilience," *Chemical Engineering Science,* vol. 40, no. 7th, pp. 1229-1237, 1985.

[23] E. Zafiriou, Integration of Process Design and Control, 1st ed., Baltimore: IFAC, 1994.

[24] K. Trickett, "QUANTIFICATION OF INVERSE RESPONSE FOR CONTROLLABILITY ASSESSMENT OF NONLINEAR PROCESSES," University of London, London, 1994.

[25] P. Daoutidis and C. Kravaris, "Structural Evaluation of Control Configurations for Multivariable Nonlinear Processes," *Chemical Engineering Science,* vol. 47, no. 5th, pp. 1091-1107, 1992.

[26] M. Morari, "Effect of Design on the Controllability of Chemical Plants," *Interactions Between Process Design and Process Control,* vol. 1, no. 1st, pp. 3-16, 1992.

[27] E. N. Pistikopoulos and N. A. Diangelakis, "Towards the integration of process design, control and scheduling: Are we getting closer?," *Computers and Chemical Engineering,* vol. 91, no. 1st, pp. 85-92, 2015.

[28] Y. Y. H. Vira Chankong, Multiobjective Decision Making: Theory and Methodology, Mineola: Dover Publications, 1983.

Criteria for Determining the Need for Installation of Fixed Fire and Gas Detectors in Petroleum Refineries

Laurence Siqueira
PETROBRAS
28, Henrique Valadares Av. Rio de Janeiro-RJ– 20231-030
laurencelima@petrobras.com.br

Bruno Ferreira Barsotti
PETROBRAS
bruno.barsotti@petrobras.com.br

Socrates Fofano
PETROBRAS
fofano@petrobras.com.br

Prepared for Presentation at
American Institute of Chemical Engineers
2017Spring Meeting and 13thGlobal Congress on Process Safety
San Antonio, TX
March 26 – 29, 2017

Criteria for Determining the Need for Installation of Fixed Fire and Gas Detectors in Petroleum Refineries

Laurence Siqueira
PETROBRAS
28, Henrique Valadares Av. Rio de Janeiro-RJ– 20231-030
laurencelima@petrobras.com.br

Bruno Ferreira Barsotti
PETROBRAS
bruno.barsotti@petrobras.com.br

Socrates Fofano
PETROBRAS
fofano@petrobras.com.br

Keywords: Safety, Detectors, Flame, Gas, Frequency, Probability, Ignition, Flammable, Toxic

Abstract

For determining of what accidental scenarios need the installation of fixed fire and gas detectors was necessary the preparation of a criterial and a technical guideline that could determine the probability of ignition of a released flammable mass, the probability of toxic gas plumes, the equipament failure rate and maximum tolerable frequency.

The probability of a flammable mass release due to leakage takes into account the operating conditions of the process plant such as pressure, temperature and physical state of the most significant flammable substance after the leakage. The physico-chemical data of the substance are also considered, such as molecular weight, auto ignition temperature, LEL (Lower Explosion Limit), MIE (Minimum Ignition Energy), Flash Point and Boling Point.

With the ignition probability of the leakage and the failure rate of the considered systems, it was possible to determine the probable frequency of its occurrence. For those scenarios with results above the tolerable frequency, it was considered that they were eligible for installation of fire and gas detectors.

The technical guideline has been applied in more than 200 process units of refineries and has allowed not only to reevaluate the need of installation of fire and gas systems, but also to increase its effectiveness, resulting in a reduction of consequence severity from leakage of flammable mass. Additionally it has contributed to the optimization of installation and reduction of the maintenance costs of the plants.

1. Introduction

Fixed fire and gas detection systems have been used in industrial installations where any leakage of gas or vapors (flammable or toxic), as well hydrocarbon and hydrogen fires represent significant hazards to people, assets and/or environment.

Significant variations in fire and gas systems installed in refining process units of the same type have been observed, like as some units have few, if any, detectors, while others have dozens of detectors installed. Due to this discrepancy, it was noted that different criteria have been adopted for installation of fixed fire and gas detectors.

This present paper aims to present criteria for the use of fire and gas systems focused on oil refineries. It sought to adopt criteria using a risk-based approach, observing the specific characteristics and limitations of these systems.

In this context, it was used the probability of ignition of a released flammable mass, the failure rates per type of process equipment, calculation of the risk, as well defining the tolerable frequency for the selected scenarios.

2- Criteria for Scenarios Selection

As initial tasks, qualitative risk studies was carried out, identifying the accidental risk scenarios of each process unit. In this stage, the respective hazards, causes and consequences associated with each scenario are related.

The next step was select among the identified scenarios, which should be evaluated as the need for detection. The scenarios of interest were scenarios whose danger was leakage to the environment, the cause was loss of containment, and the consequence was formation of flammable / toxic cloud, fire and/or explosion.

2.1 - Criteria for Consequences of Scenarios

Observing the mitigation (and non-prevention) characteristic of fire and gas systems, a possible approach to be adopted is to consider for evaluation only the critical scenarios, that is, scenarios with a high consequence severity (critical or Catastrophic). See risk tolerance matrix in Figure 1. Scenarios of lower severity (medium or marginal severity) will not be considered for evaluation as the need for detection.

Consequency Severity				Frequency of Occurrence				
	People	Assets / Production	Environment	Extremely Remote	Remote	Not likely	Probable	Frequent
Catastrophic	Multiple fatalities	Catastrophic demages	Catastrophic demages	M	M	NT	NT	NT
Critical	One fatality	Severe demages	Severe demages	T	M	M	NT	NT
Medium	Severe injuries	Moderate demages	Moderate demages	T	T	M	M	NT
Marginal	Light injuries	Light demages	Light demages	T	T	T	M	M
Negligible	No injuries	Insignificant demages	Insignificant demages	T	T	T	T	M

T	Tolerable
M	Moderate
NT	Not Tolerable

Figure 1: Risk Tolerance Matrix

2.2 - Scenario Frequency Criteria

Each of the critical scenarios has associated a likely frequency of occurrence. Therefore, as an additional criterion, it is also necessary to establish a maximum tolerable frequency of occurrence, below which the use of detection is not considered necessary.

In the present work the concept of ALARP (As Low As Reasonably Practicable) developed by Health and Safety Executive, UK was used. This concept, which is well accepted in the oil and gas industry, including regulatory agencies, presents a range of risk values divided into three regions, namely:

a) "Bottom-up" region, in which risk values are clearly acceptable, and there is no need to adopt additional actions and/or measures to reduce these values. Reference: frequency of occurrence values equal to or less than 10^{-6}/year;

b) "Top-of-the-range" region, in which risk values are clearly unacceptable, and additional actions and/or measures have to be taken to reduce these values. Reference: frequency of occurrence values equal to or greater than 10^{-2}/year; and

c) "Mid-scale" region, called the ALARP region, in which the operating company has to demonstrate that it applies actions and/or measures to reduce the so-called "practicable" risk values, and the adoption of additional actions and/or measures are not "practicable". Reference: frequency values occurring in the range of 10^{-5}/year up to 10^{-3}/year.

It is up to the operating company to define the region of ALARP that it considers suitable to apply in its installations, especially the maximum frequency value of this interval. Applying the concept of ALARP in the risk tolerance matrix presented (see figure 2), the value of 10^{-4}/year can be adopted for maximum tolerable frequency (not likely frequency).

Thus, only the critical (leakage) scenarios, whose frequency of occurrence are greater than or equal to 10^{-4}/year, are then selected for evaluation as to the need for detection.

Consequency Severity				Frequency of Occurrence				
	People	Assets / Production	Environment	Extremely Remote	Remote	Not likely	Probable	Frequent
Catastrophic	Multiple fatalities	Catastrophic demages	Catastrophic demages	M	M	NT	NT	NT
Critical	One fatality	Severe demages	Severe demages	T	M	M	NT	NT
Medium	Severe injuries	Moderate demages	Moderate demages	T	T	M	M	NT
Marginal	Light injuries	Light demages	Light demages	T	T	T	M	M
Negligible	No injuries	Insignificant demages	Insignificant demages	T	T	T	T	M

T	Tolerable
M	Moderate
NT	Not Tolerable

Figure 2: Risk Tolerance Matrix

3- Evaluation of Selected Scenarios

It is considered a basic leak events tree presented in Figure 3.

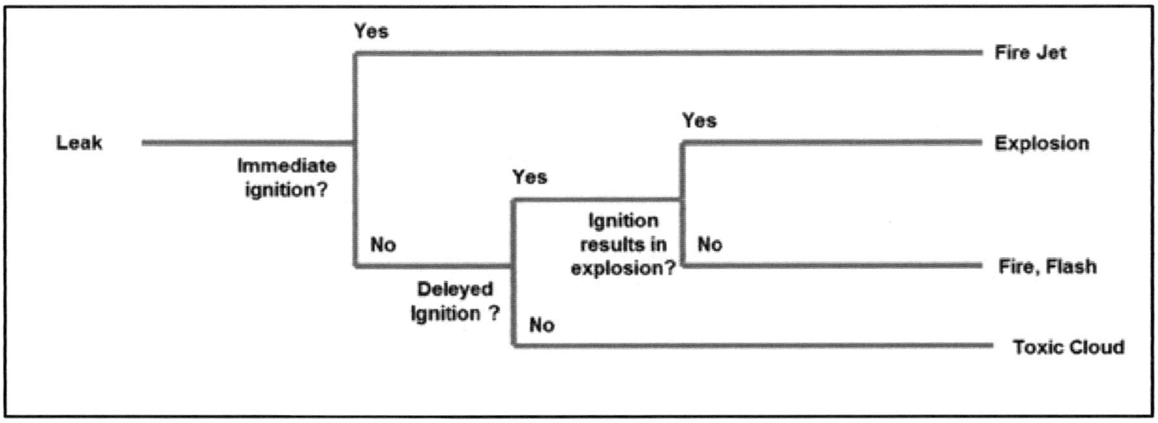

Figure 3: Leak events tree

For each scenario selected for evaluation, it is necessary estimate the respective frequency of leakage (loss of containment). For this, values of failure rate that were used are indicated in table 1, and it was necessary to correlate the type or set of equipment to the associated scenario.
The failure rates frequency adopted has considered data from the following sources: DNV - Failure Frequency Guidance, DNV - A Guide to Quantitative Risk Assessment of Offshore Installations, Energy Institute – Ignition Probability Review, Model Development and Look Up Correlations and HSE - Off Shore Ignition Probabilities

Table 1: Equipment failure rates per year adopted

Reciprocating Compressors (e.g. Make-up HDT)	**2,0 x 10^{-2}/year**
Pumps	**5,0 x 10^{-3}/year**
CentrifugalCompressors (ex: Gas, Recycling)	**2,5 x 10^{-3}/year**
Exchangers / Pipes and Valves in Preheating Batteries	**2,0 x 10^{-3}/year**
Scenarios with Other Static Equipment	**1,0 x 10^{-3}/year**
Scenarios with Pipes and Accessories only	**3,3 x 10^{-4}/year**

For all leak scenarios, the hole diameter of 10 mm can be adopted, since this value characterizes small leaks (suitable for detection by detectors), has higher frequency of occurrence, and is difficult to detect by process monitoring instruments, such as flow or pressure meters. This hole diameter adopted has used as reference the CCPS literatures: Guidelines for Determining the Probability of Ignition of a Released Flammable Mass, Continuous Monitoring for Hazardous Material Releases, Guidelines for Fire Protection in Chemical, Petrochemical, and Hydrocarbon Processing Facilities and Estimating the Flammable Mass of a Vapor Cloud

As a next step, it was necessary estimate, for each scenario selected for evaluation, the following probabilities of occurrence:

- Probability of immediate ignition;
- Probability of delayed ignition; and
- Probability of ignition with explosion.

For estimate all of these probabilities the literature that has been used was CCPS - Guidelines for Determining the Probability of Ignition of a Released Flammable Mass. Once these probabilities of occurrence (and non-occurrence) values were estimated, the selected scenarios were then evaluated as follows:

a) The values of the leakage frequencies were multiplied by the probabilities values of immediate ignition, delayed ignition and explosion;

b) The obtained results were the scenarios frequencies of occurrence, which were then compared with the maximum tolerable frequency of occurrence adopted (10^{-4}/year);

c) Scenarios which frequency of occurrence value were greater than the maximum tolerable were considered scenarios that are eligible for detection;

d) The detection type to be used in each scenario (flame, flammable gas or toxic gas) will depend on the numerical results of the calculations performed, as exemplified in Figure 4.

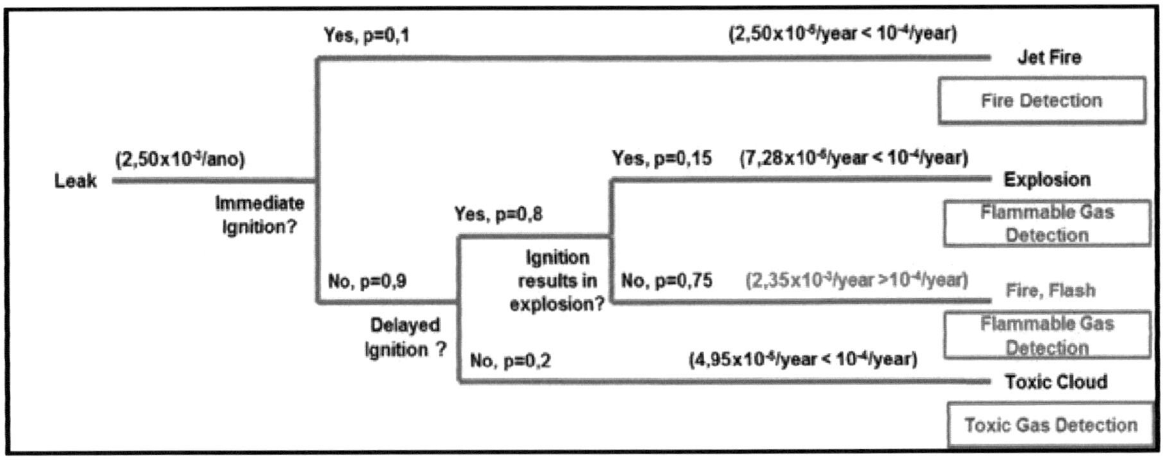

Figure 4: Leak events tree and examples of calculations performed

4- Ignition Probability Estimates

In order to estimate the ignition probabilities, it is necessary to obtain process and facility data for the scenarios selected for evaluation.

The basic process data required are pressure, temperature and flow in the operating conditions of the leaked product. However, it is also necessary to know various physicochemical properties of the leaked products, such as, physical state, molecular weight, density, Cp/Cv, vapor pressure, auto ignition temperature, flash point, boiling point, lower explosive limit and minimum ignition energy.

One way of obtaining such data is correlating the leaked products with a reference substance, considered as more representative, as exemplified in table 2.

Table 2: Examples of reference substance

Leaked Product	Reference Substance	Molecular weight	Auto ignition temperature (°C)	Flash Point (°C)	Boling Point (°C)	LEL - Lower Explosive Limit (%)	MIE (mJ)
FUEL GAS	Methane	16	600	-187,2	-162,5	5	0,28
GASOIL	n-decane	142,3	210	46	174	0,8	0,2
PETROLEUM	n-decane	142,3	210	46	174	0,8	0,2
GASOLINE	n-hexane	86	225	-23	69	1,6	0,24
DIESEL OIL	n-nonane	128	205	31	150	0,8	0,2
NAPHTHA	n-pentane	72	243	-40	36,1	1,5	0,28
LPG	Propane	44	450	-102	-42	2,6	0,25

If the critical scenario contains process equipment that runs with toxic substances, it was also necessary define the concentration of this substances in the gas phase. In this way, it was verified if the concentration is above the level of IDLH - Immediately Danger to Life and Health.

For those scenarios where the streams contains only toxic substances and the concentration is above of IDLH, then it is recommend to use toxic detectors.

For those scenarios where the streams contains both toxic and flammable substances and the concentration of toxic substance is above of IDLH, then it is recommend to use toxic detectors prior to flammable detectors, because in these cases, the detection is rated in ppm, while for flammable gas in % of LEL (lower explosive level).

The most common toxic substances considered for this study were H_2S concentration above 100 ppm, CL_2 above 10 ppm, NH_3 above 300 ppm, CO above 1200 ppm, HF above 30 ppm, Phenol above 250 ppm and ClO_2 above 5 ppm.

5 - Additional Criteria for Detection

The following are some additional criteria that can be applied to rationalize the scenarios selected for installation of fixed gas and flame detection:

5.1 - The detection of flammable gases/vapors may be dispensed for those scenarios where the leaked liquid has operating temperature lower than the Boiling Point value.

5.2 - The Hydrocarbon flame detectors can be dispensed for scenarios where there are passive protection (fireproofing) for the equipment inside the fire envelope.

5.3 - For the cases of the process units where the results of frequencies of occurrence are above the 10^{-4}/year (maximum tolerable frequency of occurrence adopted) that already has deluge systems installed and it can be interlocked with the detection system, than hydrocarbon flame detection is recommended, otherwise, it can be dispensed.

5.4 – For the cases where the frequency of occurrence results are above the 10^{-4}/year (maximum tolerable frequency of occurrence adopted) but the process unit doesn´t have any automatic or semiautomatic active protection, the decision for installation of the fixed hydrocarbon flame detectors by itself needs to be evaluated in terms of reducing the consequences of injuries to persons, assets and/or environment

5.5 – The use of fire detection can then be limited to scenarios that produces hydrogen or methanol flames, because these flames are not "visible" to the human eye. Thus, only leak scenarios with a frequency of occurrence greater than 10^{-4}/year, and with high potential of occurrence of fire involving hydrogen or methanol, were indicated for fire detection.

Examples considered:

- Compressor operating with H_2 gas;
- Equipment operating with H_2 gas and having a temperature above 500 °C; and
- Pump operating with methanol.

6 – Diagrams and formulas used to determine and calculate the risks

In order to maximize the understanding of the work that has been done, refer to the diagrams as follow:

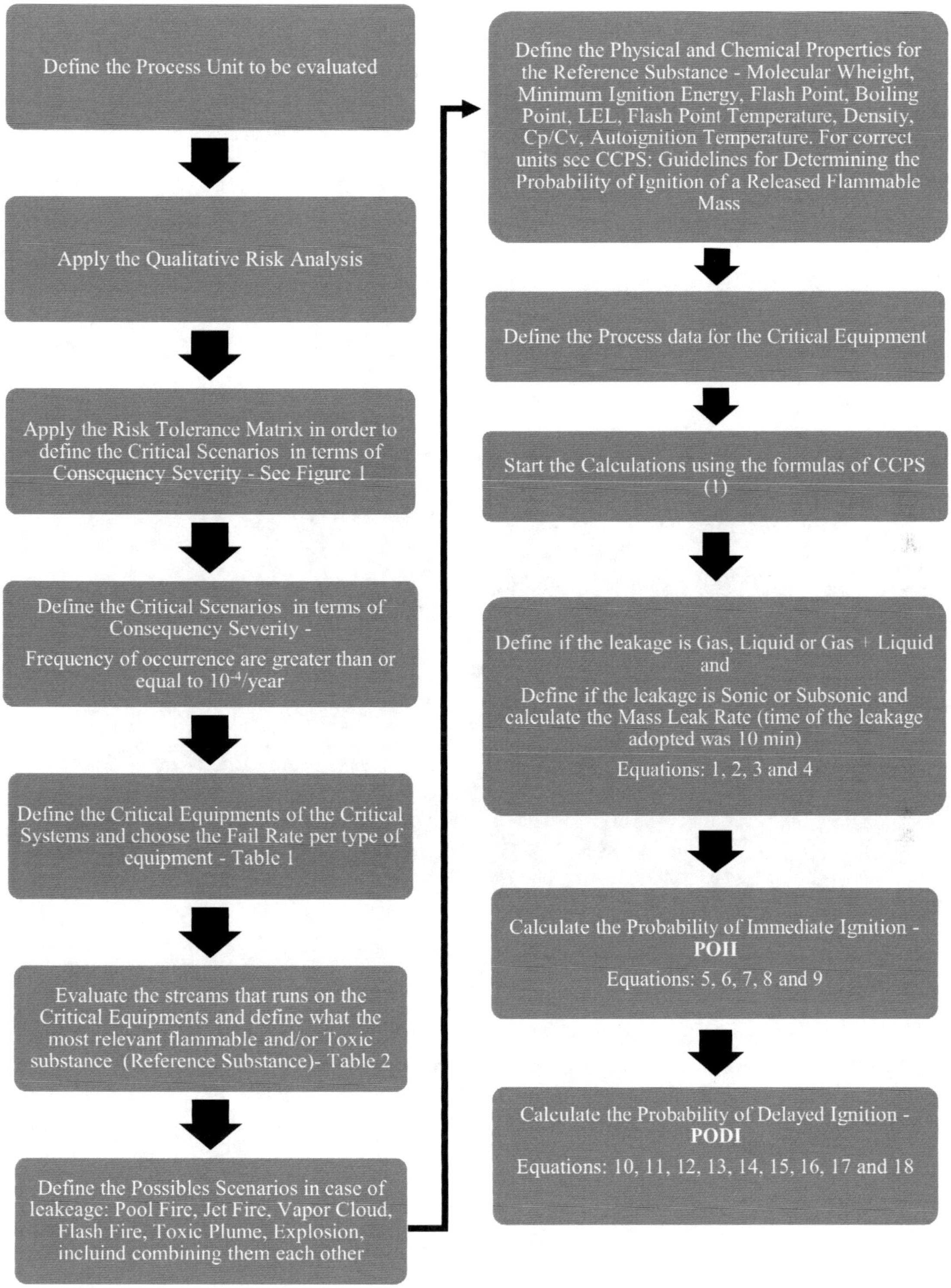

Diagram continues next page

Fire Analysis

If the POII x Failure Rate of Critical Equipment is greater or equal than Maximum Tolerable Frequency of occurrence of 10^{-4}/year, so the scenario is Elegible for Flame Detection.

Define techonology for Detection

(See criteria on items 4 and 5 of this paper)

↓

Calculate the Probability of Explosion Given a Delayed Ignition - POEGDI

Equations: 19, 20 and 21

↓

Explosion Analysis

If the POEGDI x Failure Rate of Critical Equipment is greater or equal than Maximum Tolerable Frequency of occurrence of

10^{-4}/year, so the scenario is Elegible for Flammable Gas Detection.

Define techonology for Detection

↓

Toxic Plume Analysis

If the reference toxic substance is above the IDLH and the probability of toxic plume multiplied per failure rate of critical equipment is greater or equal than Maximum Tolerable Frequency of occurrence of

10^{-4}/year, so the scenario is Elegible for Toxic Gas Detection.

Define techonology for Detection

7- Equations

The criteria presented in this paper have used the following equations and the descriptions, special notes as well the meaning of each formula presented bellow can be found on literature of CCPS - *Guidelines for Determining the Probability of Ignition of a Released Flammable Mass* and *Guidelines for Fire Protection in Chemical, Petrochemical, and Hydrocarbon Processing Facilities*.

$$W_n = C_d \cdot K_{v,n} \cdot \rho_l \cdot \frac{A_n}{C_1} \sqrt{\frac{2 \cdot g_c \cdot (P_s - P_{atm})}{\rho_l}} \quad (1)$$

$$P_{trans} = P_{atm}\left(\frac{k+1}{2}\right)^{\frac{k}{k-1}} \quad (2)$$

$$W_n = \frac{C_d}{C_2}\cdot A_n \cdot P_s\sqrt{\left(\frac{k\cdot MW\cdot g_c}{R\cdot T_s}\right)\left(\frac{2}{k+1}\right)^{\frac{k+1}{k-1}}} \quad (3)$$

$$W_n = \frac{C_d}{C_2}\cdot A_n \cdot P_s\sqrt{\left(\frac{MW\cdot g_c}{R\cdot T_s}\right)\left(\frac{2\cdot k}{k-1}\right)\left(\frac{P_{atm}}{P_s}\right)^{\frac{2}{k}}\left(1-\left(\frac{P_{atm}}{P_s}\right)^{\frac{k-1}{k}}\right)} \quad (4)$$

$$MIE_v = MIE_{reported} \times (10{,}000/P_{liquid})^{0.25} \quad (5)$$

$$MIE_{adj} = MIE_v \times \exp[0.0044(60 - T)] \quad (6)$$

$$P_{AIT} = 1 - 5{,}000e^{-9.5(T/AIT)} \quad (7)$$

$$POII_{static} = 0.003 \times P^{1/3} \times MIE_{adj}^{-0.6} \quad (8)$$

$$POII_{Level\ 2} = P_{ai} + (1 - P_{ai}) \times POII_{static} \quad (9)$$

$$PODI_{S/D} = 1 - [(1 - S^2) \times e^{-St}] \quad (10)$$

$$M_{MAG\_Amount\ Released\ (liquid)} = (\text{Amount Released}/5{,}000)^{0.3}$$

$$M_{MAG\_Amount\ Released\ (vapor)} = (\text{Amount Released}/1{,}000)^{0.5}$$

$$M_{MAG\_Hole\ Diameter\ (liquid)} = (\text{Hole Diameter})^{0.6}$$

$$M_{MAG\_Hole\ Diameter\ (vapor)} = (\text{Hole Diameter})$$

(11,12, 13 and 14)

$$M_{MAT} = 0.5 - 1.7 \log(MIE) \quad (15)$$

$$M_T = 0.4 + (T - 1.3 \times FP)/230 \quad (16)$$

$$PODI_{Level\ 2} = PODI_{S/D} \times M_{MAG} \times M_{MAT} \times M_T \times M_{IN/OUT} \quad (17)$$

$$PODI_{final} = PODI_{calc}\,(1 - POII) \quad (18)$$

$$POEGDI_{Level\ 2} = 0.3 \times M_{CHEM} \times M_{MAGE} \times M_{IN/OUT} \quad (19)$$

$$M_{MAGE} = (PODI\ M_{MAG})^{0.5} \quad (20)$$

$$POEGDI_{FINAL} = PODI_{FINAL} \times POEGDI \quad (21)$$

8- Results

The criteria presented for selection and evaluation of scenarios were applied in a set of 80 different process units, covering a total of 10 refineries. The results obtained, in terms of average amounts of scenarios, are presented in table 2, grouped by type of refining process.

Table 2: Scenarios by type of process units

PROCESS UNIT	Scenarios A	Scenarios B	Scenarios C	Scenarios D
Catalytic Cracking	119	88	57	35
Delayed Coking	74	61	46	25
Atmosp. and Vacuum Distillation	67	53	35	13
Diesel Hydrotreating	52	40	23	15
Catalytic Reform	51	37	34	22
Sulfur Recovery	33	21	20	11
Hydrogen Generation	25	15	11	5

a) Scenarios A - Total amount of risk scenarios studied, with preliminary risk analysis technique.

b) Scenarios B - Subset of Scenarios A. Those are scenarios of leaks (to the environment) that have the potential to be evaluated for the need for flame or gas detection.

c) Scenarios C - Subset of Scenarios B. Those are scenarios B which the severity of the consequence is critical or catastrophic (see Figure 1).

d) Scenarios D - Subset of Scenarios C. Those are scenarios C which frequency of leakage, combined with the probability of ignition (see Figure 5), is greater than the tolerable frequency of occurrence (see Figure 2). The additional criteria (see item 5) are also considered.

9 - Conclusions

The use of the presented criteria for selection and evaluation of scenarios, especially the use of the probability of ignition, leads to an expressive reduction of scenarios indicated for detection. Considering a hypothetical refinery, that has the process units listed in Figure 7, the total amount of scenarios indicated for detection is 126 (Figure 7, scenarios D), against a total of 226 (Figure 7, scenarios C). In this case, the use of the presented criteria represents a reduction of 100 (or 44%) in the indicated scenarios for detection.

Considering an average value of 1.4 detectors per scenario, and the cost of US$ 8,000 for a detector installation, and the cost of US$ 1,000.00 per year for a detector maintenance (checks, calibrations and replacements), we have for the hypothetical refinery:

- Reduction of investment value (CAPEX) in fixed fire and gas detection system: US$ 1,120,000.00;
- Reduction of the maintenance costs value (OPEX) in fixed fire and gas detection system: US$ 140,000.00/year.

10 - Bibliography

(1) CCPS – *Guidelines for Determining the Probability of Ignition of a Released Flammable Mass*, Center for Chemical Process Safety, American Institute of Chemical Engineers, New York, NY, 2014.

(2) CCPS – *Consequence Analysis of Chemical Releases.* Center for Chemical Process Safety, American Institute of Chemical Engineers, New York, NY, 1999.

(3) CCPS - *Guidelines for Fire Protection in Chemical, Petrochemical, and Hydrocarbon Processing Facilities.* Center for Chemical Process Safety, American Institute of Chemical Engineers, New York, NY, 2003.

(4) CCPS – *Guidelines for Evaluating the Characteristics of Vapor Cloud Explosions, Flash Fires and BLEVEs.* Center for Chemical Process Safety, American Institute of Chemical Engineers, New York, NY, 1994.

(5) CCPS – *Estimating the Flammable Mass of a Vapor Cloud.* Center for Chemical Process Safety, American Institute of Chemical Engineers, New York, NY, 1998.
(6) CCPS - *Guidelines for Chemical Process Quantitative Risk Analysis – Second Edition.* Center for Chemical Process Safety, American Institute of Chemical Engineers, New York, NY, 2000.
(7) CCPS - *Guidelines for Process Equipment Reliability Data – Second Edition.* Center for Chemical Process Safety, American Institute of Chemical Engineers, New York, NY.
(8) Lee´s Loss Prevention in the Process Industries – Hazard Identification, Assessment and Control Volume 1 – Fourth edition, Texas, USA, 2012
(9) Electrical Installations in Hazardous Areas – Eur Ing McMillan C Eng. FIEE FInstMC, Oxford OX2 8DP, Woburn, MA, 1998
(10) DNV - Failure Frequency Guidance, Process Equipment Leak Frequency Data for Use in QRA
(11) DNV - A Guide to Quantitative Risk Assessment of Offshore Installations
(12) Handbook Failure Frequencies – Safety Report, Flemish Government, LNE Department, Environment, Nature and Energy Policy Unit – Safety Reporting Division. 2009
(13) Energy Institute – Ignition Probability Review, Model Development and Look Up Correlations
(14) HSE – Health and Safety Executive, London, UK. Off Shore Ignition Probabilities, Report number HSL/2005/50, Harpur Hill, Buxton, Derbyshire SK17 9JN
(15) HSE – Health and Safety Executive, Guidance on Risk Assessment for Offshore Installations, Sheet N.3/2006
(16) IOGP – International Association of Oil & Gas Producers Risk Assessment Data Directory – Report N.434-6.1 – Ignition Probabilities, 2010
(17) IOGP - International Association of Oil & Gas Producers Risk Assessment. Data Directory – Report N.434-6.1 – Safety Performance Indicators Process Safety Events, 2014 Data
(18) IOGP - International Association of Oil & Gas Producers - Risk Assessment Data Directory – Report N.434-6.1 – Safety Performance Indicators Process Safety Events, 2015 Data
(19) IOGP – International Association of Oil & Gas Producers – Risk Assessment Data Directory, Storage Incident Frequencies. Report N.434-3, March 2010.
(20) ANSI/ISA - 5.1 Instrumentation Symbols and Identification
(21) ANSI/ISA - 92.00.02 - 2013 Installation, Operation and Maintenance of Toxic Gas-Detection Instruments;
(22) API 580 – Risk Based Inspection
(23) API 581 – Part 3, Consequence Analysis in an API RBI Assessment;
(24) API 1996 - Final Report 1996 American Petroleum Institute/National Petroleum Refiners Association Survey of Refining Operations and Product Quality First Edition
(25) API 2021- Fire Protection in Refineries
(26) API 2218 – Fireproofing Practices in Petroleum and Petrochemical Processing Plants
(27) API 752 – Management of Hazards
(28) NFPA 497 – Recommended Practice for the Classification of Flammable Liquids, Gases or Vapors and of Hazardous (classified) Locations for Electrical Installations in Chemical Process Areas

Stratification During Vapor or Gas Freeing of Storage Tanks

Michael L. Hanks, Ph.D., P.E.
ESI
146 2nd St. N, #205
Saint Petersburg, FL 33701
mlhanks@engsys.com

Amy E. Gray, Ph.D.
ESI
3310 Green Park Circle
Charlotte, NC 28217
aegray@engsys.com

Ronald Pape, Ph.D., P.E.
ESI
4215 Campus Drive
Aurora, IL 60504
rpape@engsys.com

Prepared for Presentation at
American Institute of Chemical Engineers
2017 Spring Meeting and 13th Global Congress on Process Safety
San Antonio, TX
March 26 – 29, 2017

Stratification During Vapor or Gas Freeing of Storage Tanks

Michael L. Hanks, Ph.D., P.E.
ESI
146 2nd St. N, #205
Saint Petersburg, FL 33701
mlhanks@engsys.com

Amy E. Gray, Ph.D.
ESI
3310 Green Park Circle
Charlotte, NC 28217
aegray@engsys.com

Ronald Pape, Ph.D., P.E.
ESI
4215 Campus Drive
Aurora, IL 60504
rpape@engsys.com

Keywords: Vapor Freeing, API RP 2016, Stratification.

Abstract

API RP 2016 provides guidelines for vapor freeing and degassing of tanks. The instructions include warnings and methods to recognize and avoid stratification or channeling during vapor freeing, and to control and prevent ignition sources from occurring in the tank during vapor freeing. A recent incident occurred where vapor in a tank ignited shortly after vapor freeing. The investigation revealed that the vapor freeing method used most likely resulted in stratification of the vapor in the tank, and the atmosphere testing method used to test the vapor level in the tank was not capable of detecting stratification. The low vapor reading resulted in shutting down the vapor freeing operation after only 5 hours when the planned duration was 48 hours. By following API RP 2016, this event would likely not have occurred. This paper discusses methods for estimating when stratification is likely to occur during vapor freeing of tanks based on jet mixing correlations, and uses a CFD model to examine the likelihood of stratification in several cases. These methods can be used to develop criteria for vapor freeing of tanks and for investigating incidents involving stratification of flammable vapors in an enclosed space.

1 Introduction

ANSI/API Standard 2015 Requirements for Safe Entry and Cleaning of Petroleum Storage Tanks applies to atmospheric and low-pressure tanks used in all sectors of the petrochemical industry [1]. It provides requirements for safe entry for tank cleaning operations from decommissioning through recommissioning. API RP 2016 Guidelines and Procedures for Entering and Cleaning Petroleum

Storage Tanks is a recommended practice that supplements ANSI/API 2015 [2]. Section 5 of API RP 2016 provides guidance for vapor and gas freeing of storage tanks. It defines vapor freeing as:

> The removal of flammable or toxic vapors and gases from a tank by displacement or the reduction of the percentage of vapors and gases in the tank's atmosphere to a safe level by dilution with fresh air.

According to API RP 2016, "mechanically introducing fresh air into a tank is the preferred method of removing vapors or gas from a storage tank."

API RP 2016 recommends two methods for vapor freeing using fresh air: dilution and displacement. The dilution method assumes the vapor and air mix thoroughly during vapor freeing, and the guide recommends a minimum air flow rate of 5 tank volumes of air per hour to accomplish this. The displacement method assumes the air and vapor do not mix, and relies on buoyancy of the vapor compared to air (either positive or negative buoyancy) to stratify the vapor and air. This way, essentially pure vapor is discharged from the tank as air flows into the tank. The guide recommends that the air be introduced at a low flow rate to ensure the vapor in the tank stays stably stratified.

The flow rate recommended in API RP 2016 for the dilution method would be expected to mix the air and vapor in the tank once sufficient vapor has been removed that there is insufficient buoyancy remaining to maintain stratified layers. However, the air flow rate into a tank may be restricted by equipment limitations, restrictions on discharges to atmosphere by environmental boards, or by other considerations. For example, the lead author is aware of a case where the local air quality board restricted the flow rate through a thermal oxidizer used for vapor freeing to less than 2500 CFM. This means there is a maximum tank volume for which the thermal oxidizer could be run at the recommended flow rate for vapor freeing.

For cases where the flow rate through the tank must be less than the recommended amount, there is no guidance in API RP 2016 for estimating whether the air and vapor will be well mixed during the vapor freeing operation or whether the tank contents will be stratified. Also, the API RP 2016 guide does not provide information on how low of a flow rate is required to ensure the contents do remain stratified for a displacement operation other than using natural ventilation to displace the vapor.

The dilution method of vapor freeing can generally be modeled using the Perfect Mixing Model (PMM) if the jet is sufficiently strong to overcome stratification. This model is a mass balance on flows in and out of the tank and assumes the tank contents are instantaneously perfectly mixed. The model constants relate to known parameters and provide estimates of the time required to complete the vapor freeing operation. The displacement method cannot be modeled as easily, but the time required to free a tank using the displacement method can be bounded.

Air streams used for vapor freeing are generally induced in the tank by fans or eductors placed in openings on the side or top of the tank. The resulting air stream into the tank can be reasonably well modeled as a buoyant free jet, at least until the airstream encounters the tank wall, roof or base. Chow (2009) modelled vapor freeing of tanker compartments and found that published research on jets mixing in enclosed volumes was useful for comparing to vapor freeing operations [3].

This paper uses correlations developed for jet mixing in tanks to examine conditions for which the air and vapor will mix completely, and conditions for which they will remain stratified. The correlations used correspond most closely to vapor freeing of cone roof tanks that do not have significant flow obstructions inside the tank (e.g. baffles, partitions, etc.) The concepts can be applied in general to vapor freeing of a floating roof tank, but the specific correlations may not be valid for these tanks because of the extreme aspect ratio of the headspace in a floating roof tank.

2 Jet Mixing in Tanks

When fans are used to introduce fresh air into a tank during vapor freeing, the airstream for the fan can be treated as a free jet at least until the jet impacts the tank walls or other significant internal obstructions. The jet can follow three different flow regimes depending on the density of the vapor or gas in the tank and the initial momentum of the air jet: a plume, a pure jet, or a fountain. For a plume, buoyancy dominates the initial momentum of the jet and the jet momentum acts generally in the same direction as the buoyant force. A pure jet occurs when the density difference between the air and the vapor or gas is sufficiently small that the initial jet momentum drives mixing. A fountain is a buoyant jet in which the initial momentum acts in the direction opposite to the buoyant force.

The density of the vapor or gas in the tank can be less than air (e.g. methane); comparable to air (e.g. ethane, nitrogen); or greater than air (e.g. propane or other heavier volatile molecules.) The manways or ports where air is blown into or drawn from the tank are generally located in the tank sidewall near the ground, or on the top of the tank. This means there are 8 general cases for how an air stream will flow in the tank headspace during vapor freeing as shown in Table 1.

Table 1. General cases governing airstream flow in the tank during vapor freeing.

Vapor or Gas Density	Side Air Inlet	Top Air Inlet
Less than air	(A) Horizontal momentum jet	(B) Vertical momentum jet or plume
Comparable to air	(C) Horizontal pure jet	(D) Vertical pure jet
Greater than air	(E) Horizontal momentum jet possibly transitioning to a vertical plume	(F) Fountain
Combination (e.g. high density layer at the bottom of the tank, low density layer at the top)	(G) Horizontal momentum jet, possibly transitioning to a plume and/or a fountain.	(H) Vertical momentum jet or plume, possibly transitioning to a fountain.

Whether the jet will mix with all the vapor in the tank or whether a stratified layer will form depends on the above factors and on the jet flow rate, the tank geometry, and the location of the discharge from the tank.

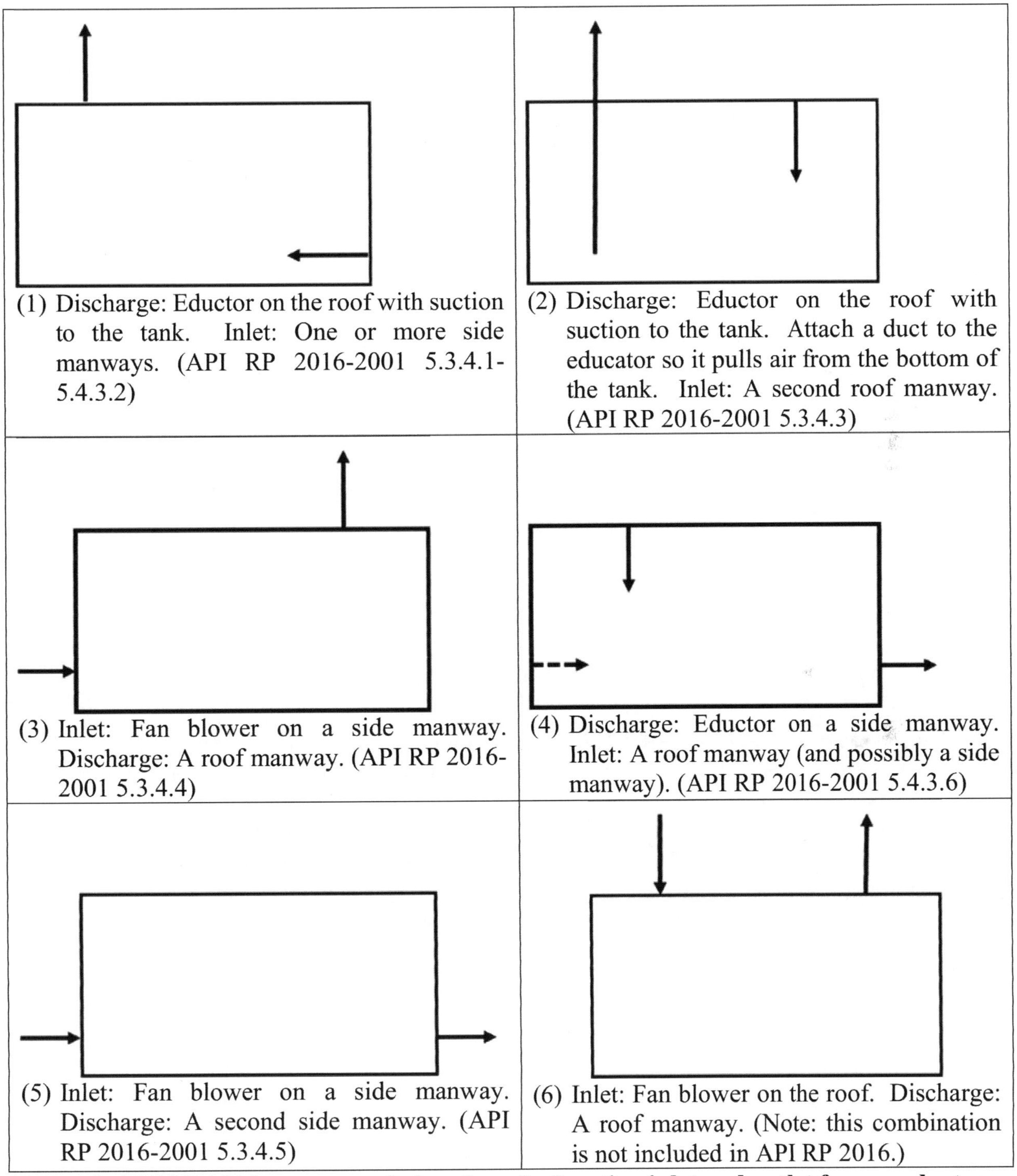

(1) Discharge: Eductor on the roof with suction to the tank. Inlet: One or more side manways. (API RP 2016-2001 5.3.4.1-5.4.3.2)

(2) Discharge: Eductor on the roof with suction to the tank. Attach a duct to the educator so it pulls air from the bottom of the tank. Inlet: A second roof manway. (API RP 2016-2001 5.3.4.3)

(3) Inlet: Fan blower on a side manway. Discharge: A roof manway. (API RP 2016-2001 5.3.4.4)

(4) Discharge: Eductor on a side manway. Inlet: A roof manway (and possibly a side manway). (API RP 2016-2001 5.4.3.6)

(5) Inlet: Fan blower on a side manway. Discharge: A second side manway. (API RP 2016-2001 5.3.4.5)

(6) Inlet: Fan blower on the roof. Discharge: A roof manway. (Note: this combination is not included in API RP 2016.)

Figure 1. API RP 2016 recommended options for locating inlet and outlet fans or eductors.

API RP 2016 recommends the combinations for positioning the air inlet and vapor discharge during vapor freeing of a cone roof tank shown in Figure 1.

Methods 1 and 3 are similar to each other in terms of air flow in the tank. Methods 2 and 4 are also similar to each other. Method 6 is not included in API RP 2016 but is the other obvious possible combination of inlet and outlet locations.

Comparing the possible vapor density vs. air density conditions from Table 1 to the recommended combinations of inlet and discharge locations in Figure 1, it is clear there are preferred combinations for dilution vapor freeing and for displacement vapor freeing. The most obvious combinations are shown in Table 2.

Table 2. Preferred combinations of inlet and outlet locations for various combinations of vapor density.

Inlet and Outlet Location Method	Dilution (High air flow rate)	Displacement (Low air flow rate)
(1) and (3)	(A), (C), (G) (E) possible	(A)
(2) and (4)	(D), (F) (B), (H) possible	(F)
(5)	(E)	(E)
(6)	(B)	(B)

Combinations that take advantage of both the momentum of the air jet and of the buoyancy of the vapor relative to air will be the most efficient and effective for vapor freeing. For example, when freeing a low-density vapor from a tank, Methods (1) or (3) will be better than would (2), (4), or (6). Therefore, (A) would be preferred over (B) for a low-density vapor and (F) would be preferred over (E) for a high-density vapor. For freeing buoyant vapors, the natural combinations with inlet and outlet location are (1 or 3)(A) and (2 or 4)(F). For cases (C) and (D) where the vapor or gas in the tank is neutrally buoyant with respect to air, the location of the air inlet and vapor discharge are less important and displacement is not possible. For cases (H) and (G) either (1 or 3) or (2 or 4) can be effective.

Several combinations would not be recommended because they could result in short circuiting of the air. For example, Method (5) with a low-density vapor and a low air flow rate could result in air short-circuiting to the discharge without freeing vapor from the upper part of the tank headspace. A similar situation could happen for Method (6) with a heavy vapor and low air flow rate.

Correlations for jet mixing can provide guidance for whether the air flow rate into the tank will result in displacement or dilution for combinations (1)(A), (2)(F), and (4)(H); and whether the air jet has sufficient momentum to adequately mix with the vapor for cases (C) and (D).

The jet mixing correlations can be divided into two parts. The first part examines whether the vapor and air will be stratified during the vapor freeing operation or whether the air jet will break the stratification. The second part examines the time to mix the tank contents for cases or conditions where the air and vapor will not be stratified.

Vapor freeing by displacement occurs when the air flow rate is low enough that stratification occurs or existing stratification remains intact. For an ideal displacement operation where there is no mixing between the air and vapor, the time to free vapor from the tank is the time required to flow one tank volume of air into the tank. The actual time required will be longer than this due to partial mixing.

Vapor freeing by dilution occurs when the air flow rate is sufficiently high to break up existing stratification, prevent stratification from occurring during the vapor freeing operation, and to mix the air and vapor fast enough to avoid short-circuiting of air through the tank. This later condition can be evaluated by comparing the mixing time to the residence time in the tank.

2.1 Stratification

Stratification can occur either because the vapor density is sufficiently different than air that introducing air results in stratification, or because there are multiple vapors in the tank that result in the tank contents being stratified before vapor freeing begins. Stratification is unlikely to occur if the relative density difference between the two fluids is:

$$\left(\frac{\rho_2-\rho_1}{\rho_2}\right) \leq 0.05. \tag{1}$$

for $\rho_1 < \rho_2$.

2.1.1 Side-Entry Jet

For a side-entry jet mixing in a tank, when the relative density difference exceeds 0.05, there is a critical jet velocity, u_c, above which the jet may be able to mix the fluids so no stratification occurs [3]. This is estimated by:

$$u_c = \left[\frac{2gGH\left(\frac{\rho_2-\rho_1}{\rho_2}\right)}{\sin^2\theta}\right]^{0.50} \tag{2}$$

where $\theta = (\beta+5)^{\circ}$, β is the angle of the jet from horizontal, and G is a stratification constant that is a function of the relative density ratio and the ratio of the tank height to the jet discharge diameter, H/d_j. The value of G can be estimated by:[1]

$$G = 0.8\left(\frac{\rho_2-\rho_1}{\rho_2}\right)^{-0.286} \tag{3}$$

Equation (3) is for cases where $H/d_j > 80$. When $H/d_j < 80$, the value of G is lower for a given relative density ratio. This means the calculated value of u_c tends to be higher than is actually required to mix the two fluids.

This configuration naturally corresponds with the combination of (1 or 3)(A): a side jet inlet, a roof outlet and a vapor in the tank that is less dense than the jet.

Figure 2 shows the estimated critical velocity (the velocity necessary for breaking up the stratified layers) is a strong function of the jet angle. Pointing the jet upward into the tank significantly reduces the air jet velocity needed to break up stratified layers.

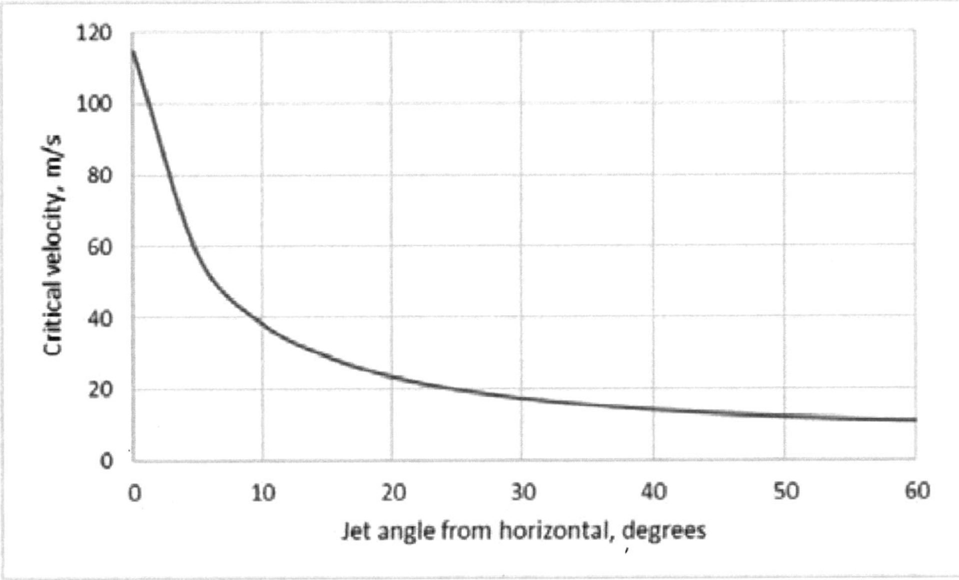

Figure 2. Estimated critical velocity from equation (2) for a side-jet of air and the tank initially filled with methane. Tank dimensions are those in API RP 2016-2001 example 5.1: tank height = 40 ft; Tank diameter = 125 ft.

2.1.2 Vertical Entry Jet or Plume

Hunt, et al. (2001) studied a filling box - an enclosed, rectangular tank with a vertical, buoyant jet but with no other inlets or outlets [4]. For the purposes of this paper, the interesting results from the filling box studies are information on the stability of a stratified layer as a function of the jet momentum and the geometry of the volume. The vapor freeing geometry corresponding to this

---

[1] Linear least squares fit to values scaled from [3] Figure 9.6, p. 173.

correlation is a turbulent air jet entering vertically downward from the roof into a less-dense vapor, which is the combination (6)(B).

For a single vertical jet, they report a correlation between critical jet length (the distance over which momentum of the jet dominates the flow) and the volume aspect ratio (height over width or length):

$$\left(\frac{L_j}{H}\right)_{crit} = \frac{1}{K}\left[\frac{1}{3\alpha^{0.5}}\left(\frac{H}{R}\right)^{-1} - 1\right] = 0.863\left[1.157\left(\frac{H}{R}\right)^{-1} - 1\right] \quad (4)$$

where $\frac{L_j}{H}$ is the dimensionless jet length, H is the tank height, R is the characteristic length of the base (diameter or length) and $K = \frac{1.057}{2^{1.25}\alpha^{0.5}\pi^{0.25}} = 1.159$, for $\alpha = 0.083$. For values of $\frac{L_j}{H} > \left(\frac{L_j}{H}\right)_{crit}$ the momentum of the jet is sufficient to overturn the entire volume so no stratification occurs, while for $\frac{L_j}{H} < \left(\frac{L_j}{H}\right)_{crit}$ a stable stratified layer forms.

The dimensionless jet length is a function of the jet momentum and buoyancy. The relevant factors are:

Dimensionless jet length: $$\frac{L_j}{H} = \frac{M_0^{0.75}}{HB_0^{0.5}} \quad (5)$$

Jet momentum flux: $$M_0 = \frac{Q_0^2}{A_j} \quad (6)$$

Jet buoyancy flux: $$B_0 = g_0'Q_0 = \frac{g|\Delta\rho|}{\rho_v}Q_0 \quad (7)$$

where Q_0 is the initial jet volume flow rate, A_j is the cross-sectional area of the jet at the discharge, g is the gravitational constant, $|\Delta\rho|$ is the absolute value of the density difference between the jet fluid and the vapor, and ρ_v is the density of the vapor.

For the same tank dimensions given above and using an air jet directed vertically down from the roof into a tank filled with methane, with an air jet nozzle area of 1 m^2, the correlation predicts an air flow rate of approximately 18 ACH (air changes per hour) would be required to prevent stratification during vapor freeing. It is unlikely that this high of an air flow rate is feasible. This means the vapor freeing should be expected to start as displacement and transition to dilution once most of the methane has been removed.

2.1.3 Fountain Jet

The fountain jet case is likely a common occurrence in vapor freeing of storage tanks because there will generally be a heel of liquid in the bottom of the tank emitting a vapor with heavier components that can form a layer near the bottom of the tank. Lighter components or the protective gas pad can form a layer in the top portion of the tank. In this situation, a downward vertical air jet introduced through the tank roof would flow through the light layer due to negative buoyancy and initial momentum, and then form a fountain when it encountered the heavy layer. Similarly, a horizontal side-entry jet at a low flow rate could become buoyant in the heavy layer, turn vertically upward, and become a fountain when it encountered a lighter vapor layer.

Correlations for fountain rise height can be used to estimate whether the air jet is sufficiently strong to break down the stratification, or whether the air jet will erode the layer by mixing at the interface between the heavy and light layers. It is unlikely that the location of the interface will be known, so some arbitrary rise height must be chosen for this estimate. The most conservative value is to use the tank height.

Burridge & Hunt (2015) developed correlations for rise height of fountains [5]. For turbulent air jets ($Re_0 > 2000$, where $Re_0 = \frac{4\rho_j Q_0}{\mu_j \pi d_j}$) the rise height of a fountain can be estimated by:

$$Z = \begin{cases} 2.46 \cdot L_M \; for \; Fr_0 \geq 4.0 \\ 2.8 \cdot L_M - 2.1 \cdot L_Q \; for \; 2.0 \leq Fr_0 \leq 4.0 \\ 0.82 \cdot r_0 Fr_0^{2/3} \; for \; 1.0 \leq Fr_0 \leq 2.0 \\ 0.81 \cdot L_{Hyd} \; for \; 0.3 \leq Fr_0 \leq 1.0 \end{cases} \tag{8}$$

where $L_M = M_0^{3/4}/B_0^{1/2} = r_0 Fr_0$, $L_Q = Q_0/M_0^{1/2} = r_0$, $L_{Hyd} = r_0 Fr_0^{2/3}$, and $Fr_0 = \frac{M_0^{5/4}}{Q_0 B_0^{1/2}}$. They make the constraint that $\frac{\Delta\rho}{\rho_v} \ll 1$. This means that the correlation may not be accurate for freeing very low density or very high density vapors with air. Their data were measured using liquids and cover a range of *Re* and *Fr* that do not necessarily correspond to those for gases and vapors, so the correlations may not be accurate for vapor freeing. However, the correlation does at least give a general idea of how far a fountain will penetrate the stratified layer.

It is unlikely that the location of the interface of a heavy vapor would be known. The interface layer location and the initial jet flow rate can be varied to estimate the air flow rate required to penetrate through the entire tank regardless of where the interface is located. If the jet penetrates through the entire layer, it is likely the jet will have sufficient momentum to break down the stratified layer and mix the contents of the entire tank.

If the jet is unable to break up the stratified layers, then a complex situation can result where the heavy and light layers remain intact, and air mixed with one or both vapors can form a third layer between them. The natural approach to remove both layers would be to have two outlets – one at the side to remove the heavy layer and one at the roof to remove the light layer. However, this approach may not be feasible and should only be used after a hazard review.

After sufficient time, the air jet can erode each layer and eventually remove both the heavy and light vapors as long as the air flow rate is sufficient to mix the tank headspace when buoyancy is no longer a significant factor.

2.1.4 Summary of Stratification Effects

If the air jet will not break up the stratified layers, then the vapor freeing operation will initially be by displacement. The locations of the inlets and outlets should be selected to take best advantage of the jet momentum and the vapor density. Initially, the outlet will be essentially pure vapor. As the vapor freeing proceeds, the concentration of vapor in the discharge will drop and the vapor remaining in the tank will be mixed with air. As a consequence, the relative density difference between the air jet and the vapor/air mixture in the tank will decrease.

At some point the density difference will become small enough that the air jet will break up any remaining stratification and buoyancy will no longer significantly affect the flow regime in the tank. Once this condition is reached, the vapor freeing will naturally transition to the dilution method.

2.2 Dilution Method

When stratification is not an issue or is no longer an issue, then the consideration of importance is whether the air jet will adequately mix the tank vapor space. Two factors are of importance in making this determination. First, the distance the jet travels in the tank before hitting the tank walls must be less than ~400 nozzle diameters: $L_j < 400 \cdot d_j$ [4]. Second, the tank contents must be mixed rapidly enough to prevent significant concentration fluctuations in the outlet. This can be evaluated using correlations for characteristic mixing time. If the mixing time is much less than the characteristic residence time in the tank, then the jet will likely be capable of thoroughly mixing the vapor space.

The characteristic residence time can be estimated as $t_r = \frac{V}{Q_0}$, where V is the volume of the vapor space. In the case where there are multiple inlets, Q_0 is the total flow rate of all inlets.

2.2.1 *Mixing Time Correlations*

Correlations for mixing time exist for several configurations relevant to vapor freeing [3]. Fox (1956) developed a correlation for mixing time for a side entry jet discharging into a cylindrical tank. This correlation is for fluids with the same density and viscosity [6]:

$$t_m = \frac{ADH}{u_0 d_j}\left(\frac{d_j}{H}\right)^{0.50} \tag{9}$$

where

$$A = \begin{cases} \dfrac{7.8x10^5 Fr_j^{0.17}}{Re_0^{1.33}} \; for\ 250 < Re_0 < 2000 \\ \dfrac{120 Fr_j^{0.17}}{Re_0^{0.17}} \; for\ 2000 < Re_0 < 160{,}000 \end{cases} \tag{10}$$

where $Fr_j = \frac{u_0^2}{gd_j}$. For a jet discharging upward from the bottom center of a cylindrical tank, mixing time can be approximated using equation (9) with the following values for *A* [7]:

$$A = \begin{cases} \frac{7.08x10^5 Fr_j^{0.17}}{Re_0^{1.33}} \; for\; 100 < Re_0 < 2000 \\ \frac{145 Fr_j^{0.17}}{Re_0^{0.17}} \; for\; 2000 < Re_0 < 100{,}000 \\ 22.8 Fr_j^{0.17} \; for\; 100{,}000 < Re_0 < 140{,}000 \end{cases} \tag{11}$$

Jets pointing down have mixing times about 90% longer than upward pointing jets, so for an air inlet on the roof, multiply *A* by 2.

The following design guidelines provide limits for using jets to mix tank contents [3]:

- Roof mounted jets (axial jet):
    - A single axial jet is acceptable if the tank height to diameter aspect ratio is in the range $0.75 \leq H/D \leq 3.0$.
    - Otherwise, use more than one jet.
- Side entry jet:
    - A single side-entry jet is acceptable for $0.25 \leq H/D \leq 1.5$
    - Use multiple side-entry jets if $H/D \leq 0.25$ or $H/D \geq 3.0$

3 Perfect Mixing Model

If the residence time is long compared to the mixing time (e.g. $t_r >> t_m$) then the vapor concentration in the outlet will not fluctuate significantly, and the perfect mixing model (PMM) may estimate the component concentration quite well. The PMM can be used to estimate the vapor concentration versus time, and can be used to estimate the time required to reach a specific concentration target.

If the inlet flow rate is fast compared to the rate at which blending occurs (e.g. $t_m >> t_r$) then the component concentration in the outlet will fluctuate. At the extreme, a component in an inlet stream could essentially short circuit the enclosure and flow to the outlet without mixing significantly. In this case, the perfect mixing model will not be a good estimate of the component concentration in the enclosure [8].

The PMM can be written as:

$$x(t) = x(0) \cdot e^{-t/t_r} \tag{12}$$

where $x(t)$ is the volume fraction of vapor in the tank, and $x(0)$ is the initial volume fraction of vapor. In the case where vapor freeing starts with the displacement method, the value of $x(0)$ is the vapor fraction at the time stratified layers break up and the freeing transitions to the dilution method, and the value of *t* is set to 0.

Solving equation (12) for t allows an estimate of the time required to reach a specified volume fraction of the component, $x(t)_s$, for $0 < x(t)_s < x(0)$. This is given by:

$$t = -t_r \ln\left(\frac{x(t)_s}{x(0)}\right) \tag{13}$$

A common final vapor concentration target is to be less than 25% of LEL. Assuming LEL = 1% by volume and the initial volume fraction of vapor is 1.0, the time required is:

$$t_{0.25\cdot LEL} = -t_r \ln\left(\frac{0.25\cdot 0.01}{1.0}\right) \approx 6\cdot t_r = 6\cdot\frac{V}{Q_0} \tag{14}$$

With appropriate units, V/Q_0 is the number of air changes per hour (ACH) in the tank vapor space. Therefore, equation (14) can be used to determine the number of air changes required to reach the target vapor concentration. In the example above, 6 air changes are required.

If the vapor freeing operation begins with stratified layers and uses the displacement method, the total time to free the tank headspace will be approximately the time to flow one tank volume of air plus the time calculated to get the remainder of the way to the target vapor concentration, using equation (13) with x(0) set to the vapor concentration after one air change.

If the jet is initially capable of breaking up any stratification and if $t_r >> t_m$, then the time needed for vapor freeing can be estimated by equation (13) with x(0) = 1.

4 Examples

There is little published data on vapor freeing of tanks, so computer simulations were used to compare the mixing correlations given above to vapor freeing. The Fire Dynamics Simulator (FDS) software package was used to model velocity of flows and the concentration of vapor in the tank with time. FDS was developed to model compartment fires, and so is well suited to model buoyant flows and the concentration of components in the compartment. It is fundamentally a computational fluid dynamics software package. It has the added advantage that it is freely distributed software so it can be used by anyone with the requisite background and training. For the specific application to vapor freeing, FDS has weaknesses that may reduce the accuracy of the simulation. For example, FDS uses rectangular boxes for the mesh so the walls of a cylindrical tank are not smoothly modelled. Similarly, a round fan or educator on a manway must be modelled as a square vent.

Therefore, the FDS simulation results provide a qualitative analysis of the specific cases modeled, and should not be viewed as providing a quantitatively reliable prediction without comparison to relevant experimental data.

The examples given below compare FDS simulation results to the correlations given above for the side-entry jet and the vertical jet. The tank dimensions used are from Example 5.1 in API RP 2016: 125 ft diameter, 40 ft tall fixed (cone) roof tank. The roof was modeled as a flat surface rather than as a cone. The side vents were modeled as 1-meter squares. The FDS mesh used 0.5-

meter cubes. The bottoms of side vents were located 1 meter above the base of the thank, and roof vents were located 2 meters in from the edge of the tank. The geometry for one case is shown in Figure 3.

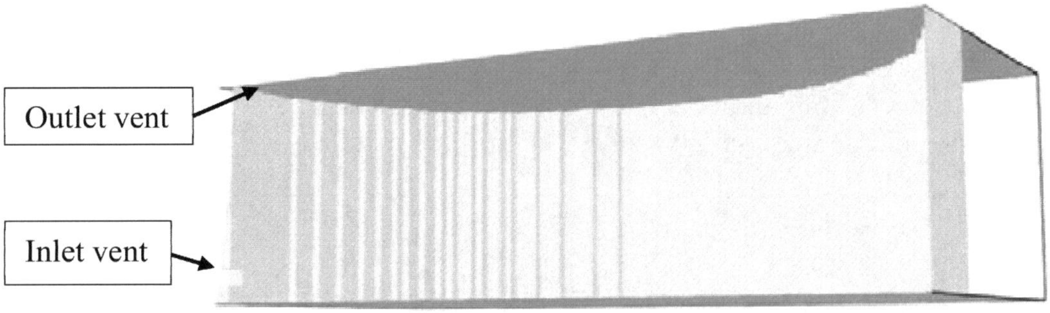

Figure 3. Cutaway view of the tank geometry used for FDS simulations. This view shows the half of the tank divided through a side-inlet vent.

4.1 Side-Entry Jet with Low-Density Vapor

The condition modeled for this case is the Combination (1)(A): an eductor on the roof drawing air from the tank with a side-entry vent for air into the tank, and the tank initially filled with methane. Simulations were run for three flow rates, as shown in Table 3.

Table 3. Combination (1)(A) Simulations (Methane).

Case Identification	Flow Rate (m^3/s)	Flow Rate (CFM)	ACH	Purpose
(1)(A) Low	2.0	4,230	0.52	Develop and maintain a stratified layer.
(1)(A) Mid	19.3	40,800	5.0	Match API RP 2016 recommended flow rate.
(1)(A) High	114	241,300	29.5	Match the critical velocity from equation (2).

Figure 4 compares the methane concentration from the FDS simulation at the tank vertical axis in 0.5 m increments to the concentration calculated by the PMM.

(a)

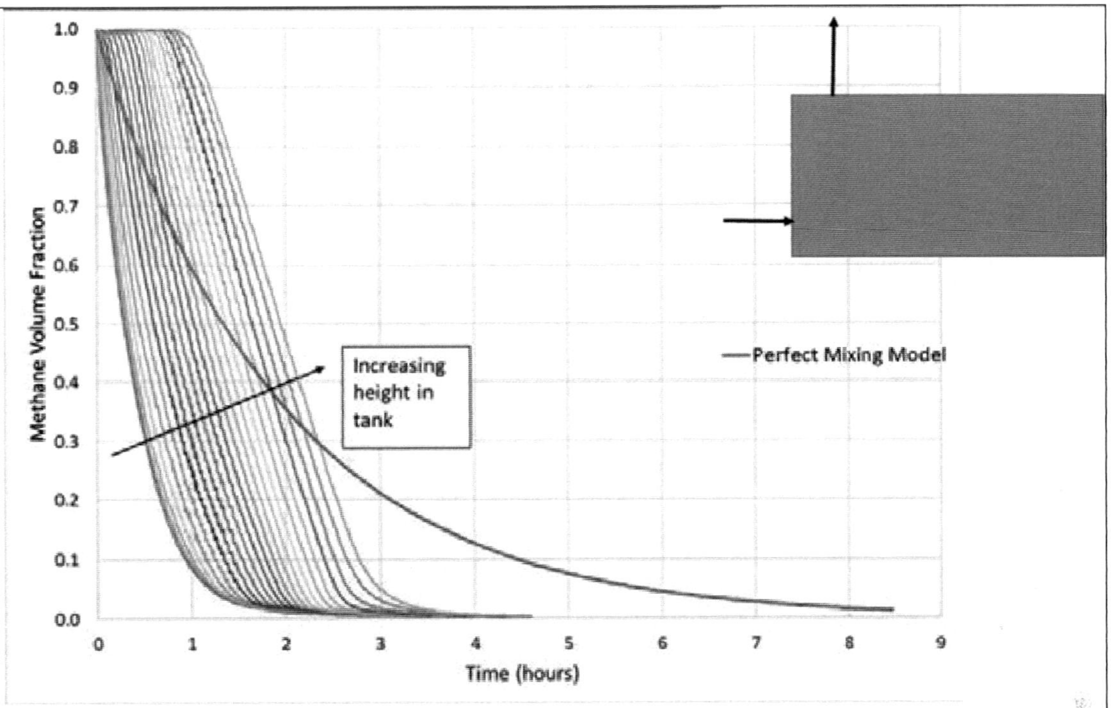
Methane Volume Fraction
Time (hours)
Increasing height in tank
Perfect Mixing Model
1.0
0.9
0.8
0.7
0.6
0.5
0.4
0.3
0.2
0.1
0.0
0
1
2
3
4
5
6
7
8
9

(b)

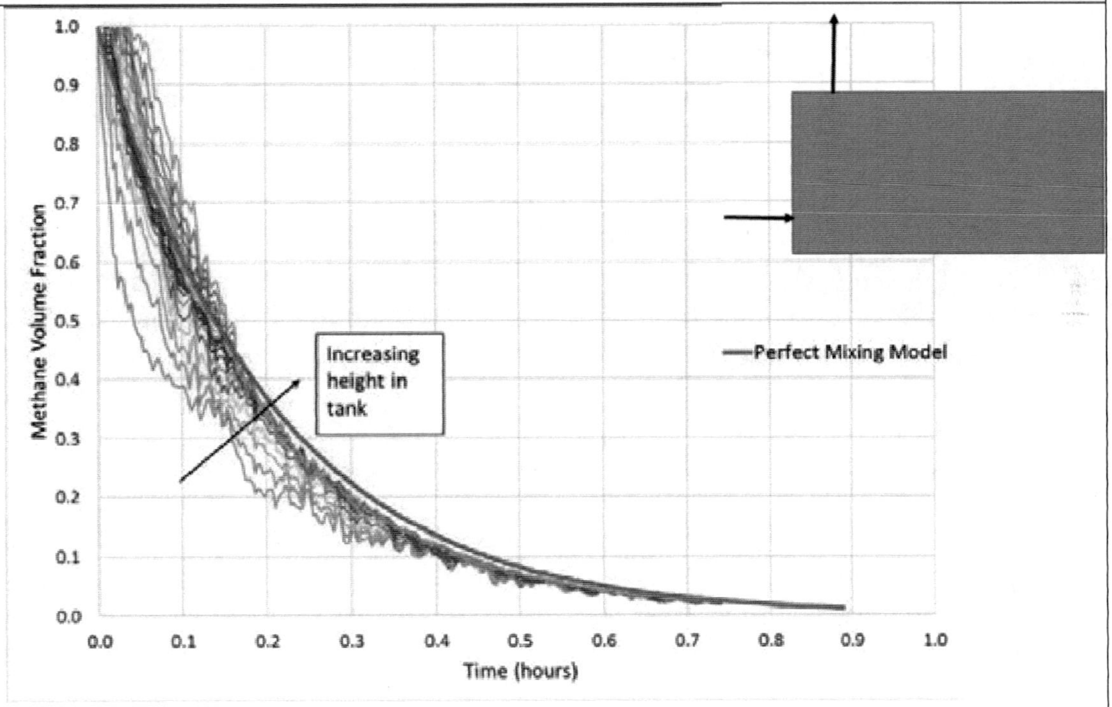
Methane Volume Fraction
Time (hours)
Increasing height in tank
Perfect Mixing Model
1.0
0.9
0.8
0.7
0.6
0.5
0.4
0.3
0.2
0.1
0.0
0.0
0.1
0.2
0.3
0.4
0.5
0.6
0.7
0.8
0.9
1.0

(c)

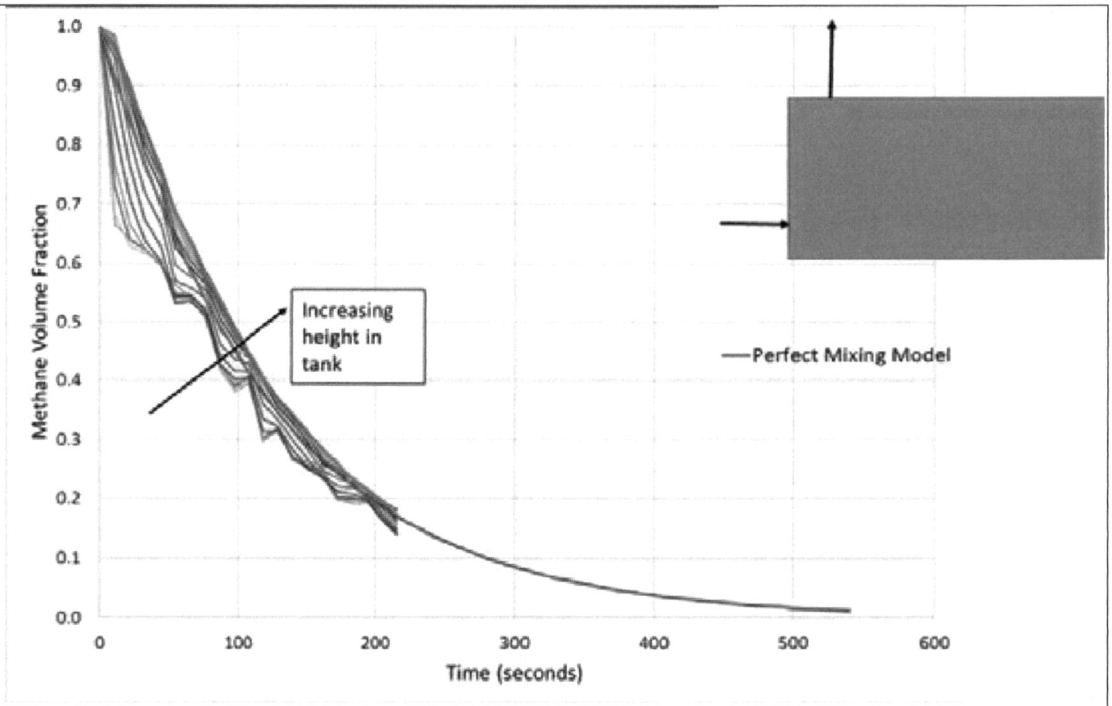

Figure 4. Comparison of methane concentration in the tank from the FDS simulation to the Perfect Mixing Model (PMM) for Combination (1)(A): (a) 0.5 ACH; (b) 5 ACH; (c) 29.5 ACH. The arrows on the blue rectangle indicate the location of the inlet and outlet vents on the tank.

The FDS results for the 0.5 ACH case of Combination (1)(A) show that the air flow rate is sufficiently low that a stratified layer forms during the vapor freeing process. This can be seen by the spread of the curves from the PMM line. These results also show that if the air flow rate that can be used is this low, and the vapor density is significantly different than that of air, then using displacement to remove vapor is faster than using dilution (e.g. the PMM line).[2]

The 5 ACH flow rate recommended by API RP 2016 results in the tank contents following the PMM reasonably well throughout the freeing process. It also shows that the time required to reach the target concentration using dilution (e.g. low air flow rate) is significantly lower than for using displacement (e.g. higher air flow rate). This is true because there is a specific number of air changes required to reach the target concentration. Even though displacement requires fewer air changes than dilution, the time to do each air change in displacement is so much longer than for dilution that displacement is a slower process.

The 29.5 ACH rate results show that the critical velocity for breaking up a stratified layer does result in much less variation of the methane concentration vertically through the tank. However, the flow rate is so high it is unlikely that any gas freeing process would come close to this flow

[2] In order to use dilution in this case, some mechanism for mixing the tank contents would need to be added. For example, placing a fan inside the tank as a mixer.

rate. The results for the 5 ACH case show that a flow rate within ~20% of the critical velocity predicted by the correlation is sufficient to allow use of the PMM.

4.2 Vertical Entry Jet With Low-Density Vapor

The condition modeled for this case is the Combination (6)(B): an eductor on the roof drawing air from the tank with a second roof vent on the opposite side of the tank for the air inlet, and the tank initially filled with methane. The results are shown for simulations with three flow rates, as shown in Table 4.

Table 4. Combination (6)(B) Simulations (Methane).

Case Identification	Flow Rate (m^3/s)	Flow Rate (CFM)	ACH	Purpose
(6)(B) Low	2.0	4,230	0.52	Develop and maintain a stratified layer.
(6)(B) Mid	19.3	40,800	5.0	Match API RP 2016 recommended flow rate.
(6)(B) High	70.3	148,900	18.2	Match the critical jet length from equation (4).

Figure 5 compares the FDS simulation results to the PMM for the three cases. Again, the PMM does not model the 0.5 ACH case well. As with the case shown in Figure (2a), displacement results in slightly faster vapor freeing than does dilution for the same air flow rate. Case (6)(B) does take best advantage of initial jet momentum and buoyancy in that the initial jet momentum acts opposite to buoyancy. This results in increased mixing during the displacement part of the vapor freeing process. The result is that it takes longer to reach the vapor fraction target for Case (6)(B) Low than it does for Case (1)(A) Low.

The PMM does model Case (6)(B) Mid reasonably well, indicating once again that the API RP 2016 recommended flow rate of 5 ACH is likely to break up stratification sufficiently that the vapor freeing will be by dilution for most of the process.

Figure (5c) for Case (6)(B) High shows that the flow rate required to achieve the critical jet length predicted by equation (4) results in little variability of the vapor fraction from the PMM line, indicating no significant stratification occurs during this process. As in Case (1)(A) High, the air flow rate is so high that it may not be feasible for any real vapor freeing process. A flow rate ~25% of the predicted critical value appears to be sufficient to allow use of the PMM to model the vapor freeing process.

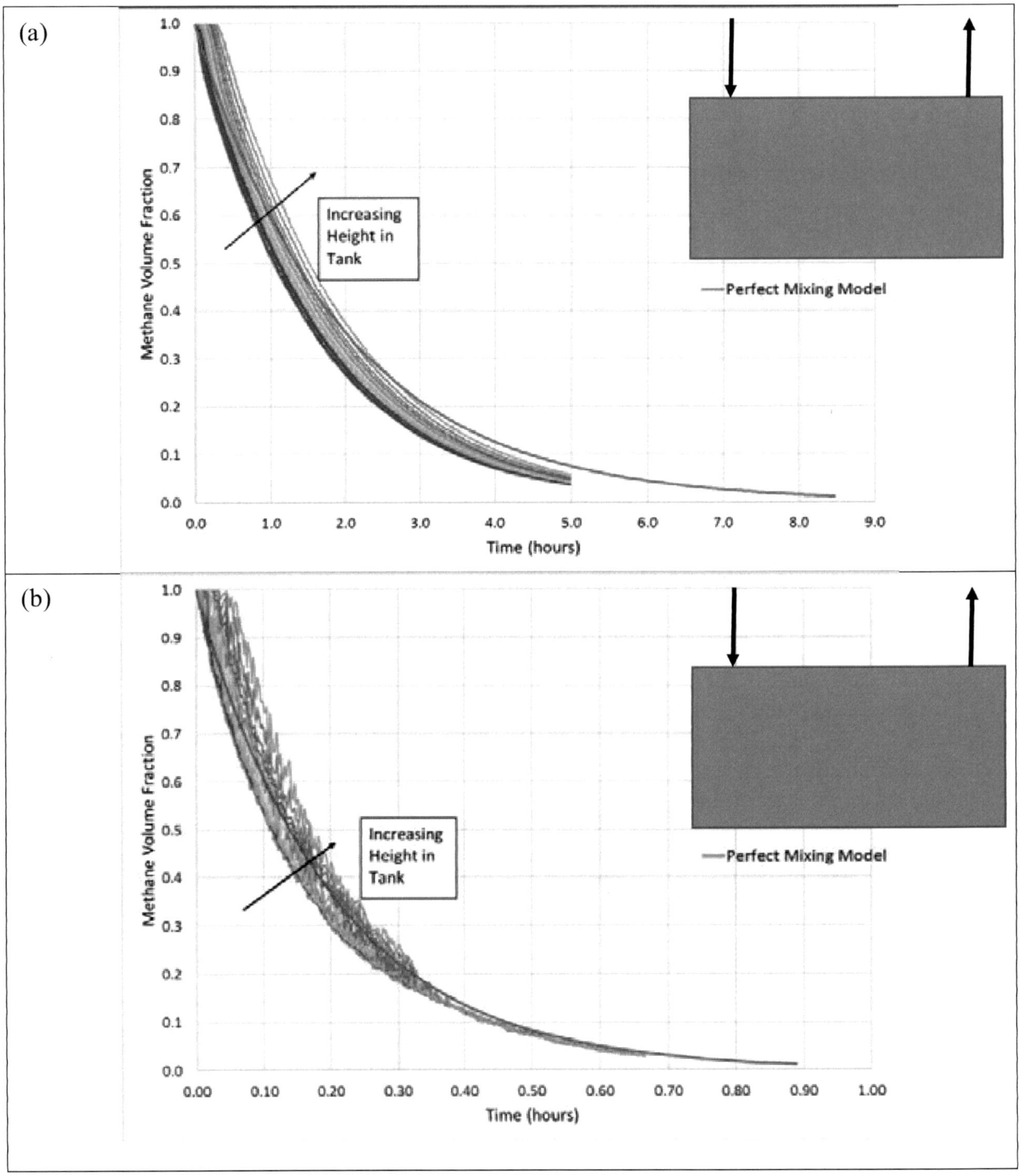
(a)
Methane Volume Fraction
Time (hours)
Increasing Height in Tank
Perfect Mixing Model
(b)
Methane Volume Fraction
Time (hours)
Increasing Height in Tank
Perfect Mixing Model

(c)

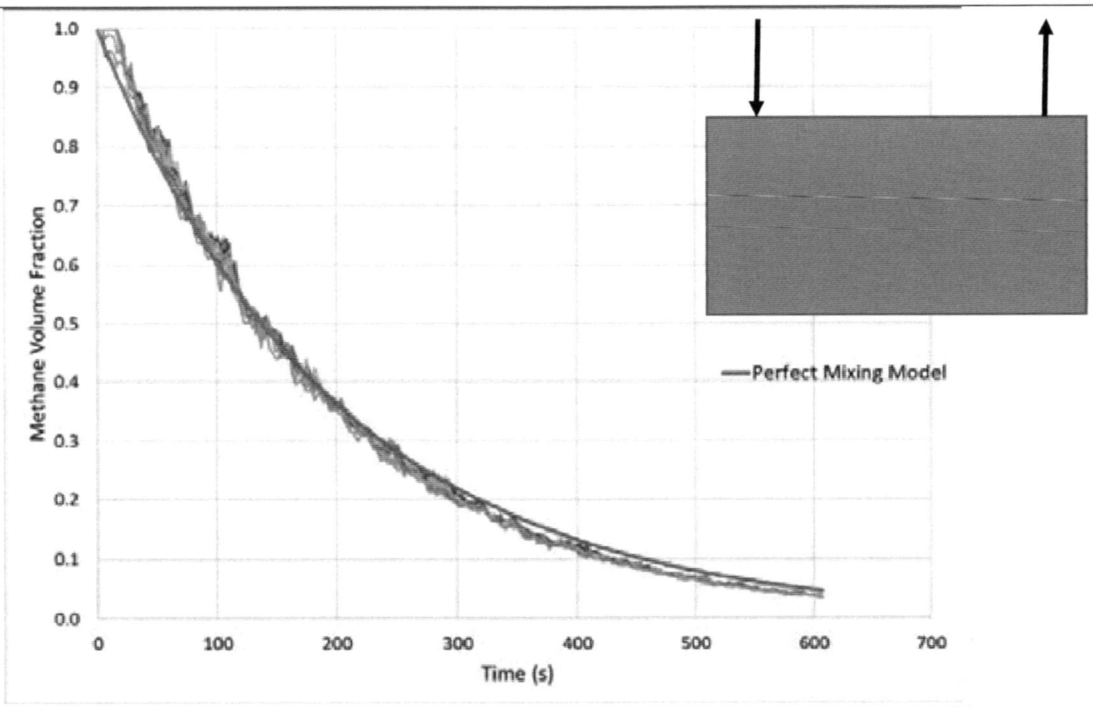

Figure 5. Comparison of methane concentration in the tank from the FDS simulation to the Perfect Mixing Model (PMM) for Combination (6)(B): (a) 0.5 ACH; (b) 5 ACH; (c) 29.5 ACH. The arrows on the blue rectangle indicate the location of the inlet and outlet vents on the tank.

4.3 Fountain Flow for Side-Entry Jet into Two Layers With Different Densities

The condition modeled for this case is the Combination (1)(G): an eductor on the roof drawing air from the tank with a side-entry vent directly below the outlet. The tank is initially filled half with methane and half with propane. The air inlet is in the propane layer. The results are shown for simulations with two flow rates, as shown in Table 5.

Table 5. Combination (1)(G) Simulations (Half Methane/Half Propane).

Case Identification	Flow Rate (m^3/s)	Flow Rate (CFM)	ACH	Purpose
(1)(G) Low	2.0	4,230	0.52	Maintain the stratified layers.
(1)(G) Mid	19.3	40,800	5.0	Match API RP 2016 recommended flow rate.

The results of the FDS simulations are shown in Figure 6 compared to the PMM model. In Figure (6), the buoyancy of the air jet in the two gases results in the total gas volume fraction remaining relatively high near the roof and floor of the tank, while the gas volume fraction near the middle of the tank decreases. This means that there is no clear trend with height in the tank for the profiles shown in Figure (6).

For the low flow rate case in Figure (6a), the stratified layers remain intact for a significant fraction of the vapor freeing process. The location of the air inlet and outlet does not take advantage of the buoyancy of air in propane. The effect of this can be seen in Figure (6a) in that the total gas volume fraction remains above the PMM line for the entire vapor freeing process. This occurs because the buoyancy of the air jet results in the air jet deflecting upward into the methane layer before it mixes well with the propane layer. Therefore, the air jet erodes the propane layer rather than mixing with and diluting the propane layer. The higher flow rate case shown in Figure (6b) is sufficient to overcome this effect.

Figure (6c) shows the FDS simulation results for a case with an air flow rate of 0.5 ACH and with two outlets in the tank: one in the roof opposite the air inlet, and one in the side opposite the air inlet. The total gas volume fraction profiles show significant variability from the PMM indicating that much of the vapor freeing process is occurring by displacement. Adding a second outlet on the side takes advantage of buoyancy for both the methane and propane. The result is that the total vapor volume fraction reaches the target sooner than for the case with no side outlet (Figure (6a)).

(a)

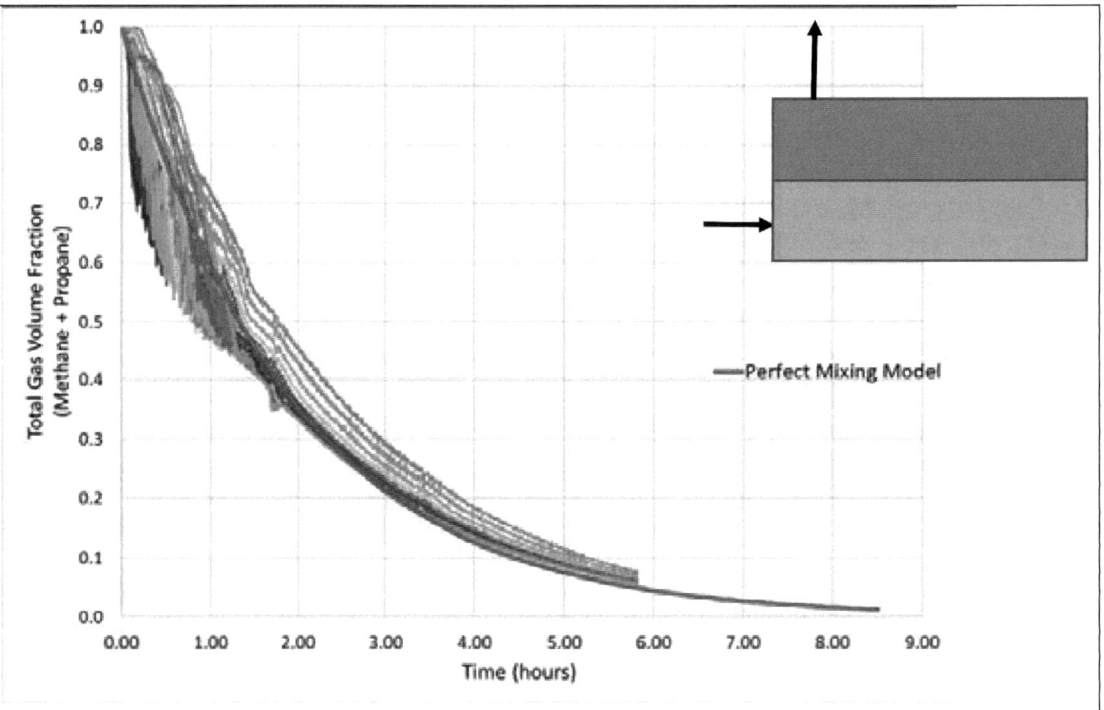

(b)

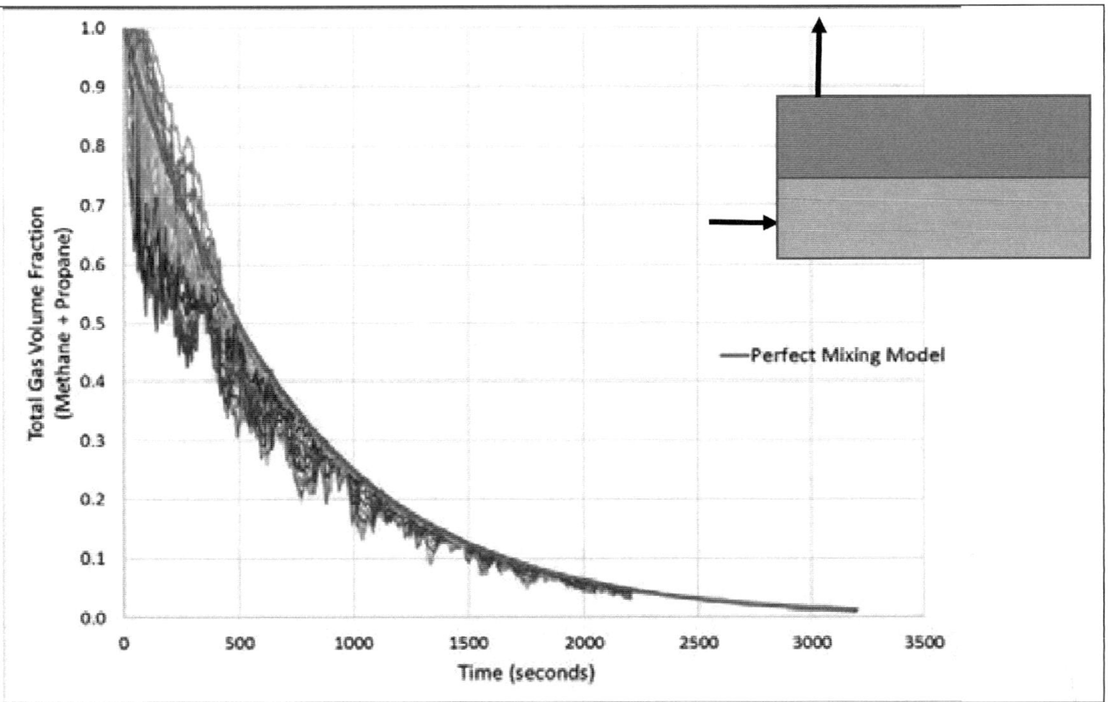

(c)

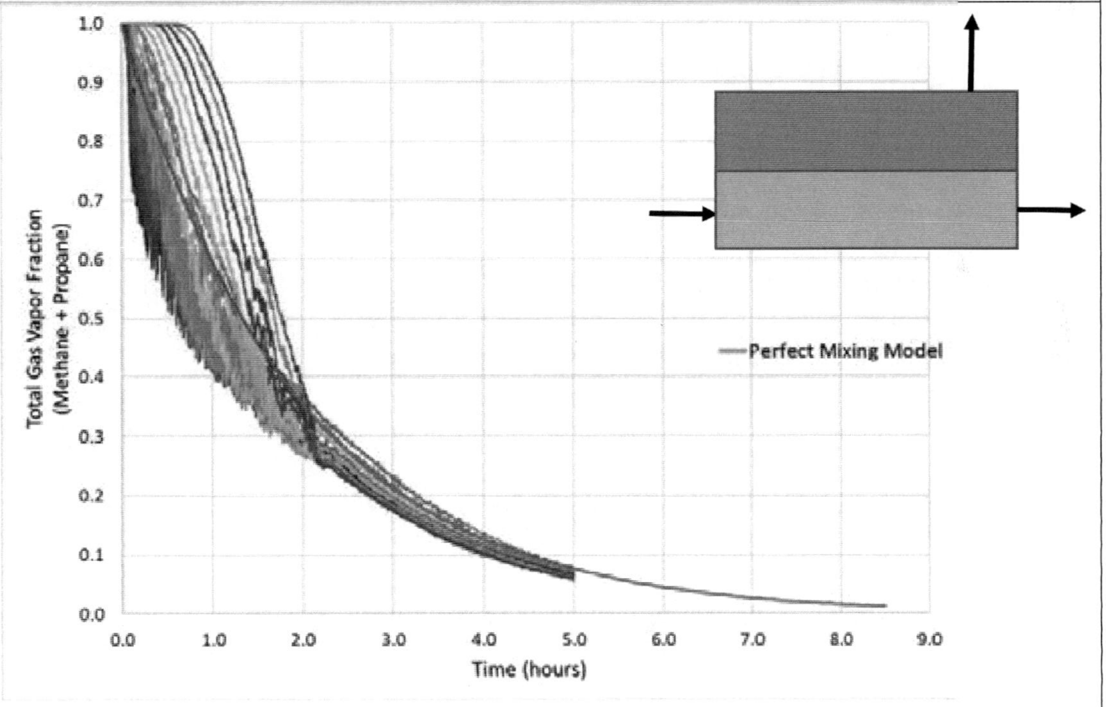

Figure 6. Comparison of total gas volume fraction in the tank from the FDS simulation to the Perfect Mixing Model (PMM) for Combination (1)(G): (a) 0.5 ACH; (b) 5 ACH ; (c) 0.5 ACH with two outlets. The rectangle with arrows indicates the location of the inlet and outlet vents on the tank. The blue area in the rectangle indicates methane, the green area indicates propane.

5 Conclusion

API RP 2016 provides recommended methods for vapor freeing storage tanks. Following these recommendations should result in safe and effective freeing of tank headspaces of toxic or flammable vapors. When using air to remove vapors from a tank headspace, it may not be feasible to achieve the 5 ACH air flow rate recommended in the guide. In some situations, air at the recommended flow rate may not break up stratified layers in the tank. In such cases, it is important to orient the air inlet and vapor outlet to take advantage of both the buoyancy of the vapors and the jet momentum.

The correlations given above provide a way to estimate whether the air jet will break up stratified layers, and provide some guidelines on when a single air inlet may be sufficient or when multiple inlets may be required to provide adequate mixing in the tank.

The FDS simulations indicate that the correlations for critical velocity of a side-entry jet and for a vertical-entry jet are conservative for the purposes of vapor freeing. It appears that a flow rate greater than approximately 20% of the critical values calculated using the correlations will be sufficient to break up stratification, so that the vapor freeing process will be by dilution. For flow rates below this, one should anticipate that at least the first air change in the tank will be mostly by displacement.

For low air flow rates, consider angling the air jet so that it traverses the maximum distance possible through the tank before encountering a wall. This will increase mixing efficiency when buoyancy effects are reduced or are nonexistent.

Safety issues posed by flammable mixtures in the tank during vapor freeing are beyond the scope of this paper. However, the Perfect Mixing Model and the FDS results do provide qualitative information on where in the tank and at what times flammable mixtures may be present. For vapor freeing by dilution (e.g. when the air flow rate is sufficiently high that the PMM can reasonably be applied) the PMM can be used to estimate the time range during which the entire tank headspace is likely to contain a flammable mixture. For vapor freeing by displacement (e.g. when the air flow rate is sufficiently low that the vapor volume fraction profiles show significant variation from the PMM model), a conservative approach is to consider that a flammable mixture exists in some part of the tank headspace for virtually the entire vapor freeing process. The region where the air jet enters the tank and begins to mix with the vapor should also be considered to contain a flammable mixture for virtually the entire vapor freeing process. Care should be taken to consider any potential mixture variations throughout the tank during any vapor freeing operation.

6 Nomenclature

B_0	Initial jet buoyancy flux, m^4/s^3
D	Tank diameter, m
d_j	Jet nozzle diameter, m
Fr_0	Froude number for jet based on initial jet properties
G	Stratification constant
g	Gravitational constant, 9.8 m/s^2
g'_0	Reduced gravity, g

H	Tank height, m
K	Constant = 1.159
L_{hyd}	Hydraulic length scale, m
L_j	Jet length scale, m
M_0	Initial jet momentum flux, m^4/s^2
Q_0	Initial jet volume flow rate, m^3/s
R	Characteristic tank base length scale, m
Re_0	Jet Reynolds number based on the nozzle diameter
r_0	Jet nozzle radius, m
t	Time, s
t_m	Characteristic mixing time, s
t_r	Characteristic residence time, s
u_0	Jet initial velocity, m/s
u_c	Critical jet velocity, m/s
V	Tank volume, m^3
x(t)	Vapor volume fraction in the tank
Z	Fountain jet penetration distance. m

Greek Symbols:

α	Constant = 0.083
β	Angle of a side-entry jet from horizontal, degrees
θ	Defined as $\beta+5^o$
π	Constant = 3.14
ρ	Density, kg/m^3

7 References

[1] ANSI/API Standard 2015, *Requirements for Safe Entry and Cleaning of Petroleum Storage Tanks,* Washington, D.C.: API, 2014.

[2] API RP 2016, *Guidelines and Procedures for Entering and Cleaning Petroleum Storage Tanks,* Washington D.C.: API, 2016.

[3] K. Chow, *Simulation and analysis of gas freeing of oil tanks,* Coventry University: Ph.D. Thesis, 2009.

[4] B. K. Revill, "Jet Mixing," in *Mixing in the Process Industries, 2nd ed.*, M. F. E. A. W. N. N. Harnby, Ed., Oxford, Butterworth Heinemann, 1992, pp. 159-183.

[5] G. R. Hunt, P. Cooper and P. F. Linden, "Thermal stratification produced by plumes and jets in enclosed spaces," *Building and Environment,* vol. 36, pp. 871-882, 2001.

[6] H. C. Burridge, A. Mistry and G. R. Hunt, "The effect of source Reynolds number on the rise height of a fountain," *Physics of Fluids,* vol. 27, p. 047101, 2015.

[7] E. A. Fox and V. E. Gex, "Single-phase blending of liquids," *AIChE Journal,* vol. 2, no. 4, pp. 539-544, 1956.

[8] A. G. C. Lane and P. Rice, "Comparative assessment of the performance of the three designs for liquid jet mixing," *Industrial & Engineering Chemistry Process Design & Development,* vol. 21, pp. 650-653, 1982.

[9] P. L. Lagus and K. H. Lie, "A numerical investigation of the constant tracer flow technique," in *Air Change Rate and Airtightness in Buildings*, M. H. Sherman, Ed., Philadelphia, ASTM, 1990, pp. 21-30.

Refinery PHAs – Revalidation or Re-Dos? Do They Ensure Safe Operations?

Kumar (Chris) Israni
Principal Consultant
ERM
840 W Sam Houston Pkwy N, Suite 600, Houston, TX
chris.israni@erm.com

Joshtin Cooper
Senior Consultant
ERM
3838 Causeway Blvd, Metairie, LA
joshtin.cooper@erm.com

Radhika Bompelly
Consultant
ERM
840 W Sam Houston Pkwy N, Suite 600, Houston, TX
radhika.bompelly@erm.com

Prepared for Presentation at
American Institute of Chemical Engineers
2017 Spring Meeting and 13th Global Congress on Process Safety
San Antonio, TX
March 26 – 29, 2017

Refinery PHAs – Revalidation or Re-Dos? Do They Ensure Safe Operations?

Kumar (Chris) Israni
Principal Consultant
ERM
840 W Sam Houston Pkwy N, Suite 600, Houston, TX
Chris.israni@erm.com

Joshtin Cooper
Senior Consultant
joshtin.cooper@erm.com

Radhika Bompelly
Consultant
radhika.bompelly@erm.com

Keywords: PHA, OSHA, Refinery, Revalidation, Redo

Abstract

OSHA 1910.119 Process Safety Management (PSM) has been the industry standard for over 20 years. Its elements have been clearly rolled out and adopted by refineries throughout the USA. However, the most recent grassroots refineries in the country were built over 45 years ago. There have been various initiatives from regulators and operating companies by way of compliance audits, emphasis programs and due diligence, to ensure that the process safety management aspects are adhered too, while following the path of compliance and conformance to the 14 elements, one of these being the every five year occurring Process Hazard Analysis (PHA) cycle.

The success of these PHAs rely on various factors – choice of methodology, scope inclusive of changes from the last few years, addressing new and earlier unidentified scenarios, tracking of previous actions and incorporation of findings from incidents and lessons learned. While this should be the scope of the PHA, it is often misrepresented with a check the boxes type of approach. The key is ensuring compliance as well as satisfying the requirements of a revalidation or redo – a choice made by a few operating companies.

It is commonly observed that while sites have completed this five year compliance requirement, some of effectiveness of the PHA is lost due to non-adherence to scope. This paper highlights the experiences of the authors while conducting PHAs using its various methodologies in the revalidation and re-do approaches.

Satisfying the every five year requirement, addresses the compliance need. The effectiveness of the PHA (irrespective of approach – revalidation / re-do) is depends on adherence to the claims made in the PHA. The operating risk stays low, when the identified scenarios and their associated safety mechanisms (claims) are continuously implemented and practiced.

1 Introduction

Petroleum refining has evolved continuously in the US for different products. Currently, there are 141 refineries in the US with a total capacity of more than 18 million barrels per day [1, 2, 10]. Apart from transportation fuels, petroleum refineries also produce liquids that are used by the petrochemical industry to make a variety of chemicals and plastics. Refineries have to ensure compliance with government regulations such as Occupational Health and Safety Administration (OSHA) 1910.119 [2] and Environmental Protection Agency (EPA) 40 CFR Part 68 [4].

2 Process Hazard Analysis and Methodology

Center for Chemical Process Safety (CCPS) Guidelines for Hazard Evaluation Procedures defines hazard as "an inherent physical or chemical characteristic that has the potential to do harm (to people, property, or the environment)" [4]. A Process Hazard Analysis (PHA) is defined as a systematic effort designed to identify and analyze hazards associated with the processing or handling of highly hazardous materials and a method to provide information which will help workers and employers in making decisions that will improve safety. The OSHA 1910.119 regulation [3], amongst its other requirements, requires the sites to perform a PHA on each process.

The method selected for PHA will be influenced by many factors such as the type of process, complexity of the process, time required, etc. The process hazard analysis methodology selected must be appropriate to the complexity of the process and must identify, evaluate, and control the hazards involved in the process. Proper selection of the methodology is one of the most basic factors that can aid in making a good PHA. The most commonly used techniques of PHA are What-If, Checklist, What-If/Checklist, Hazard and Operability Study (HAZOP), Failure Mode and Effect Analysis (FMEA) and Fault Tree Analysis (FTA) [5]. All PHA methodologies are subject to certain limitations [9]. So regardless of the method used, the PHA shall address all of the following issues:

- The hazards of the process;
- Previous incidents which could have been catastrophic;
- Engineering and administrative controls;
- The consequences of failure of engineering and administrative controls;
- Facility siting;

- Human factors; and
- The range of possible safety and health effects caused by the failure of controls.

The process hazard analysis shall be performed by a team with expertise in engineering and process operations, and the team shall include at least one employee who has experience and knowledge specific to the process being evaluated. Without adequate knowledge and experience specific to the process, a PHA team may not identify unique hazards or appropriately assess potential causes, consequences, or the effectiveness of safeguards to control process hazards [5, 6, 9].

3 PHA Revalidation

As mentioned, at least every five years after the completion of the initial PHA, the companies are required to update or revalidate their PHA as per the OSHA process safety management (PSM) regulation, 29 CFR 1910.119. Revalidation is a process of examining the initial PHA to ensure that it adequately addresses process hazards as they are currently identified. This includes modifying the previous PHA accordingly to reflect the necessary changes in the facility since the previous analysis. Also, current understanding of the process hazards might differ from time to time – the initial PHA may have a certain type of understand, but due to changes in technology, regulation and personnel, the current group may have a different understanding.

There could be many factors that would have introduced new hazards such as process changes, changes in occupancy patterns either on-site or off-site, knowledge in better understanding the potential hazards, incidents occurred and changes in the safeguards identified in previous PHA. So the PHA revalidation must address the following items [6]:

- Changes subsequent to previous PHA
- Identification and correction of PHA omissions and deficiencies
- Ensure that Process Safety Information (PSI) is complete, current and accurate and all the training procedures are up to date
- Review the new process technology or information since the previous PHA
- Update the PHA according to new or modified regulations (OSHA and EPA)

One of the advantages of revalidation is time effectiveness. Conducting an initial PHA requires considerable amount of time and effort. The revalidation process significantly reduces the time and effort as the PHA scope is less since revalidation only reviews the modified sections. The prior PHA serves as the starting point for the revalidation process. The new content is added to the PHA and results will be documented as a new PHA.

PHA Revalidation Approach

PHA revalidation is an essential and critical activity. The first step in any PHA revalidation is to consider the issues before determining the appropriate revalidation approach. The most appropriate approach depends upon several factors including the quality of previous PHAs, information available from the previous PHA, the complexity of the changes, the number of undocumented changes and the quality of Management of Change (MOC). The next step would be to define the scope of the revalidation; the scope should include all the affected changes but not be limited or dictated by the scope of prior PHA.

The procedure for PHA revalidation study is similar to that of the initial PHA and should follow the steps listed below:

- Confirm on the scope
- Select the study team and the leader
- Select the appropriate methodology
- Assemble the required information such as previous PHA data, updated P&IDs, equipment data, incident investigation reports, MOC and revised operating procedures
- Review of actions from previous PHA
- Prepare a list of items to be reviewed during revalidation
- Conduct sessions for revalidation
- Establish new recommendations and action plan
- Issue revalidated PHA report

Selection of team members for revalidation is as critical as selection of team members for an initial PHA. A team comprising a mix of both new and previous team members would be efficient as this helps in providing consistency of approach, implementing lessons learned from previous PHA and providing fresh insight and objectivity. Just as conducting an initial PHA, the team should be trained in advance on the objective of the revalidation to make the sessions effective. Training may include providing the process overview, purpose of PHA revalidation, methodology, etc.

Revalidation requires more time to be spent on collecting the information, reviewing the previous and preparing the information to be reviewed during the sessions. Selection of an efficient study leader results in proper documentation of the revalidation process and generating a clear, concise report for implementation of the new findings and retention of record for future use.

Essential Tasks for PHA Revalidation

In order to optimize the PHA revalidation and to avoid pitfalls it is important to follow essential tasks as described below [7]:

1. Review of modifications: involves reviewing all the MOCs, audit reports, comparing current P&IDs with those used for previous PHA, operating procedures, etc.
2. Review of previous incidents: involves gathering incident investigation reports and reviewing with the full team to address any potential issues that are likely to cause catastrophic consequences.

3. Review previous PHA recommendations: involves reviewing the previous PHA recommendations to check the status of closure and to address unresolved actions.
4. Update facility siting/human factors: involves preparing a checklist to address the modifications in the facility siting or human factor analysis that could affect the risk scenarios.
5. Evaluate non-routine operations: involves evaluation and addressing the hazards such as start-up, shutdown, maintenance, etc.
6. Evaluate the standards and regulations: involves reviewing of the new regulations and comparing with those used in previous PHA to determine the additional information to be added in the new PHA to ensure that the new PHA is compliant with the new regulations.

4 PHA Re-do

A PHA redo can be defined as a process to perform a complete, new PHA using one or more recognized techniques as described in the sections above that meet all the regulatory, industry and company requirements. A complete redo would be considered a feasible and cost effective option depending on various factors.

A PHA redo may be a most appropriate choice in situations where [7]:

- Gaps or deficiencies exist in the manner in which the prior PHA was conducted indicating that significant hazards might have been overlooked or not controlled appropriately as a result of improper use of process safety information.
- Uncontrolled changes were occurred in the process or equipment since the previous PHA and the complexity of the process may require a redo.
- Inappropriate use of technique or methodology in the previous PHA.
- Occurrence of various incidents since the previous PHA
- Previous PHA documentation do not meet the current standards of regulatory requirements due to new interpretations of codes and standards

A redo is a time and resource effective option if the facility has undergone a large number of MOCs since the previous PHA as reviewing all MOC documents may require more time than redoing a PHA to ensure that all the hazards were assessed adequately. A redo also provides an opportunity to conduct a PHA using an appropriate method if a different method was used in the previous one. A redo is also considered a better option after a certain number of revalidations in order to ensure that no potential hazards were overlooked in the revalidating process over the period of time. Also document retention after an indefinite period of time becomes unmanageable with the addition of large number of MOCs to the PHA revalidation reports.

5 Comparison between Revalidation and Redo

The primary aim to either revalidate or redo a PHA is to identify, evaluate, and control the hazards involved in the process. The decision to revalidate or redo a PHA for the facility depends on the number of factors. Both revalidation and redo has its own limitations as well as benefits. The information in the paragraphs below are based on the authors experiences with facilitating PHAs in refineries, using either of the two approaches. These were based on their interactions with refinery sites, at a national level, and hence covered the various regions within the United States.

In the case of PHA revalidations, the limitations observed were:

- Inadequate documentation of the previous PHA(s) results in inefficient revalidation process,
- Limitations in defining the scope as it might be dictated by the previous PHA,
- The same team that performed the prior PHA may not be available for revalidation,
- Improper or poor documentation of the MOCs may result in inadequate evaluation of the hazards,
- Incorrect or missing PSI for revalidation,
- Unavailability of P&IDs, node information, operating procedures used in the prior PHA to compare with the new changes,
- Missing information on the closed out actions which forms the basis for revalidation,
- Noncompliance of the previous PHA with the regulations and standards,
- Amount of time spent in collecting and reviewing the previous PHA could be more than actual revalidation session,
- May require more time than an actual PHA to incorporate each MOC when there are a number of changes since the prior PHA,
- Data could become unmanageable after several revalidations, and
- Some hazards may be overlooked over a period of time.

At the same time, the benefits gained were:

- The approach was a huge time saver,
- Lessons from previous PHA can be implemented during revalidation,
- Experience and on-the-job training of the team aids in identifying the new issues that were not evaluated in the previous PHA,
- The team with a mix of new and old team members provide consistency and new insight for the session, and
- Revalidation offers an opportunity to pull multiple PHAs together.

In the case of PHA redos, the limitations observed were:

- Time required to redo a PHA is more compared to that of revalidation,
- Redo might just be a repetition of previous PHA in case of small changes,
- The team may not be familiar with the previous PHA, and
- The method selected for redo may not be effective in evaluating certain hazards identified in the initial PHA.

At the same time, the benefits gained were:

- Filling the gaps or deficiencies identified in the prior PHA,
- Selection of an appropriate method if the method used in the previous PHA was inefficient for the process,
- Defining the scope according to the new changes in the process, and
- Incorporating the current standards and regulations.

6 Conclusion

The two approaches have their benefits, and neither of them replace the other.

The revalidation approach is preferred due to time savings implying budgetary savings. But this is only successful if:

- The previous PHA is of a good quality and is used in the revalidation,
- Updates to the action items from the previous PHA are available,
- Incident reports and MOCs since the previous PHAs are discussed in the revalidation,
- The study still uses the most current set of PSI,
- Incorporates any regulatory changes, and
- During the revalidation, thought is given to changes to the process and the team also contributes towards any missed scenarios.

The re-do approach, essentially means beginning the PHA from scratch and will be successful:

- As long as the drawings are current, and brainstorming is followed with an experienced team, the study will cover all aspects and
- Even if the previous PHA has any deficiencies, they are covered by the complete re-do approach.

It is our observation, that however time and labor intensive the re-do process is, many companies are choosing it on their ten or fifteen year cycles. That covers for a lot of gaps otherwise noticed in:

- The previous PHAs,
- Any skipped or missed scenarios, and
- Changes in personnel, thereby becoming a training tool for new team members.

Hence, the choice is up to the operating company, on the approach they take, the key is that irrespective of route taken, the PHA must always be on the current state of the process, and provide adequate information on hazards identified, to minimize risks to personnel, environment and the business, during operations.

7 References

[1] U.S. Energy Information Administration, http://www.eia.gov/energyexplained/index.cfm?page=oil_refining#tab1Accessed on February 17, 2017.

[2] U.S. Energy Information Administration, https://www.eia.gov/tools/faqs/faq.cfm?id=29&t=6, Accessed on February 17, 2017.

[3] OSHA 3132 Process Safety Management Informational Booklet, 2000.

[4] American Petroleum Institute, API Recommended Practice 75, Recommended Practice for Development of a Safety and Environment Management Program for Offshore Operations and Facilities, 3rd edition, Washington, DC: API, May 2004.

[5] CCPS Guidelines for Hazard Evaluation Procedures, Center for Chemical Process Safety, American Institute of Chemical Engineers, New York, NY, 1992.

[6] Nigel Hyatt, Guidelines for Process Hazard Analysis, Hazards Identification & Risk Analysis, Dyadem, 1st Edition, 2004

[7] Walter L. Frank and David K. Whittle, CCPS Revalidating Process Hazard Analysis, Center for Chemical Process Safety, American Institute of Chemical Engineers, New York, NY, 2001

[8] Kevin E. Smith and David K. Whittle, Six Steps to Effectively Update and Revalidate PHAs, Chemical Engineering Progress, 2001

[9] Israni, Chris and Cooper, Joshtin, "Technical Risk – It's Not a One-Size-Fits-All Proposition", paper presented at AICHE 2013 Spring Meeting, 9th Global Congress on Process Safety, San Antonio, Texas, April 28 – May 1, 2013

[10] US energy Information Administration https://www.eia.gov/dnav/pet/pet_pnp_cap1_dcu_nus_a.htm, last accessed March 6, 2017

How Small of a Cloud Do We Want to Detect with our Gas Detectors?

Vinicius Simoes
Genesis
11750 Katy Freeway Suite 100
Houston, TX, 77079
Vinicius.Simoes@genesisoilandgas.com

Massimiliano Kolbe, Sarah Chambliss
Genesis
11750 Katy Freeway Suite 100
Houston, TX, 77079
MKolbe@genesisoilandgas.com, Sarah.Chambliss@genesisoilandgas.com

Prepared for Presentation at
American Institute of Chemical Engineers
2017 Spring Meeting and 13th Global Congress on Process Safety
San Antonio, TX
March 26 – 29, 2017

How Small of a Cloud Do We Want to Detect with our Gas Detectors?

Vinicius Simoes
Genesis North America
11750 Katy Freeway Suite 100
Houston, TX, 77079
Vinicius.Simoes@genesisoilandgas.com

Massimiliano Kolbe, Sarah Chambliss
Genesis North America

Keywords: Gas Detectors, Cloud Dispersion, Explosion, VCE

Abstract

There is a big discussion in the industry about how small is really small. The objective of this paper is to cover a series of cloud sizes and their impact to determine how small of a cloud we should really be detecting in order to avoid personnel injury and structural damage. There are a lot of different discussions in the industry about what clouds we should be detecting.

1 Introduction

The purpose of this paper is to determine a cloud size or release rate that one should focus on when defining their gas detector placement. It is well known there are several different methodologies when it comes to cloud detection or release detections. There are several criteria which are applied by different companies or even industries (e.g., chemical, pharmaceutical). Those criteria could be based on dispersion, cloud volume, or separation distances. It is of paramount importance to determine the target for our analysis, e.g., flammable or toxic materials. Those materials will have very distinct thresholds which would directly impact the study. This paper focuses on the flammable releases, and more importantly on the explosion effects coming from those clouds that could be generated from such releases.

The criteria could be, for example, based on release rate ranges. This could be using very small releases (0.1 kg/s) up to small releases (5 kg/s), depending on what criteria are applied. There are a series of questions that could still be unanswered after the gas detector layout has been designed. For example, have we optimized the number and location of our gas detectors? Did we detect all the clouds needed? The idea of the study was to identify if we really need to detect clouds being formed from very small releases or if there is a certain range that one should focus on when it comes to overpressure analysis resulting from vapor cloud explosions (VCE).

2 Modeling Approach

2.1 Scenario 1

The scenario was conducted using a series of releases of Ethane under 120 °F and 500 psi. The modeling of the discharge, dispersion and explosion was done using Safeti (Software for the Assessment of Flammable, Explosive and Toxic Impact) Version 6.7. Safeti is well-recognized and a very comprehensive quantitative tool available for assessing process plant risks. It is designed to perform all the data processing, generation of results and presentation elements of a Quantitative Risk Assessment (QRA) within a structured framework. Safeti went through an extensive validation process.

The Safeti QRA software is designed to comply with the regulatory requirements of many countries. For example, specific modules have been included to ensure compliance with the Dutch Yellow Book, the Reference Manual Bevi Risk Assessments version 3.1 and UK HSE regulations.

The explosion model adopted for this analysis was TNO Multi-Energy. A set of TNO strengths were used for the assessment, ranging from 3 to 7. It is understandable that detonations are possible, but for the purpose of this exercise this was not included in the assessment. The study focuses on the far-field assessment of the explosions. TNO is well accepted on the far-field but it is important to be cautious when applying it to the near-field. For a near-field exercise the authors do believe that a more sophisticated model, such as Computational Fluid Dynamics (CFD) should be applied. Based on the tool and approach used, it is important to highlight that the focus of this paper has been the onshore industry.

The study used a sample layout, which can be seen in Figure 1. This layout was created to define several congested regions, structures and buildings. In total, five congested regions and three monitor points were defined in the model, and two buildings were assessed for overpressure. The east wall of the two buildings and the roof of the control room building were monitored.

Several ignition sources were placed along the congested regions to ensure proper coverage of the different cloud volumes and ignition locations.

Two weather conditions were used for this study, including 2 m/s with a class F for the Pasquill Stability and 5 m/s wind speed with a class D for the Pasquill Stability. This was done to ensure we properly covered both stable and unstable conditions. Table 1 presents the other atmospheric parameters used for the study.

Table 1: Atmospheric Parameters

Parameter	Value
Atmospheric Temperature	68 °F

Parameter	Value
Atmospheric Pressure	14.7 psi
Surface Roughness	183.2 mm
Humidity	70 %
Solar Radiation Flux	250 $Btu/ft^2 \cdot hr$

In addition, four release locations were assumed and are indicated in Figure 1 by the red marks. For the purpose of this study, the authors covered a series of locations to ensure proper representation of the possible explosive cloud locations.

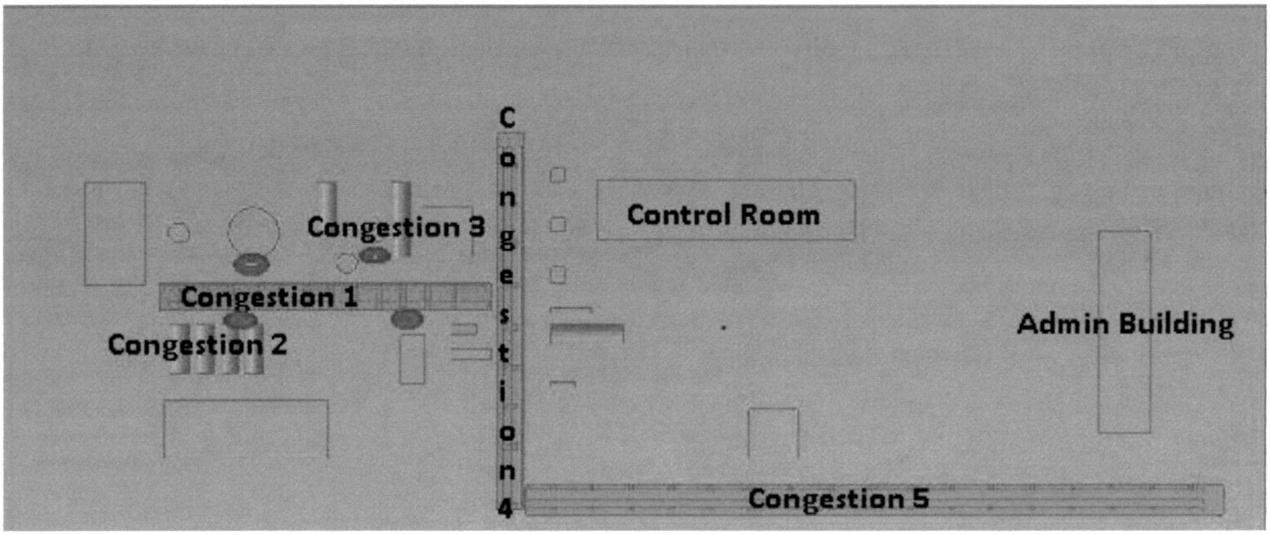

Figure 1: Facility Layout

Table 2 shows the list of Ethane release scenarios and their parameters.

Table 2: Ethane - Scenarios Parameters

Release Rate (kg/s)	Location	Pressure (psi)	Temperature (°F)
0.1	Location 1,2,3,4	500	120
0.5	Location 1,2,3,4	500	120
1	Location 1,2,3,4	500	120
2	Location 1,2,3,4	500	120
3	Location 1,2,3,4	500	120
4	Location 1,2,3,4	500	120
5	Location 1,2,3,4	500	120
7	Location 1,2,3,4	500	120
9	Location 1,2,3,4	500	120
10	Location 1,2,3,4	500	120
15	Location 1,2,3,4	500	120

Release Rate (kg/s)	Location	Pressure (psi)	Temperature (°F)
20	Location 1,2,3,4	500	120

2.2 Scenario 2

The scenario was conducted using a series of releases of Propane under 160 °F and 250 psi. The scenario used the same sample layout, congested regions, ignitions and weather stability class as described in section 2.1. Table 3 presents the other atmospheric parameters used for the study.

Table 3. Atmospheric Parameters

Parameter	Value
Atmospheric Temperature	68 °F
Atmospheric Pressure	14.7 psi
Surface Roughness	183.2 mm
Humidity	70 %
Solar Radiation Flux	250 Btu/ft2·hr

In addition, four release locations were assumed, and are indicated in Figure 1 by the red marks. For the purpose of this study, the authors covered a series of locations to ensure proper representation of the possible explosive cloud locations.

Table 4 provides the list of propane release scenarios and their parameters.

Table 4: Propane - Scenarios Parameters

Release Rate (kg/s)	Location	Pressure (psi)	Temperature (°F)
0.1	Location 1,2,3,4	250	160
0.5	Location 1,2,3,4	250	160
1	Location 1,2,3,4	250	160
2	Location 1,2,3,4	250	160
3	Location 1,2,3,4	250	160
4	Location 1,2,3,4	250	160
5	Location 1,2,3,4	250	160
7	Location 1,2,3,4	250	160
9	Location 1,2,3,4	250	160
10	Location 1,2,3,4	250	160
15	Location 1,2,3,4	250	160
20	Location 1,2,3,4	250	160

3 Results

3.1 Scenario 1 – Ethane Releases

A series of Ethane simulations were conducted as described in Section 2.1. Those simulations were conducted with different release rates that led to different cloud volumes and explosion locations.

From these simulations, explosions were modeled in the congested regions, and the flammable mass that could reach the congested volumes was determined. The analysis was conducted explicitly to determine how much of the release could be found in the congested region and contribute to the explosion energy and lead to overpressure that were then evaluated for each of the monitor points. The main findings are listed in Table 5.

Table 5: Results Summary for Ethane Scenarios

Release Rate (kg/s)	Monitor Points	Overpressure (psi)	Overpressure (bar)	Impulse ($N \cdot s/m^2$)	Flammable Cloud Volume (m^3)
0.1	East Wall CR	NA	NA	NA	0.07
0.1	Center CR Roof	NA	NA	NA	0.07
0.1	Admin Building	NA	NA	NA	0.07
0.5	East Wall CR	NA	NA	NA	0.35
0.5	Center CR Roof	NA	NA	NA	0.35
0.5	Admin Building	NA	NA	NA	0.35
1	East Wall CR	NA	NA	NA	4
1	Center CR Roof	NA	NA	NA	4
1	Admin Building	NA	NA	NA	4
2	East Wall CR	NA	NA	NA	6
2	Center CR Roof	0.51	0.04	10.12	6
2	Admin Building	NA	NA	NA	6
3	East Wall CR	0.65	0.05	15.56	12
3	Center CR Roof	0.35	0.02	8.98	12
3	Admin Building	NA	NA	NA	12
4	East Wall CR	0.74	0.05	19.56	21
4	Center CR Roof	0.41	0.03	11.23	21
4	Admin Building	NA	NA	NA	21
5	East Wall CR	0.90	0.06	22.69	32
5	Center CR Roof	0.45	0.03	12.98	32
5	Admin Building	NA	NA	NA	32
7	East Wall CR	1.36	0.09	40.47	60
7	Center CR Roof	0.59	0.04	19.39	60
7	Admin Building	NA	NA	NA	60

Release Rate (kg/s)	Monitor Points	Overpressure (psi)	Overpressure (bar)	Impulse (N·s/m²)	Flammable Cloud Volume (m³)
9	East Wall CR	1.75	0.12	58.21	92
9	Center CR Roof	0.71	0.05	27.36	92
9	Admin Building	0.25	0.02	12.3	92
10	East Wall CR	1.90	0.13	65.64	108
10	Center CR Roof	0.77	0.05	30.6	108
10	Admin Building	0.28	0.02	13.74	108
15	East Wall CR	2.44	0.17	92.5	221
15	Center CR Roof	0.81	0.06	42.8	221
15	Admin Building	0.33	0.02	19.0	221
20	East Wall CR	3.42	0.24	155.2	361
20	Center CR Roof	1.17	0.08	62.48	361
20	Admin Building	0.38	0.03	22.82	361

Another important part of the analysis was defining generic overpressure contours generated from specific leak rates in order to define the threshold for cloud size generated for the overall facility rather than for a specific building.

Figures 2 through 9 provide an overall guidance on the vicinity of the congested regions and were used to determine the cloud size that would impact the area around the explosion source. The figures provide overpressure contours for the simulations conducted.

Figure 2 presents the results for the 0.1 kg/s release rate and demonstrates that rates this small impact only the areas directly around the release and around congested regions. This simulation did not generate overpressure above 2 psi.

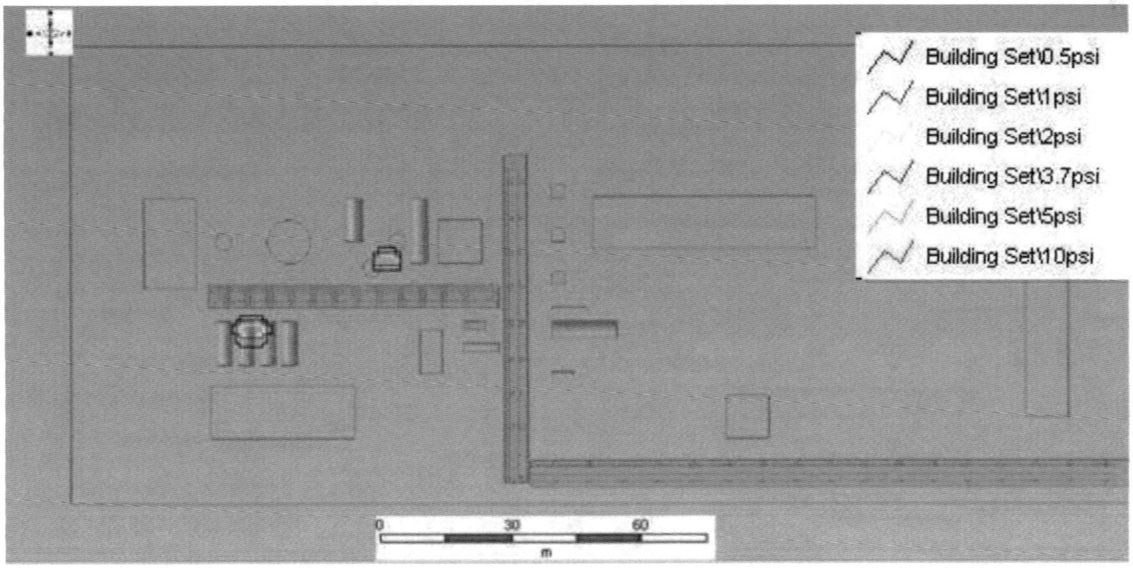

Figure 2: 0.1 kg/s release Ethane

Figure 3 presents the results for the 1.0 kg/s release rate and demonstrates that rates as small as those impact the areas around the explosion region. Higher overpressures were registered for this simulation, but the impact was still only around the explosion source, which would likely require a more refined model to produce robust results.

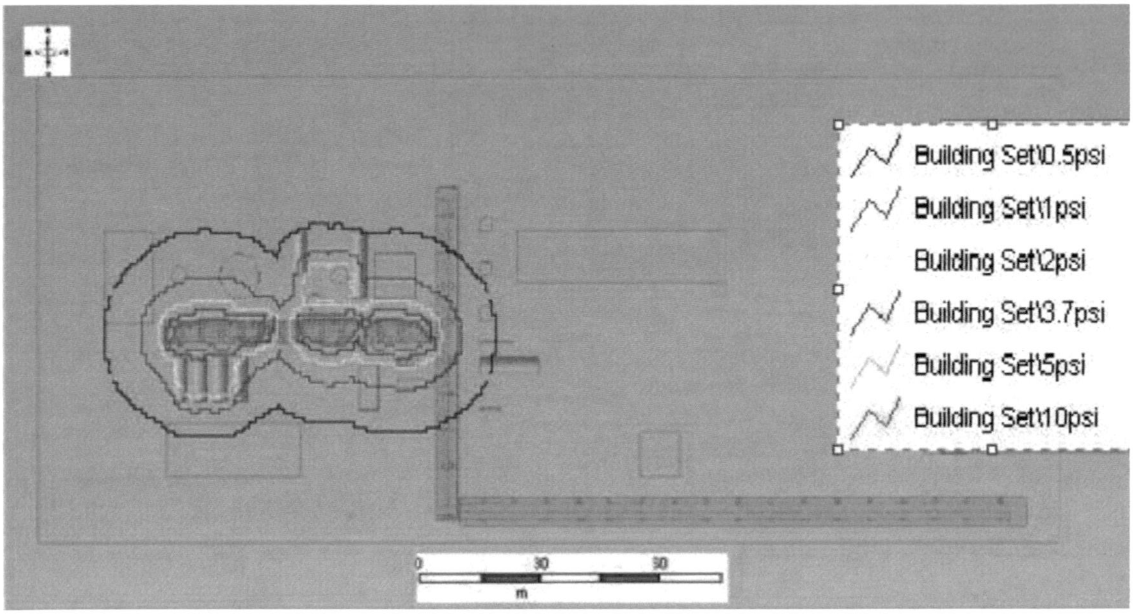

Figure 3: 1kg/s release Ethane

Figure 4 presents the results for the 3.0 kg/s release rate and demonstrates that those rates start to produce some overpressure on the building of concern, but the overpressure and associated impulse is not significant enough to cause damage to the structure and compromise the personnel located inside.

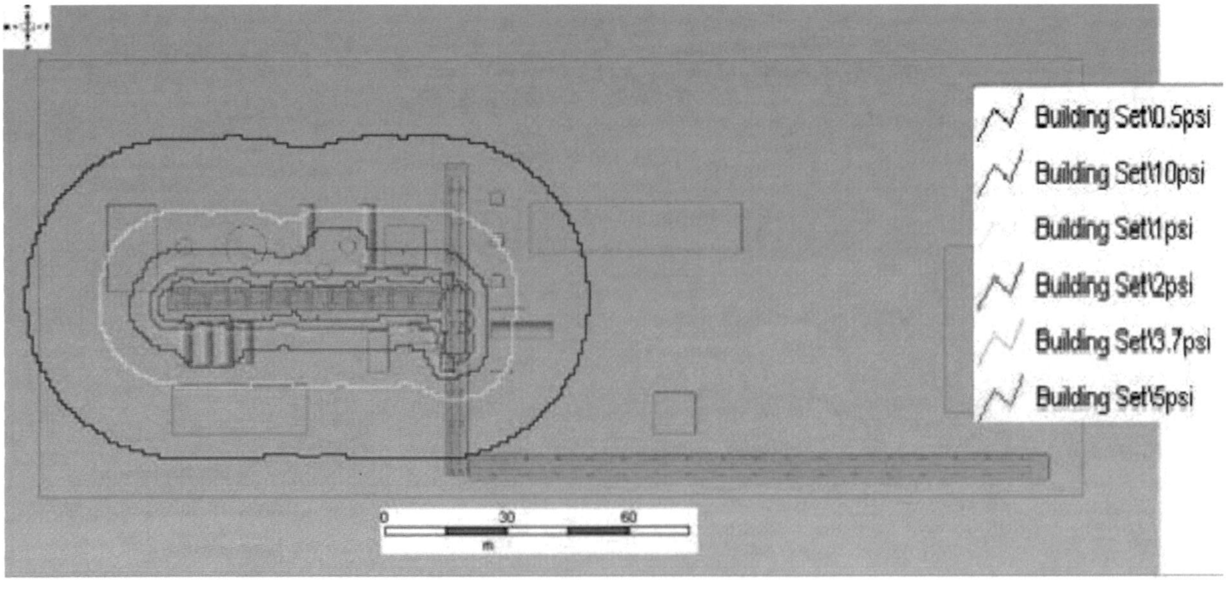

Figure 4: 3 kg/s release Ethane

Figure 5 presents the results for the 5.0 kg/s release rate and demonstrates that those rates start to produce overpressures above 1 psi on the building of concern. These values are still not an issue for the occupants, but the building would need to have appropriate windows, frames and other reinforcements.

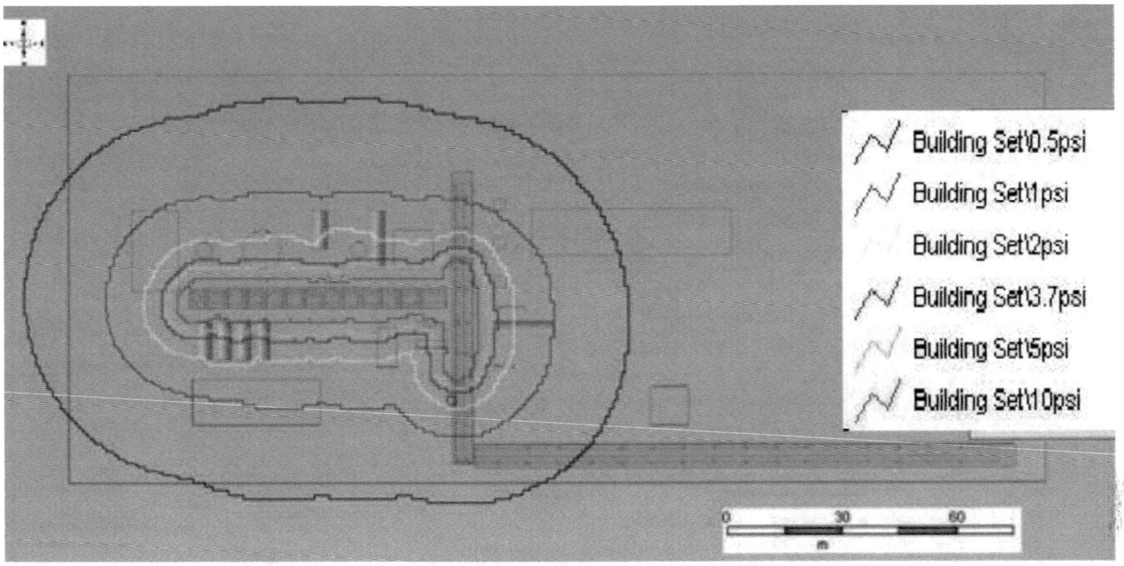

Figure 5: 5 kg/s release Ethane

Figure 6 presents the results for the 7.0 kg/s release rate and demonstrates that those rates start to produce overpressures around 2 psi on the building of interest. These buildings would require reinforcement, at a significant cost, in order to sustain overpressures generated from those scenarios.

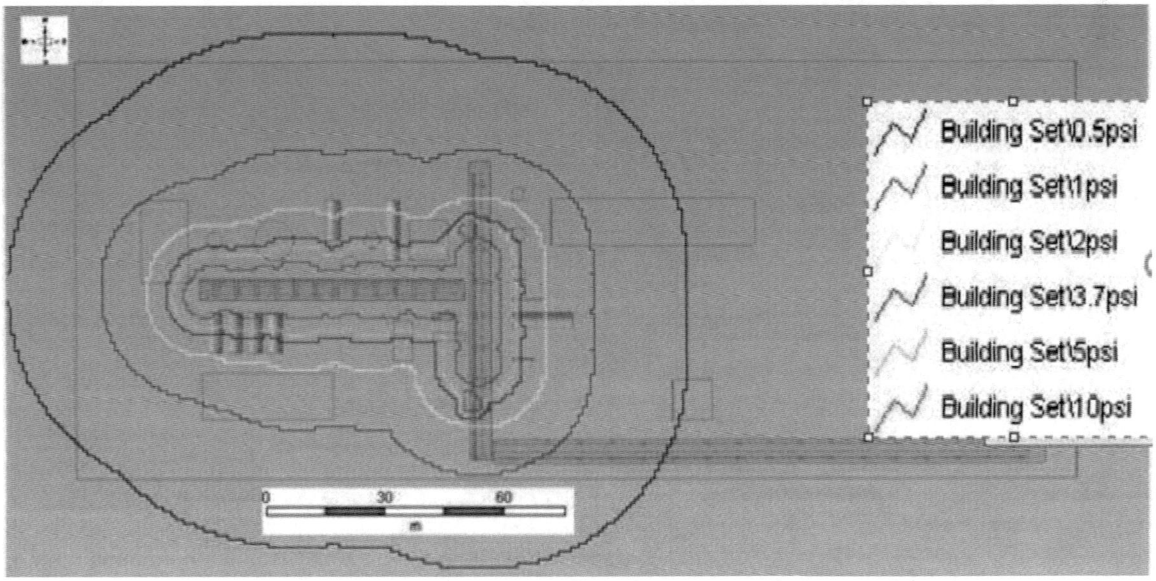

Figure 6: 7 kg/s release Ethane

Figure 7 presents the results for the 9.0 kg/s release rate and demonstrates that those rates also produce overpressures around 2 psi on the building of concern. Again, these buildings would require reinforcement, at a significant cost, in order to sustain overpressures generated from those scenarios.

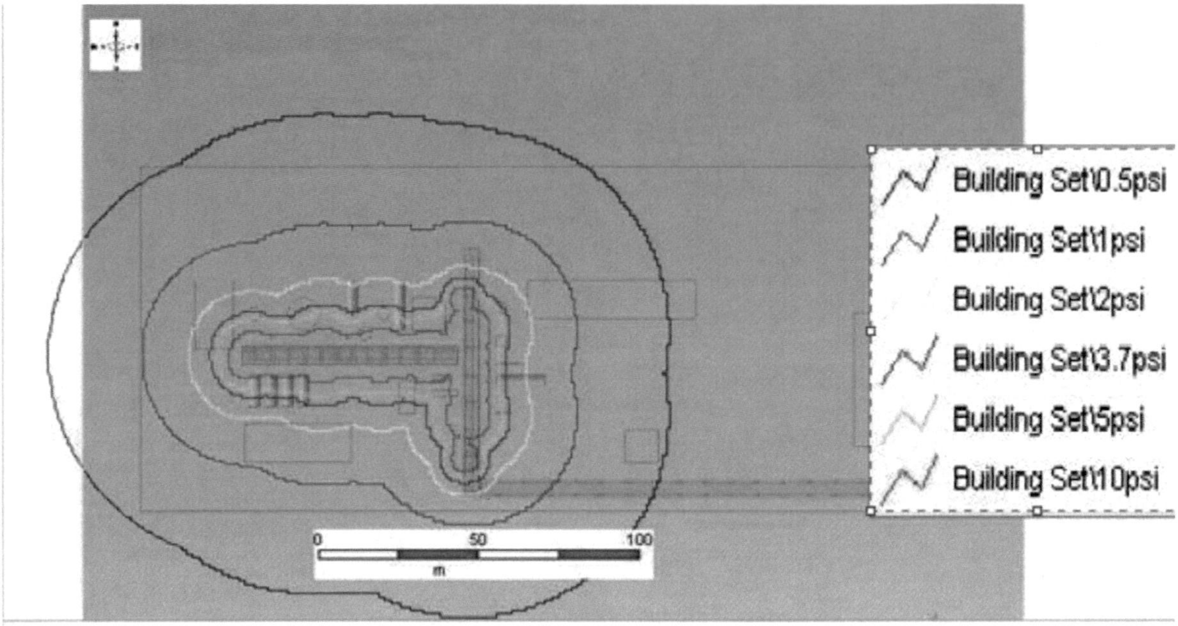

Figure 7: 9 kg/s release Ethane

Figures 8 and 9 present results for release rates of 10 kg/s and 15 kg/s, respectively. These demonstrate severe impact to the buildings and a proper retrofit would be required. These releases would produce overpressure and impulse significant enough to cause severe damage, dependent on the building type used in the facility.

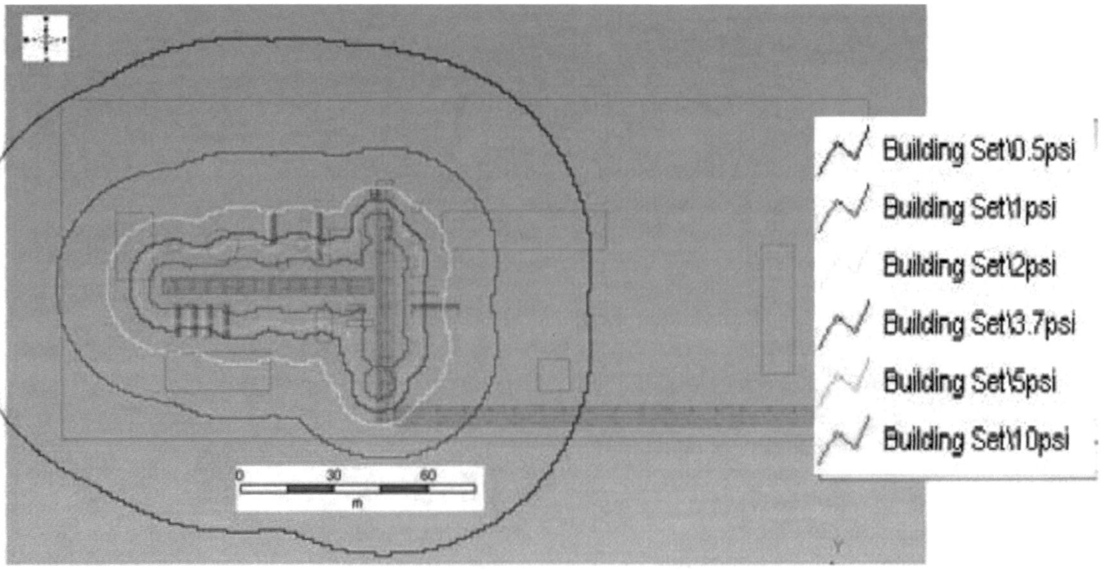

Figure 8: 10 kg/s release of Ethane

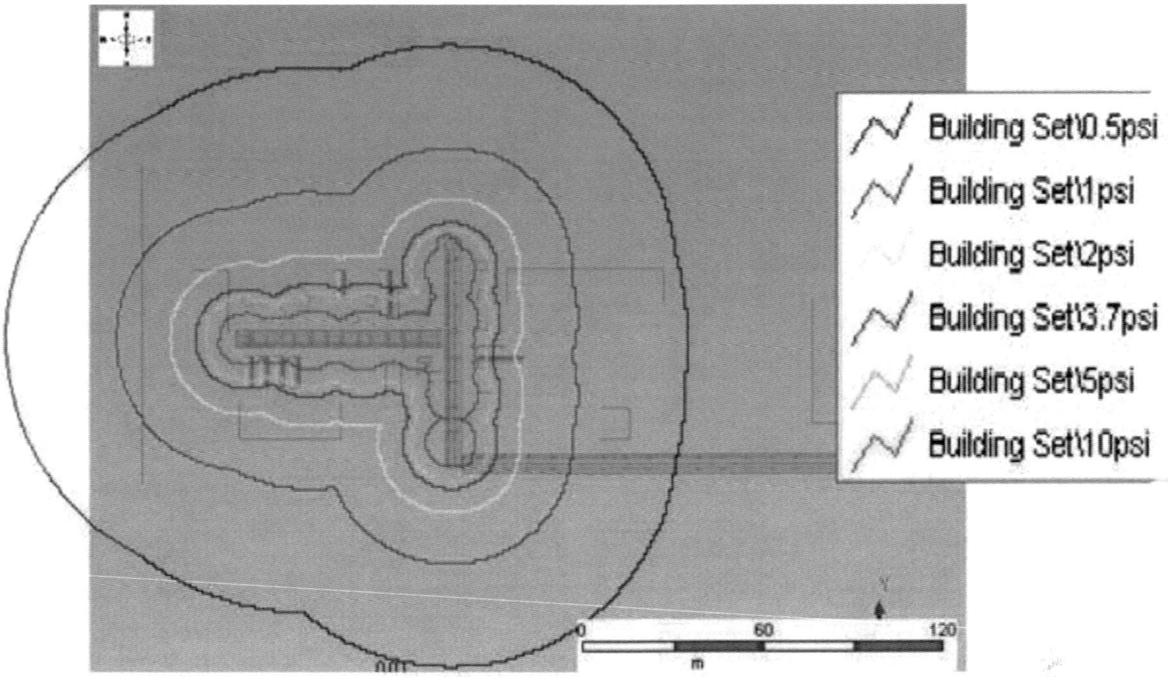

Figure 9: 15 kg/s release Ethane

3.2 Scenario 2 – Propane Releases

A series of Propane simulations were conducted as described in Section 2.2 with different release rates that led to different cloud volumes and explosion locations.

From these simulations, explosions were modeled in the congested regions and the flammable mass that could reach the congested volumes was determined.

The analysis was conducted explicitly to determine how much of the release could be found in the congested region and contribute to the explosion energy and lead to overpressure. The main findings are listed in Table 6 for each monitor point.

Table 6: Results Summary for Propane Scenarios

Release Rate (kg/s)	Monitor Points	Overpressure (psi)	Overpressure (bar)	Impulse (N·s/m^2)	Flammable Cloud Volume (m^3)
0.1	East Wall CR	NA	NA	NA	0.04
0.1	Center CR Roof	NA	NA	NA	0.04
0.1	Admin Building	NA	NA	NA	0.04
0.5	East Wall CR	NA	NA	NA	0.2
0.5	Center CR Roof	NA	NA	NA	0.2
0.5	Admin Building	NA	NA	NA	0.2
1	East Wall CR	NA	NA	NA	3
1	Center CR Roof	NA	NA	NA	3

Release Rate (kg/s)	Monitor Points	Overpressure (psi)	Overpressure (bar)	Impulse (N·s/m²)	Flammable Cloud Volume (m³)
1	Admin Building	NA	NA	NA	3
2	East Wall CR	0.55	0.04	11.59	6
2	Center CR Roof	0.42	0.03	6.75	6
2	Admin Building	NA	NA	NA	6
3	East Wall CR	0.70	0.05	17.2	12
3	Center CR Roof	0.38	0.03	9.9	12
3	Admin Building	NA	NA	NA	12
4	East Wall CR	0.81	0.06	21.2	21
4	Center CR Roof	0.42	0.03	12.15	21
4	Admin Building	NA	NA	NA	21
5	East Wall CR	1.05	0.07	26.95	33
5	Center CR Roof	0.49	0.03	14.52	33
5	Admin Building	NA	NA	NA	33
7	East Wall CR	1.60	0.11	49.64	61
7	Center CR Roof	0.65	0.05	23.03	61
7	Admin Building	0.23	0.02	10.43	61
9	East Wall CR	1.91	0.13	66.2	94
9	Center CR Roof	0.77	0.05	31.02	94
9	Admin Building	0.28	0.02	13.91	94
10	East Wall CR	2.03	0.14	72.87	110
10	Center CR Roof	0.81	0.06	34.1	110
10	Admin Building	0.29	0.02	15.26	110
15	East Wall CR	2.84	0.20	117.6	234
15	Center CR Roof	1.02	0.07	48.8	234
15	Admin Building	0.33	0.02	20.5	234
20	East Wall CR	3.77	0.26	178.8	387
20	Center CR Roof	1.26	0.09	70.87	387
20	Admin Building	0.39	0.03	25.29	387

Another important part of the analysis was defining generic overpressure contours generated from specific leak rates in order to define the threshold for cloud size generated for the overall facility rather than for a specific building.

Figures 10 to 17 provide an overall guidance on the vicinity of the congested regions and were used to determine the cloud size that would impact the area around the explosion source. The figures will provide overpressure contours for the simulations conducted.

Figure 10 presents the results for the 0.1 kg/s release rate and demonstrates that rates this small impact only the areas around the release and around congested regions. This simulation did not generate overpressure above 2 psi.

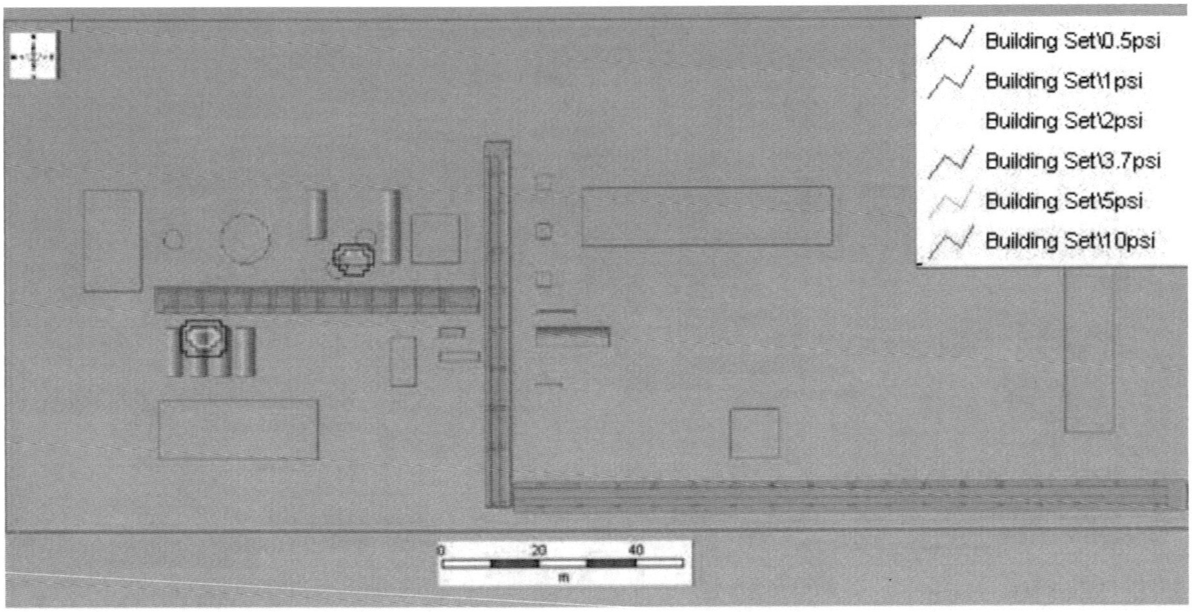

Figure 10: 0.1 kg/s release Propane

Figure 11 presents the results for the 1.0 kg/s release rate and demonstrates that rates as small as those are impacting the areas around the explosion region. Higher overpressures were registered for this simulation, but the impact was still only around the explosion source, which as previously described would likely require a more refined model in order to produce robust results.

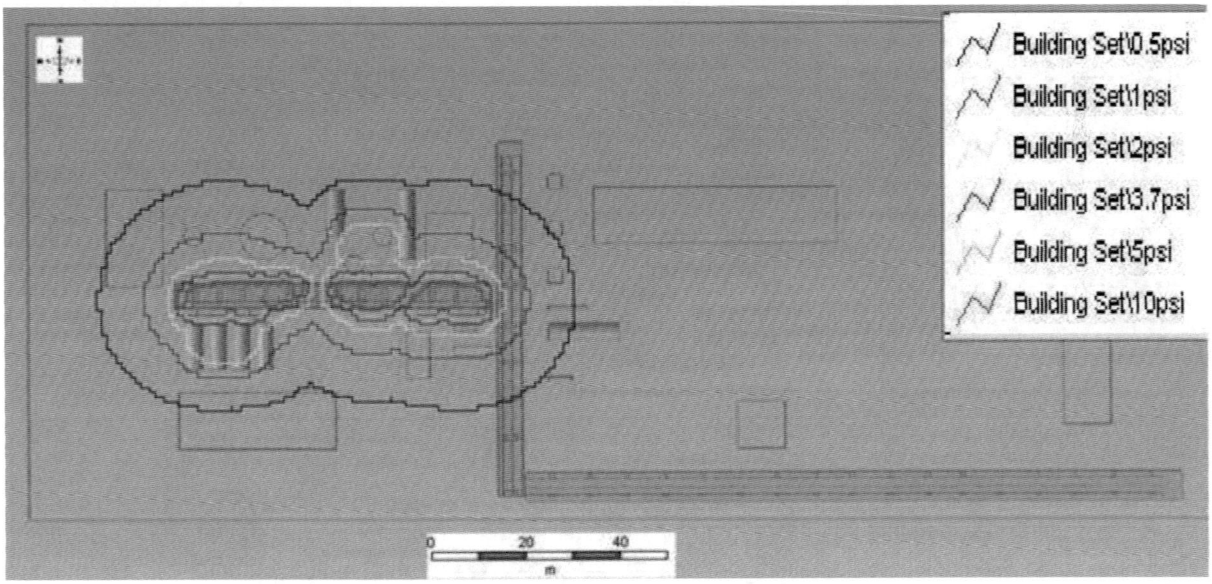

Figure 11: 1kg/s release Propane

Figure 12 presents the results for the 3.0 kg/s release rate and demonstrates that those rates start to produce some overpressure on the building of concern, but the overpressure and associated impulse is not significant enough to cause damage to the structure and compromise the personnel located inside.

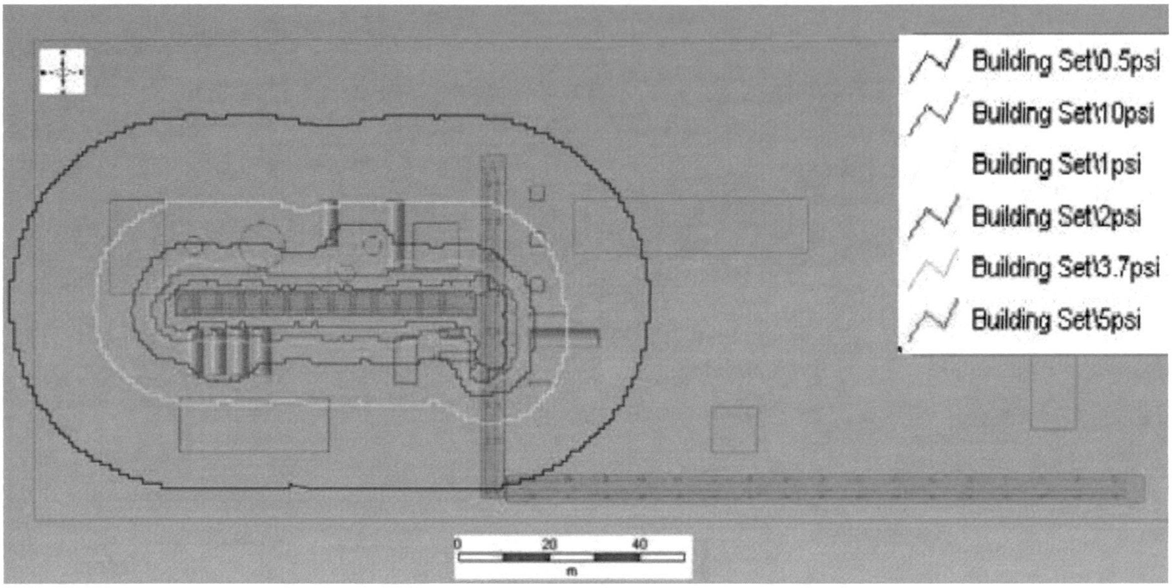

Figure 12: 3 kg/s release Propane

Figure 13 presents the results for the 5.0 kg/s release rate and demonstrates that those rates start to produce overpressures above 1 psi on the building of concern. These values are still not an issue for the occupants, but the building would need to have appropriate windows, frames and other reinforcements.

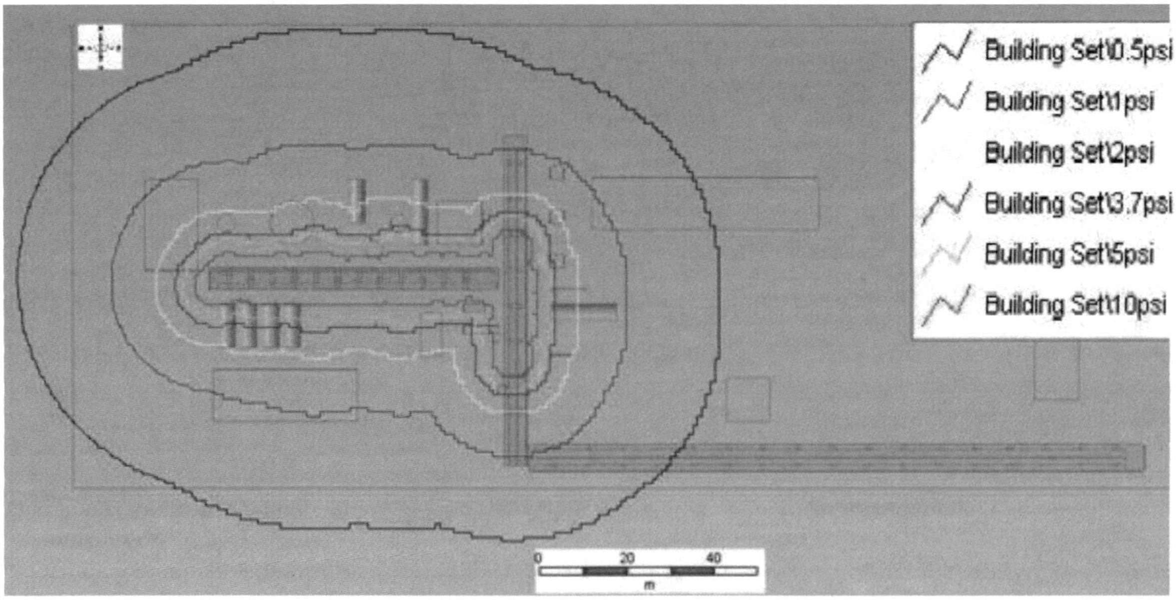

Figure 13: 5 kg/s release Propane

Figure 14 presents the results for the 7.0 kg/s release rate and demonstrates that those rates start to produce overpressures around 2 psi on the building of concern. These buildings would require reinforcement, at a significant cost, in order to sustain overpressures generated from those scenarios.

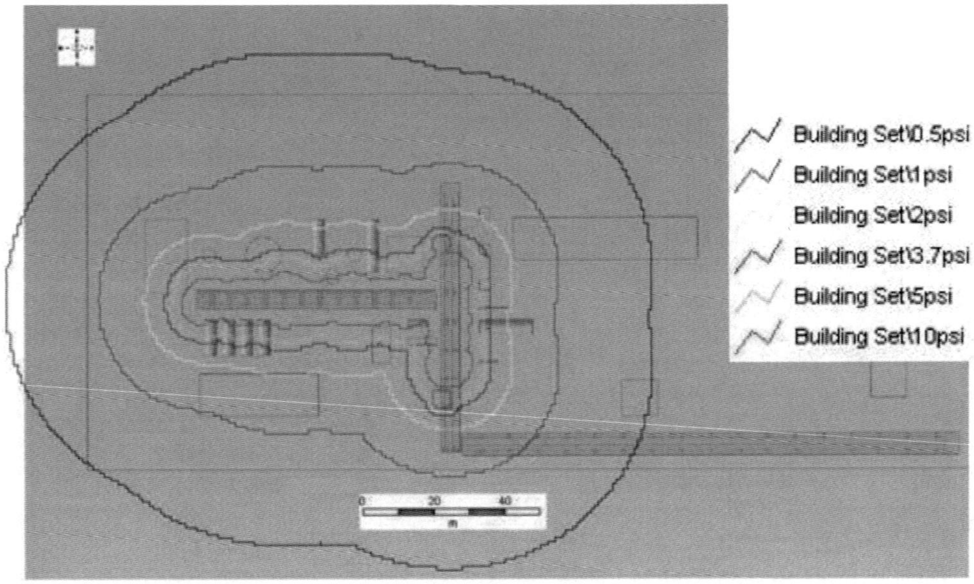

Figure 14: 7 kg/s release Propane

Figure 15 presents the results for the 9.0 kg/s release rate and demonstrates that those rates also produce overpressures around 2 psi on the building of concern. Again, these buildings would require reinforcement, at a significant cost, in order to sustain overpressures generated from those scenarios.

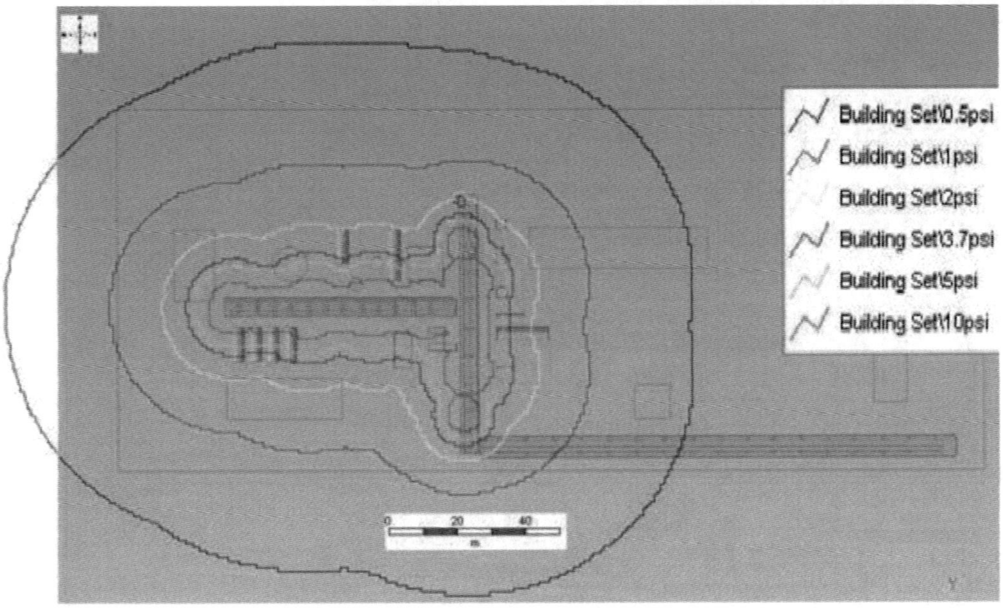

Figure 15: 9 kg/s release Propane

Figures 16 and 17 present results for the release rates of 10 kg/s and 15 kg/s, respectively. Those demonstrate severe impact to the buildings and a proper retrofit would be required. Those releases would produce overpressure and impulse significant enough to cause severe damage dependent on the building type used on the facility.

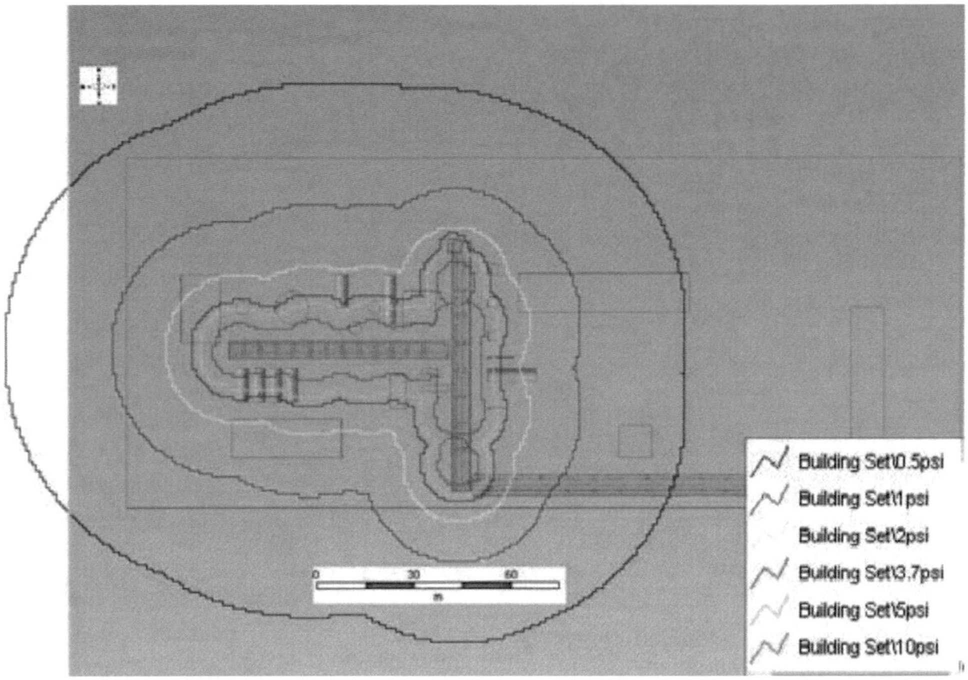

Figure 16: 10 kg/s release of Propane

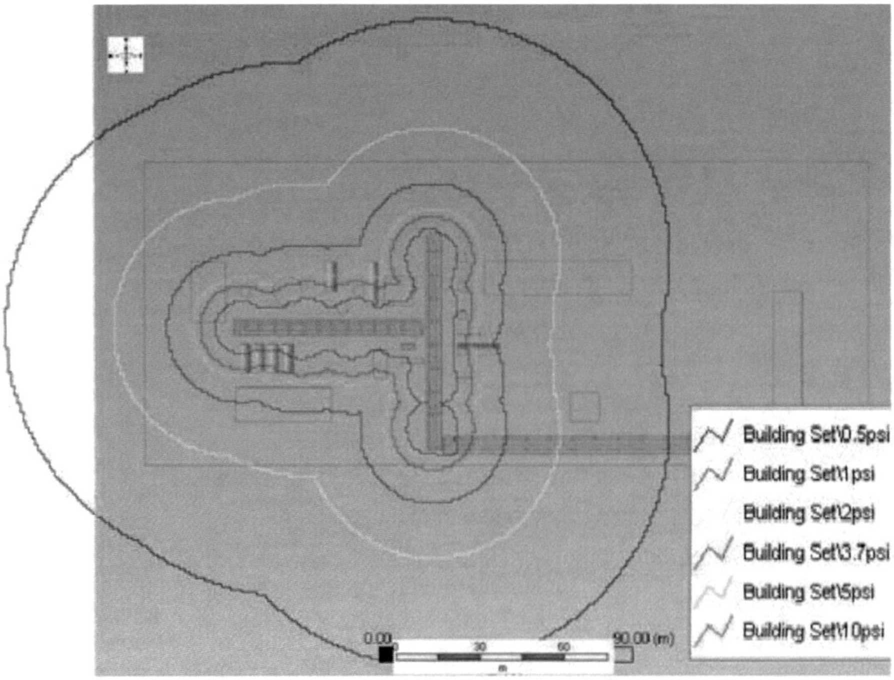

Figure 17: 15 kg/s release Propane

4 Discussion

Several conclusions could be drawn from the study. The first conclusion is that the material being released will be of importance. Propane releases seem to be forming larger clouds and lead to larger impacts to the occupied buildings studied, as expected. Another important conclusion is that there seems to be a threshold for the cloud sizes that we should be focusing on for our detection system when it comes to overpressure impact on buildings. The major conclusion that can be drawn from this study is that releases under 5 kg/s or releases which will form flammable clouds with volumes below 35 m^3 will have a local impact on the congested regions and not the areas outside of the explosion source. The releases below 5 kg/s will produce overpressure below 1 psi and impulses below 27 $N \cdot s/m^2$. Most of the regular constructions currently used in the onshore facilities will be able to withstand this load without significant damage to property or personnel. Furthermore, the releases that would lead to cloud sizes above the 35 m^3 threshold are produced from releases of 5 kg/s and above.

To determine the load that would damage the occupied buildings, the authors have assessed two typical building construction types widely used in the Unites States. This consists of Wood Frame and Unreinforced Concrete Masonry Unit (CMU). The authors have plotted the results from the simulations, described on Sections 3.1 and 3.2, using Pressure and Impulse (P-I) curves for each of those buildings types. The results can be found in Figures 18-21 for both types of buildings and releases.

The authors have used the Building Damage Level (BDL) 1 curve to determine the damage threshold of interest, which means the buildings exposed to this level will not present significant damage and will not require reinforcement. The definitions used in Table 7 are taken from a previous publication – Vulnerability Model for Occupants of Blast Buildings [1].

Table 7: BDL Definitions

Building Damage Level (BDL)	Damage Description
1	Onset of visible damage to reflected wall of building.
2.0 (2A)	Reflected wall components sustain permanent damage requiring replacement. Other walls and roof have visible damage that is generally repairable.
2.5 (2B)	Reflected wall components are collapsed or very severely damaged. Other walls and roof have permanent damage requiring replacement.
3	Reflected wall has collapsed. Other walls and roof have substantial plastic deformation that may be approaching incipient collapse.
4	Complete failure of the building roof and a substantial area of walls.

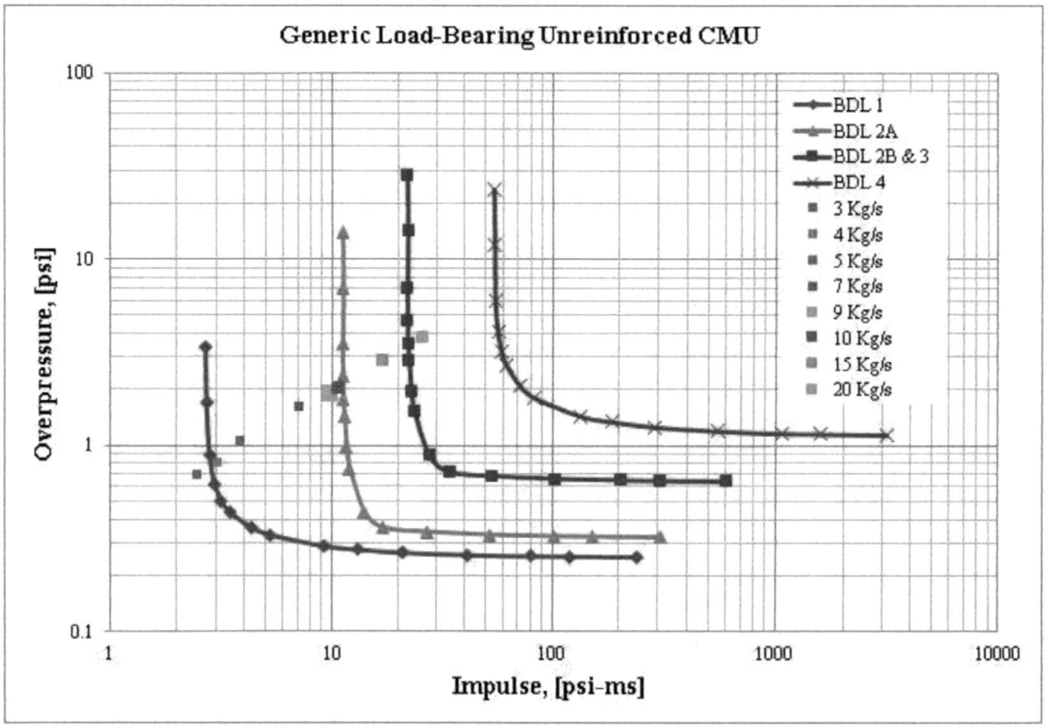

Figure 18: P-I Results for Propane – Unreinforced CMU

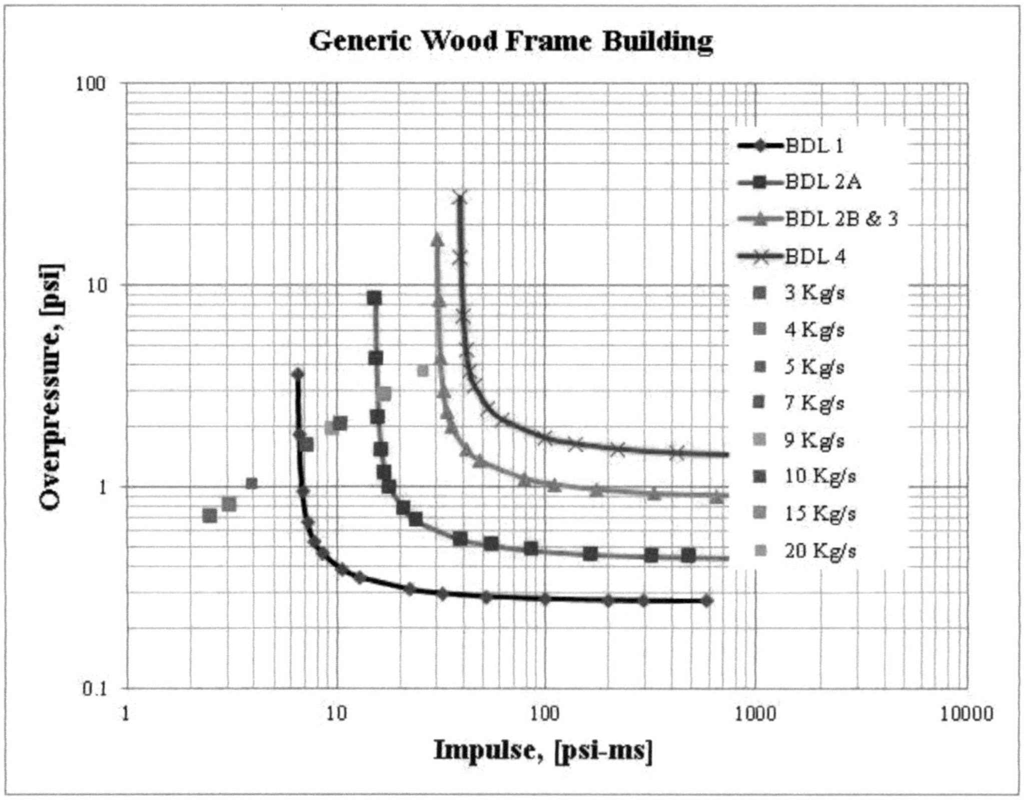

Figure 19: P-I Results for Propane – Wood Frame

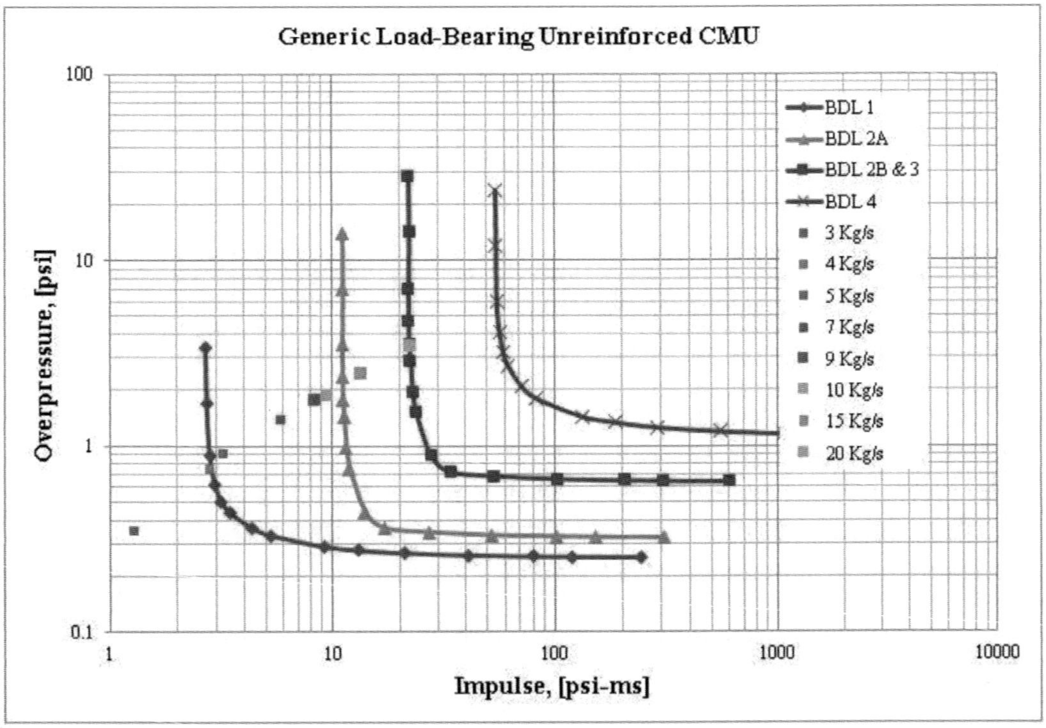

Figure 20: Results for Ethane – CMU

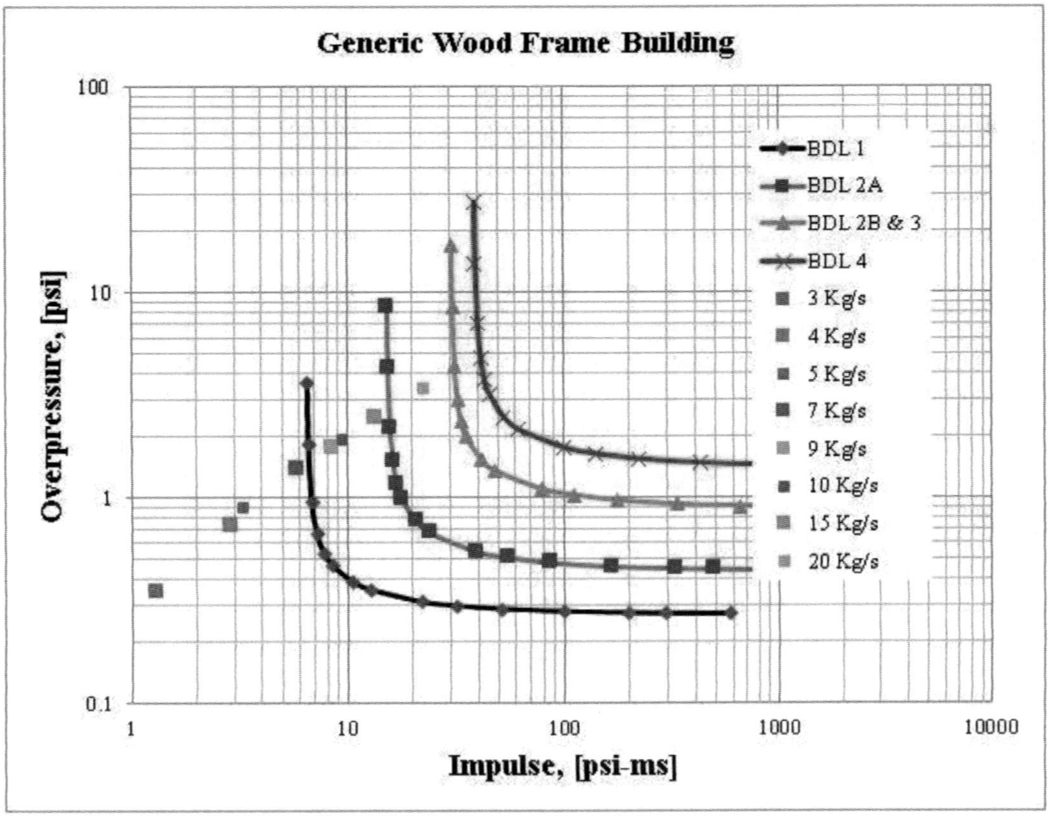

Figure 21: Results for Ethane – Wood Frame

Using the defined building damage levels for building assessments, the conclusion is that any cloud smaller than 35 m^3 would not produce damage to the typical construction types normally used for onshore facilities and would not present a concern to the building integrity and personnel inside the building itself.

An important point to be highlighted is that if one is concerned with damage to equipment within or very close to the explosion source, the study should be reevaluated with a more robust and refined model. The authors also conclude that the reinforcement of the buildings will need to be drawn from the Facility Siting Study, and this could allow, depending on the facility, a higher load on their occupied buildings.

As stated initially, the models used are normally reasonably accurate for far-field explosion and dispersions, but they should be used with caution on the near-field. The authors understand the model limitations for near-field explosions and highlight the importance of a more accurate model to be used for those predictions.

The authors would also like to highlight two very important factors regarding Computational Fluid Dynamic (CFD) cloud sizes when compared with Gaussian Dispersion Models and Gas Detector Placements.

First the authors would like to emphasize their understanding that a more refined simulation could lead to smaller flammable clouds. This is discussed in further detail in a previous publication titled “Evaluation beyond a consequence-only approach: When is it really needed?” [2]. In this publication the authors demonstrated that CFD modeling reduces the size of the clouds being estimated in the Gaussian models, even without significant obstacles. A similar facility was simulated and the modeling was carried out and explained in the paper. Figure 22 presents the difference in the cloud volumes obtained when comparing CFD to Phast results.

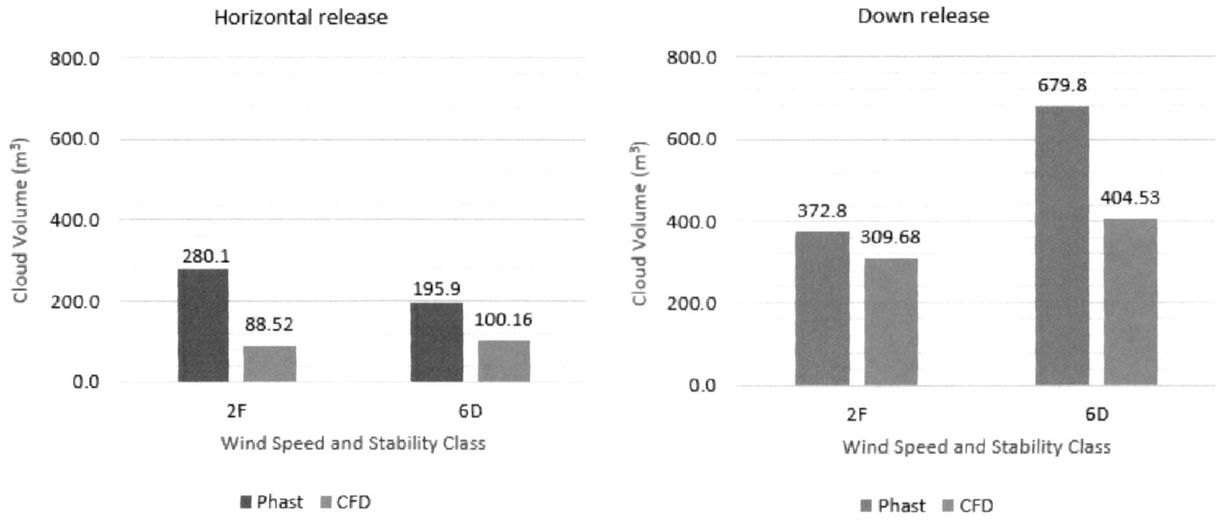

Figure 22: Cloud Volumes Phast vs CFD

This could lead others to question the need to detect even 5 kg/s releases, as the clouds would probably be reduced using CFD models. This topic is addressed in a separate publication, "Optimi – A Novel 3D Computational Tool for Gas Detector Optimization" [3]. In this separate publication, the authors ran a series of CFD simulations with 5 kg/s release rates which led to leak locations and cloud sizes to determine the characteristic length (or grid cells size) that should be applied for the gas detector optimization. For this approach, the authors have proposed that from the range of clouds generated we want to ensure detection of the smaller cloud threshold for the series of clouds being generated.

As the authors have highlighted in the Gas Detector Optimization publication [3], there were a series of CFD simulations with several wind and release directions that were modeled. The series of releases provide a wide range of cloud volumes ranging from 9 m^3 to 59 m^3, depending on the wind and release direction. For this analysis, it is critical to highlight that a very large recirculation zone was created behind the process area which causes some of the clouds to be accumulated depending on the direction of the release and wind conditions. Based on the results from the Gas Detector Optimization publication [3], the placement of detectors using a 5 kg/s release rate seems to provide the optimal point of detection, which aligns with the proposed approach of this paper based on the cloud volumes for the explosion assessment. In the next couple of figures, the layout and cloud volumes from the Gas Detector Optimization paper [3] are presented.

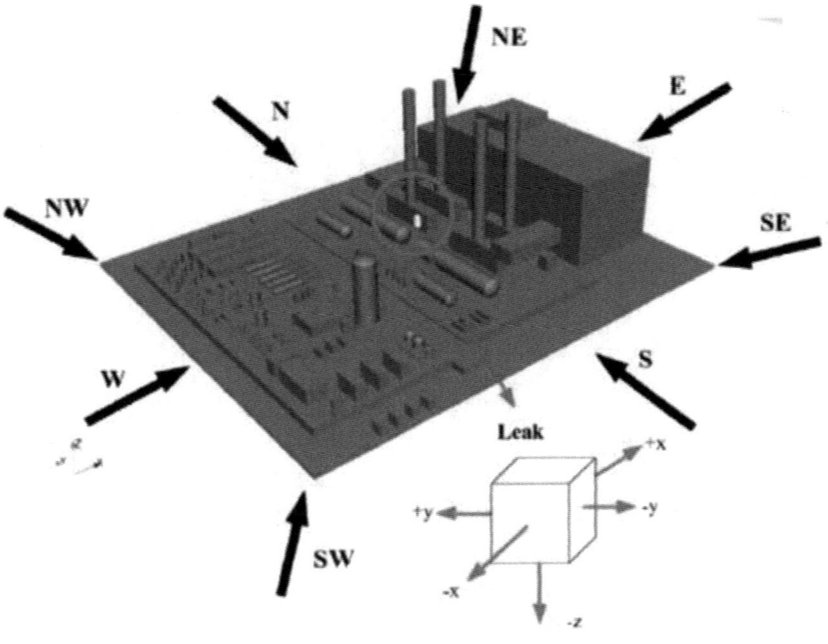

Figure 23: Sample Facility Layout

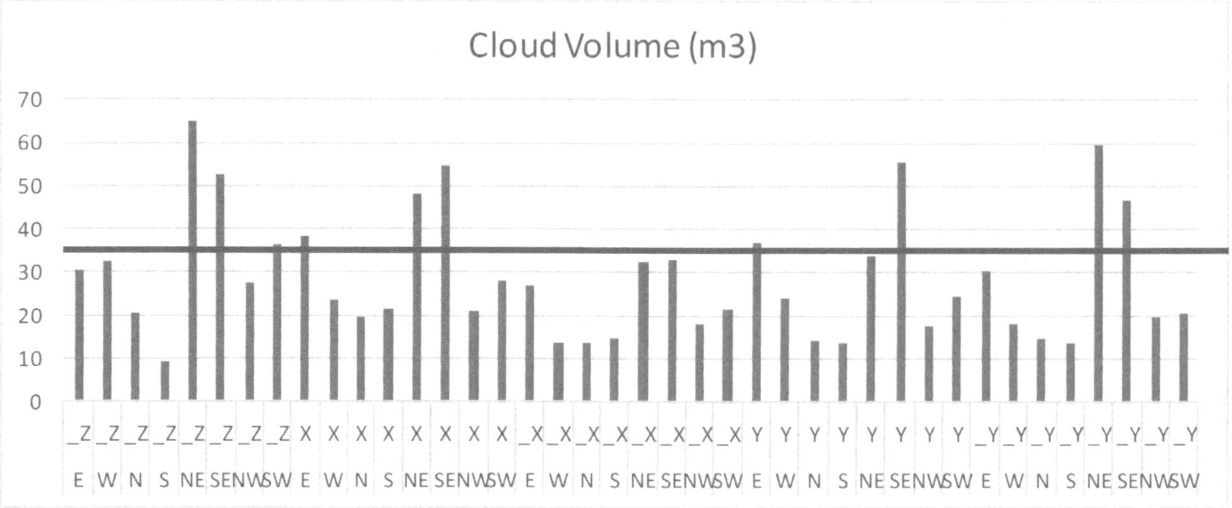

Figure 24: 5 kg/s CFD Cloud Volume

The red line on the graph above indicates the threshold proposed for this paper of 35 m^3. As indicated in Figure 23 and 24, some of the releases would have clouds above the suggested threshold of 35 m^3. It should be noted these releases are upwind releases, which were not modeled by the authors for this analysis due to limitations in the model. In addition, the cloud would be concentrated around the area of the release and upwind of the release rather than towards the congested regions.

5 Conclusion

From the simulations that were carried out, it's reasonable to conclude that there is certainly a threshold that one should focus on in order to place their gas detectors based on cloud volumes that produce significant damage to structures and buildings. The release rates of 5kg/s, or more specifically the cloud volume of 35 m^3, were determined to be the minimum threshold that would cause damage from a vapor cloud explosion, which would compromise facility structures and possibly lead to casualties. It is concluded that this minimal threshold should also be used to determine the optimal location for the gas detector.

It's expected that other materials will have similar or even larger cloud size thresholds based on their average or low reactivity, e.g. Methane, Butane. It's also important to note that the simulations conducted were specific to this facility. So it's very important to highlight that the authors are suggesting that caution should be used when applying a spacing criteria, and the main idea was not to determine a single point to be used but rather focus on a specific range to which one should look into. The conclusion is that very small releases will not significantly impact the structures and/or personnel, and the focus was exactly to try to determine a range in which one should focus their simulations on in order to achieve the proper spacing required for their detectors. The authors do believe a proper consequence or QRA model should be used in order to determine the appropriate separation for the gas detector spacing in each facility based on the importance of both the material being released and the configuration of the facility.

6 References

[1] Oswald, C. J., Baker, Q. A., "Vulnerability Model for Occupants of Blast Damaged Buildings", 34th Annual Loss Prevention Symposium, 2000.

[2] Vianna, S. S. V., Sofyan S., Chambliss S., Simoes V., Kolbe M., "Evaluation beyond a consequence-only approach: When is it really needed?", Mary Kay O'Connor Process Safety International Symposium, 2016.

[3] Vianna, S.S. V., Fiates, J., Simoes, V., "OPTIMI – A Novel 3D Computational Tool for Gas Detectors", Mary Kay O'Connor Process Safety International Symposium, 2015.

479060 API-RP521 Psv Capacity Calculation Compared Against Universal Bernoulli's Homogeneous Equilibrium Method

Abstract:

Bernoulli's Homogeneous Equilibrium (HEM) Method is used to compute relief valve capacity for liquids, vapors and two-phase flows. HC liquids, vapor, water, steam and two phase flow of water and water vapor as well as HC vapor and HC liquid are considered at set pressures ranging from 50 psig to 1500 psig. The calculated coefficents of discharge by HEM method are compared against industry practise coefficients of discharge that are recommended by API-RP521. While the coefficients are comparable, there are differences that are noteworthy.

479316 Using Combined Failure Data Sources to Assess Equipment Failure Rates

Abstract:

Understanding asset reliability is of the utmost importance to manage and maintain your assets. This is even more crucial when the equipment is critical to operations and/or related to safety systems. The challenge in estimating failure rates of equipment is to identify the proper and most relevant data to support the assessment. Examples of available data sources include generic data, and operator and vendor failure data. Generic data has the advantage of being proven and widely accepted, but is criticized for not being specific enough to the asset under evaluation or to the application. Alternatively, vendor and operator data is praised for being specific, but the cleanliness of the data and the sample size can be challenged. This presentation will discuss the pros and the cons of the different sources of failure data and approaches for assessing failure rates. The presentation will illustrate a methodology for combining different data sources, as a trade-off for failure rate estimations by using a Bayesian framework. The methodology may be used for estimating an asset's failure rate, which will in turn allow the stakeholders to make informed decisions about asset selection and maintenance, and to manage equipment failures and related risks. Ultimately, the benefits of the proposed methodology are the decisions made based on the analysis that can lead to increased equipment reliability, system availability, and production uptime.

Industry/Academic Safety Partnership - An IChemE Case Study

Trish Kerin
Director, IChemE Safety Centre
Level7, 455 Bourke St
Melbourne, VIC, 3000
Australia
tkerin@icheme.org

Prepared for Presentation at
American Institute of Chemical Engineers
2017 Spring Meeting and 13th Global Congress on Process Safety
San Antonio, TX
March 26 – 29, 2017

Industry/Academic Safety Partnership - An IChemE Case Study

Trish Kerin
Director, IChemE Safety Centre
Level7, 455 Bourke St
Melbourne, VIC, 3000
Australia
tkerin@icheme.org

Keywords: process safety, education

Abstract

There are standards around the world defining what must be included in engineering degrees. These standards are vital in ensuring that students all over the world understand fundamental principles. The overarching principles are defined in the Washington Accord, which boasts 17 countries as signatories, providing a stable basis for engineering education. In addition, various other bodies facilitate accreditation processes to ensure that the necessary topics are addresses. For Chemical Engineering, these include the Institution on Chemical Engineers (IChemE) and the Accreditation Board for Engineering and Technology (ABET). The IChemE Safety Centre (ISC) has commenced a project to look beyond these accreditations and explore with industry and academia what is needed in process safety education. This paper summarizes some elements of the project and discusses the way forward.

1 Introduction

The ISC is a not-for-profit industry-led consortium focused on advancing process safety. This is achieved by developing projects that involve industry and academia working together. These projects are selected, by the ISC members, to address specific concerns. One such project looked at how to improve undergraduate engineering education in process safety. This was not an attempt to produce process safety graduates, but to ensure that chemical engineers have sufficient awareness of process safety management prior to entering the workforce.

The various accrediting bodies address process safety in a number of ways. ABET covers the need to address safety [1] and has recently added reactive chemistry to its requirements for example and IChemE has requirements including understanding inherent nature of safety and loss prevention as well as risk assessment [2]. The ISC is not an accrediting body in itself but seeks to provide additional guidance and assistance to universities. This will start to address the identified gaps and in turn produce engineering graduates with a higher process safety understanding.

2 The project

The project consisted of seven phases, with the final phase (to develop recommendations for improvement) underway at the time of writing. The phases can be described as follows:

- Review of existing standards and accreditation guidelines
- Determining what industries requirements for process safety were
- Survey of the trial universities on what and how they teach process safety
- Development of the learning outcomes based on what industry required
- Delivery of a masterclass for trial university academics and presenting the proposed learning outcomes by industry practitioners
- Feedback and comment from trail and non trial universities on the learning outcomes
- Develop recommendations for improvement

A key initial assumption was that industry had a duty to assist in the education of future engineers: change could not be requested without a willingness to assist universities in making those changes.

In order to ensure that learning outcomes were written as required for university education, the project team included at least two team members with an academic background and the learning outcomes were developed using Blooms Taxonomy [3].

Although the project was initially run as a trial with Australian and New Zealand based universities, the work was reviewed by a number of other universities from other countries, including Canada, the United Kingdom and the USA.

This paper outlines the learning outcomes that were developed by the project team, summaries the feedback received from academia and discusses the proposed recommendations for improvement.

3 The learning outcomes

The project developed learning outcomes under four topics:

A. Process safety overview
B. Process safety practice - design
C. Guidelines for process safety risk assessment
D. Process safety practice - operations

The proposed learning outcomes are described in Table 1 [4].

Table 1. Proposed learning outcomes

Topic	Learning outcome
A. Process safety overview	
A1	Define the main concepts of process safety and describe the benefits of process safety to an organization and to society.
A2	Describe and identify the similarities and differences between process and personal safety.
A3	Describe the local regulatory framework in relation to process safety and list differences in global process safety legislation.
A4	List typical Process Safety Key Performance Indicators and describe the role of auditing and assurance in PSM (subset of PSM Elements).
A5	Describe the main elements of a Process Safety Management System.
A6	List typical ways that an engineer is exposed to process safety on a daily basis.
B. Process safety practice - design	
B1	Define the concepts of inherent safety and list typical approaches to inherently-safer process design.
B2	Describe the benefits of multiple barriers and list typical barrier types for various process excursions.
B3	Compare the advantages and disadvantages of risk-based design versus codes and standards based design in the overall PSM framework.
C. Guidelines for process safety risk assessment	
C1	Describe the steps involved in the risk management process; and define the main concepts related to risk management including hazards, failed protection layers, threats, consequences and effects.
C2	Identify various process 'hazards' and develop 'risk' phrases in terms of loss of control of hazards. Including identification of: • process hazards • the potential consequences of a failure to keep them contained/controlled • threats to containment of hazards • prevention and mitigation controls
C3	Define the main concepts of common process hazard analysis tools. Describe qualitative and quantitative risk assessment techniques, their uses, benefits and limitations. Perform a hazard identification and risk assessment study on a process safety scenario.
C4	Describe the concepts of ALARP/SFARP and describe the steps required to demonstrate that a risk has been reduced So Far As is Reasonably Practicable.
D. Process safety practice - operations	
D1	List typical factors that contribute to barrier effectiveness. Explain the role of critical

Topic	Learning outcome
	activities like monitoring, inspection and maintenance in barrier effectiveness.
D2	List typical Safety Systems of Work (SSOW) and identify key attributes of SSOW systems in managing Process Safety.
D3	List typical Human Factors concepts (e.g. competence, fitness for work) and briefly describe how they affect process safety management.
D4	Define and describe the concepts that relate to Operating Integrity.
D5	Identify key attributes of management of change (MoC) processes in relation to process safety management.
D6	Describe the steps in risk-based decision making involving both cost and safety implications, taking into account the hierarchy of controls and principles of inherently safer design.
D7	Critically analyze the causes, consequences and effects of a process safety incident, including identification of recommendations to prevent future incidents.

From an industry perspective, it was considered that whilst existing undergraduate courses broadly cover topics A, B and C, opportunities exist to improve the teaching of topic D, process safety in operations.

4 The feedback

Feedback was sought from both industry and universities to understand the value and practicality of implementing the proposed learning outcomes. The feedback varied greatly; however, there were some common concerns regarding the following items:

1. Legislative frameworks: as legislation varies between jurisdictions, there is no value in teaching any specific legislative framework.
2. Limited teaching time: there is a limited number of hours that can be assigned to teaching any topic in an undergraduate degree.
3. Process safety in operations: difficult to teach aspects of operations in a university setting.

5 Recommendations for improvement

The following recommendations, which were being drafted at the time of writing, propose some options which may be considered as part of the final project:

1. Noting the similarities between various laws covering the management of major hazard industries, a general overview of process-safety related legislation from around the world could be provided.
2. Process safety education could be incorporated into other subjects, e.g. covering risk assessment and inherently-safer design concepts within the final year design project.

3. Recognizing the potential for collaboration between industry and academia to support the teaching of process safety, two options are proposed to facilitate education towards achieving the learning outcomes for Topic D:

   a. Industry passport
   Before conferring an undergraduate engineering degree, there is an expectation that students will undertake plant visits and work experience. It is proposed that, during their time in these activities, students could record, in a personal passport, the activities they undertake and what they have learned about managing process safety in operations. This could include, e.g., specific hazards identified during induction, process-safety-related training undertaken or how the organization addresses human factors. Students would need to ask questions of the organization to understand all elements and the awareness gained would assist them in applying process safety concepts in their final year design projects. The passport could be used as an assessable part of the work experience report.

   b. Management system for university laboratories
   Feedback on the proposed learning outcomes noted that management systems only make sense once the graduate is in the workplace. Noting that a university laboratory is not dissimilar to an industrial application (both in terms of process management and the risk of injury and damage), there is potential for students to appreciate the intent and application of management systems by developing and/or applying a basic university laboratory safety management system. This would have the additional benefit of creating a more-structured approach to managing laboratory safety, noting the occurrence of injuries and fatalities in university laboratories. A management system would highlight elements that implicitly relate to experimental practice, such as writing of procedures for conducting experiments, identification of the hazards involved in an experiment, evaluation of risk, determination of the required safety procedures, use of protective equipment, application of basic safe systems of work and management of change to control variations in experimental parameters and the effects thereof. Whilst most universities already include variations of these requirements, bringing them together in an integrated management systems (as is done within industry), will provide students with the opportunity to appreciate how industry formally manages safety in operations.

6 Way forward

The ISC has developed a number of case studies that consist of simple, short videos and engagement of the audience in making decisions within the context of the incident and the events that lead to it. This gives the experience of an incident, without actually suffering the consequences. This concept was delivered as a paper and oral presentation at the 12th Global Congress on Process Safety in Houston in 2016 [5]. Some universities have started to use these to highlight different elements of learning, such as how much information is seen by operators on any given day and how risk assessments can be applied when making design decisions. The ISC will continue to promote the use of the case studies in the education of undergraduate engineers.

The ISC intends to continue developing useful resources for universities to use to improve undergraduate education in process safety, as well as bringing academics and industry together to learn from each other. Future developments may include the industry passport and the simplified laboratory safety management system that could be easily adopted by universities.

Future improvements in process safety are dependent on our ability to adequately design and operate facilities. This project will assist our future engineers to be better prepared for the workforce.

7 References

[1] ABET. (2017). "Accreditation policy and procedure manual (APPM) 2017-2018". Baltimore: ABET.

[2] IChemE. (2015). "Accreditiation of chemical enigneering programmes". Rugby, UK: IChemE.

[3] Anderson L.W. and Krathwohl, D. (2001). “A taxonomy for learning, teaching, and assessing : a revision of Bloom's taxonomy of educational objectives”. New York: Longman.

[4] ISC. (2016). "Improve process safety in engineer undergraduate education". Melbourne: IChemE Safety Centre.

[5] Kerin, T. (2016). "Improved learning through interactive case studies". *12th Global Congress on Process Safety*. Houston: AIChE.

8 Acknowledgements

The author would like to acknowledge the contribution of the following people in the project team:

Ivica Ninic - Origin Energy (Sponsor)
Stuart Chia - Sherpa Consulting
Catharina Grant - Safety Solutions

Rao Vasantharao - Woodside
Morgane Le Cossec - MMI Engineering
Yogesh Koirala - Mary Kay O'Connor Process Safety Center
Pierre Le Clech - University of New South Wales
Simon Casey - Simon Casey Consulting (Facilitator)

13TH GLOBAL CONGRESS ON PROCESS SAFETY

Waste Solvents to Trash Haulers: Lessons Learned from Hazardous Waste Accidents

Delmar "Trey" Morrison, Ph.D., P.E.
Exponent, Inc.
4580 Weaver Parkway, Suite 100, Warrenville, IL 60555, USA
tmorrison@exponent.com

Michael Stern, Ph.D., P.E.
Exponent, Inc.
Natick, MA, USA
mstern@exponent.com

Carmen H. Osorio-Amado, Ph.D.
Exponent, Inc.
Los Angeles, CA, USA
cosorio@exponent.com

Prepared for Presentation at
American Institute of Chemical Engineers
2017 Spring Meeting and 13th Global Congress on Process Safety
San Antonio, TX
March 26 – 29, 2017

Waste Solvents to Trash Haulers: Lessons Learned from Hazardous Waste Accidents

Delmar "Trey" Morrison, Ph.D., P.E.
Exponent, Inc.
4580 Weaver Parkway, Suite 100, Warrenville, IL 60555, USA
tmorrison@exponent.com

Michael Stern, Ph.D., P.E.
Exponent, Inc.

Carmen H. Osorio, Ph.D.
Exponent, Inc.

Keywords: Hazardous waste, incident investigation, chemical reactivity

Abstract

Process safety management (PSM) of hazardous chemicals is a core tenet of the chemical, oil, and gas industries. Organizations in these industries have increasingly woven the elements of PSM into their strategic vision, day-to-day operations, and interactions with contractors and vendors. This cultural evolution has, in part, been driven by major accidents in these fields. In these industries, the chemicals, their hazardous properties, and their potential interactions, are usually either well-defined or capable of being well-defined. Control and mitigation strategies for those chemicals can be developed based on this available knowledge. In contrast, companies that handle wastes have experienced catastrophic accidents due to the ill-defined nature of the hazards. This industry handles the cast-offs, dregs, and refuse, which makes identification of the hazards and prevention of incidents more challenging. How do you implement a PSM program when your input streams can vary due to processes at a plant you do not control or monitor? Does the EPA (Environmental Protection Agency) Resource Conservation and Recovery Act (RCRA) provide sufficient guidance of safely handling of wastes? The overriding theme in the incidents is that PSM is more complex to apply to unknown, mixed, and uncontrolled chemicals, both solid and liquid. To assist companies that handle wastes, this paper provides several case studies of fires and explosions involving multiple waste material handling processes at different facilities. The incidents involved numerous matrices from solid industrial wastes, EPA RCRA hazardous wastes, to waste solvents. An analysis of the lessons learned will be presented and used to provide realistic guidelines to reduce safety risks based on a pragmatic implementation of process safety management principles. The paper will also compare OSHA (Occupation Safety and Health Administration) PSM with EPA RCRA approaches to characterizing wastes versus managing the hazards, and will emphasize those elements that played critical roles in the accident case studies.

1 Introduction

This paper discusses several case histories of process incidents occurring at waste handling or processing facilities and compiles lessons learned from those accidents. Depending upon the source, the industry, and the regulatory classification of the waste, industrial waste may pass through a municipal waste transfer station to a landfill or be collected, aggregated and then incinerated by a RCRA (Resource Conservation and Recovery Act) [1] permitted hazardous waste Treatment, Storage, and Disposal Facility (TSDF). Managing waste is inherently more challenging than managing typical chemical process hazards where the identity and properties of the process chemicals can be effectively defined. While many types of hazardous waste facilities test for the properties of incoming waste streams, these tests can be challenging if the waste is phase separated, if the waste is still reacting and changing over time, or if the waste is incompletely identified by the generator. Similarly, investigations of accidents are complicated by the fact that wastes involved in the accident could have contained unknown chemicals with unknown properties.

Several case studies are presented in this paper. The accident locations range from waste transfer stations, waste storage facilities, waste treatment facilities, and a cement manufacturing facility using waste as a fuel source. The wastes involved in the accidents ranged from municipal waste, unregulated solid and liquid wastes, to regulated hazardous solid and liquid wastes. Significant fires and explosions occurred in these incidents. The overarching causes of these incidents were due to the gap between the true properties and the recognized properties of the wastes. While the task of characterizing waste streams may be challenging, the overall process safety and incident investigation principles remain similar. In this paper, we start with a typical decision tree analysis for screening chemical reactivity hazards and modify it for use for waste processing facilities. We will then discuss several case studies involving fires and explosions in waste handling and treatment and provide lessons learned.

2 Background

Application of Process Safety Management (PSM) principles in the chemical industry has been codified for highly hazardous chemicals above threshold quantities through Title 29 Part 1910.119 of the Code of Federal Regulation since 1992 [2]. The Center for Chemical Process Safety (CCPS) subsequently expanded on the process safety model presented by the regulations to develop twenty elements of risk-based process safety [3]. The overlap between the CCPS process safety model and the OSHA PSM regulatory elements (bolded) is represented in the list below (Table 1).

Table 1. Overlap between OSHA (bold) and CCPS PSM elements

1. Process Safety Culture	2. Contractor Management **(h) Contractors**
3. Compliance with Standards	4. Training and Performance Assurance **(g) Training**
5. Process Safety Competency	6. Management of Change **(l) Management of change**
7. Workforce Involvement **(c) Employee participation**	8. Operational Readiness
9. Stakeholder Outreach	10. Conduct of Operations
11. Process Knowledge Management **(d) Process safety information**	12. Emergency Management **(n) Emergency planning and response**
13. Hazard Identification and Risk Analysis **(e) Process hazard analysis**	14. Incident Investigation **(m) Incident investigation**
15. Operating Procedures **(f) Operating procedures**	16. Measurement and Metrics
17. Safe Work Practices **(k) Hot work permit**	18. Auditing
19. Asset Integrity and Reliability **(j) Mechanical integrity**	20. Management Review & Continuous Improvement

Deficiencies in many of these elements can be traced back to the root causes of hazardous materials incidents, as will be illustrated in the case studies in this paper. However, it is uncommon for TSDFs and other waste handling facilities to formally apply PSM to their processes. Usually, facilities follow a prescriptive approach to waste characterization and handling that is based upon the EPA (Environmental Protection Agency) regulations and relies heavily upon the waste generator's characterization of the waste properties [4]. This reliance is not unusual when considering the quantity of distinct and unique waste streams that an individual municipal facility or TSDF may handle and dispose.

The primary hazard identified in the case studies is delayed reaction leading to pressure generation, heat generation, and unintended ignition of packaging or the waste material itself. This is commonly referred to as spontaneous ignition or spontaneous combustion since a piloted ignition source is not necessary. Trash and municipal waste handling facilities are frequently susceptible to delayed fires due to spontaneous combustion. Such ignition events may be caused by unintentional mixing of incompatible chemicals or exothermic oxidation of self-heating materials, such as vegetable-derived oils. For example, the National Fire Protection Association (NFPA) estimated that 4,460 annual fires involving spontaneous combustion of trash occurred between 2005 and 2009 [5].

Recent studies have identified several trends in hazardous waste TSDF fires and explosions [6], [7], [8]. Incidents can occur in any stage of the process from initial storage, waste repackaging, or transfer to the final treatment stage (i.e., incinerator, calciner, or kiln). The consequences of a fire or explosion can accelerate rapidly, depending upon the nature of the initial waste and its proximity to surrounding waste materials, surrounding waste materials' properties, and the adequacy of fire protection and emergency response actions. The causes of fires and explosions typically derive from unknown or unrecognized chemical reactivity issues with the waste material in question.

2.1 Waste generation in the United States

According to the US EPA, 26,237 generators produced 33.6 million tons of hazardous waste, of which over 6 million tons were non-wastewater [9]. Of the 6 million tons, 26% is combusted for energy recovery, 19% is put into landfills, and 13% is incinerated. Over 76% of the non-wastewater hazardous waste is treated off-site by a TSDF.

The EPA statistics also show that 258.5 million tons of non-hazardous Municipal Solid Waste (MSW) was generated in the United States in 2013. While the per capita waste has slightly decreased since the year 2000, the total annual generation continues to climb due to population growth [10]. Approximately 89.4 million tons (35%) of the MSW were recycled in 2013. Another 13% of the total MSW was combusted with some amount of energy recovery. The remaining 52.6% was placed into landfills.

2.2 Resource Conservation and Recovery Act

A hazardous waste can be any waste material with a potential to harm human health or the environment, either by itself or through interaction with other factors. However, for regulatory purposes in the U.S., a waste material must meet specific criteria to be considered as a hazardous waste. These conditions, summarized in Figure 1, are defined by the EPA under the RCRA in Title 40 of the Code of Federal Regulations [1].

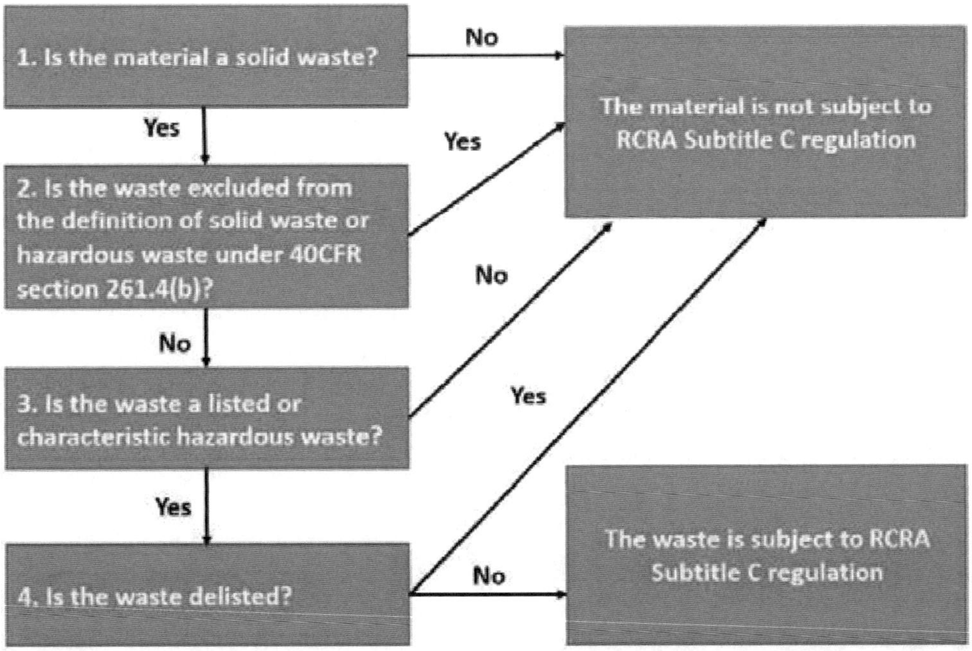

Figure 1. EPA's Hazardous waste identification process [11]

Per the EPA, the first condition of a hazardous waste is to be a solid waste, which paradoxically includes not only wastes that are physically solid, but also wastes that are liquid, semi-solid, or contained gaseous material. A hazardous waste is also required to meet certain characteristics based on its toxicity, ignitability, reactivity, and corrosivity, and needs to be listed in either the F, K, P, or U lists found at the 40CFR section 261 [12].

Some solid wastes are specifically excluded from RCRA regulatory compliance due to different reasons such as impracticability, economy impact, lack of information to justify applying the regulation, regulation by other laws, or they were removed (delisted) from this category. These exclusions refer to conditions (2) and (4) from the hazardous waste identification process shown in Figure 1. For example 40 CFR 261.4(b) excludes household waste and solid waste from commercial or industrial sources (that don't contain solid hazardous waste). It also excludes fly ash waste. These types of waste material may pose physical and chemical hazards, but they are not regulated under RCRA.

2.3 RCRA chemical compatibility testing

As part of the RCRA regulations, hazardous waste generators are required to classify their waste. When TSDFs receive waste, they must perform pre-acceptance analysis to confirm the waste's characteristics and identify incompatible wastes [13]. The EPA states "Compatibility testing may include determining water reactivity (e.g., foaming, heat generating, explosivity), oxidation-reduction potential, spot tests for cyanide and sulfide, among others" [14]. The EPA provides a list of some common potentially incompatible wastes in Appendix V to Part 264. When analysis is required, the EPA has published guidance on how to qualitatively evaluate the compatibility of wastes [15]. The guidance document uses a two-part approach: step-by-step compatibility analysis procedures and then a compatibility chart that considers 41 reactivity groupings. To simplify this analysis when chemical testing is not required, free computer software packages are available for performing quick assessments of chemical compatibility [16].

2.4 Guidelines for evaluating chemical reactivity hazards

According to the U.S. Chemical Safety Board (CSB), a chemical reactive hazard can be defined as "*a sudden event involving an uncontrolled chemical reaction with significant increases in temperature, pressure, or gas evolution that has the potential to, or has caused serious harm to people, property or the environment*" [17]. In order to decide whether a chemical can be considered a reactive hazard or not, the CCPS published a preliminary screening test based on a series of questions summarized in Figure 2 [18].

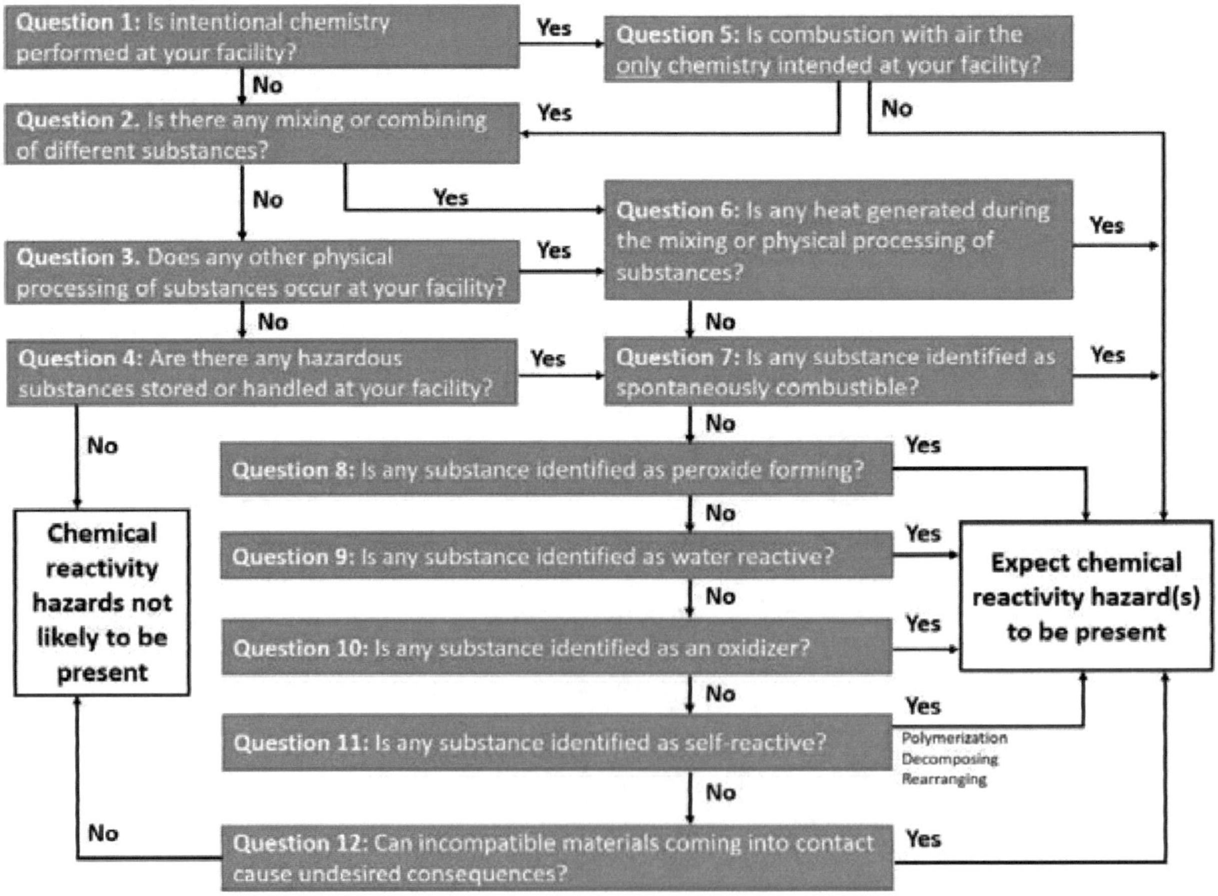

Figure 2. Preliminary screening for chemical reactivity hazards [18].

This test should be conducted on an ongoing basis, in particular when there have been changes in the process/conditions or a new chemical is introduced into the facility. If a chemical reactive hazard is expected to be present at the facility, the CCPS also suggests the implementation of a chemical reactivity management program structured consistent in Figure 3.

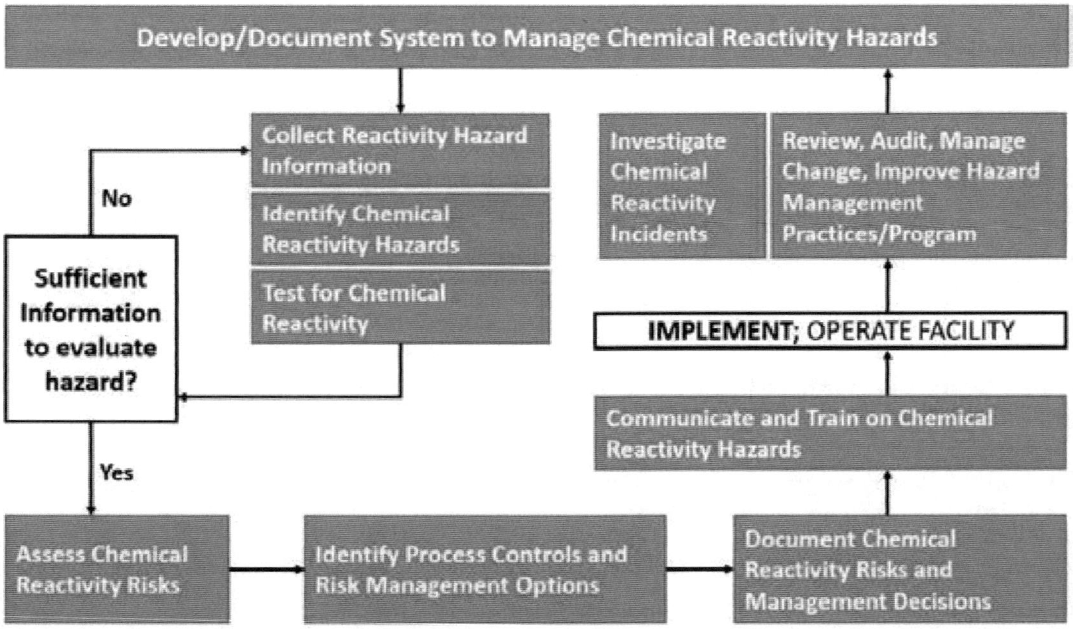

Figure 3. Flowchart for implementing chemical reactivity hazard management [18].

2.5 Why and how an event happened? Protocols for investigating reactive chemical incidents

Given the nature of fires and explosions, especially when involving hazardous waste materials, necessary evidence may be destroyed or lost during the incident or subsequent emergency response. Thus, conclusive determination of the chemistry and proximate causes may be impossible. The investigation process must rely upon the scientific method to develop hypothetical causes, test those causes, and determine the most likely causes and contributing factors.

Two commonly relied upon resources for performing incident investigations per the scientific method are NFPA 921 *Guide for Fire and Explosion* Investigations [19] and the CCPS publication Guidelines for Investigating Chemical Process Incidents [20]. Figure 4 shows the progression of an investigation as described in Chapter 4 of NFPA 921. Two essential concepts, inductive reasoning and deductive reasoning, are used to develop and test hypotheses, respectively. Inductive reasoning is using the analyzed data to develop a consistent hypothesis. Deductive reasoning is used to test the hypothesis by comparing it to all known facts of the accident and scientific literature. NFPA 921 cautions that "[t]he investigator does not have a valid or reliable conclusion unless the hypothesis can stand the test of a careful and serious challenge." Recent editions of NFPA 921 have also added cautionary language regarding the use of *negative corpus* also known as the process of elimination. Investigators should not confuse the lack of an alternative hypothesis for affirmative proof of a hypothesis with no supporting evidence (but that cannot be ruled out with the available information).

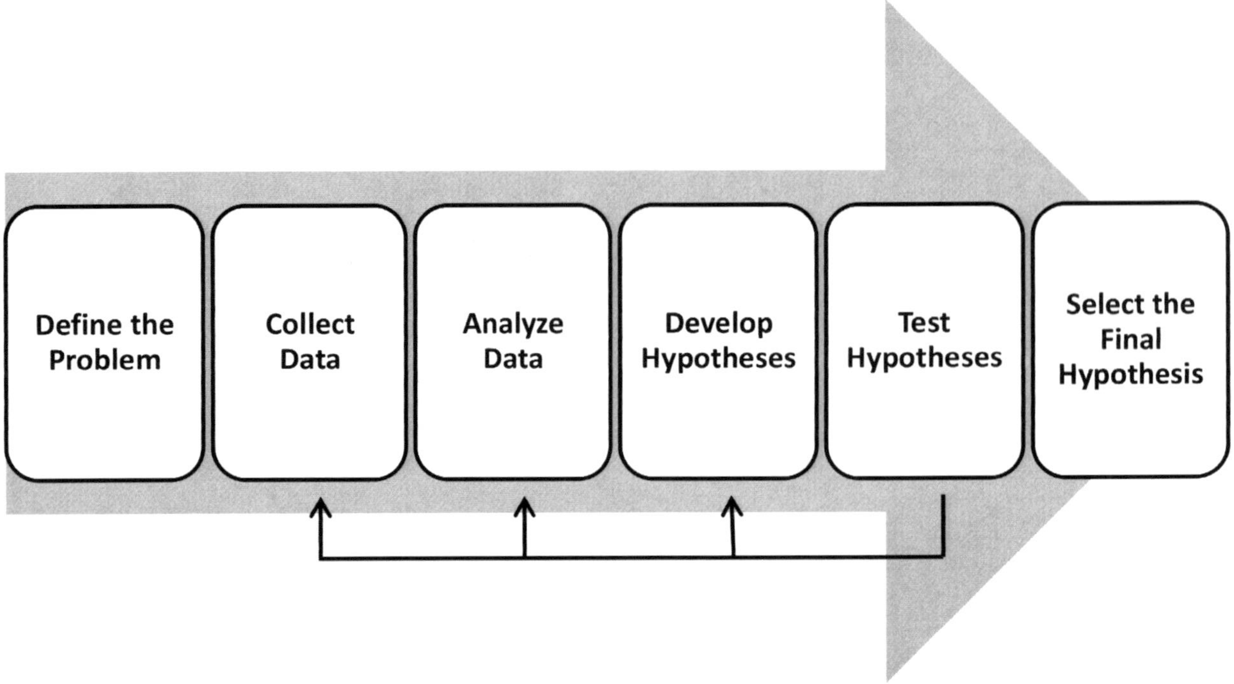

Figure 4. The scientific method of investigation.

The CCPS Guidelines text provides a similar scientific methodology for accident investigation but also offers information specific to industrial accident investigation. For example, information for designing incident management systems within a corporation is provided. Specific investigation and root cause analysis tools commonly employed for investigations, organized by the level of structure, are shown in Figure 5.

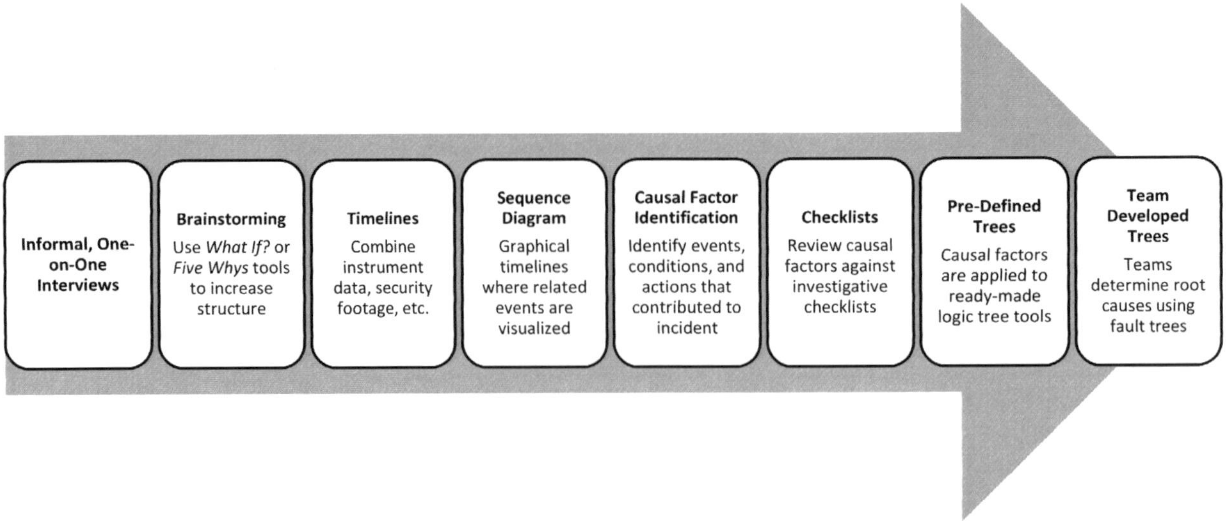

Figure 5. Investigation tools for process incident investigations.

The case studies below describe the causes and lessons learned from MSW handling, solvent reclamation, and TSDF incidents. All of the incidents were characterized by fires or explosions, which caused significant damage and posed potential for catastrophic injuries or fatalities to onsite personnel. Many of the incidents were investigated by external public agencies in addition to the company investigations. The authors also participated firsthand in the investigations of the incidents. Where possible, lessons learned from the investigation process as one element of PSM are shared.

3 MSW Transfer Facility Fires

Two fires, separated by three days, occurred at a MSW transfer station. The first fire involved a trailer of waste awaiting transfer to the landfill, and the second fire involved a different trailer and a large portion of the building and waste contained within. The facility was unsophisticated when considering the element of process safety management, which is understandable because it did not handle reactive materials. As a consequence, early investigation of the causes of the fires relied upon supposition and poor science to conclude that one waste stream was the cause of the fire. The incorrect conclusion adopted by the facility meant that the real cause went unaddressed until a later, thorough engineering investigation occurred.

The waste transfer station was a large, open warehouse structure with two sunken loading bays. The facility handled trash classified as municipal solid waste (MSW) and construction and demolition (C&D) debris. Waste materials were piled on the concrete and moved around using a front end loader. Different materials were commingled by the facility operators, both in piles and in the open trailers located in the loading bays. The commingled waste was periodically hauled to a municipal landfill. The facility did not accept RCRA hazardous waste or liquid wastes. The local regulations and the permit for the facility both acknowledged the reasonable possibility of "hot loads" - in other words, the receipt of loads of waste material that are undergoing self-heating or are already smoldering.

The first fire involved the contents and the rear of a trailer (Figure 6), which was located in a loading bay overnight. The trailer contained a mixture of bottom ash from a fluidized bed coal combustor and other unknown MSW and C&D materials. No formal investigation was conducted to attempt to determine the cause of the trailer fire. The waste material in the trailer was not fully extinguished or overhauled by the fire department, and the facility located the trailer away from the building so that it could eventually burn itself out. The fire witness statements indicated that the contents of the trailer continued to burn for the next several days. The fire damage patterns to the trailer are consistent with a fire originating at the rear wheels, perhaps due to overheated brakes. Three days after the initial fire, a second fire was discovered after a different trailer was loaded with similar waste materials from within the building and held overnight in the undamaged loading bay. The second fire burned to completion and destroyed the building.

Figure 6. Photograph of the trailer damage resulting from the first fire.

An early investigation of the second fire concluded that the combustor bottom ash was the equivalent of pure lime (CaO), caused both fires due to exothermic heat of hydration, and should have been classified as a RCRA hazardous waste. Pure CaO is referred to as "lime" or "quick lime." The reaction of CaO with water vapor results in solid calcium hydroxide, $Ca(OH)_2$, while mixing with excess water creates an aqueous solution, according to the chemical reaction:

$$CaO_{(s)} + H_2O \rightarrow Ca(OH)_2$$

This reaction is exothermic, meaning that it generates heat, with a heat of reaction of -65.27 kJ/mole of CaO reacted [21]. The quantity of heat released is directly proportional to the quantity of CaO and H_2O undergoing the reaction.

The transfer station's incorrect initial conclusion was based on the description of the waste by the generator as "agricultural lime," which is commonly used to describe calcium carbonate ($CaCO_3$), hydroxide ($Ca(OH)_2$), and oxide (CaO) used in farming. Further, the early investigation concluded that the mischaracterization of the waste was a direct cause of the fires. A subsequent engineering investigation analyzed that conclusion and also identified several plausible causes including an electrical fire, spontaneous combustion of MSW or C&D, or re-kindling of the smoldering remains of the first fire after being transported across the facility.

Chemical analysis of the combustor bottom ash and a review of the literature indicated that the ash did not contain sufficient CaO to cause spontaneous combustion through hydration and was excluded under 40 CFR 261.4(b). The residual CaO in combustor tram will undergo hydration by reacting with water (H_2O) in either a vapor (e.g., atmospheric) or liquid form. The hydration reaction for residual CaO would run largely to completion within a few days when exposed to the outdoor atmosphere [22]. Comparison samples of fresh combustor bottom ash revealed that it contained approximately 10% CaO by weight.

If concentrated CaO is mixed with liquid water, the hydration reaction can lead to very high temperatures capable of igniting combustible materials. However, it is unlikely that these conditions existed in the current matter. The maximum temperature reached by the combustor ash via this reaction, if any unreacted CaO is present, will be tempered by the presence of liquid water and surrounding solid material (i.e., $CaCO_3$, $Ca(OH)_2$, $CaSO_4$, sand, and onsite waste materials). The water will absorb the released heat and act to cool the material as the water evaporates; the other solid species will act to absorb the liberated heat and increase in temperature.

The false conclusion that the ash waste was CaO, thus causing the fires, highlights the importance of precise descriptions for waste streams even if they are not bound to the RCRA regulation. Effective description can help investigators, who may be unaware of the precise waste compositions, more effectively assess the important properties of the waste stream after an accident. Similarly, investigators must be careful when examining the properties of non-RCRA waste streams as their descriptions may be more colloquial. Conclusions should not be drawn until a more detailed composition of the waste can be obtained through chemical analysis or discussion with the generator.

4 Solvent Recovery Plant Vapor Cloud Explosion

In 2009, a release of ignitable liquid and vapors, followed by catastrophic vapor cloud explosion and subsequent fires, took place in a solvent recovery plant. The explosion severely injured two employees, destroyed part of the plant (see Figure 7), and broke windows in neighboring facilities and residences. The explosion was investigated by the Chemical Safety and Hazard Investigation Board (CSB) [23].

The solvent recovery plant (Figure 8) was located at a facility that handled hazardous and non-hazardous waste materials that were received via tanker truck, drums, totes, and lab packs. The facility performed fuel blending, solvent recovery, and waste consolidation. Incoming dirty solvents were unloaded into tanks in the dirty tank farm. Clean solvent was recovered from the dirty solvents via one of three distillation units (Unit 1, Unit 3, and Unit 4). The recovery process involved the circulation of distillation bottoms back into the source tank in the dirty tank farm, while purified "clean" solvent was condensed and transferred to the clean solvent tank farm. Final waste liquids and sludge were shipped offsite to a waste incinerator as blended fuels.

At the time of the incident, the three Processing Units were being run by a single outside operator. Unit 1 was used to recycle a paint thinner material whose primary compound was acetone, while Unit 3 was processing a tetrahydrofuran (THF)/water mixture. Unit 4 worked as a closed circuit reflux-distillation unit where a mixture of THF and water near the azeotropic concentration was processed.

Figure 7. Aerial view of the solvent recovery plant after the incident from the CSB report [23].

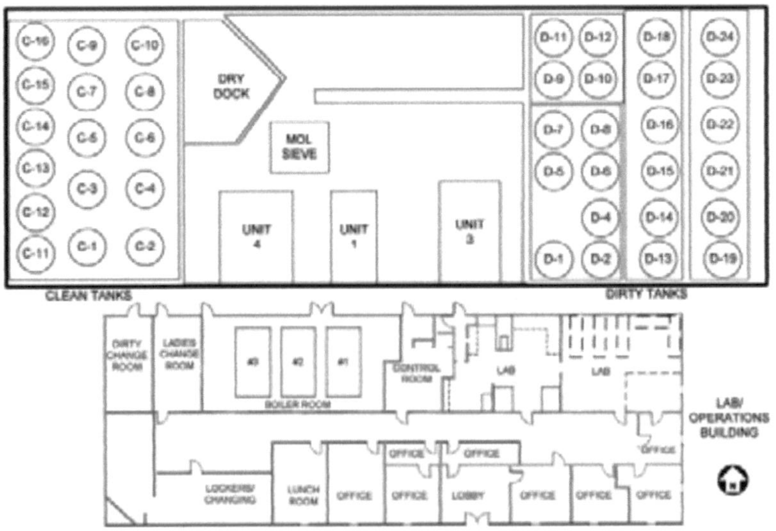

Figure 8. Facility layout (not to scale) [23].

Three of the six workers onsite reported hearing a loud noise like a pressurized release, accompanied by a strong solvent odor and a visible aerosol/or vapor cloud coming from the Dirty Tank Farm prior to ignition. Security camera footage from onsite and a nearby business supported the conclusion of a pressurized release, likely of Tetrahydrofuran (THF) solvent mixture; however, the exact composition and source of the flammable material released could not be determined.

The incident investigation concluded that the vapor cloud spread out from the Dirty Tank Farm, entered the laboratory/Operations Building, and ignited inside the Boiler Room causing, an initial explosion. Consequently, additional flammable material within the Dirty Tank Farm was also ignited causing a blast pressure wave (vapor cloud explosion) and subsequent pool fire that burned for several hours.

The CSB's findings from this incident included the following facility siting concerns:

- Lack of hazard analysis to determine the risks associated to the proximity of occupied buildings to processes: The most critical injured employees were at the Laboratory/operations building at the time of the incident. The nearby building was occupied with non-essential personnel during the day shift.
- Inappropriate area classification: Two boilers were in service in the Laboratory/Operation Building, where the ignition source was identified. The flammable vapor cloud was ingested and ignited by one of the boilers.

Although the facility was covered by the OSHA PSM standard, the CSB found that the TSDFs did not have industry-specific guidance for fire safety and facility siting.

5 Fires and Explosions in Storage Operations at TSDFs

Several fire incidents have occurred during storage of hazardous materials at TSDFs. The ultimate fate of hazardous materials is often destruction through incineration. The type of incinerator and local environmental emission limits dictate handling activities with the hazardous waste. To manage the limits on the heat of combustion for the incinerator, heavy metal emissions, and other factors, TSDFs commonly repackage wastes by either combining similar waste streams in the same container or by breaking up larger containers into smaller container sizes. For example, if the bulk heat of combustion of the 55-gallon drum of waste received by a TSDF is too high for the incinerator's operating limit, then the 55-gallon drum must either be metered (if a liquid) into the incinerator, or the drum must be subdivided into smaller containers (e.g., 30-gallon drums).

Thus, hazardous materials often pass through the following stages at a TSDF:

1. Receipt and characterization
2. Temporary storage
3. Repackaging – either bulk up or break down
4. Storage awaiting disposal
5. Final disposal

The waste generator must provide specific information on the hazardous waste that they are providing to the TSDF. The TSDF oftentimes works with the generator to help characterize the waste stream as part of the initial contracting process. The TSDF typically samples and performs basic characterization of individual wastes as they are received. However, this process is imperfect because the TSDF cannot be guaranteed that a set of samples from distinct locations in a waste container represent the bulk of the waste in that container. The potential deviation of actual properties from the results of this sample may grow significantly for larger-sized containers (e.g., a liquid tote or a 1 m^3 box), wastes that are heterogeneous, wastes with layers or pockets of different composition, or large volume wastes that are not completely sampled (e.g., sampling a subset of containers).

After initial contracting and waste specification, a TSDF will develop a plan for handling that waste stream, which will include sampling, characterization, and verification that as-received waste meets the contract specification. This process falls under Process Knowledge Management in the 20 elements of CCPS PSM. Deviations of as-received waste from the specification should be addressed under a Management of Change element, but the TSDF may not have a robust change management policy. Additionally, changes from the specification may not be identified during the characterization process as discussed above.

Incident investigation after a hazardous materials incident at a TSDF can be very challenging. A TSDF storage area may contain hundreds or even thousands of individual waste containers (see Figure 9). A fire or explosion in the storage area will destroy or damage the waste containers, displace waste containers, and mix the residue from involved containers. All these aspects make determination of the cause of the incident difficult. The likelihood of determination is usually directly related to whether the incipient stages of the fire or explosion were witnessed. The following three case studies provide incident summary and investigation results for a progressively larger TSDF fire incidents in order to characterize the difficulty in cause determination and lessons learned.

Figure 9. Example of TSDF storage warehouse.

5.1 Delayed Ignition - No Repackaging – Cause Undetermined

An investigation was performed on a fire that occurred in a small TSDF's storage bay. The facility handled a wide variety of wastes including flammables materials and acids. The facility included storage bays, laboratories, treatment equipment and office areas. In the storage bay where the fire occurred, several different types of containers were present including:

- Plastic buckets
- Plastic, steel, and paper fiber drums
- Steel encaged plastic totes known as intermediate bulk containers (IBCs)

The 55-gallon drums were placed on wooden pallets in sets of four. The pallets of 55-gallon drums and IBCs were stacked two or three levels high. A photograph of the bay after the fire is shown in Figure 10.

Figure 10. Photograph of storage area after fire.

Based on witness statements, the fire in the storage bay was preceded by a loud hissing sound and a pressurized release of a reddish dust/mist from a container near the center row of the image. The witness did not see the exact container where the release originated, but was able to place it within a few pallet stacks in the area. Immediately after the pressurized release, a fire ensued in the area. Due to fire suppression efforts, the fire damage was contained to the storage bay where it originated. The investigation focused on the adjacent rows containing the waste drums that were damaged. The exact location of each waste container in the neighboring pallet stacks was reconstructed based upon storage records, remaining containers, damage patterns, and witness statements about container placement. The origin was confirmed to be one pallet stack in the center row.

The containers in the center row had not been repackaged or disturbed from the time of receipt until the incident; therefore, commingling of incompatible wastes or mixing of a reactive pocket within the source drum were refuted as potential causes. Delayed decomposition of waste material in one of the 55-gallon drums is the likely source of the pressurization. Delayed decomposition may be the result of chemical incompatibility or thermal instability leading to a temperature rise, gas generating reaction, or both. Eventually, the pressure increased to the level that the source container ruptured. It was most likely that this release occurred from either a steel drum or a plastic drum given their ability to contain some degree of pressurization compared to a fiber drum. After the initial pressurized release and fire, several other containers were damaged and ruptured as a result of the growing fire.

Based on the photo-documentation of the scene, it was possible to identify several potential drums that showed evidence of pressurization and/or rupture. However, due to the extent of the damage, it was not possible to identify the serial number on all of the containers, and therefore, their origin and contents. Examination of the fire damage showed evidence of burned away paint, burned pallets, and melting of non-metal containers. Based on the damage patterns, it was possible to rule out some waste containers including a set of drums containing phosphoric acid. A ground-level area containing four drums was identified as the likely area of origin, but further specificity was not possible due to the extent of the damage and several other drums that could not be conclusively eliminated. In a regular facility, the containers of all the stored containers would be known, providing valuable information to incident investigators. In a waste handling facility, significantly less certainty is available about the exact contents of each container. The ability for many of the waste container to have been the source of the pressure release significantly complicated this specific investigation. The final determination as to which container was the source could not be determined given the degree of damage.

The extent of damage does not always preclude determination of the likely cause of a fire in storage. A similar TSDF facility experienced a fire in a storage bay, which consumed all of the hazardous waste and combustible containers in the area (see Figure 11). The point of origin of the initial fire could be more precisely placed than the incident just discussed because a nearby operator observed the location of the incipient fire at ground level at a specific pallet stack. The source of the fire was most likely a leak of fuming red nitric acid that spilled onto a wood pallet holding the drums. The drums were stacked on combustible wood pallets, and fuming nitric acid is a strong acid capable of igniting such materials. The facility did not identify the potential consequences of a spill of the acid; thus, a noncombustible pallet (i.e., steel) was not selected. Use of a what-if strategy for packaging and storage selection may have prevented the fire (but not necessarily the leak).

Figure 11. Another TSDF storage bay after extensive fire damage.

5.2 EQ Industrial Services Warehouse

A hazardous waste TSDF was complete destroyed due to a fire for which the initial cause was undetermined (see Figure 12). The magnitude of this incident forced thousands of local residents to be evacuated in two days. The CSB investigated the cause of the incident, and identified several hypothetical causes. Their recommendations focused on eliminating those hypothetical causes.

Figure 12. Facility after the incident [24].

Based on witness statements, a strong chlorine odor and plume of smoke came from the TSDF on the night of the incident. Initial firefighters responding to the scene reported that a small fire started in one of the hazardous waste storage bays and rapidly spread to the other locations of the facility, including a bay containing many 55-gallon drums of flammable solvents. The incident investigation revealed that the fire originated in a storage bay where containers of solid chlorine-based pool chemicals and unspent aircraft oxygen generators awaiting for final disposal.

Many incidents have occurred involving storage of chlorine-based pool chemicals (see for example [25]-[28]), and around the time of the incident, the NFPA was pursuing research into fire protection considerations for pool chemicals in warehouse settings [29]. Concentrated calcium hypochlorite or sodium dichloroisocyanurate containing pool chemicals pose potential thermal runaway concerns if contaminated with organic material or if stored at elevated temperature. Given their nature as solid oxidizers, pool chemicals would also serve to accelerate fire growth. The witness observations of a chlorine-odor associated with the smoke likely indicate that these chemicals were present near the fire origin, but the investigators could not conclude that these were cause of the fire.

Although the first material involved could not be determined due to the extent of damage, the CSB investigation findings focused on chemical oxygen generators [30]. The CSB concluded that chemical oxygen generators most likely contributed to the intensification of the fire, which quickly spread to an adjacent bay where flammable materials were stored. The investigation found that the unspent oxygen generators, coming from an aircraft maintenance facility, were not properly packaged by the aircraft maintenance personnel, and were misidentified as general oxidizer waste on shipping paperwork. A lesson learned from this incident was to consider oxygen generators that have exceed their service life as still posing a risk of heat and oxygen generation if accidentally actuated. Therefore, the CSB recommended that unspent oxygen generators should be actuated before they are transported [30].

5.3 Teris El Dorado Warehouse

This incident occurred in a warehouse at a TSDF that contained 4000 to 5000 drums of different wastes. The facility operated two rotary kiln incinerators for solid disposal and thermal oxidation for liquid incineration [31]. The incident was witnessed by an operator who saw a drum rupture, which quickly spread fire throughout the warehouse contents (see Figure 13). The entire contents of the warehouse were consumed, damaged, and/or released during the incident. No injuries or fatalities were reported; however, nearby residents were evacuated during the emergency response.

The investigation found that the fire was caused by spontaneous combustion of sodium chlorate solid waste recovered from a rail car accident site. This material was brought to the facility in $1m^3$ boxes to be incinerated shortly before the incident. However, the waste generator incorrectly profiled this material as containing only sodium chlorate in soil; the waste actually contained vegetative matter, rail car components, and portions of railroad ties. The organic contaminants are incompatible with the strong oxidizer and would act as fuel sources, allowing spontaneous ignition of the mixture. The material was repackaged into smaller drums and a portion was moved into the warehouse prior to the fire. The remainder of the material had not been relocated at the time of the fire. This residual material was isolated in a trailer onsite. The trailer containing the residual material underwent spontaneous ignition several days after the original fire incident, which supported the conclusion that the sodium chlorate waste material was the cause of the fire.

Figure 13. Aerial image during fire event [31].

Given the non-homogeneous nature of the waste material received, grab sampling of a single container was insufficient to fully characterize the mixed material. In contrast to liquid sampling, a thorough visual examination of the mixed, non-homogeneous waste material may have identified the presence of organic materials in the oxidizer and soil waste. Further, the use of "soil" in the waste analysis plan submitted by the generator may have triggered a further review by the TSDF since soil typically contains a significant organic component. It is likely that the repackaging activity mixed reactive sodium chlorate with fresh organic fuel materials that led to the delayed spontaneous combustion event. After the incident, the facility enhanced their material analysis and handling plans for wastes containing solid oxidizers.

6 Fires & Explosions during Incineration Processes

6.1 Explosion during Solid Waste Homogenization

An explosion occurred at an incineration facility that processed hazardous and non-hazardous solid and liquid wastes through thermal destruction in a rotary kiln. Drums of waste were homogenized in a nitrogen-inerted shredder. Drums were scheduled to be incinerated and assembled onto pallets based upon the waste characteristics. The system automatically moved individual pallets into an airlock, purged the airlock with nitrogen, then dumped the pallets into the already inerted shredder chamber. The shredder was used to shred steel drums as well as plastic drums. After shredding, the homogenized waste was conveyed to the rotary kiln via a screw conveyor.

The pallets, packaging materials, and waste materials are combustible materials, and flammable materials may also be processed based on the incinerator waste schedule. Due to the potentially flammable nature of the waste and the likelihood of mechanical sparks and heat generation from the shredder, the system was inerted with nitrogen gas. The nitrogen inerting function was performed by using a fixed timing of nitrogen addition. Although the system did not use an oxygen measurement control loop to dictate the duration of purging, it did use oxygen measurement as a permissive control for the automated conveyor movements and door opening between chambers in the system.

A central incinerator control room at the facility monitored several operating parameters, including the oxygen concentration and temperature within the inerted system. Since one of the primary safety mechanisms for the unit was a reduced oxygen atmosphere, the unit was not intended for processing oxidizing materials including solid oxidizers, oxygen containers/cylinders, explosives, and/or water reactive materials. During one waste load, an explosion occurred in the shredder chamber, which damaged the unit and caused explosion panels to open on the shredder chamber and airlock.

An approximation of the oxygen concentration data and shredder chamber temperature is provided in Figure 14. There was a lag noted in the oxygen sensor response due to the delay for gas to traverse the sampling leg to the sensor. The temperature measurement had less of a response lag; thus, the sudden temperature rise in the shredder chamber actually occurred coincidentally with the explosion and the opening of the explosion vents. On the night of the incident, an additional waste load was fed into the unit, and shredding commenced less than one minute after it was added. Approximately one minute after shredding began, the oxygen concentration started to increase from 2% to over 13% over a period of 20 seconds. According to the recorded data, the temperature in the shredder chamber suddenly increased at approximately the same time as the oxygen concentration increased. This temperature increase is consistent with the heat generated by the explosion within the shredder chamber. An engineering analysis concluded that the most likely source of the rapid oxygen influx was from an item in the waste stream that released oxygen when shredded. All potential sources of air via external sources were examined and rejected.

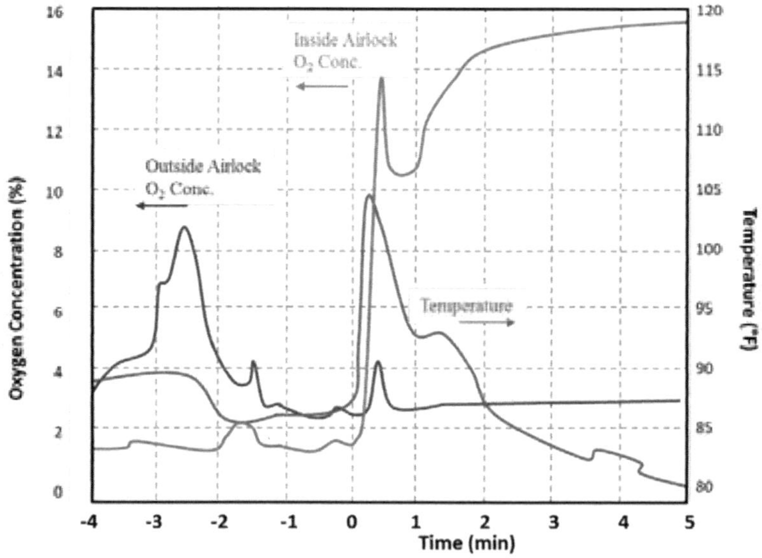

Figure 14. Process oxygen concentration and temperature during incident.

The explosion in the shredding system demonstrated the serious consequences of processing waste with unintended properties. In this case, a system that was intended to operate while inerted inadvertently contained compressed oxygen, resulting in a damaging explosion. While the waste was screened prior being fed into the shredder, the screening process did not identify and divert the oxygen-containing waste. In this case, the explosion mitigation system worked as intended by relieving the explosion pressure through explosion vents. The system was damaged, but the damage was not catastrophic, and there were no injuries.

6.2 Spontaneous Combustion at a Calciner

A fire occurred at a cement production facility that used an engineered solid fuel composited from MSW and other industrial wastes to co-fire in the calciner. The engineered fuel was a blended mixture of solid wastes including paper, plastic, and sawdust as well as non-hazardous waste liquids, slurries, and sludge. Because the waste streams possessed variable energy content, a qualifications process (including laboratory analysis) was used to determine the appropriate mixture for creating a consistent engineered feed. A block flow diagram of the shredding, blending, and mixing process is shown in Figure 15. The engineered fuel had the potential for self-heating due to containing oxidizing materials such as polymerizing agents and vegetable-derived oils. Material handlers inspected the stockpiles for external signs of high temperature due to self-heating and excavated the piles to identify internal high temperatures. Infrared pyrometers were used for surface temperature measurements. The temperatures were monitored with the handheld pyrometers and logged prior to the loading the material into trucks for transport to the cement production facility.

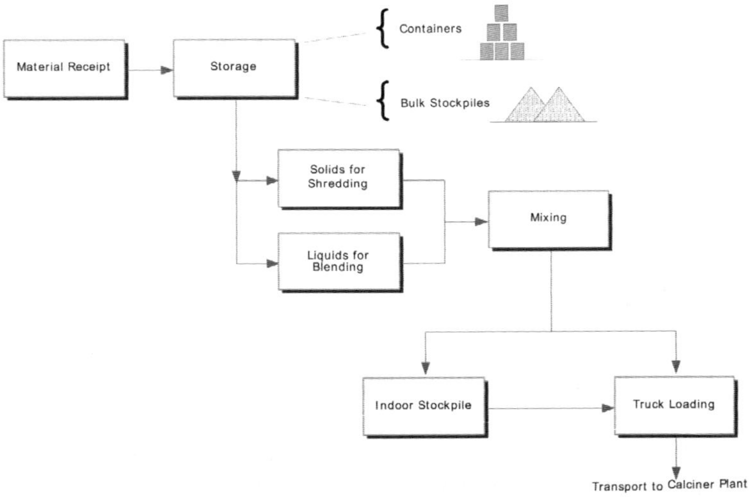

Figure 15. Block flow diagram of the engineered solids feed preparation process.

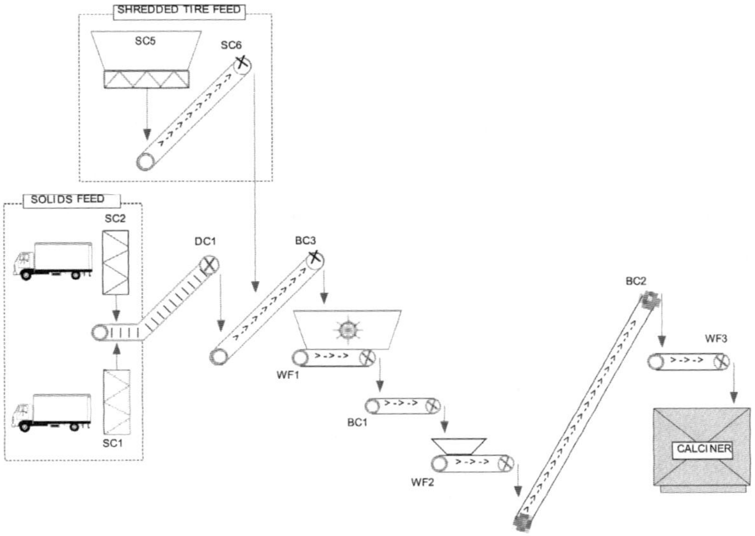

Figure 16. Conveyor flow diagram between truck delivery and the calciner.

The engineered fuel was transported to the cement production facility in tractor trailers for eventual feed into the calciner unit. A conveyor flow diagram is provided in Figure 16. The engineered fuel was unloaded into enclosed hoppers that discharged into horizontal screw conveyors. The horizontal screw conveyors discharged into a central drag conveyor that delivers the fuel to a belt conveyor which also received a separate shredded tire feed. The combined feed then traveled through an additional series of hoppers and conveyors before being deposited on a 1,600 feet long enclosed belt conveyor that climbed from ground level to the upper levels of the calciner structure. The belt of the long conveyor was originally a flame-resistant belt material but had been subsequently patched with non-flame resistant material. The total travel time of the engineered fuel was estimated to be 20-30 minutes by the cement facility's personnel.

A fire was discovered overnight soon after the conveyor system was restarted after a two-day outage. During that two-day period, the belts were left loaded with the engineered fuel and combined feed. When the kiln was restarted, the system was operated for approximately one hour then shut down due to an upstream clog during which a fire was discovered near the head section of the long conveyor. The belt failed and broke as a result of the fire.

An investigation into the fire cause determined that, while various friction-based ignition sources (belts, bearing, etc.) did exist within the conveyor system, the most likely cause of the fire was spontaneous ignition of the engineered feed. This was substantiated because three days after the fire, a stockpile of the engineered feed that had been fed to conveyor prior to the fire enflamed at the storage warehouse. Although staff had monitored stockpiles at the feed mixing facility for external signs of ignition, external methods can be insufficient to detect self-heating inside the pile. After exposing the self-heating pile to oxygen during transport and handling, self-heating may accelerate and lead to spontaneous ignition. The facility implemented a strategy that avoided leaving materials on conveyors for long periods of time. Through the use of conveyor interlock strategies and external dump locations, the system can be operated to empty conveyors in the event of an unintended shutdown.

6.3 TDI Waste Rail Tank Car

An explosion occurred at another TSDF cement plant in the liquid fuels rail tank car unloading area [6], [32]. As consequence, the tank ruptured and was propelled about 750 feet away. Other tank cars, as well as storage tanks located nearby, were also heavily damaged (Figure 17). No injuries or fatalities were reported; however, total property losses were estimated at nearly $8.2 million.

This facility incinerated liquid hazardous waste streams as supplementary fuel in rotary kilns, and used to receive several tank cars containing column bottoms waste from a toluene diisocyanate (TDI) manufacturing process. TDI is a monomer feedstock used in the manufacture of polyurethane polymers. The waste generator had stored tank cars containing this type of waste material for several years, and reprocessed them as the opportunity arose. Prior to the incident, several tank cars were shipped to the TSDF for incineration, and had been successfully emptied, with contents incinerated in the kilns. Despite the TDI waste was partially polymerized in most tank cars, the car's contents were non-homogeneous and did not have identical properties between individual tank cars.

Figure 17. Locations (1) and (2) indicate positions of the tank car's vessel before and after the explosion, respectively [33].

In order to decrease the viscosity of the polymerized TDI, so that it was pumpable to the kilns, the facility had successfully used external steam heating through the built-in steam jackets on the tank cars.. A schematic of the approximately 30,000 gallon tank car is provided in Figure 18. While attempting to unload the incident TDI material, the tank car vessel was steam-heated in two opportunities to ensure the waste mixture was sufficiently fluid to be pumped. After three unsuccessful unloading attempts, the shipment was rejected, but the tank car was left in the unloading location. The explosion incident took place two days after the last heating attempt.

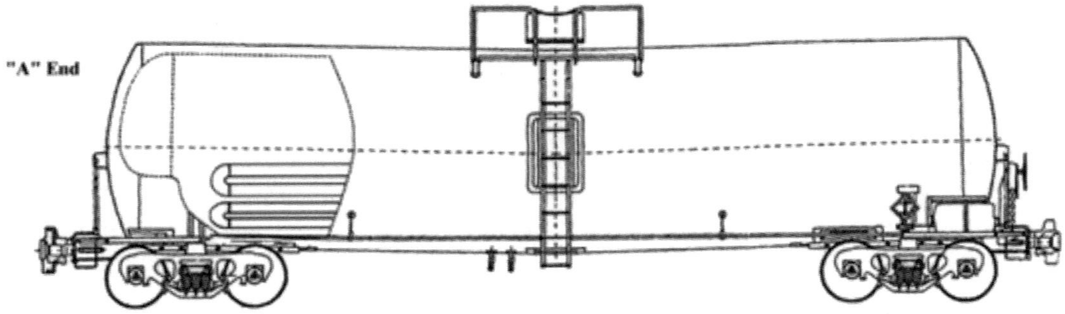

Figure 18. TDI tank car schematic with cutaway showing the external steam heating coils. [20]

Retained samples of the original material and residue ejected during the incident were analyzed, and the presence of self-polymerized TDI, TDI monomer, and ureas were determined. Thermal analysis of the materials indicated that the polymerized TDI and TDI monomer exhibited an onset temperature for thermal runaway that was much higher than possible with the steam heating system. A chemical hazard assessment revealed that the cause of the explosion was the runaway thermal decomposition of a contaminant mixed with the TDI waste. The TSDF operator relied on the waste generator for information regarding the waste composition and safe heating procedures. In this case, the waste generator supplied incomplete waste characterization data and inadequate instructions and warnings for heating and handling the TDI waste.

Four critical factors and lessons learned were identified from this incident [6] [20]:

(1) Use of inappropriate tank car design that was unsuitable for this service, creating difficulties for unloading of TDI solid waste.

(2) Storing of TDI waste for several years in a warm climate

(3) Providing inadequate instructions and warnings for unloading the TDI waste.

(4) Commingling an unknown contaminant with the TDI waste; this last factor was the most serious error in this incident.

7 Discussion and Conclusions

In the case studies presented above, hazardous properties that were not known to exist led to accidents. In several cases, the accidents were the result of a chemical reaction, leading to a release of heat or pressure that caused a storage vessel to fail. Chemical reactions due to the mixing of incompatible chemicals, storage of unstable chemicals, or heating of thermally unstable chemicals occurred in several of the case studies described. While characterization is a significant preventative safeguard for these types of failures, it may not always be possible to determine if a waste stream is completely stable based on testing; therefore, facilities may still need to consider the unlikely event of an exothermic or pressure-generating reaction within their waste streams. This may be accomplished by using separation distance or barriers between certain waste streams, fire protection systems, or fast turnaround from receipt through final processing.

In several incidents, oxidizers including oxygen tank, oxygen generators/tanks and sodium chlorate, are involved in the initiation of the incident or contribute to the severity of the incident. Because many waste streams contain flammable or combustible materials, enriched oxygen atmospheres or the presence of chemical oxidizers increases the likelihood and severity of a fire or explosion. While in many process operations, inerting atmospheres may be an effective and robust protective measure, due to the possibility of oxidizer ingress in mis-identified waste, additional protections such as explosion vents may be required. The following flowchart was created based on the cases presented in this paper, and may serve as a guideline for investigating suspected waste streams and their role on initiating a fire or explosion event.

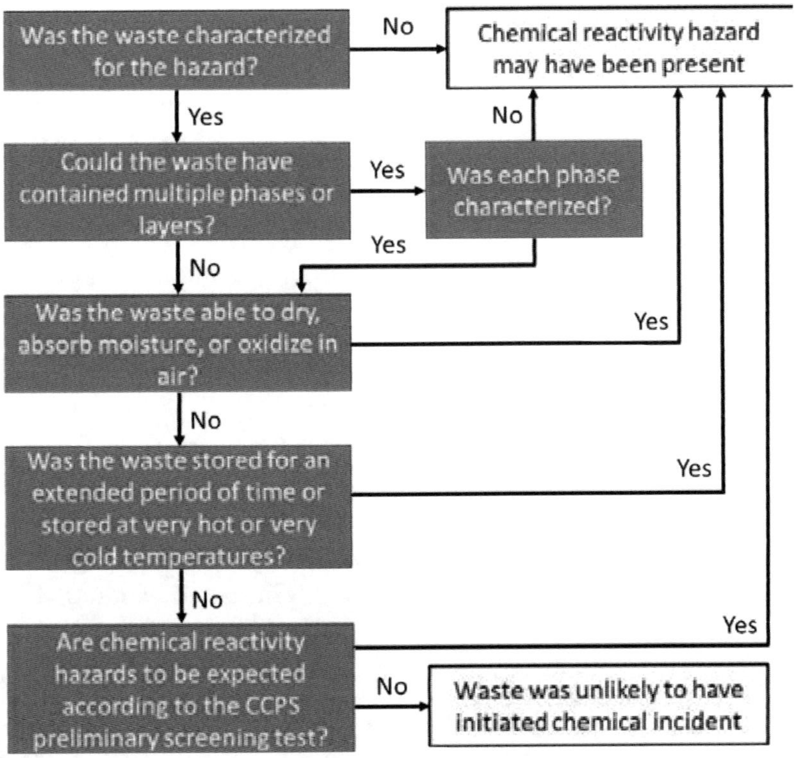

Figure 19. Role of suspected waste stream on initiating a fire or explosion.

8 References

[1] U.S. Environmental Protection Agency (EPA) Resource Conservation and Recovery Act, Title 40 of the Code of Federal Regulations.

[2] 29 CFR §§1910.119

[3] Center for Chemical Process Safety (CCPS). Guidelines for Risk Based Process Safety. Center for Chemical Process Safety, American Institute of Chemical Engineers, New York, NY, 2007.

[4] U.S. Environmental Protection Agency: Resource Conservation and Recovery Act (RCRA) Laws and Regulations, available at https://www.epa.gov/rcra. Accessed on February 15, 2017.

[5] Evarts, B. "Fires Caused by Spontaneous Combustion or Chemical Reactions," National Fire Protection Association, November 2011.

[6] Cox BL, Carpenter AR, Ogle RA. Lessons learned from case studies of hazardous waste/chemical reactivity incidents. Process Safety Progress 2014; 33(4):395–398.

[7] E.C. Buc. "Hazardous Waste Treatment, Storage, and Disposal Facility Fire Code Gap Analysis, Final Report," The Fire Protection Research Foundation, 2015.

[8] Environmental Protection Agency. Report on Emergency Incidents at Hazardous Waste Combustion Facilities and Other Treatment, Storage, and Disposal Facilities (TSDFs). EPA 1999

[9] U.S. Environmental Protection Agency (EPA): 2015 Biennial Report Summary, available at: https://rcrainfo.epa.gov/rcrainfoweb/action/modules/br/national/view. Accessed January 15, 2017.

[10] U.S. Environmental Protection Agency: Advancing Sustainable Materials Management: 2014 Fact Sheet. US EPA, November 2016.

[11] U.S. Environmental Protection Agency, Available at https://www.epa.gov/hw. Accessed on February 15, 2017.

[12] 40 CFR §§261

[13] 40 CFR §§264/265.17

[14] U.S. Environmental Protection Agency. Waste Analysis at Facilities that Generate, Treat, Store, and Dispose of Hazardous Wastes – Final: Guidance Manual (EPA 530-R-12-001). April 2015.

[15] Hatayama H.K., J. J. Chen, E. R. de Vera, R. D. Stephens, and D. L. Storm. A Method for Determining the Compatibility of Hazardous Wastes. U.S. Environmental Protection Agency, Washington, D.C., EPA/600/2-80/076 (NTIS PB80221005), 1980.

[16] National Oceanic and Atmospheric Administration (NOAA), "Chemical Reactive Worksheet," NOAA, available at http://www.aiche.org/ccps/resources/chemical-reactivity-worksheet-40. Accessed January 15, 2017.

[17] Davis E.M., D. Silva, and J. Murphy. The CCPS Chemical Reactivity Evaluation Tool. 8th Global Congress on Process Safety. April 2012.

[18] Center for Chemical Process Safety (CCPS). Essential Practices for Managing Chemical Reactivity Hazards. Center for Chemical Process Safety, American Institute of Chemical Engineers, New York, NY, 2013.

[19] National Fire Protection Agency. NFPA 921: Guide for Fire and Explosion Investigations

[20] Center for Chemical Process Safety (CCPS). Guidelines for Investigating Chemical Process Incidents. 2nd Ed. 2003

[21] Anthony, EJ, L Jia, M Caris, F Preto, and S Burwell, "An examination of the exothermic nature of fluidized bed combustion (FBC) residues," Waste Management Journal, 19 (1999).

[22] Serris, E, L Favergeon, M Pijolat, M Soustelle, P. Nortier, RS Gartner, T Chopin. "Study of the hydration of CaO powder by gas–solid reaction," Cement and Concrete Research 41(2011) 1078-1084.

[23] U.S. Chemical Safety Board (CSB). Veolia ES Technical Solutions, L.L.C. Case Study. Report 2009-10-I-OH. Washington, D.C. 2010

[24] U.S. Chemical Safety Board (CSB). EQ North Carolina Case Study. Report 2007-01-I-NC. Washington, D. C. 2008.

[25] Gray B.F. and B.W. Halliburton. 2000. "The thermal decomposition of hydrated calcium hypochlorite (UN 2880)." *Fire Safety Journal*. 35.3: 223-239.

[26] Gray B.F. 2002. Spontaneous Combustion and Self-Heating. In SFPE Handbook of Fire Protection Engineering, eds. P.J. DiNenno, D. Drysdale, C.L. Beyler, W.D. Walton, R.L.P. Custer, J.R. Hall and J.M. Watts, Quincy, Massachusetts: National Fire Protection Association, 211-228.

[27] Ogle R. A, Carpenter A. R. Toxic gas release caused by the thermal decomposition of a bulk powder blend containing sodium dichloroisocyanurate. Process Safety Progress 2003; 22:75–82.

[28] Ogle RA, Morrison DR, Carpenter AR, Su YS. Missed opportunities in reactive chemical hazard evaluations. Process Safety Progress, March, 2006, 25(1):2-7.

[29] Buc E.C. and D.J. Hoffman, "Development of an Enhanced Hazard Classification System for Oxidizers Research Project," The Fire Protection Research Foundation, One Batterymarch Park, Quincy, Massachusetts, April, 2006

[30] U.S. Chemical Safety Board (CSB). Safety Advisory Dangers of Unspent Aircraft Oxygen Generators (2007-I-NC-01-SA). 2007

[31] U.S. EPA, "Teris El Dorado Explosion," downloaded on March 13, 2017, https://response.epa.gov/site/site_profile.aspx?site_id=NRC745957.

[32] Ogle R.A, Carpenter AR, Morrison DR. Explosion of a railcar containing toluene diisocyanate waste. Process Safety Progress 2004 Dec; 23(4):316-320.

[33] National Transportation Safety Board. Rupture of a Railroad Tank Car Containing Hazardous Waste Near Clymers, Indiana. Hazardous Materials Accident Report NTSB/ HZM-01/01. Washington, DC. 2001

479039 ExxonMobil Torrance Refinery FCC ESP Explosion

Abstract:

On February 18, 2015, an explosion occurred in the electrostatic precipitator (ESP) at the ExxonMobil Refinery in Torrance, California near Los Angeles. The explosion injured four workers caused significant property damage to multiple process units within the refinery, and resulted in an off-site accidental release of catalyst dust. As a result of the incident, the FCC unit was down for over one year. The explosion occurred when hydrocarbons from the FCC unit reached the energized ESP while the unit was idling to perform maintenance on equipment that had failed. Multiple failures contributed to the incident. During the ESP explosion, there was also a near miss release of hydrofluoric acid when a large piece of ESP debris fell within feet of a storage vessel storing thousands of gallons of modified hydrofluoric acid in the neighboring alkylation unit. If the storage vessel had failed due to impact from ESP debris following the explosion, hydrofluoric acid would have been released. This presentation will detail key findings from the CSB investigation into the incident.

Overfill to Overpressure – Investigation Findings and Contributing Events to the CAPECO Explosion

Scott Davis
GexCon US
Bethesda, MD
sgdavis@gexcon.com

Tom DeBold
Claudio Marsegan
Bryant Hendrickson
GexCon US
Bethesda, MD

Prepared for Presentation at
American Institute of Chemical Engineers
2017 Spring Meeting
13th Global Congress on Process Safety
San Antonio, Texas
March 26-29, 2017

Overfill to Overpressure – Investigation Findings and Contributing Events to the CAPECO Explosion

Scott Davis
GexCon US
Bethesda, MD
sgdavis@gexcon.com

Tom DeBold
Claudio Marsegan
Bryant Hendrickson
GexCon US
Bethesda, MD

Keywords: CAPECO, petroleum, overfill, bulk storage, explosion

Abstract

On October 23, 2009, a vapor cloud explosion occurred after an above ground gasoline storage tank was overfilled at the Caribbean Petroleum Corporation (CAPECO) tank farm facility in Bayamón, Puerto Rico. The blast registered a 2.9 on the Richter scale and sparked several fires throughout the facility that burned for another three days and damaged a total of 21 tanks before emergency responders were able to extinguish the flames.

Operations on the day of the incident included a ship transfer of unleaded gasoline to the CAPECO facility and the transfer was scheduled for four smaller capacity tanks. At approximately 10:00 pm, all of the unleaded gasoline was being transferred to tank 409 and a visible vapor cloud was noted during the walk around at 12:00 am on October 23, 2009. It was determined that tank 409 was overfilled by approximately 5,000 barrels (210,000 gallons) of gasoline, which caused gasoline to flow out of the tank vents near the top of the tank and cascade down the side forming a vapor cloud and liquid pool. This is equivalent to overfilling the tank by over 20 gasoline tanker trucks, which if 10% of the liquid formed the vapor cloud it would cover an area larger than 30 American football fields of a rich flammable mixture. The gasoline vapors mixed with the air forming an ignitable vapor cloud spanning a vast portion of the facility, which subsequently ignited neighboring tanks that burned for three days.

US regulations do not consider CAPECO and other bulk petroleum storage terminals as high hazard facilities, yet a large vapor cloud formed. This paper will present investigation findings and key lessons learned into the massive explosion and fire. This includes: effects of confinement and congestion in a seemingly open space that contributed to the incident; ship offloading and tank filling procedures that revealed significant operational deficiencies; failures to secure the confined drainage system; and level controls.

1 Introduction

On October 23, 2009, a vapor cloud explosion occurred after an above ground gasoline storage tank was overfilled at the Caribbean Petroleum Corporation (CAPECO) tank farm facility in Bayamón, Puerto Rico. The blast registered a 2.9 on the Richter scale and sparked several fires throughout the facility that burned for another three days and damaged a total of 21 tanks before emergency responders were able to extinguish the flames. The explosion and ensuing fires did not cause any fatalities; only a few minor injuries were reported. However, approximately 300 homes and business within 1.25 miles of the site reported damages [1].

GexCon was retained to perform an independent and objective investigation into the cause and origin of the explosion. Due to the nature of the investigation, some information provided to GexCon was confidential and could not be provided. Therefore, information provided in the reports of other agencies were used in the present analysis, and include: the ATF Investigation 764045-10-0011 Report [1]; U.S. Chemical Safety and Hazard Investigation Board Final Investigation Report – Caribbean Petroleum Tank Terminal Explosion and Multiple Tank Fires [2]; and the Puerto Rico Fire Department Fire Marshal Report [3] . This paper will present investigation findings and key lessons learned into the massive explosion and fire that occurred at the CAPECO facility. This includes: effects of confinement and congestion in a seemingly open space that contributed to the incident; ship offloading and tank filling procedures that revealed significant operational deficiencies; failures to secure the confined drainage system; and level controls.

We will first present a brief description of the facility, events leading up to the incident, and details of the explosion. Next, we will evaluate the conditions and contributing factors that lead to the explosion.

CAPECO Facility

The Caribbean Petroleum Corporation (CAPECO) facility was a petroleum storage terminal located in Bayamón, Puerto Rico. At the time of the incident, the petroleum storage terminal employed approximately 65 people. Prior to the gasoline overfill and ensuing explosion, the CAPECO facility (Figure 1.1) consisted of a tank farm of above ground storage tanks, offices, decommissioned refinery, waste water treatment plant and distribution piping that serviced the San Juan Bay. The CAPECO facility stored gasoline, diesel, fuel oil, and jet fuel at the time of the incident. The tank locations along with what fuel was stored in each can be found in Figure 1.2.

The tank farm was split by roads into three areas; the northern, central, and southern tank farms. The tanks were separated by dikes or berms designed to hold 110% of the largest tank capacity in that berm (see Figure 1.2) [3]. The southern tank farm contained eleven tanks, which included diesel and fuel oil tanks, divided by six berms. Jet fuel, diesel, and fuel oil were contained in the central tank farm. The northern tank farm held gasoline, diesel, and fuel oil. CAPECO estimated that 613,816 barrels (25,780,272 gallons) of gasoline, 428,260 barrels (17,986,920 gallons) of diesel, 360,496 barrels (15,140,832 gallons) of fuel oil, and 132,724 barrels (5,574,408) of jet fuel were present at the time of the incident [4].

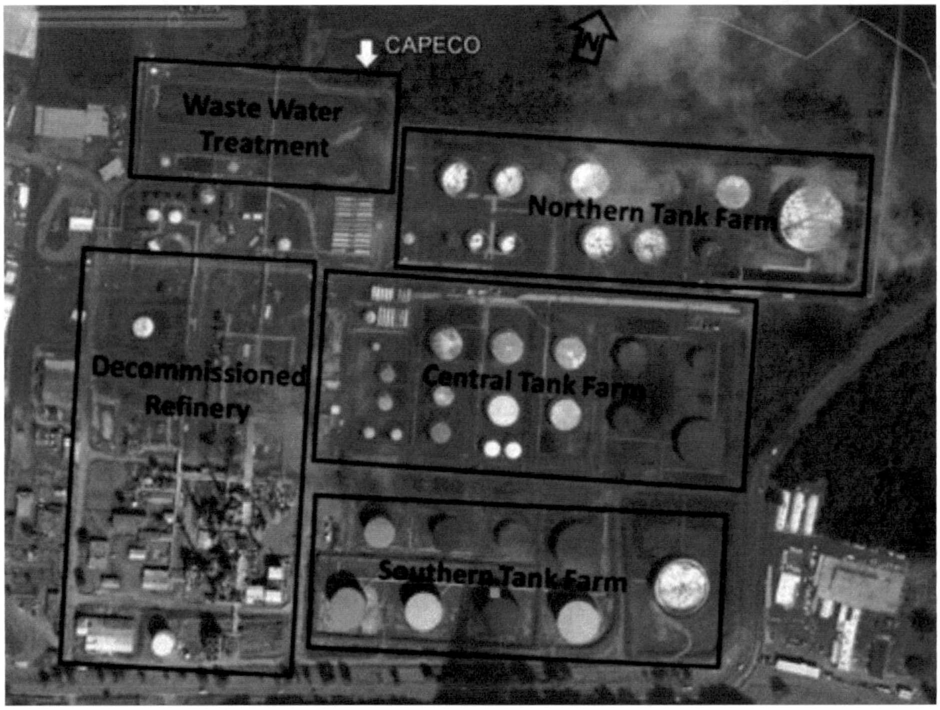

Figure 1.1: Facility Layout.

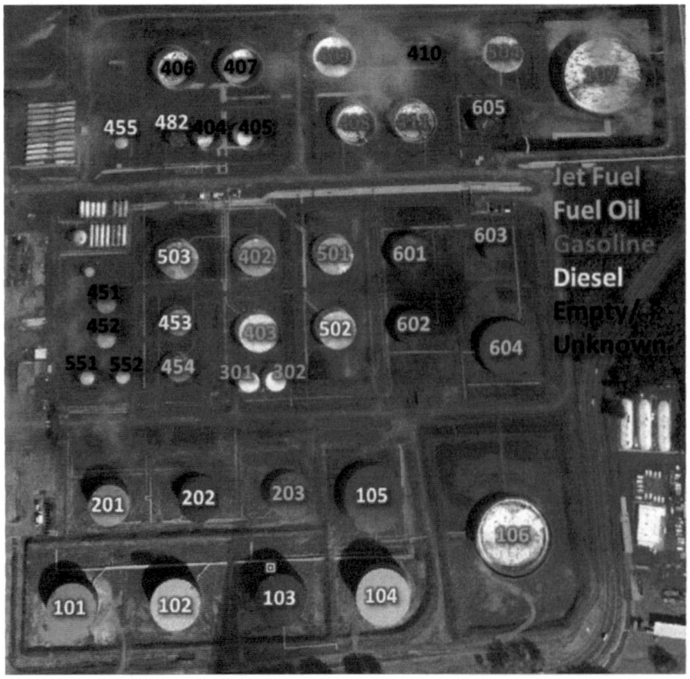

Figure 1.2: CAPECO tank location and inventory.

During fuel transfer operations, vessels would load/unload through the facility's 2 ½ mile pipeline which ended at a loading dock located in the San Juan Bay [2]. The terminal dock could accommodate one tanker of up to 75,000 tons. The transfers were made through six pipelines that connected to the facility transferring crude oil, FCC stock, fuel oil, gasoline, diesel, jet fuel and kerosene. The pipelines were equipped with remote operating valves that shutdown through: an electrical actuator; manually closing the valve; thermal shut-off switch set to 221°F (105°C); five push buttons in the operator's office, guards office, stairway to the main platform, fire pump building and Chiksan platform [5].

Events Prior to the Incident

Operations the day of the incident included: 1) A transfer of diesel fuel to the Puerto Rico Electric Power Authority Plant in Puerto Nuevo from tank 202; 2) A transfer of fuel oil to the Puerto Rico Electric Power Authority Plant in Palo Seco from tank 105; and 3) A transfer of gasoline from the Cape Bruny vessel to the CAPECO facility. Due to CAPECO's largest tank 107 (capacity 500,000 bbls – 21,000,000 gal) being used for a different quality gasoline, the transfer was scheduled for four smaller capacity tanks (see Figure 1.3) [6] .

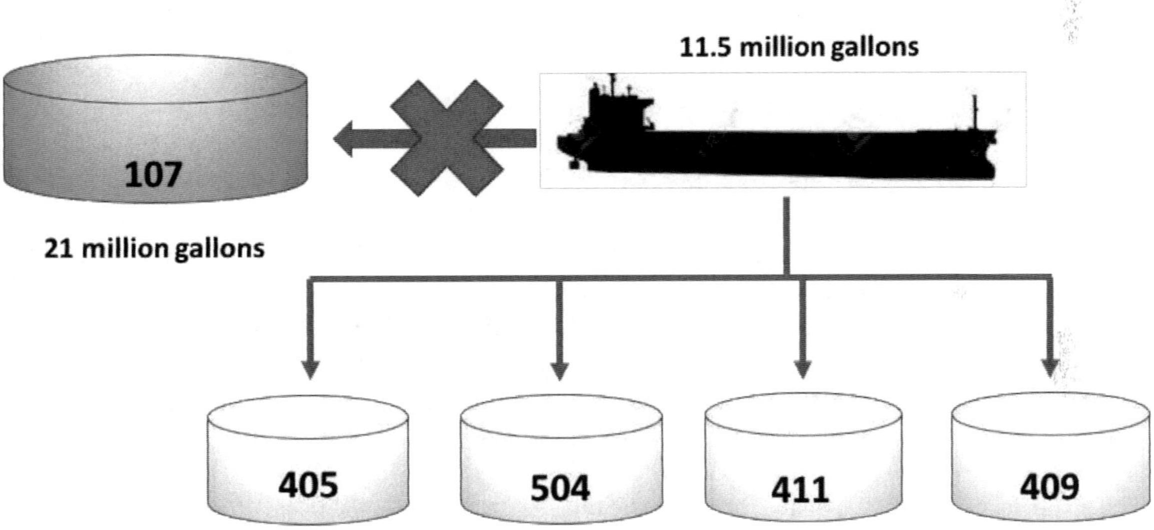

Figure 1.3: Cape Bruny transfer plan.

Pumping from Cape Bruny to the CAPECO facility began on October 21, 2009. Four tanks (Tank 405, 504. 409, and 411) in the northern tank farm received gasoline during the transfer. Initial pumping started at approximately 8:47 p.m. A sequence of operations can be found in Figure 1.4 [2].

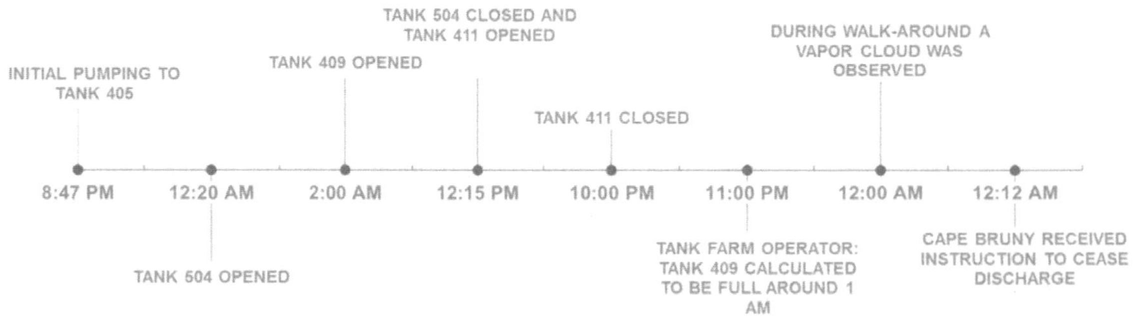

Figure 1.4: Transfer Operations [6].

Shortly after 11:00 p.m., the tank farm operator advised the supervisor in charge of the operations that night that the inspector found a difference between the side gauge, monitor console reading and the physical status of Tank 411 (already closed), and advised that the discrepancy was to be investigated [1].

ATF interviews determined that numerous operators and employees observed a vapor cloud starting at midnight. Tank 409 began to overfill between the 11:00 p.m. and 12:00 a.m. tank checks while the average pumping rate was approximately 11,134 bbls/hr (467,628 gal/hr) [7]. The overfilling of tank 409 caused gasoline to flow out of the tank vents near the top of the tank and cascade down the side forming a vapor cloud and liquid pool (as will be discussed in detail later). Tank 409 was contained in a berm with Tank 410 having a capacity of approximately 132,000 barrels (5,544,000 gallons) based on the 110% capacity requirement [3] . As the overfilling continued, gasoline (liquid and vapor) began to fill the containment area and expand towards the waste water treatment facility resulting in the white mist or vapor cloud that was seen in the facility by employees [6]. It was determined that approximately 5,082 barrels (213,444 gallons) of gasoline overflowed from tank 409 [7]. This is equivalent to overfilling the tank by over 20 gasoline tanker trucks, which if 10% of the liquid formed the vapor cloud it would cover an area larger than 30 American football fields of a rich flammable mixture.

The employees saw a white cloud extending approximately two to three feet above the ground by tank 404, and the white vapor cloud could be seen throughout the area between tanks 453 and 301, and near tanks 201 and 409 (Figure 1.5). The vapor cloud could only be seen in areas that were illuminated and extended approximately 200 yards wide by 400 yards long (183 m by 366 m) [1].

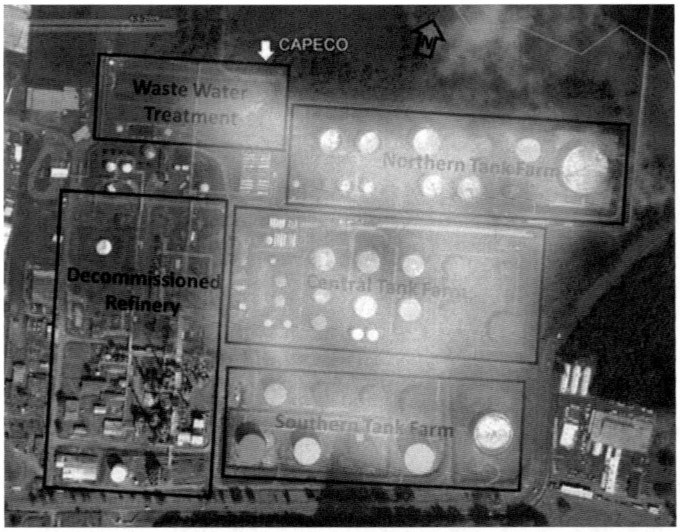

Figure 1.5: Approximate migration area of the vapor cloud.

Incident

At approximately 12:23 a.m. on October 23, 2009, the resulting vapor cloud from the overfill of gasoline ignited on the western side of the facility, away from the tank farm area. As the flame propagated (Figure 1.6 and Figure 1.8) towards the above ground storage tanks, a large explosion (Figure 1.8) occurred in the tank farm area that registered 2.9 on the Richter scale. Due to ignition of the large vapor cloud and ensuing explosion, many of the facility storage tanks were consumed by fire, which burned for another three days before emergency responders extinguished the flames [3].

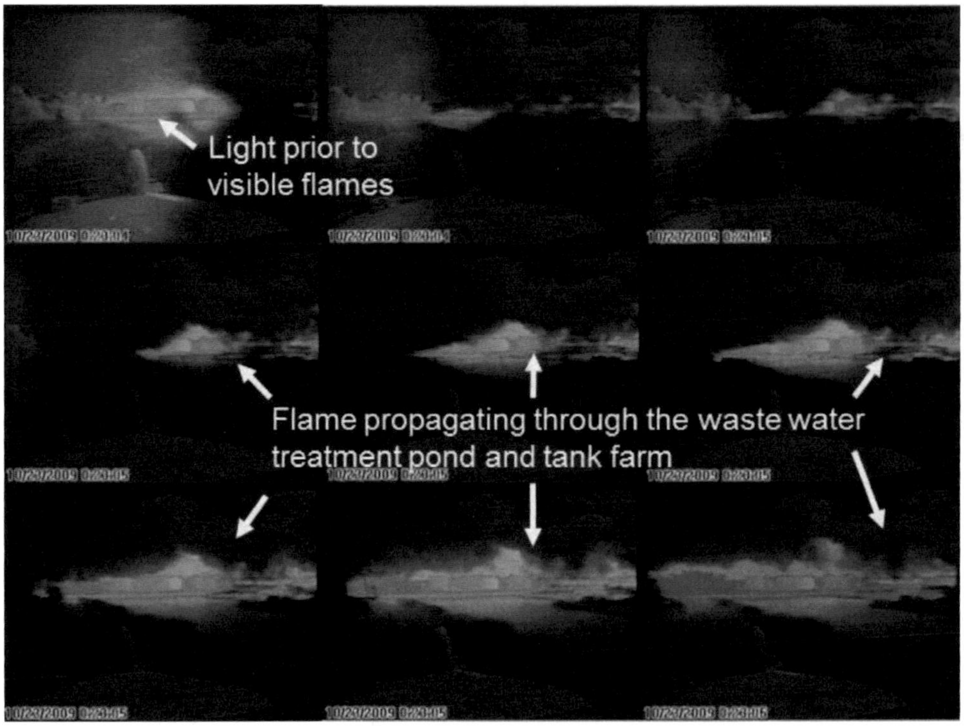

Figure 1.6: CAPECO security footage from camera 4. Visible light is seen prior to the flame propagation through the waste water treatment pond and tank farm.

Figure 1.7: Approximate propagation of the flame front prior to the explosion in the underground drainage as indicated by arrows. Camera locations and angles (red-camera 3 and blue-camera 4) at the first sign of ignition.

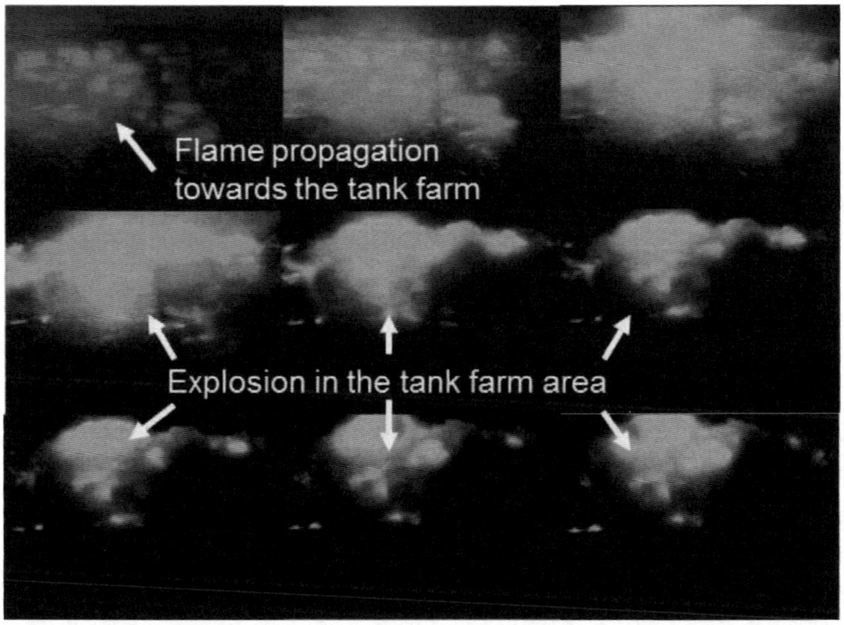

Figure 1.8: CAPECO security footage from camera 3 as the flash fire approached the tank farm from the west, leading to an explosion within the tank farm area (top-left to bottom-right).

Post Incident and Blast Damage

Emergency crews were alerted to the incident when a nearby citizen heard a large explosion and saw flames at the CAPECO facility [1]. At approximately 12:30 a.m. on October 23, 2009 the Bayamón and Cataño Fire Departments arrived at the CAPECO facility where they waited for the site fire chief to arrive approximately 45 minutes later. Eleven of the CAPECO tanks were damaged by the initial fire and explosion.

Firefighting efforts, as shown in Figure 1.9, lasted almost three days until approximately October 25, 2009 at 11:40 a.m. when the fires were extinguished [6]. Additional tanks were damaged by ensuing fires (see Figure 1.10 and Figure 1.11) [3] and when the fire was completely extinguished, the damage caused by the large gasoline vapor cloud explosion and fire included 21 tanks onsite. Both blast and thermal damage were observed on the tanks at the CAPECO facility (see Figure 1.12), consistent with the 11 tanks initially damaged by the large vapor cloud explosion and ensuing gasoline fire, and the others falling victim to the resultant fire [1].

Figure 1.9: Firefighting efforts after the incident [8].

Figure 1.10: Aerial view of the CAPECO facility after the initial gasoline vapor cloud explosion.

Figure 1.11: Secondary tank fires caused by the initial vapor cloud explosion and fire from the gasoline vapors.

Figure 1.12: Post incident damage to the tank farm.

2 Physics of vapor cloud explosions

Prior to the explosion at the CAPECO facility, tank 409 was overfilled and gasoline was being discharged from the overflow vents at the top of the tank. Recall that ignition and burning of liquid hydrocarbons occurs when the hydrocarbon is in the vapor state. A large vapor cloud developed resulting in the 'white cloud' described by CAPECO employees. As the vapor cloud grew from the continuous overflow, the gasoline vapors continued to mix with air forming a flammable mixture throughout the facility. This cloud found a competent ignition source in the western side of the facility and flames propagated towards the tank farm leading up to the main explosion event. This phenomenon is referred to as a vapor cloud explosion (VCE). The specifics on how the vapor cloud was formed and on the VCE will be discussed next.

Formation of vapor clouds

When evaluating the potential for large vapor cloud formation, which may lead to a massive explosion event, it is important to consider the mechanism by which a vapor and vapor clouds are formed. Recent experimental and modeling research has been performed to explore the generation of vapor clouds from overfilling gasoline storage tanks [9], [10]. The research determined that during a tank overfill, the gasoline would flow out of the ventilation holes at the top of the tank and over the deflector plate. The deflector plate channeled some of the escaped fuel onto the tank wall, but the rest ran over the top of the plate, fragmenting into droplets that cascaded through the air. Most of the fuel running down the wall hit a wind girder (a structural stiffening ring) and detached from the tank wall, creating a second cascade of droplets.

These conditions promoted the evaporation of the lighter components of gasoline (e.g., butanes, pentanes and hexanes). The free-fall of droplets leads to entrainment of air and mixing between the air and fuel vapor, and the formation of a rich fuel/air mixture. In addition, the wind conditions at the time of the overfill were calm, and had the wind speeds been higher, the evaporating gasoline vapor would be quickly diluted as it escaped the berm. Therefore, the overfill led to conditions that increased the break-up of gasoline into droplets, which significantly increased the evaporation rate of the lighter components of gasoline as compared to that from the pooled gasoline. Therefore, the overfill rate coupled with low wind speeds generated very large vapor clouds capable of filling a significant region of the CAPECO tank farm as seen in modeling shown in Figure 2.1.

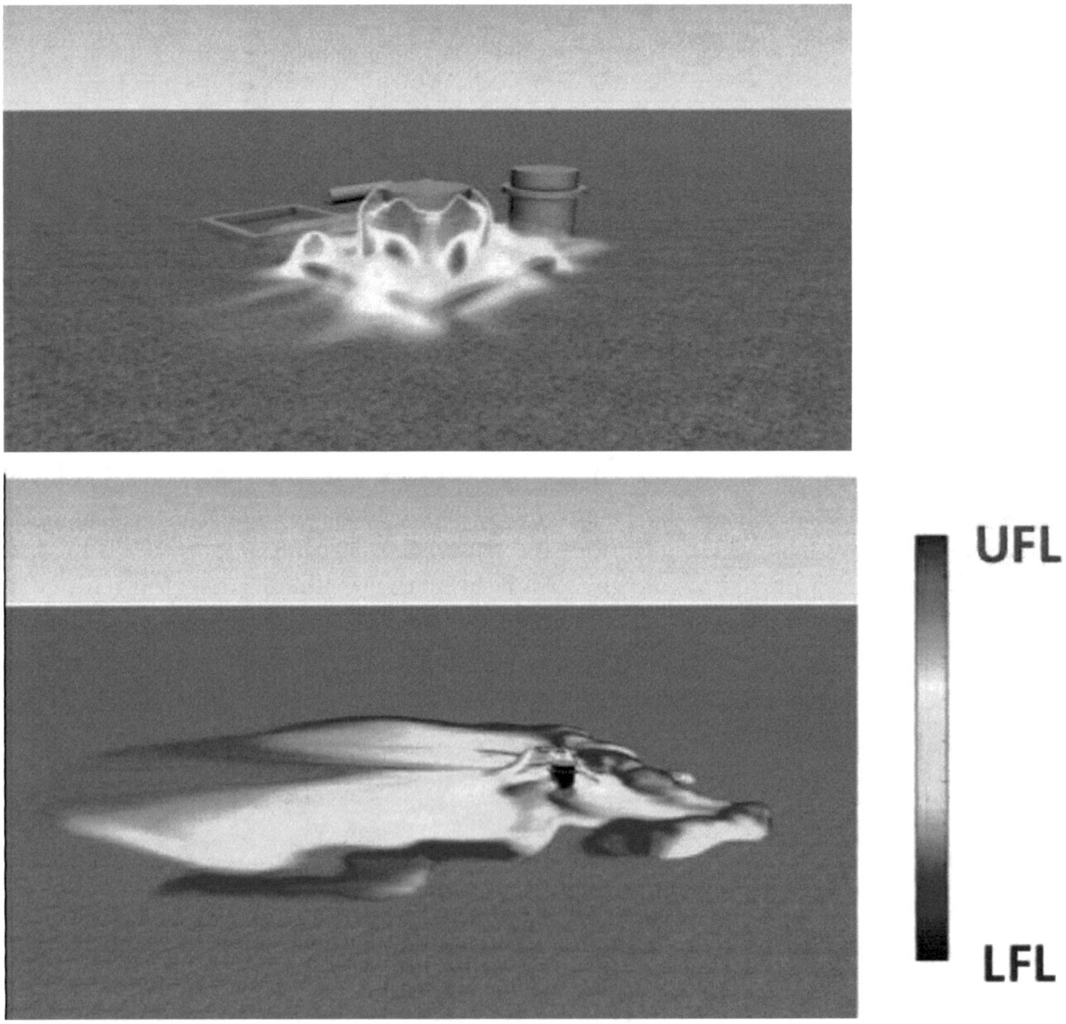

Figure 2.1: Vapor cloud formation and growth as gasoline initially cascades down the side of a storage tank and is dispersed.

Vapor cloud explosions

Vapor cloud explosions (VCEs) are oftentimes understood as "confined explosions", where a flammable mixture of gas (or vapor) and air is ignited within a confined structure or vessel. The heat released from the burning mixture causes the gases to expand and the pressure within the confinement increases above normal atmospheric pressure; this is referred to as "overpressure". This occurs until the envelope of the structure ultimately fails, similar to an ignition of a flammable mixture within an empty tank causing the roof to blow off. The sudden failure of the structure/vessel causes a shock or blast wave to form. However, conditions at CAPECO were mostly unconfined and open, except for the underground drainage system. An open and unconfined environment presents little to no resistance of an expanding flame front when a mixture is ignited; therefore, an overpressure is not expected if the gases are allowed to freely expand. The formation of a shock or blast wave in an open area cannot be explained in the same manner, and in fact, requires a unique set of circumstances.

Overpressures can indeed be created for unconfined vapor cloud explosions and require mechanisms that accelerate the combustion or burning process. The combustion rate for burning vapor/air mixtures is measured by how fast the flame propagates (flame speed) relative to the unburned mixture. For quiescent fuel/air mixtures the flame propagates at approximately 0.5 m/s and minimal overpressure is observed as the hot gases can expand in the unconfined environment. However, the flame propagation rate is extremely sensitive to turbulence in the flammable mixture into which the flame is burning. This turbulence can be caused by repetitive obstacles, (e.g. pipes, equipment, trees) within the vapor cloud itself. This results in a feedback loop, which has been extensively studied by Gexcon (see Figure 2.2), whereby the turbulence causes the flame to burn faster, which in turn pushes the unburned gases past the obstacles even faster creating higher turbulence levels, which in turn causes the flame to burn even faster. This results in the flame accelerating and can lead to very high propagation rates.

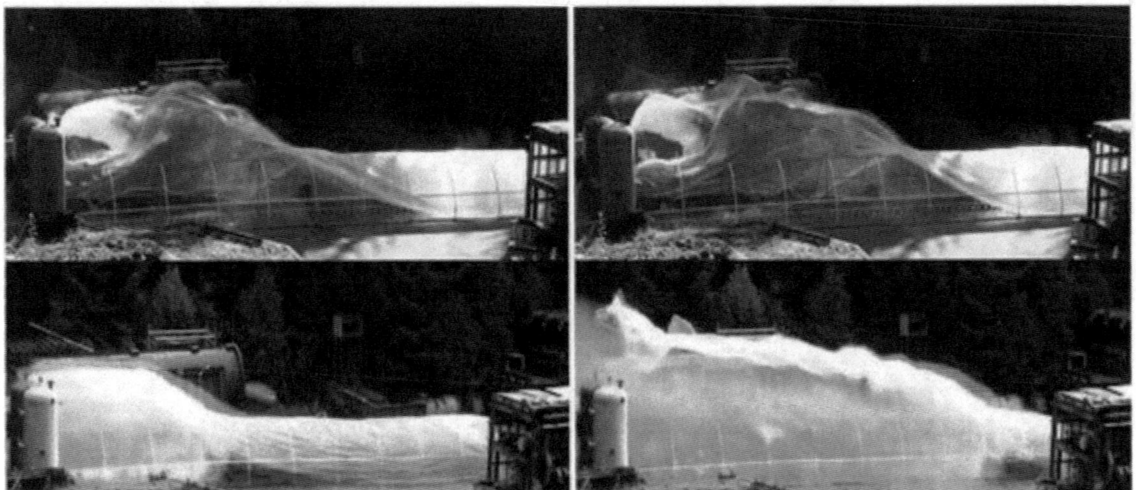

Figure 2.2: Flammable fuel air experiments for "unconfined" vapor clouds with no obstacles (top) and with trees (bottom). Clouds ignited on the left and maximum flame speeds with trees are six times larger than without.

The flame may propagate at such high speeds that the expanding gas ahead of the flame is actually compressed, which can ultimately result in a shock or blast wave. The strength of the blast wave will depend on the flame speeds achieved during the acceleration process.

If the congestion in the cloud is low, the flame speeds will remain low and will not generate a shock or blast wave (i.e., a "flash" or "whoosh" type sound), while relatively higher flame speeds can occur in high congestion areas, resulting in shock wave and "boom" type sound. Flame speeds between 100-150 m/s can generate significant overpressures, and those above 400 m/s can generate pressures well in excess of 1 barg, and can even transition to detonations. This phenomenon was precisely what occurred at the Buncefield explosion in 2005, whereby the flame accelerated into highly congested tree lines that extended over 100 meters. No confinement was found at the Buncefield site and the flame accelerations, and extremely high overpressures, were exclusively caused by the tree density [11].

In order to generate the massive explosion observed at the CAPECO facility, the resulting large vapor cloud needs to migrate into a congested (as described above) or confined region to accelerate the flame. Similar to Buncefield, the CAPECO facility had congestion that included numerous low-lying pipelines connecting the storage tanks, and vegetation to the north and east of the facility. During the post-incident scene investigation, thermal damage was found to extend into the tree line approximately 50 yards past the east property fence [1]. Unlike Buncefield, the CAPECO facility had additional long channels of confined drainage pipes that were not properly closed. Recall that the vapor cloud was ignited western side of the plant propagating towards the waste water treatment (see Figure 1.7) and it propagated east towards the tank farm and vegetation. Video analysis confirmed (see Figure 2.3) that while the main flame propagated above ground through the tank farm, a faster flame front traveled via the underground drainage pipes where a fireball exited the sump pump encasement at the northeast corner of the central tank farm well before the above ground flame.

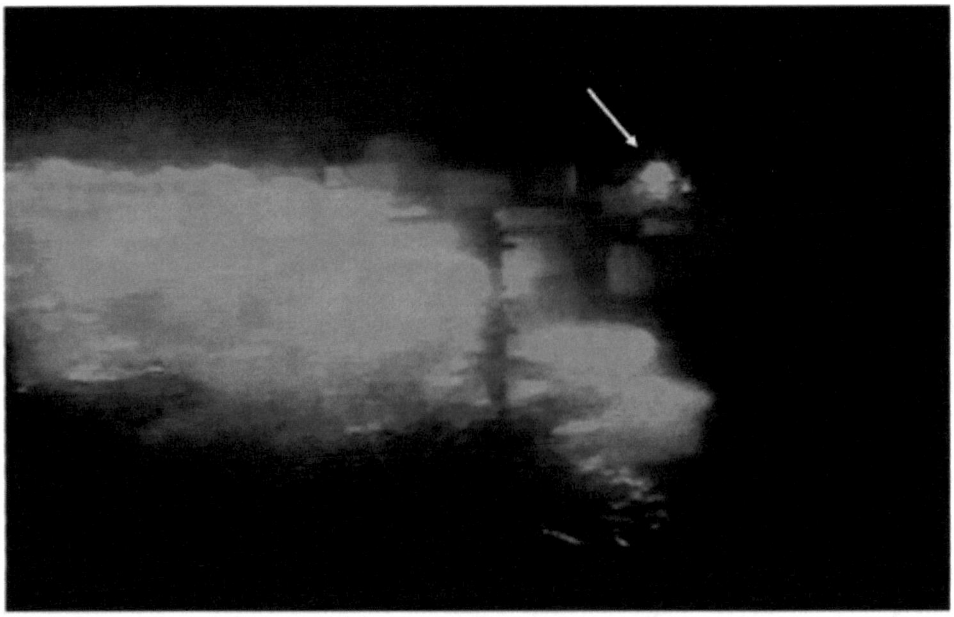

Figure 2.3: View from camera 3 showing the flame front in the underground drainage pipes reaching the northeast corner of the central tank farm before the above ground flame front. Ignition occurred on the western side of the facility resulting in flame propagation from left to right in the image above.

In fact, the 10 ft x 10 ft concrete lid on the sump pump, weighing about 12,000 pounds, was found over five feet from its original location [1]. Furthermore, blast damage to the above ground storage tanks was generally in the direction of the sump pump encasement and eastern vegetation and not the western side of the facility where the cloud was first ignited [1]. This confirms that the explosion pressures were exacerbated due to the confinement caused by the underground drainage pipes, which resulted in higher overpressures and faster accelerating flames that also entered the congested region. To evaluate the effect of the underground drainage system, FLACS simulation were performed to evaluate the resulting overpressures and blast damage both with and without the presence of the drainage system.

The CAPECO tank farm area (both with and without the drainage system) was modeled in FLACS, a computational fluid dynamics (CFD) tool predominantly used in the Oil & Gas and chemical industries to predict the consequences of hydrocarbon deflagrations. A large vapor cloud (e.g., stoichiometric cloud 4 meters tall) was placed throughout the tank farm and water treatment areas, and ignited in the water treatment area. Figure 2.4 shows the tank farm with the large vapor cloud and drainage system.

The flame propagation rates were evaluated both through the above ground cloud and through the drainage system. Figure 2.5 shows that the vapor cloud entering the drainage system resulted in flame speeds of over 400 m/s, significantly higher than above ground flame propagation. In fact, the confinement and acceleration of the flame in the drainage system resulted in overpressures that exceeded 2 barg (see Figure 2.6), which could possibly result in condition where the exiting deflagration transitions to a detonation. In contrast, overpressures generated by the flame burning through the above ground portion of the cloud did not exceed 0.03 barg. Therefore, the confined underground drainage system resulted in flames burning faster than the speed of sound and with pressures exceeding 2 barg, which ignited the remaining unburned vapor cloud in the tank farm. This high-pressure, supersonic flame front resulted in a significantly more devastating explosion than the "flash" or "whoosh" that would have been observed had the drainage system been properly closed prior to the overfill.

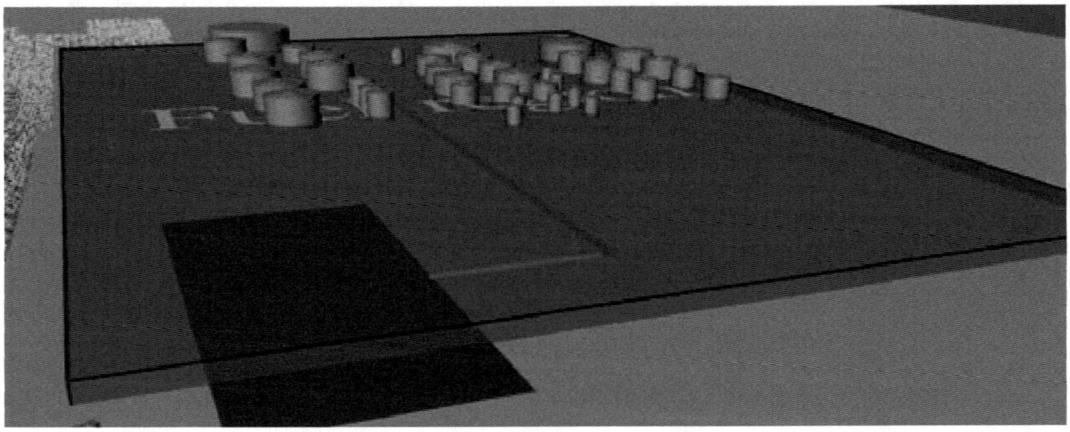

Figure 2.4: FLACS model of the CAPECO underground drainage system.

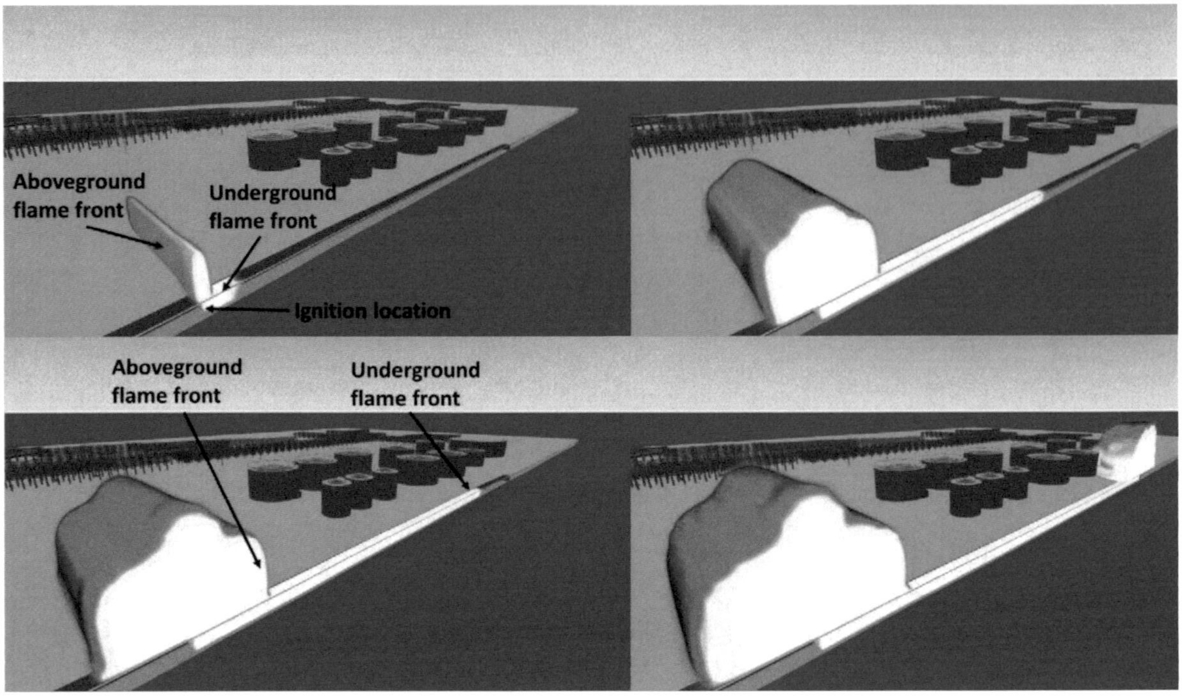

Figure 2.5: Flame front above and underground.

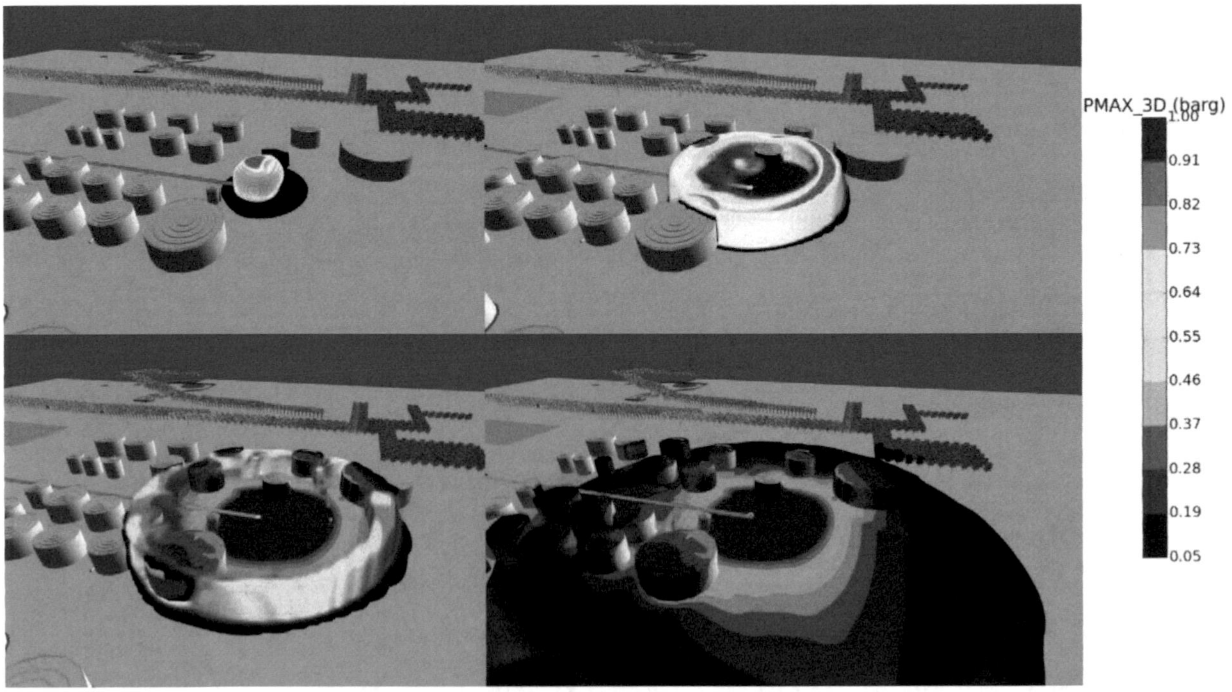

Figure 2.6: Pressure generated from the explosion in the underground drainage system.

3 Investigation findings and key lessons learned

Many of the contributors to the explosive atmosphere present at the CAPECO facility during the early morning hours on October 23, 2009 could have been avoided if the proper precautions were followed. While previous investigation reports have outlined certain contributing factors related to the incident, other major findings and lessons learned were not identified. For example, other groups have determined that the facility did not have automatic level alarms and that the automatic tank gauge system at the facility was not operational, which contributed to the event by not providing key indicators automatically and relying solely on manual measurements. However, key contributors to the incident (in addition to the automatic gauge system) were not identified and included: a failure to accurately monitor and verify the rate at which the gasoline storage tanks were being filled during the vessel offloading; and a failure to ensure "inherent safety" at the facility by not securing the dike drainage systems, which resulted high-pressure, supersonic flame fronts capable of causing the massive explosion as opposed to a "flash" or "whoosh" had the vapor cloud remained above ground. These factors will be discussed next.

Operational deficiencies that led to the Tank 409 Overfill

Recall that automatic controls were down and operations were manually monitored. Using the vessel discharge rates and tank strapping tables, it was estimated that when tanks 504 and 409 were being filled from 4:00 a.m. to approximately 12:00 noon on October 23, 2009, the transfer to each tank was approximately equal (see Figure 3.1). However, when 411 was brought online from 12:00 noon to 10:00 p.m. on October 23, 2009, it was determined that approximately 66% of the fuel was being transferred to tank 411 (see Figure 3.2), with the remaining fuel going to tank 409. This is consistent with operators' observations when preparing to take tank 411 offline: the valve to tank 409 was noted to be only open a little, confirming that tank 411 was receiving a much higher flow rate than tank 409 while they were being filled simultaneously [1].

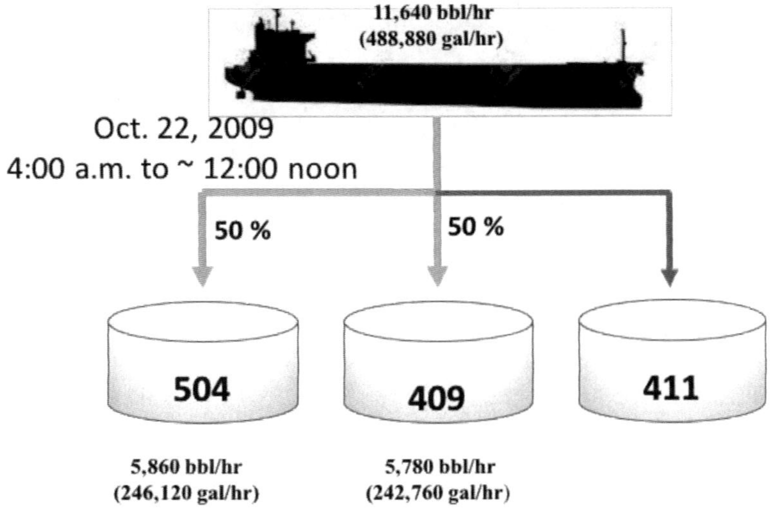

Figure 3.1: Transfer from the Cape Bruny to tanks 504 and 409 from 4:00 a.m. to approximately 12:00 noon on October 22, 2009.

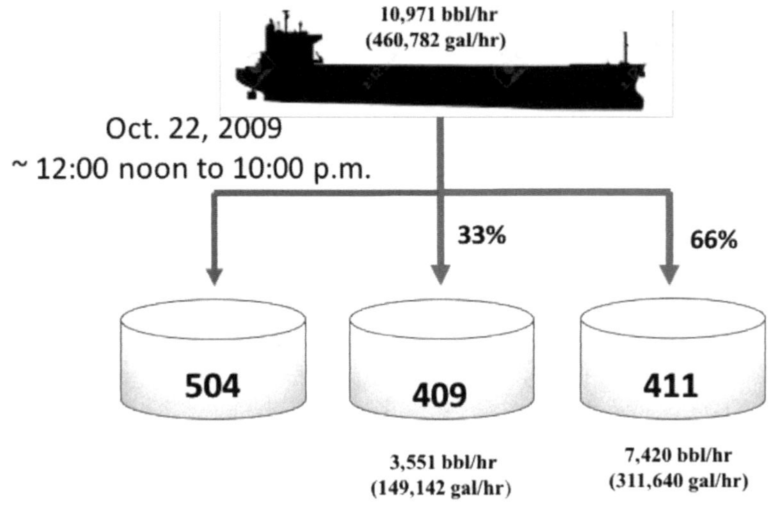

Figure 3.2: .Transfer from the Cape Bruny to tanks 411 and 409 from approximately 12:00 noon to 10:00 p.m. on October 22, 2009.

At 10:00 p.m. on October 22nd, tank 411 was closed and tank 409 was fully opened. Tank 409 contained approximately 98,716 gross barrels (4,146,072 gallons) of gasoline (overfill occurs at approximately 121,062 barrels).[1] The supervisor incorrectly estimated that the tank would be full at 1:00 a.m. by assuming only 33% of the discharge rate was going to tank 409. However, with the full vessel discharge of approximately 11,134 bbl/hr [7] going into tank 409, approximately 5,082 barrels (213,444 gallons) of gasoline overflowed from tank 409 contributing to the generation of a large flammable vapor cloud. This means that overflow occurred for approximately 27 minutes, beginning at approximately 11:45 p.m. on October 22nd until 12:12 a.m. the next morning.

Table 3.1: Estimated amounts of gasoline added to each tank and overflown at the CAPECO facility.

Tank	Estimated gasoline added Gross Barrels (Gallons)
405	4,411 (185,262)
504	62,984 (2,645,328)
411	74,198 (3,116,316)
409	114,325 (4,801,650)
Total	255,918 (255,918)
Total from Cape Bruny	261,000 (10,962,000)
Overfill	**5,082 (213,444)**

Therefore, one major cause of the incident was due to an incorrectly estimated fill time of 1:00 a.m. for tank 409, likely caused by ignorance of the increased flow rate to tank 409 after closing tank 411. Therefore, failure to compare vessel discharge rates to tank fill rates and verify that tank 409 had sufficient space to receive the scheduled amount of product resulted in tank overflow which generated a large flammable vapor cloud [12].

Drainage system

In the post incident investigation, the drainage valve in the berm surrounding tank 409 was found to be in the open position. The open valve drained to the waste water treatment facility, creating a pathway for the liquid gasoline to spread, increasing vapor generation and allowing for flammable gasoline vapor to fill the underground drainage system. The dike drainage valves should be closed during transfer operations so that in the event of an overfill, pool spread is limited. In this case, failure to close the drainage valve led to flammable vapors building up within the underground drainage system, and after ignition, resulted in a deflagration propagating within a completely confined drain and causing significant flame accelerations and overpressures. CFD analyses confirmed that these flame accelerations led to high pressure, supersonic flame fronts capable of causing the massive explosion as opposed to a "flash" or "whoosh" had the vapor cloud remained above ground. Keeping the drainage valves closed during a fuel transfer would have vastly limited the consequences and ensured inherent safety at the facility.

Level controls

The CAPECO facility has an automatic tank gauge system that transmitted the tank liquid depth to the operator's office. It would calculate the fill rate, as well as plot the flow rate and tank liquid depth versus time. Due to issues with the automatic tank gauge system (typical float and tape gauge), manual inspection of the tanks and calculation of the fill time had to occur through hourly readings.

4 Conclusions

Although US regulations do not consider bulk petroleum storage terminals as high hazards, an explosion at the CAPECO facility and ensuing fire destroyed a vast majority of the tank farm facility. This explosion was caused by a vapor cloud resulting from the overfill of a gasoline bulk storage tank, and this vapor cloud entering an improperly closed drainage system. The two main lessons learned from this incident (and previous incidents like Buncefield) include: (1) having systems and procedures in place to limit the likelihood of having such an overfill condition, and (2) providing inherent safety to such facilities to vastly limit the explosion consequences in the event such an overfill occurs (e.g. limiting confinement and congestion).

Regarding reducing the likelihood of such an event, the CAPECO facility was operating as a fully-attended facility and was in compliance with federal regulations at the time of the incident. As of today, fully-attended tank farm facilities do not require independent liquid level alarms and automatic tank gauging which allows other facilities to operate in the same fashion as CAPECO when transferring and storing flammable liquids such as gasoline. Therefore, such facilities should consider redundant controls and procedures to ensure: fill and transfer rates are constantly updated and verified throughout the transfer; and that the automatic tank gauge system is operational and complements manual measurements made at the facilities; and areas of high confinement like dike drainage systems are secured and closed prior to transfer of volatile hydrocarbons.

Inherent safety should also be evaluated at such facilities to ensure significantly lower consequences, even in the event of a significant release. CFD simulations have confirmed that physical layout of both Buncefield and CAPECO resulted in significantly higher consequences that could have resulted for identical overfill conditions. In fact, the densely coppiced tree lines at Buncefield and high confinement of the improperly opened drainage system at CAPECO resulted in explosion pressures and consequences orders of magnitude higher had such congestion/confinement not been present. Inherent safety of such facilities can be enhanced if areas of confinement and congestion are identified and mitigated in standard risk assessments. In fact, the CAPECO and Buncefield incidents would have likely resulted in large flash fires instead of the more devastating VCEs had these congestion and confinement areas been mitigated.

References

[1] ATF Investigation 764045-10-0011 Report.

[2] U.S. Chemical Safety and Hazard Investigation Board Final Investigation Report – Caribbean Petroleum Tank Terminal Explosion and Multiple Tank Fires.

[3] Puerto Rico Fire Department Fire Marshal Report.

[4] Inventory Logs.

[5] Dock Operating Manual.

[6] ATF Investigation 764045-10-0011 Interviews.

[7] Ship Transfer Logs.

[8] Caribnewspr https://www.flickr.com/photos/43578995@N03/.

[9] Davis, SG, Hinze, PC, Hansen, OR, van Wingerden, K (2010). 2005 Buncefield Vapor Cloud Explosion: Unraveling the Mystery of the Blast. Interflame 2010, Nottingham, UK..

[10] Coldrick, S., Atkinson, G.T., and Gant, S.E., Health and Safety Laboratory, Hazards XXII, 2011, pp. 560- 569..

[11] Van Wingerden, M. and Wilkins, B. (2008) Experimental investigation of the effect of flexible obstructions on flame propagation in vapour cloud explosions, GexCon report no. GexCon-08-F44110-RA-1, rev 01.

[12] API RP 2350, Appdx B.3.3 (2005).

Produced Water Tank Explosion Case Study

Presented by:
J. Kelly Thomas, Ph.D.
Baker Engineering and Risk Consultants, Inc.

2017 AIChE
Global Congress on Process Safety

BAKERRISK®

BAKERRISK®

Outline

- Purpose
- Background
- Facility Description
- Incident
  - Overview
  - Fuel
  - Oxidant
  - Ignition
- Related Incidents
- Lessons Learned
- Potential Prevention and Mitigation Approaches

2

Purpose

- Highlight commonly encountered hazard
  - Produced water tank with flammable atmosphere
  - Similar hazard associated with waste (slop) water tanks
    - Incidents involving waste water tanks in industry have occurred
- Identify lessons learned
- Identify potential prevention and mitigation approaches
  - Examples, not exhaustive compilation
  - Other approaches may provide similar level of risk reduction

3

Background

- Produced water (brine water)
  - Water separated from oil / condensate / gas streams
  - Collected in tanks and treated as waste
  - Head space may be in flammable range depending on operating parameters and separator performance (e.g., condensate carryover)
- Not typically viewed as posing significant explosion risk
  - Reasonable on a relative basis if other explosion hazards present which pose much more severe consequences and/or risk
  - Reasonable from a building siting perspective
  - Hazard is local, although debris may be thrown considerable distance
  - However, produced water tanks can, and do, explode
- Should be considered in terms of hazard/risk to personnel

4

Facility Description (1 of 6)

- Incident occurred at gas production site (West Virginia, 2013)
- 6 gas wells (no oil or condensate produced), each with own separator
  - Separators are 2.5 ft diameter by 8 ft long, and operate at 675 psig
  - Gas composition (97.2% CH_4, 2.1% C_2H_6, 0.1% C_3H_8, 0.3 CO_2, 0.3% N_2)
- Produced water tanks
  - Tanks in two groups of 3 (i.e., 6 tanks)
  - Group of 3 tanks share common overflow (3 in.) and vent (4 in.) lines
    - Common vent line not supplied by tank manufacturer
    - No flame arresters on these lines
  - Fixed roof API 12F style, 12 ft. (3.7 m) diameter, 20 feet (6.1 m) tall
  - 400 barrel (16,800 gallons, 64 m^3) & hydrotested at 1.5 psig (0.10 barg)
  - 2 access hatches (Enardo ES-660 and Jayco JT-8); relieve at 0.5 psig

5

Facility Description (2 of 6)

6

Facility Description (3 of 6)

- Fill line and access hatches

7

Facility Description (4 of 6)

- Common vent and overflow lines

- Common vent line terminates with open end at opposite end of 3-tank group

Facility Description (5 of 6)

- Sign at entrance to pad

9

Facility Description (6 of 6)

- Warning sign on each produced water tank
- Flammability rating of 3 would be indicative of gasoline

10

Incident (Overview, 1 of 7)

- Approx. 100 bbl removed from each tank every 3 or 4 days
  - Driver goes up to top of tank via catwalk, opens hatch, and gauges using wooden rod before pumping out desired liquid volume
- Driver arrived between 5:30 and 5:45 am, hooked truck up to tank manifold, and (at some point) ascended catwalk to tanks
- Explosion occurs prior to 6:45 am (February, sunrise at 7:17)
- One tank fails at shell-floor joint, ejected, and thrown 110 feet (34 m)
- One tank fails at shell-roof joint, roof thrown 165 feet (50 m)
- Catwalk damaged and displaced, throwing driver from catwalk and outside tank battery, approximately 30 feet (9 m); fatality.

11

Incident (Overview, 2 of 7)

12

BAKERRISK

Incident (Overview, 3 of 7)

13

BAKERRISK

Incident (Overview, 4 of 7)

- Ejected tank

14

BAKERRISK

Incident (Overview, 5 of 7)

- Failed tank roof
- Shell-floor joint shows uplift (was approaching joint failure)

15

BAKERRISK

Incident (Overview, 6 of 7)

- Catwalk displacement (catwalk supported by tank attachment)

16

Incident (Overview, 7 of 7)

- Tank structural capacity roughly 10 to 15 psig (gradually applied internal pressurization) based on past experience
- Trajectory analysis indicates in range of 5 to 15 psig required to achieve observed tank and tank roof throw distances
- Indicates weakly flammable mixture was present in two tanks which failed, and mixture in "middle" tank was either not flammable or barely flammable
- Post-incident flammable gas measurements
  - Tank in adjacent set ≈ 65% LFL days after incident
  - Truck ≈ 15% LFL 2 months after incident
  - Consistent with elevated fuel concentration in tanks at time of incident

17

Incident (Fuel, 1 of 3)

- Separator does not remove "all gas" from the fluid, even when functioning properly and as designed
- Produced water contains gas since at elevated pressure (675 psig), which is released when pressure dropped to atmospheric
  - Gas content function of water pressure, temperature and composition
- Can have gas "carry" over or with stuck/jammed valve in separator
  - Separators were tested post-event and found to work properly
  - Concluded gas source was solely content of produced water transferred to storage tanks from separators

18

Incident (Fuel, 2 of 3)

- Methane solubility data from Duan & Mao (2006) Geochemica, 70: 2269-86.
  - Approximately 0.043 moles/kg at relevant conditions
- See vapor space fuel concentration plot on following slide:
  - Low level to high level of 50 to 150 barrels (1/4 tank capacity)
  - Conditions present (separator pressure, fluid composition, etc.)
  - No air in-leakage (only draw in air when drop liquid level)
- Can easily exceed UFL
  - Fuel concentration would be lower at lower separator pressure
  - Minimum fuel concentration would decrease if either high level or liquid volume extracted were increased (vs. values noted above)
  - > UFL is not a "universal condition" for produced water tanks

19

Incident (Fuel, 3 of 3)

- Fuel concentration for series of 100 bbl fill / extract cycles

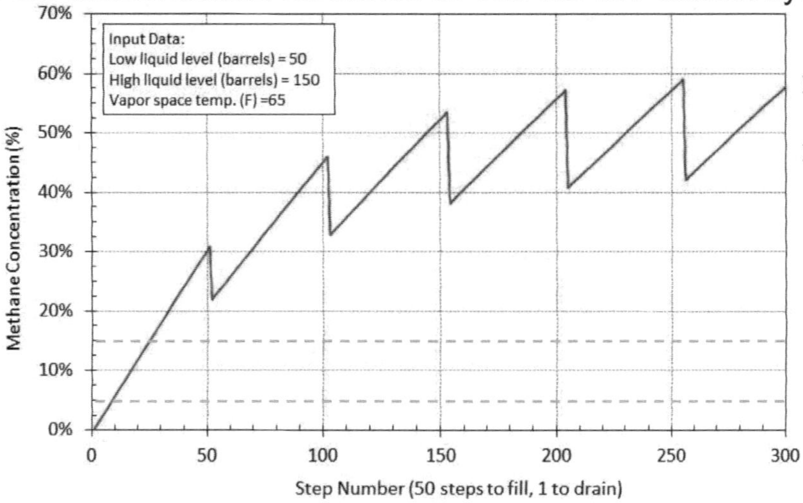

- Can easily exceed the UFL
- Issue is how get enough oxidant to be in flammable range

20

Incident (Oxidant, 1 of 4)

- Need additional air to enter flammable range (i.e., < UFL)
- Additional air can be drawn into tank via natural convection
  - Gas within tank lighter than outside air (i.e., unless outside air is hotter)
  - Light gas can flow out common vent line, pulling air in through leak points (e.g., access hatch seals)
    - Analysis subject to uncertainty since inlet flow path not well defined, but leakage rate in range of 125 scfd (2.5 lpm, 22 bbl/day) reasonable
    - Acts to limit maximum fuel concentration
    - Steady decrease in fuel concentration when tank is "stagnant" (i.e., no produced water additions or removals)
  - Much higher air in-leakage rate when hatch is open (2 to 5 scfm)
- Tanks stagnant for period before explosion & hatched opened
- See illustrate plots on following page

21

Incident (Oxidant, 2 of 4)

- Fuel concentration for series of 100 bbl fill / extract cycles, followed by stagnant period (125 scfd air in-leakage)

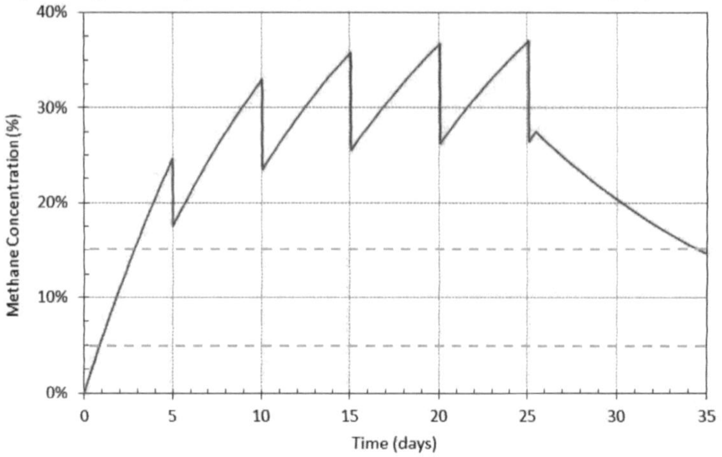

- Above UFL until enter stagnant period
- Air in-leakage during stagnant period allows fuel concentration to drop below UFL and enter flammable range
- Reach flammable condition in about 2 weeks

22

Incident (Oxidant, 3 of 4)

- Fuel concentration for series of 100 bbl fill / extract cycles, followed by stagnant period with hatch open (2 scfm air in-leakage)

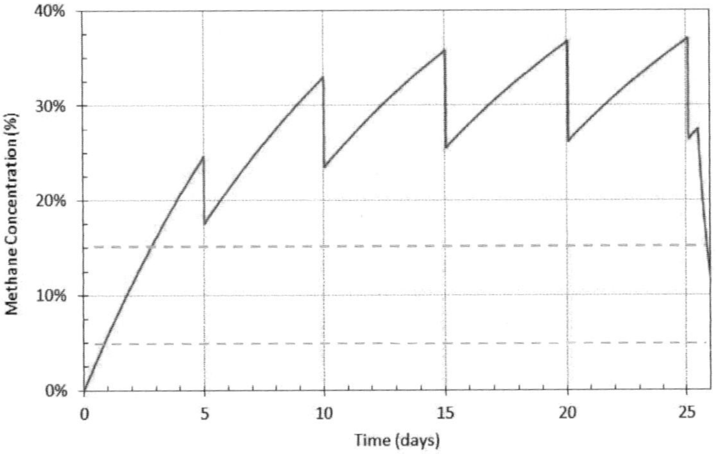

- ➢ Above UFL until hatch open for extended period
- ➢ Air in-leakage with hatch open for extended period allows fuel concentration to drop below UFL and enter flammable range
- ➢ Reach flammable condition in about 8 hours

23

Incident (Oxidant, 4 of 4)

- Natural convection limits maximum fuel concentration and can bring tank vapor space into flammable range, particularly with a tank which is stagnant for an extended period or with a hatch left open
- Note should <u>not</u> conclude that require natural convection in order to form a flammable mixture with a generic produced water tank
  - Lower separator pressure would mean less gas and hence tank could be flammable for portion of time regardless of natural convection
  - Larger volume of produced water withdrawn from tank or higher tank levels could put the tank vapor space into the flammable range
- Should assume a flammable mixture can result if the fuel concentration can exceed the LFL

24

Incident (Ignition Source)

- Driver was performing operation before sunrise
  - No headlamp or flashlight recovered from scene
  - Lighter found 25 feet from body (in same direction as thrown)
  - Driver was a smoker and one glove was off
  - Opened cigarette pack present in truck
  - Reasonable to assume driver used lighter to check liquid level
  - Driver not equipped with an LFL meter
- Static charge cannot be completely ruled out
- Flame propagation to adjacent tanks via connecting lines
- Ignition potential recognized (e.g., API RP 505)

25

Related Incidents

- Produced water tank explosions have happened in the past
- Cursory literature search
  - Texas (2013): Set of six 200 bbl produced water tanks, 2 people smoking on top of tank, tank exploded
  - Canada: Welding on tank which had contained produced water resulted in explosion
- CSB investigation report (2011-H-1):
  - "Public Safety at Oil & Gas Storage Facilities"
  - 26 incidents over 27 year period (44 fatalities and 25 injuries)
  - Includes at least one produced water tank
- Produced water SDS packages vary in flammability warnings (as does the signage on the produced water trucks)

26

Lessons Learned

- Produced water tank explosions can and do happen
  - Lesson already learned in prior incidents .....
- Hazard awareness may not be adequate among all personnel working at oil and gas production sites
- Development of flammable atmosphere in tank depends on separator pressure, tank arrangement and tank ventilation
  - Should assume tank vapor space may be flammable
  - Can also have separator malfunction leading to increased fuel concentration

27

Potential Prevention and Mitigation Approaches

- Increase hazard awareness for all personnel working at site
- Measure level at grade to eliminate routine access to most hazardous location (i.e., tank roof)
- Inert tanks using natural gas
  - Take feed from product gas line
  - Eliminate open vent paths and institute pressure control system (flow as required to maintain positive pressure and preclude air in-leakage),
  - Will result in increased emission of natural gas from facility
- Use a "lead tank" as a secondary separator
  - Inert only this tank with natural gas (limit use of natural gas)
  - Pump or gravity flow from "lead tank" to other produced water tanks
- Monitor tank vapor space

28

Questions

?

29

481704 Milestone Company Incidents and Their Impact on Air Products Process Safety Journey

Abstract:

Air Products has experienced several significant Process Safety events in our history that have helped define the Process Safety culture that exists at the company today. This presentation describes these events and their root causes, as well as the changes that they launched in the company's approach to Process Safety. The first incident was an explosion at a Nitrous Oxide plant in 1976 that resulted in four on-site fatalities. In the aftermath of this event the need for a formal Hazard Review Process was identified. The second event occurred in 1978 at an air separation plant in which a cryogenic Liquid Oxygen storage tank was ruptured and 800 tons of Liquid Oxygen was released. The Liquid Oxygen vaporized and drifted off-site, resulting in five off-site fatalities. This incident highlighted the necessity of Risk Quantification. While these milestone events served as the impetus for Air Products Process Safety journey, that journey continues to current day, and the presentation will close with a brief update on future direction.

481925 Plymouth Incident

Abstract:

The Williams Company LNG facility on their Northwest Pipeline system located in the state of Washington experience an explosion March 30th 2014. The incident occurred during a routine start-up of the liquefaction process. Directly after the facility was deemed safe the site preservation process was initiated and the investigation began. The various regulatory agencies were involved during the investigation to help determine the cause and to provide recommendations to prevent like kind occurrences elsewhere within the industry. Additional 3rd party firms were called in to review potential structural deficiencies, design flaws, metallurgy testing, and auto ignition testing.

AUTHOR INDEX

AUTHOR INDEX